Geochemistry: Principles and Applications

Geochemistry: Principles and Applications

Edited by Karolina Jensen

MURPHY & MOORE
www.murphy-moorepublishing.com

Published by Murphy & Moore Publishing,
1 Rockefeller Plaza,
New York City, NY 10020, USA
www.murphy-moorepublishing.com

Geochemistry: Principles and Applications
Edited by Karolina Jensen

International Standard Book Number: 978-1-63987-256-5 (Hardback)

Cataloging-in-Publication Data

Geochemistry : principles and applications / edited by Karolina Jensen.
p. cm.
Includes bibliographical references and index.
ISBN 978-1-63987-256-5
1. Geochemistry. 2. Earth sciences. 3. Chemistry. I. Jensen, Karolina.
QE515 .E87 2022
551.9--dc23

TABLE OF CONTENTS

PREFACE

Over the recent decade, advancements and applications have progressed exponentially. This has led to the increased interest in this field and projects are being conducted to enhance knowledge. The main objective of this book is to present some of the critical challenges and provide insights into possible solutions. This book will answer the varied questions that arise in the field and also provide an increased scope for furthering studies.

Geochemistry is referred to as the study of the phenomena that control the abundance, composition, and distribution of chemical compounds and isotopes in geologic environments. It is vital to understand the processes related to production of economic concentrations of minerals by hydrothermal, magmatic, metamorphic, hydraulic, or weathering agents. Geochemistry encompasses the topics such as the Earth's aggregate physical and chemical state, thermodynamics, solutions and simple phase relations and mineral chemistry. The chemistry of natural waters, the chemistry of igneous rocks, chemical controls on soil formation, diagenesis, metamorphism, hydrothermal ore deposition, radioactive isotope geochemistry, stable isotope geochemistry, surface sorption geochemistry, chemical kinetics, organic geochemistry and atmospheric chemistry are also undertaken within this field. This book contains some path-breaking studies in the field of geochemistry. The topics covered herein deal with the core subjects of geochemistry. This book will provide comprehensive knowledge to the readers.

I hope that this book, with its visionary approach, will be a valuable addition and will promote interest among readers. Each of the authors has provided their extraordinary competence in their specific fields by providing different perspectives as they come from diverse nations and regions. I thank them for their contributions.

Editor

1

Water-melt interaction in hydrous magmatic systems at high temperature and pressure

Bjorn Mysen

Abstract

Experimental data on the structure and properties of melts and fluids relevant to water-melt interaction in hydrous magmatic systems in the Earth's interior have been reviewed. Complex relationships between water solubility in melts and their bulk composition [Al/Si-ratio, metal oxide/(Al + Si) and electron properties of metal cations] explain why water solubility in felsic magmas such as those of rhyolite and andesite composition is significantly greater than the water solubility in basalt melts. The silicate solubility in aqueous fluid is also significantly dependent on composition with metal oxide/(Al + Si) and electron properties of the metal cations, the dominant variables. Hydrogen bonding is not important in hydrous fluids and melts at temperatures above 500°C to 550°C and does not, therefore, play a role in hydrous magmatic systems. The properties of hydrous melts and aqueous solutions are governed by how the silicate speciation (Q^n species, where n is the number of bridging oxygen in an individual species) varies with bulk composition, silicate composition, temperature, and pressure. The reactions that describe the interactions are similar in melts, fluids, and supercritical fluids. The degree of melt polymerization caused by dissolved water varies with melt composition and total water content. Silicate- and alkali-rich felsic magmatic melts are more sensitive to water content than more mafic magmas. Transport and configurational properties of hydrous magmatic melts can be modeled with the aid of the Q^n speciation variations. Liquidus and melting phase relations of hydrous systems also can be described in such terms, as can minor and trace element partition coefficients. Stable isotope fractionation (e.g., D/H) can also be rationalized in this manner. Critical to these latter observations is the high silicate concentration in aqueous fluids. These components can enhance solubility of minor and trace elements by orders of magnitude and change the speciation of H and D complexes so that their fractionation factors change quite significantly. Data from pure silicate-H_2O systems cannot be employed for these purposes.

Keywords: Hydrous magma; Aqueous fluid; Melt structure; Viscosity; Isotope partitioning; Partial melting; Water solubility; Silicate solubility; Glass transition; Solution mechanism

Review

Introduction

The principal mass and energy transport agents in the Earth are magmatic melts and water-rich fluids. Their transport properties are sensitive to their water and silicate content. Experimental characterization of solubility and solution mechanisms of water and silicate components is, therefore, central to our understanding of magmatic and metasomatic processes in the Earth's interior.

Properties of magmatic melts are sensitive to their water content as first noted by Spallanzani (1798). Perhaps the most well-known among these effects is a several hundred degree centigrade temperature depression at high pressure of solidii and liquidii of silicate phase assemblages (including natural mineral assemblages) (e.g., Kushiro et al. 1968a, b; Grove et al. 2012). Liquidus phase relations (including partial melt compositions), transport, and volume properties of hydrous melts also vary in important ways with variations in water content (Kushiro 1972; Richet et al. 1996; Ochs and Lange 1999; Grove et al. 2003).

Aqueous solutions in the Earth's interior are efficient solvents of oxide components (Zhang and Frantz 2000; Manning 2004). Several mol% of silicate components dissolve under conditions corresponding to those of the deep continental crust and upper mantle. In the upper mantle, there can be complete miscibility between H_2O and silicate (Bureau and Keppler 1999; Mibe et al. 2007).

Correspondence: bmysen@ciw.edu
Geophysical Laboratory, Carnegie Institution of Washington, 5151 Broad Branch Rd., NW, Washington DC 20015, USA

Major element solutes in aqueous fluids (silicate components) can also enhance the solubility of other components by up to orders of magnitude (Pascal and Anderson 1989; Antignano and Manning 2008; Mysen 2012a, b; Ayers and Watson 1993). Transport, volume, and mixing properties of silicate-rich aqueous fluids differ in important ways from those of pure H_2O (Audetat and Keppler 2004; Hunt and Manning 2012; Hack and Thompson 2012; Foustoukos and Mysen 2013).

The property behavior of melts and fluids in hydrous silicate systems at the high temperatures and pressures can be traced to the relationships between fluid and melt structure and their properties. Most experimental and theoretical studies have focused on the behavior in chemically simpler systems in order to isolate the effects of individual intensive and extensive variables. With the information from chemically simpler systems, we can model melt and fluid behavior in systems relevant to natural processes. In this review, these relationships will be presented and their applications to natural systems will be discussed.

Water and properties of hydrous magma

Melting phase relations

The solidus temperature depression of mantle peridotite caused by H_2O ranges between approximately 1200°C and approximately 1600°C at the upper mantle pressures (Figure 1A). The water solubility in partial melts from a peridotite parent in this pressure range is 20 to 40 wt% (Hamilton et al. 1964; Holloway and Jakobsson 1986). Furthermore, the extent of the solidus temperature depression depends on silicate composition and increases, for example, with increasing silica content (Figure 1B). In natural systems, this effect helps explain why the liquidus depression caused by H_2O of a felsic composition (e.g., granitic composition) is greater than that of peridotite (Figure 1A). An implication of these observations is that in a subduction zone water has a greater effect on solidus temperature of descending crustal material near the slab/peridotite interface than in an overlying peridotite wedge or in an underlying depleted oceanic lithosphere (200°C to 300°C temperature difference; Kushiro et al. 1968b; Grove et al. 2012).

The composition of a melt in equilibrium with silicate mineral assemblages at high pressure is affected by H_2O (Figure 2). For example, at 2 to 3 GPa, in equilibrium with enstatite + diopside + forsterite in the $CaMg_2O_6$-Mg_2SiO_4-SiO_2 system, the melt is quartz normative in the presence of H_2O and olivine normative under volatile-free conditions (Kushiro 1969). In natural peridotite systems, this difference persists so that partial melting of hydrous peridotite at least to 2 GPa is quartz normative (andesite or quartz tholeiite), whereas anhydrous peridotite melting under these conditions results in olivine tholeiite (Kushiro 1990). The silica content of partial melts is not, however, the only compositional variable affected by H_2O at high pressure and temperature. This is evident in the system $NaAlSi_3O_8$-$KAlSi_3O_8$-SiO_2 system (Figure 2B). Here, increasing water pressure (which means increasing concentration of H_2O in the melt) leads to a rapid expansion of the quartz liquidus volume relative to that of feldspar components. This evolution means increasing alumina/

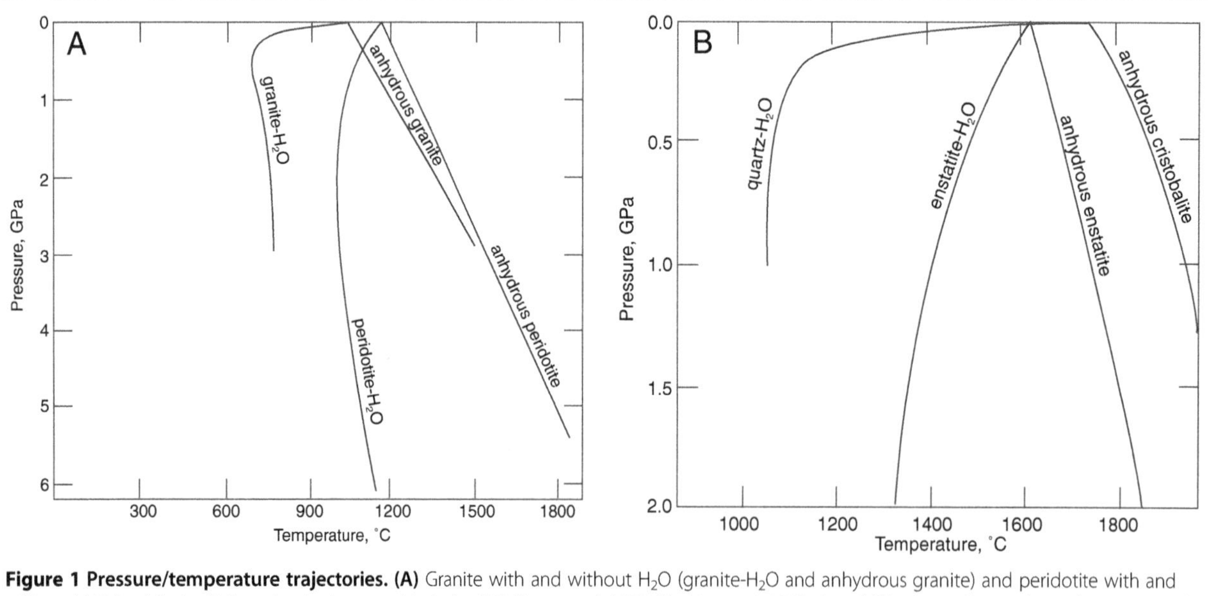

Figure 1 Pressure/temperature trajectories. (A) Granite with and without H_2O (granite-H_2O and anhydrous granite) and peridotite with and without H_2O (peridotite-H_2O and anhydrous peridotite) solidi (Stern et al. 1975; Kushiro et al. 1968a) and **(B)** silica polymorphs and enstatite with and without H_2O (cristobalite-H_2O, enstatite-H_2O, anhydrous cristobalite, and anhydrous enstatite) (Boyd et al. 1964; Kushiro et al. 1968a; Jackson 1976; Boettcher 1984).

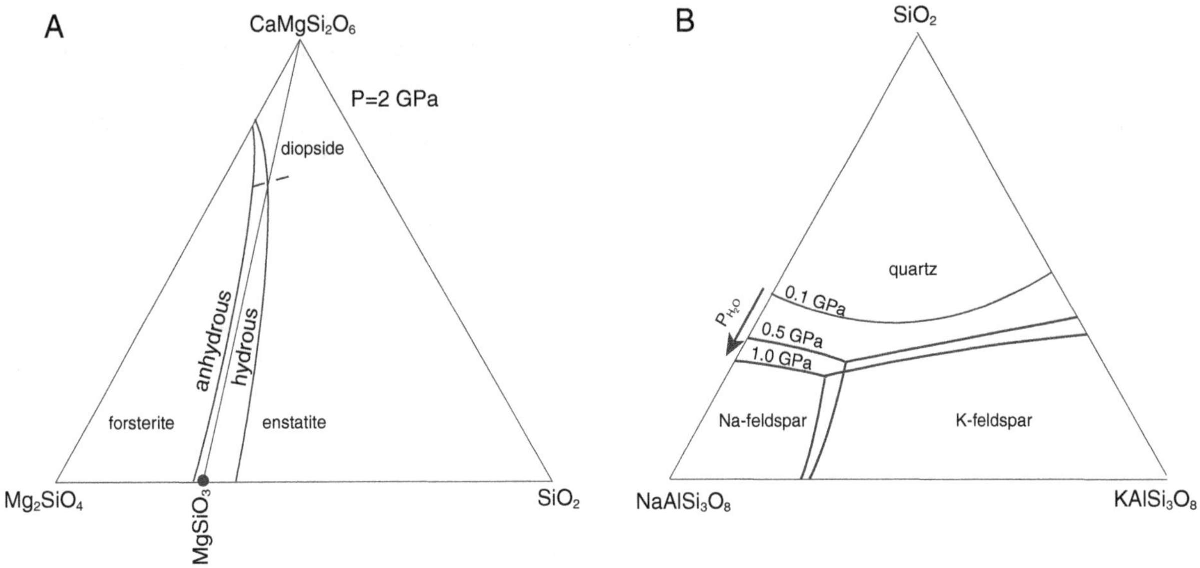

Figure 2 Liquidus phase relations in hydrous and anhydrous silicate and aluminosilicate systems. (A) Pyroxene/olivine liquidus boundary at 2 GPa in the Mg_2SiO_4-$CaMgSi_2O_6$-SiO_2 system with and without H_2O from Kushiro (1969). **(B)** Evolution of feldspar/quartz liquidus boundaries in the $NaAlSi_3O_8$-$KalSi_3O_8$-SiO_2 system with increasing H_2O pressure compared with anhydrous 0.1 MPa liquidus boundaries H_2O (data from Tuttle and Bowen 1958; Luth et al. 1964).

silica ratio in hydrous melts derived from partial melting of pelagic sediments compared with melting products of hydrous oceanic basaltic crust in the upper portion of subducting plates the higher the water content (see, for example, Poli and Schmidt 2002).

Properties of hydrous silicate melts

Transport properties of silicate melts (e.g., viscosity, diffusion, conductivity) are quite sensitive to their water content. For example, melt viscosity decreases by several orders of magnitude by a solution of as little as 1 wt% H_2O (Richet et al. 1996). This effect, however, is strongly non-linear and diminishes with increasing water content (Figure 3A). Furthermore, deviation of melt viscosity from Ahrrenian behavior is increasingly pronounced the higher the water content (Figure 3B). Water effects on viscosity are more pronounced the more felsic the melt. Water effects also are more pronounced the higher the alkali/alkaline earth ratio. This means that the viscosity of rhyolitic melts is more sensitive to water content than basalt and that alkali basalt is more sensitive to water content than olivine tholeiite.

Most other physical properties of silicate melts show analogous relations with water content and silicate composition. Examples include cation and water diffusion in melts (Watson 1994; Behrens and Nowak 1997), effects of water on glass transition temperature (Richet et al. 1996, 1997; Whittington et al. 2000), and electrical conductivity (e.g., Takata et al. 1981; Satherley and Smedley 1985). These effects exist because of the functional relationships that exist between these transport properties (Nernst 1888; Einstein 1905; Eyring 1935a, b).

Water solubility in magmatic melts

In order to understand better the solubility behavior of H_2O in silicate melts and the silicate solubility in aqueous solutions, examination of schematic phase relations in silicate-H_2O systems can be helpful (Figure 4). There exists a range of pressures and temperatures where H_2O-saturated silicate melts and silicate-saturated aqueous fluids coexist ('melt + fluid' field in Figure 4). This boundary defines the water solubility in the silicate melts. The boundary between 'aqueous fluid' and 'silicate melt + aqueous fluid' describes the silicate solubility in aqueous fluid. At temperatures above the solvus (above the critical point (c.p.) in Figure 4), there is complete miscibility between silicate melt and aqueous fluid. The temperature/pressure coordinates of the critical point ranges from about 1 GPa near 1,200°C for granitic compositions to nearly 4 GPa near 1,200°C for peridotite-H_2O (Bureau and Keppler 1999; Mibe et al. 2007). The slope of the phase boundaries at lower temperatures depends on pressure and temperature itself (Holtz et al. 1995).

It follows that water solubility in magmatic melts is temperature, pressure, and silicate composition dependent. The water solubility in melts in the simplest of silicate systems, SiO_2-H_2O, at low pressures is a linear function of the square root of water fugacity, $\sqrt{f_{H_2O}}$ (Moulson and Roberts 1961; see Figure 5A). However, with increasing pressure

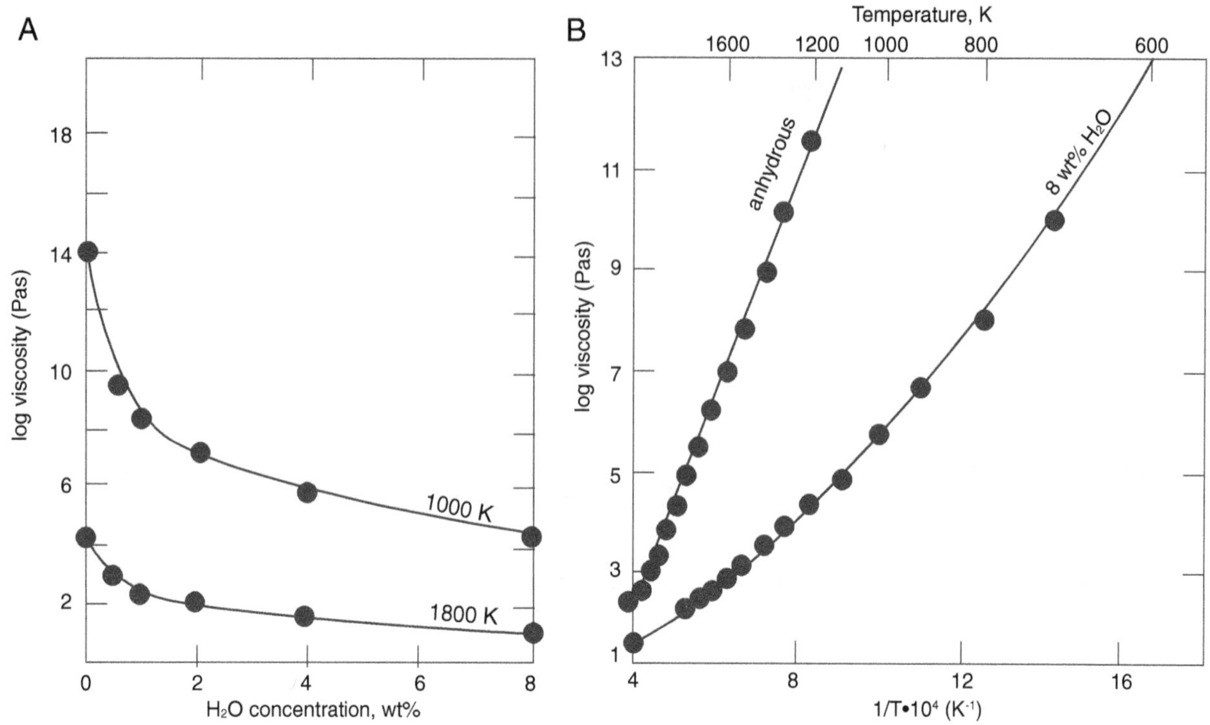

Figure 3 Viscosity of silicate melts as a function of their total H_2O content and temperature. (A) Hydrous andesite melt viscosity from Richet et al. (1996) recorded at ambient pressure by using a sample synthesized at 400 MPa and 1400°C before viscosity measurement at ambient pressure and temperatures indicated. **(B)** Temperature-dependence of melt viscosity of granite composition with (8 wt%) and without H_2O as indicated on curves from Hess and Dingwell (1996).

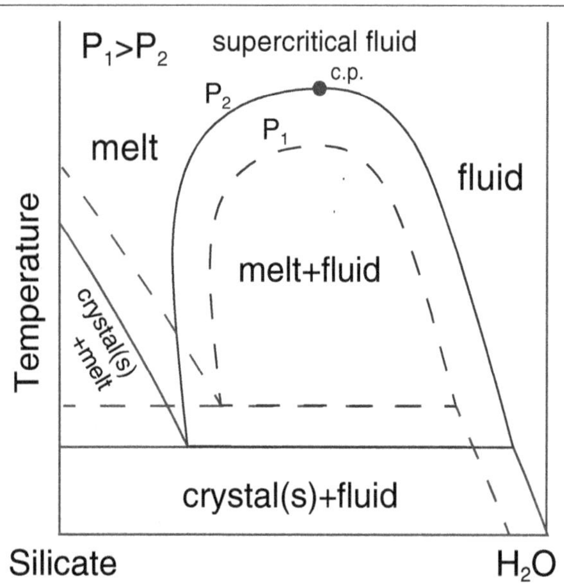

Figure 4 Schematic and simplified phase relations in silicate-H_2O systems at two different pressures, P_1 and P_2, where $P_1 > P_2$.

(and f_{H_2O}), the solubility is a non-linear function of $\sqrt{f_{H_2O}}$ (Figure 5) and where the pressure dependence itself is pressure dependent (Kennedy et al. 1962; see Figure 5B).

The features in Figure 5 notwithstanding, the pressure dependence of water solubility in more complex melts, including natural basalt and other aluminosilicate compositions, shows no simple solubility relationship with pressure (or f_{H_2O}) (Figure 5C). These complex relations likely reflect several effects on the solubility of H_2O of individual silicate components. For example, in ternary metal oxide aluminosilicate systems, water solubility is correlated with metal/Si and Al/Si ratios and electronic properties of the metal cation (Kurkjian and Russell 1958; Behrens et al. 2001; Mysen and Cody 2004; see also Figure 6). At high metal oxide/silica ratio, the water solubility decreases as this ratio increases, but this relationship reverses near composition where alkali metal/Si ratio is near 1 (Kurkjian and Russell 1958). Furthermore, whereas water solubility in peralkaline silicate melts decreases with increasing Al/Si ratio (Figure 6B), the reverse relationships have been observed for melts along SiO_2-$NaAlO_2$ (meta-aluminosilicate) joins (McMillan and Holloway 1987).

Given the complex relationships between melt composition and water solubility, it is not surprising that

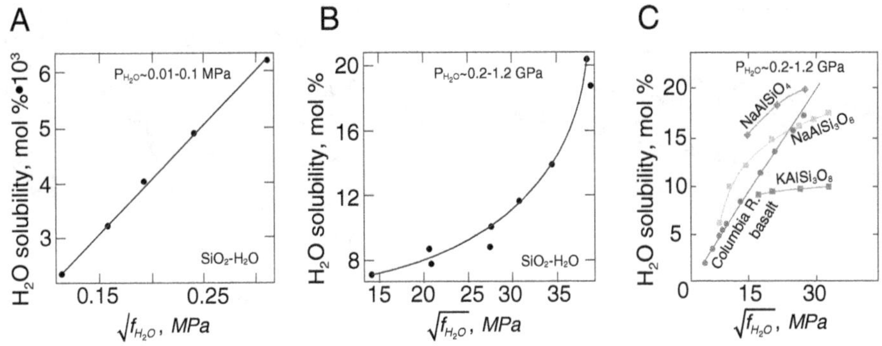

Figure 5 H_2O solubility in silicate melts at low and high pressures as a function of fugacity of H_2O, f_{H_2O}. (A) Solubility in SiO_2 melt at and below ambient pressure from Moulson and Roberts (1961). **(B)** Solubility in SiO_2 melt at high pressure as a function of f_{H_2O} from Kennedy et al. (1962). **(C)** Solubility in various silicate melts as a function of f_{H_2O} as compiled by McMillan (1994).

the solubility varies significantly in different magmatic systems. In general, water solubility in felsic magmas such as those of rhyolite and andesite composition is significantly greater than the solubility in basalt melts (Hamilton et al. 1964; Dixon and Stolper 1995; Behrens and Jantos 2001; Zhang 1999). This would be expected because of the higher alkali/alkaline earth and Si/Al ratios in rhyolite and andesite melt compared with melts of basaltic composition. These solubility relationships have been modeled with a variety of empirical models (e.g., Spera 1974; Burnham 1975; Dixon and Stolper 1995; Behrens and Jantos 2001). However, quantitative linkage between solubility behavior in chemically simple melts and more complex natural magma compositions await detailed structural characterization of the water solution mechanisms in simple and complex magmatic melts.

Silicate solubility in aqueous fluids

In the schematic phase diagram in Figure 4, the boundaries silicate melt + aqueous fluid/aqueous fluid and silicate crystals + aqueous fluid/aqueous fluid define the silicate solubility in the fluid above and below the solidus of the system. The solubility of SiO_2 in H_2O has been the subject of more extensive experimental study than other chemically more complex silicates. From a compilation of available high-pressure/high-temperature solubility data, an effective way to describe solubility is in terms of temperature and the density of pure H_2O at the temperature (and pressure) of interest (Manning 1994; see also Figure 7). It must be emphasized, however, that the empirical relationship used for this purpose,

$$\begin{aligned}\text{Log } m_{SiO_2}(\text{mol/kg}) = {} & 4.262-(5764.2/T(K) \\ & +(1.7513\cdot10^6/T^2) \\ & -(2.2868\cdot10^8/T^3) \\ & +[2.8454-(1006.9/T) \\ & +(3.5689\cdot10^5/T^2)]\cdot\log \rho_{H_2O},\end{aligned} \tag{1}$$

uses the density of pure H_2O to describe the silica solubility, and it is known that silicate components in

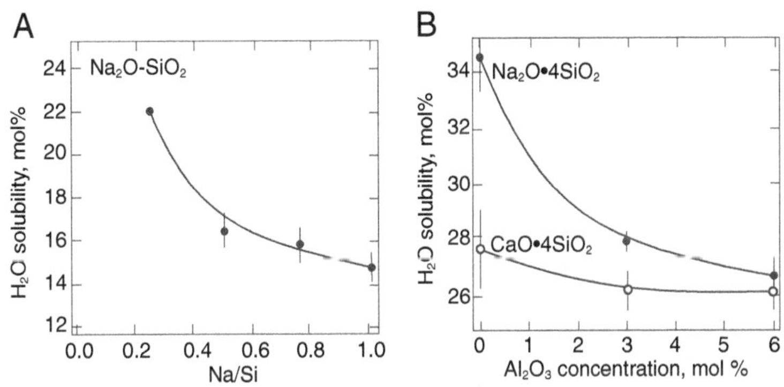

Figure 6 H_2O solubility in silicate and aluminosilicate melts with silicate composition. (A) Solubility at 1,200°C and 0.8 GPa as a function of Na/Si of the melt from Mysen (2002). **(B)** Solubility at 1,200°C and 2 GPa as a function of Al_2O_3 content in $Na_2O{\cdot}4SiO_2$ and $CaO{\cdot}4SiO_2$ melts as a function of added Al_2O_3 from Mysen and Wheeler (2000).

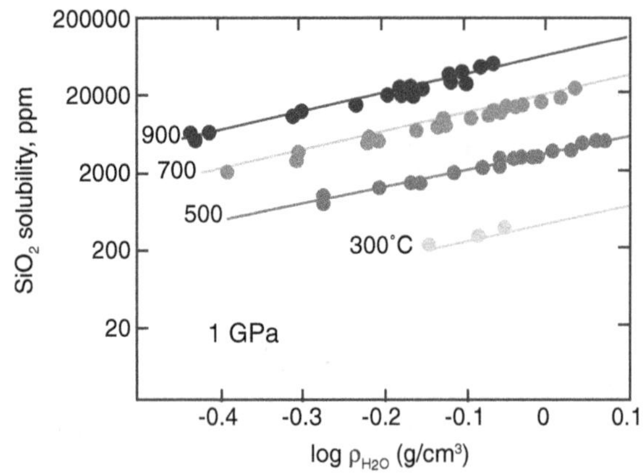

Figure 7 Silica solubility in fluid in equilibrium with quartz. As a function of the density of pure H_2O at 1 GPa and temperatures indicated calculated with the algorithm of Manning (1994) (Equation 1).

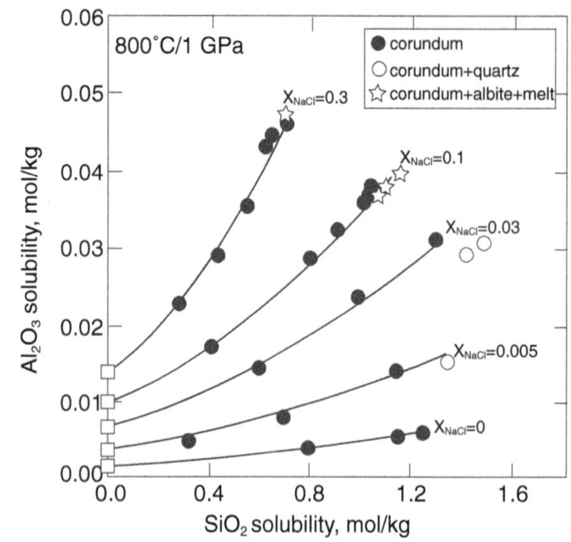

Figure 9 Solubility of Al_2O_3 in aqueous fluid in the system Na_2O-Al_2O_3-SiO_2-H_2O-NaCl. As a function of silica solubility and NaCl mol fraction in fluid. Symbols indicate the coexisting solid phases. Data from Newton and Manning (2008).

the solution affects the density of the solution (Mysen 2010a, b).

Additional components affect the silicate solubility in aqueous fluid. For example, addition of MgO to SiO_2 reduces the SiO_2 solubility (Zhang and Frantz 2000; Kawamoto et al. 2004; see also Figure 8). The Mg/Si ratio of the fluid increases with pressure (Kawamoto et al. 2004). Below about 3 GPa, MgO cannot be detected in the fluid (Zhang and Frantz 2000; Kawamoto et al. 2004). The decreased Mg/Si ratio of fluid with increasing pressure likely reflects the interaction between dissolved silica and MgO - perhaps similar to the solubility behavior of Al_2O_3 in fluids in the system SiO_2-Al_2O_3-H_2O-NaCl where increasing SiO_2 content in the fluid enhances the Al_2O_3 solubility (Figure 9). In the latter system, there is also a pronounced positive effect of increasing NaCl in the fluid (Newton and Manning 2008).

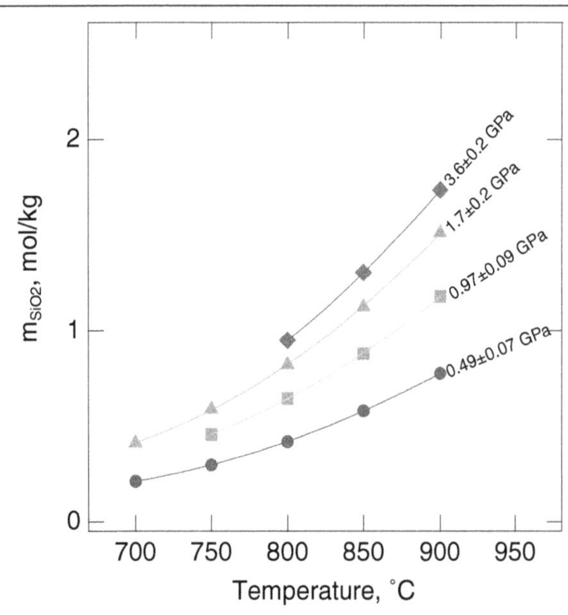

Figure 8 Solubility of silica (SiO_2) in fluid. Coexisting with enstatite + forsterite (0.49 to 1.7 GPa pressure, closed symbols) and with enstatite (3.6 GPa pressure, closed symbols) in MgO-SiO_2-H_2O fluid calculated with the algorithm of Newton and Manning (2002) (results from Mysen et al. 2013).

Oxides that are essentially insoluble in pure H_2O also can be affected quite strongly by other solutes such as silicate and aluminosilicate components. As an example, the solubility if Ti^{4+} in fluid in equilibrium with melt in the Na_2O-Al_2O_3-SiO_2-TiO_2-H_2O is a strong positive function of concentration of aluminosilicate component in addition to temperature and pressure (Antignano and Manning 2008; Mysen 2012a; see Figure 10). This is a very different solution behavior than that of TiO_2 in the simple system, TiO_2-H_2O; here, the Ti solubility is in the parts per million range (Audetat and Keppler 2005; Antignano and Manning 2008). Similar effects have been observed for P^{5+} and other high field strength cations (Mysen 2011; Bernini et al. 2013). It follows, therefore, that during dehydration of subducting slabs where the fluid is quite silicate rich, it will likely carry significantly greater proportions of nominally refractory oxides than that expected from solubility measurements with pure H_2O as the solute (see also Manning 2004).

Melt and glass

Glass transition

Most of the experimental data on solubility and solution behavior of water in silicate melts at high temperature

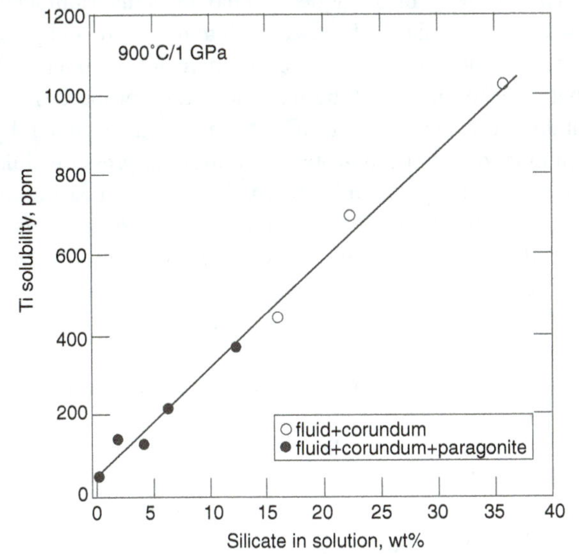

Figure 10 TiO_2 solubility in fluids in equilibrium. With rutile in the system TiO_2-Al-silicate ($NaAlSi_3O_8$ starting composition) as a function of SiO_2 in solution. Data from Antignano and Manning (2008).

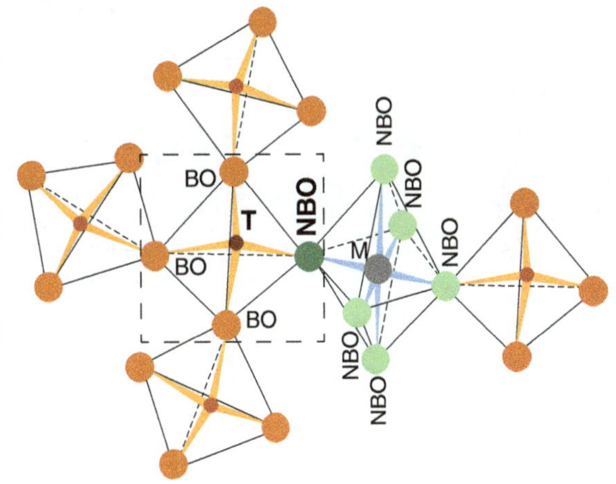

Figure 11 Schematic representation of the concept of bridging (BO) and nonbridging (NBO) oxygen.

have been obtained by analysis of the glass formed by temperature quenching from a high-temperature/high-pressure hydrous melt. Temperature-dependent structure may change during such a cooling process and eventually gets frozen in at the temperature of the glass transition (Dingwell and Webb 1990). By definition, the glass transition temperature (actually a small temperature range), therefore, is that below which the material is not relaxed on the time scale of a measurement (glass), whereas above that temperature, the material is relaxed (liquid). In the temperature interval between a liquidus and glass transition temperature, the material is a supercooled liquid with the same property behavior as that of the melt above the melting temperature. These distinctions are important when a property measured on a glass is applied to the property behavior of its melt.

Principles of silicate melt structure

Characterization of water solution mechanisms in silicate melts, including magmatic melts, requires understanding of the principles of silicate melt structure. A basic structure concept is a degree of polymerization of the silicate network of oxygen tetrahedra with their centrally located tetrahedrally coordinated cation (T-cation) (Figure 11). The dominant T-cations in magmatic melts and glasses are Si^{4+} and Al^{3+} provided that the pressure is less than that where these cations may undergo coordination transformations (coordination transformations begin to take place ≥5 to 6 GPa depending on whether it is Al^{3+} or Si^{4+} and depending on the overall melt composition; see, for example, Ohtani et al. 1985; Lee et al. 2004). There are two types of oxygen forming the corners of these tetrahedra. An oxygen shared by neighboring tetrahedra is termed a bridging oxygen (BO), and the one shared between a tetrahedron and a neighboring polyhedron that does not form tetrahedral network is termed nonbridging oxygen (NBO) (see Figure 11).

The degree of melt polymerization is defined as the proportion of nonbridging oxygen per tetrahedrally coordinated cations, NBO/T. The NBO/T of a silicate melt and glass can be calculated from its bulk composition *provided that* the type and proportion of T cations are known. This can be done by keeping in mind that electrical neutrality requires T-cations to be assigned a formal charge of 4+ and oxygen 2− so that

$$NBO/T = (4 \cdot X_T - 2 \cdot X_O)/X_T, \quad (2)$$

where X_T and X_O are the atomic proportions of tetrahedrally coordinated cations and oxygen, respectively.

The distribution of NBO/T of major rock-forming magmatic melts suggests a general correlation between aluminosilicate concentration in the magma and their degree of polymerization, NBO/T (Figure 12). The NBO/T of common magmatic melts is the 0 to 1 range (Figure 12). It is for this reason that most studies of simple system melt structure have focused on melt compositions in this NBO/T range.

Silicate glasses and melts, including those of natural magma and many commercial glasses, can be described in terms of a small number of discrete silicate polymers with integer values for their average NBO and BO (BO = 4 − NBO) between 0 and 4 (Virgo et al. 1980; Stebbins 1987; Maekawa et al. 1991; Buckermann et al. 1992). These entities sometimes have been described in terms of their silicate stoichiometries (TO_4, T_2O_7, TO_3, T_2O_5, and TO_2)

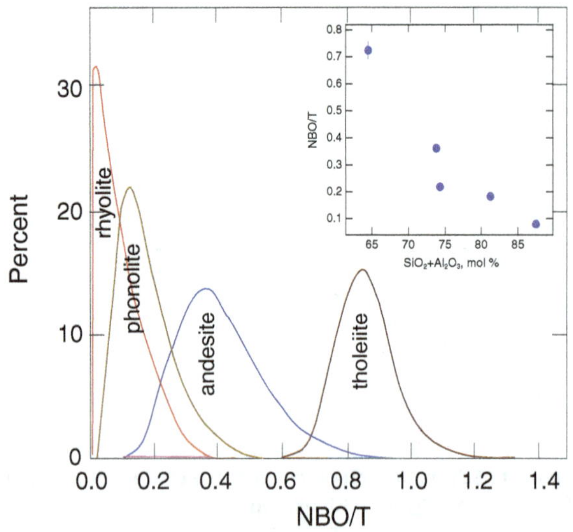

Figure 12 Distribution of NBO/T values for melts of some common igneous rocks. The inset shows the relationship between average values of $SiO_2 + Al_2O_3$ of rhyolite ($SiO_2 + Al_2O_3$ = 85.4 wt%), phonolite ($SiO_2 + Al_2O_3$ = 79.5 wt%), dacite ($SiO_2 + Al_2O_3$ = 80.8 wt %), andesite ($SiO_2 + Al_2O_3$ = 74.4 wt%), and tholeiite ($SiO_2 + Al_2O_3$ = 65.1 wt%) melts (rock composition data from http://Earthchem.org used as described by Mysen and Richet 2005, ch. 17).

and sometimes by using the Q^n species notation, Q^0, Q^1, Q^2, Q^3, and Q^4, where the superscript indicates the number of BO. Consequently, $Q^0 = TO_4$, $Q^1 = 0.5T_2O_7$, $Q^2 = TO_3$, $Q^3 = 0.5T_2O_5$, and $Q^4 = TO_2$. The proportion of these species, but not their stoichiometry, varies as a function of melt composition (metal oxide/Si and Si/Al ratios). For any melt composition, its degree of polymerization, NBO/T, can be calculated from the proportion of the Q^n species:

$$NBO/T = \sum_{n=0}^{n=4}(4-n)\cdot X_{Q^n} \tag{3}$$

where n is the number of bridging oxygen and X_{Q^n} mol fraction of individual Q^n species. However, Q^n species abundance obviously cannot be calculated from NBO/T values alone.

Solution mechanisms of silicate and water in fluids and melts in hydrous magmatic systems

Most structural data from hydrous silicate melts have been obtained from analysis of samples quenched from high temperature and pressure to ambient conditions prior to chemical and structural analysis. These data reflect, therefore, the melt structural environment near the glass transition temperature. Temperature-dependent structural features cannot be captured in such studies. That notwithstanding, many important principles have been derived from studies of quenched melt.

The behavior of an aqueous fluid in equilibrium with molten or crystalline silicates at high temperature and pressure may not be addressed by examination of quenched materials because most, perhaps all, of their properties (including the structure itself) cannot be determined by examination of the high-temperature/high-pressure fluid after quenching to ambient conditions. Fluid structure studies require, therefore, examination while the sample is at the temperature and pressure of interest. However, before addressing such experimental environments, structural data from quenched melts will be discussed.

Hydrous melts quenched from high temperature at high pressure

Water is dissolved in silicate melts in the form of water molecules, H_2O^0, and structurally bound hydroxyl groups, OH. The OH groups can form bonds with Si^{4+} and Al^{3+} as well as with other metal cations (Mysen and Virgo 1986; Xue and Kanzaki 2004; Cody et al. 2005). In either case, water dissolved in the form of OH groups in silicate melts affects their structure.

In the simplest of silicates, SiO_2, the OH formation is via breakage of bridging oxygen bonds (Wasserburg 1957; Farnan et al. 1987):

$$\text{Si–O–Si} + H_2O = 2\text{Si–OH}, \tag{4}$$

where Si-O-Si represents an oxygen bridge and Si-OH a broken bridge terminated with H^+. Formulated in terms of Q^n species and supported by ^{29}Si and ^{1}H-^{29}Si CP MAS NMR (cross-polarization magic angle spinning nuclear magnetic resonance) data for SiO_2-H_2O (Farnan et al. 1987; Cody et al. 2005), this reaction can be written as

$$2Q^4 + nH_2O = 2Q^{4-n}(H), \tag{5}$$

In chemically more complex systems, the OH-forming process also is more complex as will be discussed further below.

From the experimental studies of temperature-quenched hydrous glasses, the proportion, $X_{OH}/X_{H_2O^0}$, varies with total water concentration and with bulk chemical composition of the melt (Stolper 1982; Dixon and Stolper 1995; Zotov and Keppler 1998; see also Figure 13). Above the glass transition temperature, the $X_{OH}/X_{H_2O^0}$ is positively correlated with temperature so that a simple water speciation reaction,

$$H_2O^0 + O^{2-} = 2OH^-, \tag{6}$$

shifts to the right with temperature (Nowak and Behrens 1995; see also Figure 13B). From the temperature-dependent equilibrium, an enthalpy change, ΔH, of about 30 kJ/mol has often been reported (Nowak and Behrens 1995).

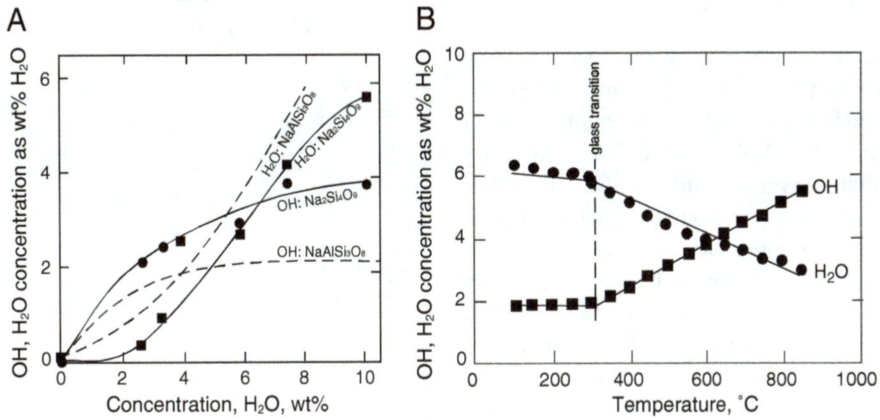

Figure 13 Abundance of OH groups and molecular H_2O as a function of water concentration and temperature. (A) Evolution of abundance of OH groups and molecular H_2O as a function of total water concentration in quenched melts (quenched from 1,100°C at 200 MPa pressure) of compositions indicated (data from Zotov and Keppler 1998). **(B)** Abundance of OH groups and molecular H_2O in haplogranite composition glass and melt as a function of temperature from Nowak and Behrens (1995) recorded *in situ* while the sample was at the temperature of interest as shown with the pressure constant at 150 MPa.

Data such as those in Figure 13 do not provide information about *how* the OH groups form bonding with the silicate melt structure. In the simple form of Equation 4 whereby Si-OH bonding is formed, there is depolymerization of the melt (NBO/T increases). By adding one or more alkali metal or alkaline earth to SiO_2, bridging oxygen bond breakage also takes place and an increase in NBO/Si as water is dissolved. The rate of NBO/Si change at constant total water concentration decreases with increasing metal oxide/silicon ratio (Na/Si in the case of the data in Figure 14; see also Cody et al. 2005). The rate of change also decreases with increasing total water content (Figure 15).

The NBO/Si decrease caused by increasing metal/silicon ratio (Figure 14) could occur either because the $X_{OH}/X_{H_2O^0}$ decreases with increasing metal/Si or it is because some of the OH groups form bonding with a network-modifying metal cation instead of breaking oxygen bridges to form Si-OH bonds. A simple example is what might be expected in an Na_2O-SiO_2 melt, where Na^+ in the anhydrous melt forms bonding with nonbridging oxygen. Here, in addition to a reaction such as shown schematically in Equation 4, a structural interaction with Na^+ can be written as

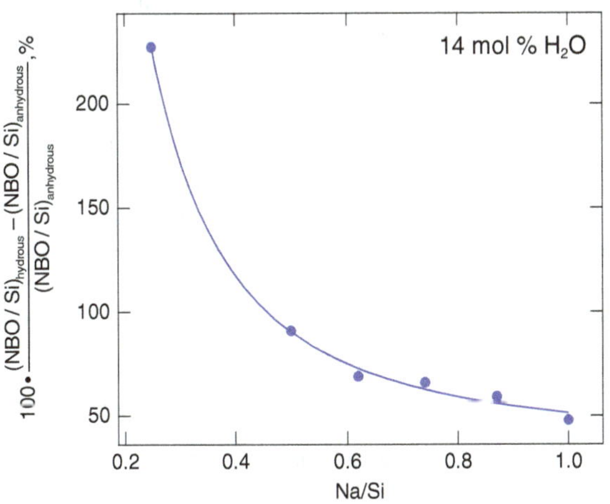

Figure 14 Effect of composition of Na_2O-SiO_2 melt on the change of NBO/Si caused by dissolved water. Samples were quenched at 100°C/s from 1,400°C at 1.5 GPa. In this system, the Na/Si ratio is equal to the NBO/Si of the anhydrous melt (data from Cody et al. 2005).

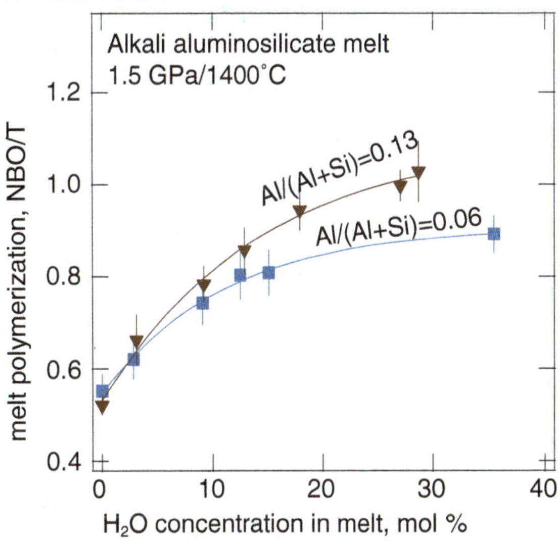

Figure 15 Relationship between degree of polymerization, NBO/T (T = Al + Si), of aluminosilicate melts. As a function of the total water content for different values of the Al/(Al + Si) of the melt at 1.5 GPa and 1,400°C. Data from Mysen (2007).

$$2Q^3(\text{Na}) + H_2O = Q^4 + 2\text{NaOH}. \quad (7)$$

In this example, the network-modifying Na^+ bonding with nonbridging oxygen in a Q^3 species (with NBO/Si = 1) in an anhydrous melt reacts with H_2O to form NaOH complexes in hydrous Na_2O-SiO_2 melts. This interaction causes the nonbridging oxygens bonded to Na^+ in anhydrous Q^3 to be transformed to bridging oxygens resulting in the formation of the more polymerized Q^4 species. Silicon-29 MAS NMR data from melts quenched from 1,400°C along the Na_2O-SiO_2 join show that this is exactly the situation and that the abundance of NaOH complexes increases with increasing Na/Si (Cody et al. 2005). Similar conclusions have been reported for alkaline earth silicate glasses wherein Ca..OH and Mg..OH groups were formed in CaO-MgO-SiO_2 melts (Xue and Kanzaki 2004).

From a compositional perspective, aluminosilicate melts are more akin to natural magmatic melts than metal oxide silicate melts. At pressures less than 5 to 6 GPa, Al^{3+} is in tetrahedral coordination where it is charge-balanced with alkalis or alkaline earths in a manner conceptually similar to that observed in crystalline aluminosilicates such as feldspars (see Lee et al. 2004; for high-pressure structural data). The aluminate groups (AlO_2^-) in aluminosilicate melts can interact with dissolved water to form either Al-OH or metal-OH bonding (metal can be alkali metal or alkaline earth), or both, in addition to Si-OH bonding (Mysen and Virgo 1986; Schmidt et al. 2001). The extent to which aluminate interaction takes place is correlated with the Si/Al ratio of the melt (Mysen and Virgo 1986). The nature of the Al^{3+} charge balance probably also affects the solution mechanism because the Al-O bond strength depends on the electronic properties of the charge-balancing cation (Roy and Navrotsky 1984).

The formation of Al-OH bonding can be illustrated schematically by using aluminate complexes, denoted as $M^{m+}{}_{1/m}Al_mO_{2m}$, to describe the silicate speciation, and $Al(OH)_3$ as a representation of Al-OH formation in the melt:

$$2M^{m+}_{1/m}Al_mO_{2m} + {}^3/_m H_2O + 2Q^n \rightleftharpoons {}^2/_m Al(OH)_3 + 2Q^{n-1}(1/mM^{m+}), \quad (8)$$

Additional complexity may exist because in melts with mixed M^{m+} cations and H^+, the H^+, because of its size for steric reasons, exhibits preference for forming OH groups in the silicate portion of the network by reacting with the nonbridging oxygen in Q^n species with the largest number of nonbridging oxygens (Cody et al. 2005).

Transformation of tetrahedrally coordinated Al^{3+} OH-bearing complexes, therefore, results in silicate depolymerization wherein the rate of depolymerization with H_2O increases the more aluminous the melt (Figure 15). This depolymerization (increasing NBO/T) takes place because as the tetrahedrally coordinated Al^{3+} interacts with H_2O to form Al-OH bonds, an equivalent proportion of the charge-balancing cation becomes a network-modifier, or charge-balancing Na^+ for Na-OH bonds leading to network-modifying Al^{3+}, or both. This depolymerization is also reflected in water concentration-dependent Q^n abundance as a function of total water content (Figure 16) where the abundance of individual Q^n species at a given total water content also varies with the Al/(Al + Si) of the melt (Mysen 2007). The example from the Na_2O-Al_2O_3-SiO_2-H_2O system shows how depolymerized species, Q^3 and Q^2, become more important as more polymerized species, Q^4, abundance decreases (Figure 16).

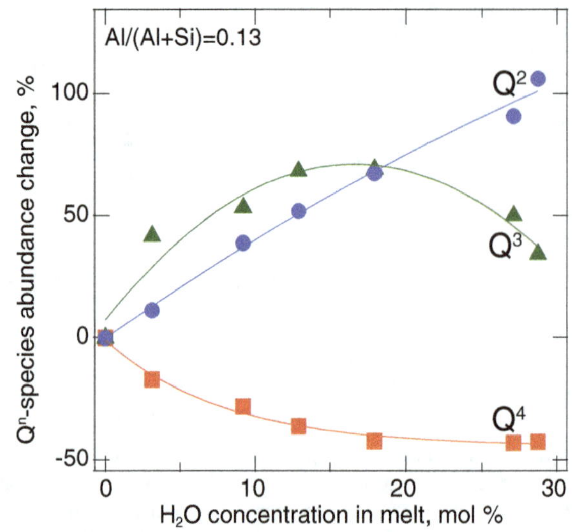

Figure 16 Evolution of Q^n species abundance. In quenched (at 100°C/s from 1,400°C at 1.5 GPa) hydrous Na-aluminosilicate melts as a function of H_2O content at fixed Al/(Al + Si) (data from Mysen 2007).

The decreasing $\partial(\text{NBO/T})/\partial X_{H_2O}$ with increasing water content, X_{H_2O}, reflects the decreasing rate of change of the abundance ratio, $X_{OH}/X_{H_2O^0}$, as the total water content increases. This evolution, in turn, reflects the diminishing rate by which the Q^n species changes with increasing concentration of water in the melt (Figure 16). These composition-dependent solution mechanisms of water in aluminosilicate melts also explain why the solubility of water in silicate melt at any temperature and pressure is significantly dependent on the bulk chemical composition of the melt itself.

Hydrous melts and aqueous fluids at high temperature and pressure

Experimental protocols have recently been implemented for examination of fluids and melts in hydrous silicate systems at deep crustal and mantle pressures and temperatures *in situ* while the sample is at the desired pressure and temperature conditions. Structural data obtained under such conditions are, therefore, increasingly available from all regions of silicate-H_2O phase diagrams (Figure 4). In considering such data, commonly obtained in so-called hydrothermal diamond anvil cells (e.g., Bassett et al. 1994), the experiments usually are conducted in such a manner that pressure is a variable dependent on temperature. This means that increasing temperature normally is associated with increasing pressure. In the following discussion, this is the case unless otherwise indicated.

High-temperature/high-pressure *in situ* examination of hydrous silicate systems includes characterization of pure H_2O, characterization of the silicate components, and mixtures between the two. For the H_2O component itself, the diminishing extent of hydrogen bonding with temperature is the only structure observable. There are, however, differences between hydrogen bonding of water dissolved in melts (typically supercooled liquids), of silicate-saturated aqueous fluid, and pure H_2O. The ΔH for hydrogen bond formation in such melts is near 10 and about 22 kJ/mol for silicate-saturated aqueous fluid (Figure 17). The ΔH for pure H_2O is slightly above 25 kJ/mol (Walrafen et al. 1986). Hydrogen bonding cannot be detected spectroscopically at temperatures above 500°C to 550°C (Foustoukos and Mysen 2012; Mysen 2013).

In hydrous silicate systems, the ratio of mol fraction of water species, $X_{OH}/X_{H_2O^0}$, in fluid, melt, and supercritical fluid varies with temperature (Mysen 2010b; see also Figure 18). The $X_{OH}/X_{H_2O^0}$ ratio of water in fluid and melt converges at the second c.p. (Figure 18). The existence of OH groups in all three phases (melt, fluid, supercritical fluid) implies structural interaction between water and silicate components. In fact, structural data obtained from vibrational spectroscopy indicate that in any silicate systems, the types of Q^n species in fluids, melts, and supercritical fluids resemble one another (Mysen 2009) although their concentration at any temperature and pressure depends on the silicate composition and whether in fluids or melts. The latter differences are evident in the NBO/T of the silicate melt being considerably lower (the melt is more polymerized) than coexisting fluid (Figure 19). The NBO/T values approach each other with increasing temperature and pressure until they merge at c.p. Interestingly, at higher temperature and pressure above those that define the c.p., the silicate in supercritical aqueous fluids becomes further depolymerized (NBO/T increases) (see Figure 19).

The NBO/T variations illustrated in Figure 19 reflect temperature- and pressure-dependent Q^n species in fluids and melts. Their partition coefficients, $K_{Qn}^{\text{fluid/melt}}$, at any temperature and pressure are sensitive functions

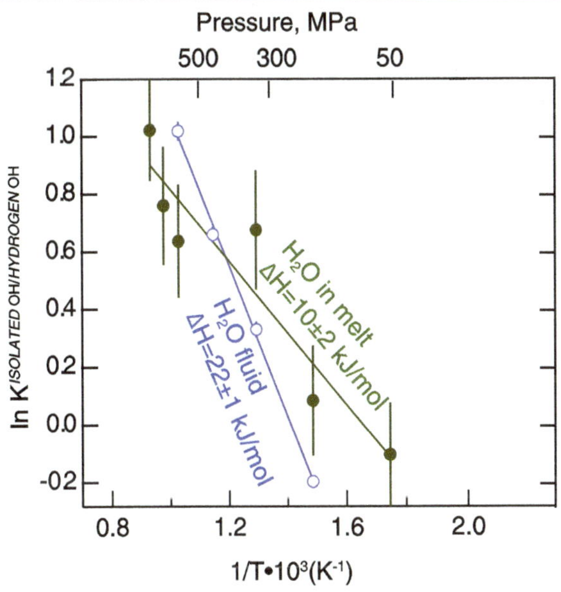

Figure 17 Evolution of ratio of hydrogen bonded to isolated OH bonds. In hydrous Na-aluminosilicate melts and in silicate-saturated aqueous fluids as a function of temperature (and pressure) (data from Mysen 2012a, b.)

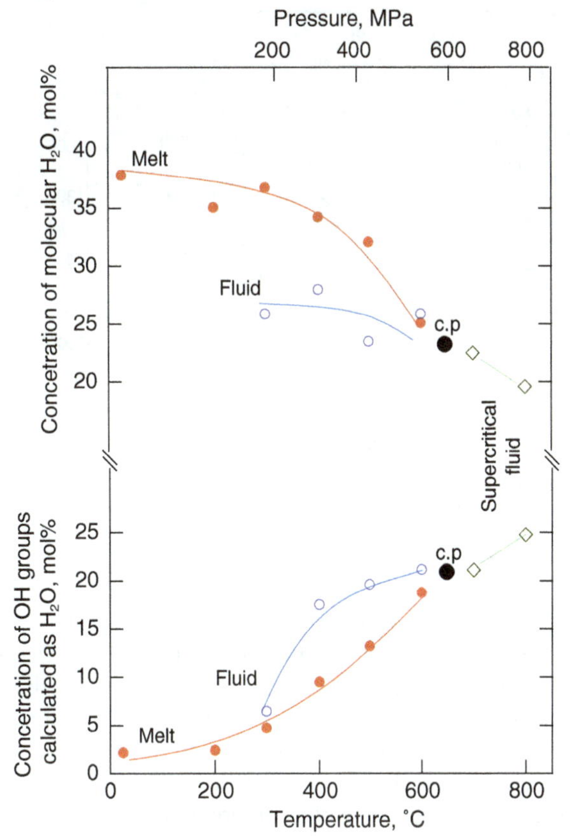

Figure 18 Abundance evolution (mol%) of OH groups and molecular H_2O. In coexisting melts and fluids and in supercritical fluid as a function of temperature (and pressure) in a hydrous Na-aluminosilicate system (data from Mysen 2010a).

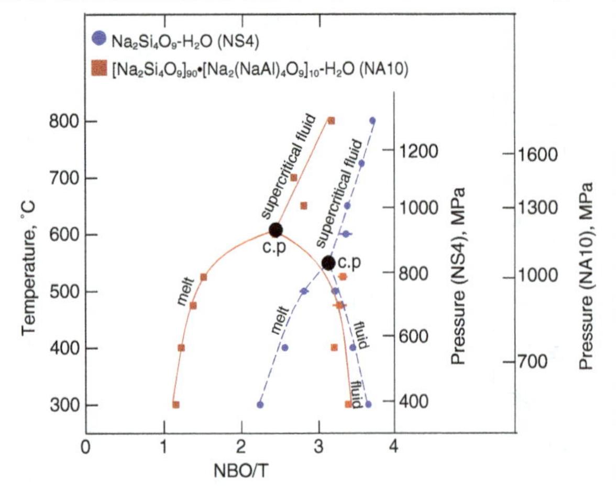

Figure 19 Degree of silicate polymerization, NBO/T. In silicate in hydrous melts, silicate-saturated aqueous fluid, and supercritical fluid in a hydrous Na-aluminosilicate system (data from Mysen 2012a, b)

of the degree of polymerization of the species (Figure 20). The $K_{Qn}{}^{fluid/melt}$ of the least polymerized Q^n species is the largest and then decreases systematically as the n value increases. Moreover, this effect becomes increasingly pronounced as the temperature and pressure decreases. The latter effect results from two factors. First, the aluminosilicate solubility in aqueous fluid decreases with decreasing pressure and temperature. The lowered silicate concentration leads to increasing abundance of less polymerized Q^n species in the fluid (Mysen 2010a). Second, the extent to which the silicate species in the hydrous melts are affected by dissolved water is a positive function of water content (Figure 16) and water solubility decreases with decreasing pressure (and temperature).

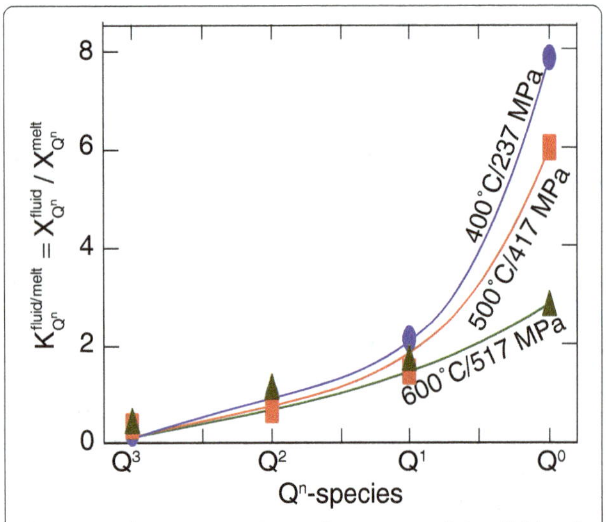

Figure 20 Q^n species partitioning between coexisting fluid and melt. As a function of type of Q^n species, temperature, and pressure in a hydrous Na-aluminosilicate system (data from Mysen 2012a, b).

Both effects would cause the $K_{Qn}{}^{fluid/melt}$ partition coefficients to increase.

In the compositionally simple silica-H_2O system in the 400°C to 900°C temperature range at pressures at and below about 1.4 GPa, only Q^0 and Q^1 species in the fluid can be detected. The equilibrium constant for the polymerization equilibrium in fluids is

$$K = X_{Q1}/X_{Qo}, \tag{9}$$

with ΔH = 13.3 ± 1.5 kJ/mol (Mysen 2010b) if it is assumed that $\Delta V = 0$ for this equilibrium, and, therefore, that the increasing pressure governed by increasing temperature does not affect the equilibrium. However, the partial molar volume of Q^0 species is smaller than that of Q^1 species (Bottinga and Richet 1995) so that with the likelihood of $\Delta V > 0$, the ΔH = 13.3 kJ/mol is a minimum value. At higher pressures (≥1.8 ± 0.2 GPa; see Mysen et al. 2013), the positive pressure- and temperature-dependent total silica concentration in fluid leads to further polymerization of the silicate species,

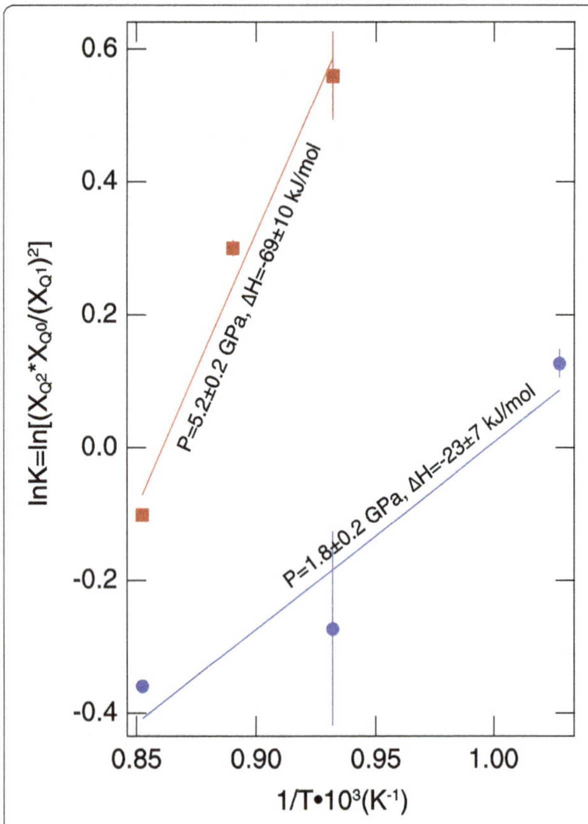

Figure 21 Temperature (1/T) vs. ln K relationship for the equilibrium, $2Q^1 \rightleftharpoons Q^0 + Q^2$, (Equation 10). From the system SiO_2-H_2O with aqueous fluid in equilibrium with quartz (lower pressure) or coesite (higher pressure). Data from Mysen et al. (2013).

and Q^2 species are also observed (Mysen et al. 2013) with the Q^n species equilibrium,

$$2Q^1 \rightleftharpoons Q^0 + Q^2, \quad (10)$$

for which the ΔH clearly is pressure dependent (Figure 21). The pressure/ΔH relationship reflects the fact the ΔV of reaction (10) is between −1 and −2 cm^3/mol (Mysen et al. 2013).

The system SiO_2-H_2O is too simple for modeling natural processes because neither Al_2O_3 nor alkali metals and alkaline earths are involved. The system Na_2O-Al_2O_3-SiO_2 is more realistic even though alkaline earths, in particular, have not yet been addressed. The Na_2O-Al_2O_3-SiO_2 system is also useful for characterization of the chemical interaction between nonbridging oxygen and important network-modifying cation (Na^+) and protons (H^+).

From NMR spectroscopy of quenched melts in this system, there is a strong preference of H^+ for nonbridging oxygen in the least polymerized of the Q^n species (Q^0), whereas Na^+ forms bonds with nonbridging oxygen in the more polymerized Q^n species (Cody et al. 2005; see also Equation 7). From the *in* situ structural characterization in alkali aluminosilicate systems, a more complex reaction (Mysen 2010a)

$$12Q^3(\mathrm{M}) + 13H_2O \rightleftharpoons 2Q^2(\mathrm{M}) + 6Q^1(\mathrm{M}) + 4Q^0(\mathrm{H}), \quad (11)$$

describes the equilibrium. Here, M denotes an alkali metal and where (M) and (H) indicate where alkali metals and protons form bonding with the relevant nonbridging oxygens, respectively. The tetrahedrally coordinated cations forming the Q^n species can be either Al^{3+} or Si^{4+}. The ΔH for this equilibrium (350 to 400 kJ/mol) is the same, within uncertainty for melt fluids and supercritical fluids (Mysen 2010a). With increasing Al/(Al + Si), equilibrium (11) likely shifts to the left (ΔH decreases) because Al^{3+} will preferably occupy the most polymerized of available Q^n species (Merzbacher and White 1991). In Equation 11, that species is Q^3, but for other more polymerization melt compositions, Q^4 species likely also would be involved. The principles outlined in equilibrium (11) may also be applied to alkaline earths, but quantitative information awaits further experiments. These are all considerations necessary for the application of the experimental data to hydrous magmatic systems. However, the necessary experimental data are at present insufficiently comprehensive for quantitative application.

Discussion

Dissolved water and melt properties

The evolution of Q^n species abundance of a melt with water content can be used to characterize how dissolved water governs phase relations and mixing properties of melts. For example, the rapid abundance increase of depolymerized species at the expense of polymerized species with water content of the melt enhances the stability of depolymerized relative to more polymerized liquidus phases. An example of this effect can be seen in the silica/pyroxene liquidus boundary of the Mg_2SiO_4-$CaMgSi_2O_6$-SiO_2 system where the silica polymorph is more polymerized than pyroxene (their NBO/Si are 0 and 2, respectively) (Figure 2A). The magnitude of liquidus boundary shifts will reflect the size of the NBO/Si difference between the minerals coexisting along a liquidus boundary. The greater the NBO/Si difference, the greater is the effect of water on the shift of the boundary as illustrated, for example, by the different shifts of olivine/pyroxene and pyroxene/silica liquidus boundaries (Kushiro 1969). Effects such as these also explain why hydrous melts in equilibrium with peridotite mineral assemblages is more silica rich (quartz normative andesitic melts) than under anhydrous conditions (olivine normative tholeiitic melts) during melting and crystallization in the upper mantle.

Configurational properties of melts can be linked to transport and thermodynamic properties of melts (Adam and Gibbs 1965; Richet 1984; Lee and Stebbins 1999). For example, melt viscosity, η, is a function of configurational entropy:

$$\eta = A_e \exp\left(B_e / T S^{conf}\right) \quad (12)$$

where A_e and B_e are constants, T is temperature, and S^{conf} is configurational entropy. The S^{conf} is described in terms of configurational heat capacity:

$$S^{conf}(T) = \left[C_p^{conf}(T_g)/T_g\right] + \int_{T_g}^{T} \left(C_p^{conf}/T\right) dT, \quad (13)$$

where T_g is glass transition temperature, T is the temperature of interest ($T > T_g$), and $C_p^{conf}(T_g)$ is the configurational heat capacity at the glass transition temperature. This configurational heat capacity of hydrous magmatic melts is a systematic function of the water content of the melt (Richet et al. 1996; Bouhifd et al. 2006).

Changes in configurational heat capacity resulting from H_2O dissolved in the melt can be calculated from the experimental data on Q^n species abundance in hydrous melts combined with the knowledge of the partial molar heat capacity of individual species (Richet and Neuville 1992; see also Mysen 2007). For example, with such information, the effects of water on the heat capacity change across the glass transition measured for a

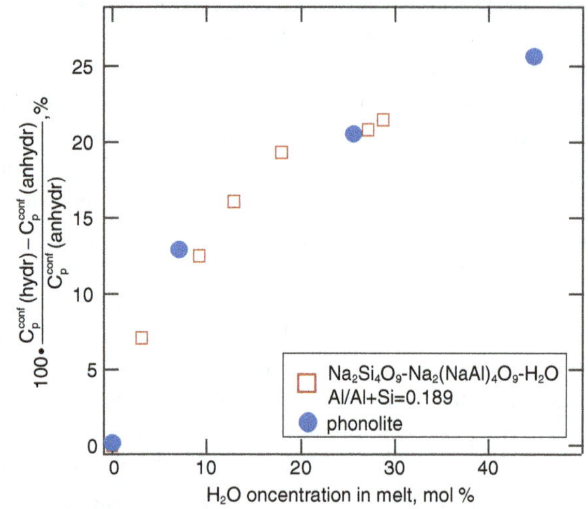

Figure 22 Calculated and observed change of configuration heat capacity. Comparison of calculated and observed change of configuration heat capacity, $C_{p,}{}^{conf}$ across the glass transition for a hydrous phonolite melt as a function of water content of the melt. Experimental data for phonolite from Bouhifd et al. (2006). Structural and thermodynamic data used to derive partial molar configurational entropy of Q^n species are from Mysen (2010a) and Richet and Neuville (1992), respectively. This calculation was originally reported in Mysen (2007).

phonolite melt can be modeled quite precisely from a hydrous haplophonolitic composition in the Na_2O-Al_2O_3-SiO_2 system (Figure 22). The NBO/T and Al/(Al + Si) of this haplophonolite melt resemble that of the natural phonolite melt composition. The good match between calculated and observed configurational heat capacity changes with water content illustrates how those compositional parameters are central to characterization of configurational properties of melts.

Water and fluid-melt equilibria

Trace and minor element partitioning is also sensitive to melt polymerization (Mysen and Virgo 1980; Kohn and Schofield 1994; Jaeger and Drake 2000; Toplis and Corgne 2002). From the relationship between NBO/T and water contents of such melts

$$NBO/T = 1.10 - 0.57 \bullet e^{-0.068 X_{H_2O}}, \tag{14}$$

and NBO/T and $K_{Mg}{}^{ol\text{-}melt}$ (Kushiro and Mysen 2002; see also Figure 23A), changes in the olivine/melt partition coefficient with water content can be calculated (Figure 23B). In the calculation in Figure 23, it is assumed that there is no structural interaction between the silicate components and the element in question (Mg). Under such circumstances, the NBO/T changes in melt have effects on partition coefficients in ways analogous to NBO/T changes affected by adding other oxide components. The 25% change in the Mg partition, which reached with near 20 mol% H_2O, is not an unreasonable situation during island arc melting and crystallization (see, for example, Grove et al. 2012, for review of data).

Stable isotope fractionation

Vibrational spectroscopy has also been used to determine D/H ratios in fluids and melts in equilibrium during experiments in diamond anvil cells at the high temperatures and pressures of interest (Foustoukos and Mysen 2012; Mysen 2013). In silicate-saturated fluids and hydrous silicate melts, D^+ and H^+ forms bonding with nonbridging oxygen in the Q^n species. From deuterium and proton MAS NMR of silicate melts, a D/H fractionation between the different structural positions has been found (Wang et al. 2011). This means that if the proportion and/or type of Q^n species vary with temperature and/or pressure so that the D/H fractionation within and between fluids and

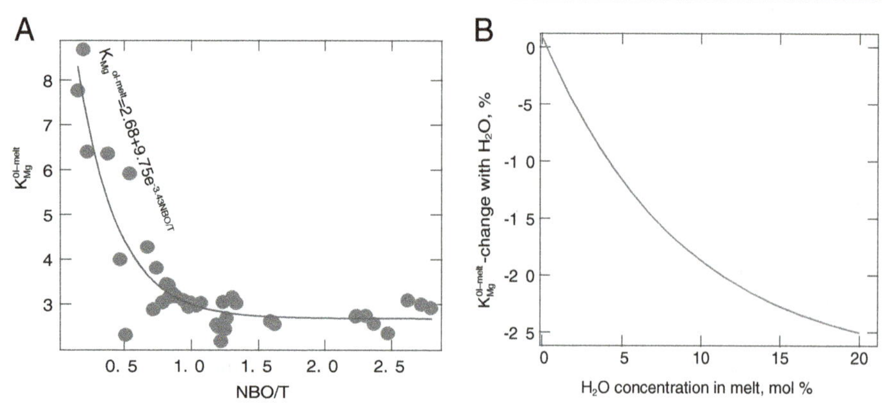

Figure 23 Olivine-melt partition coefficients for Mg. As a function of the NBO/T of the melt from experimental data in the Na_2O-K_2O-CaO-MgO-FeO-Al_2O_3-SiO_2 system (Kushiro and Mysen 2002) coupled with the effect of dissolved H_2O on NBO/T (Mysen 2007) of melts to derive effect of dissolved water in melt on the partition coefficient as discussed by Mysen 2012a, b. **(A)** The $K_{Mg}{}^{ol\text{-}melt}$ data fitted to the equation shown in this figure. **(B)** Resulting calculated change in Mg partition coefficients as a function of water content.

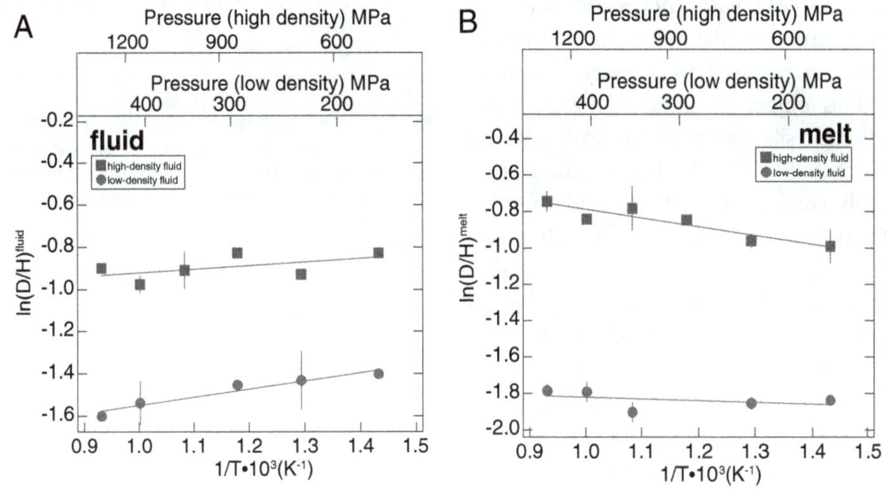

Figure 24 Evolution of D/H ratio of coexisting fluid (A) and melt (B) for (H_2O + D_2O)-bearing Na-aluminosilicate melts. As a function of different fluid densities. In these experiments, the fluid density governs the relationship between temperature and pressure as shown in the figures. Data from Mysen 2013.

melts also become temperature and pressure dependent (Figures 24 and 25).

The temperature and pressure effects in Figure 24 reflect the changes in silicate concentration and speciation in the coexisting aqueous fluid and silicate melts. Furthermore, different D/H values for different fluid densities again reflect the concentration and type of Q^n species because the different density trajectories reflect pressure/temperature trajectories of the diamond cell experiments that followed different pressure paths (Mysen 2013). It follows from the different trajectories of D/H ratios in fluids and melts that the D/H partitioning between fluids and melts will also be dependent on temperature and pressure (Figure 25). Finally, in silicate mineral/aqueous fluid or silicate mineral/hydrous melt environments, temperature- and pressure-dependent partitioning will also take place because of the temperature/pressure dependence of the D/H fractionation with the fluid and melt phases.

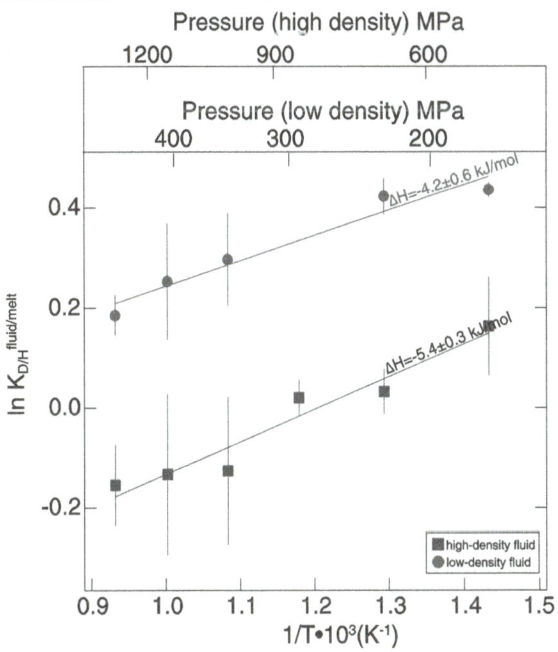

Figure 25 (D/H) fluid-melt partition coefficients for coexisting fluid and melt (Figure 24). For low- and high-density aqueous fluid from the data in Figure 24.

Conclusions

Understanding the role of hydrous melts and silicate-rich fluids in transport processes depends sensitively on how chemical composition, temperature, and pressure govern the physicochemical properties of these materials (viscosity, diffusion, thermodynamics of mixing, element partitioning between phases, phase relations, etc.; see, for example, Mysen and Richet 2005, for review of experimental and theoretical information of structure-property relations of silicate glasses and melts). Attainment of this objective requires well-determined solubilities in and partitioning between melts, fluids, and crystalline phases, detailed understanding of the oxide solution mechanisms in melts and fluids, and determination of how fluid and melt structure governs those properties.

Structural information cannot be obtained directly on chemically complex natural systems because the resolving power of spectroscopic methods employed for such purposes diminishes rapidly with increasing chemical complexity. However, structural data from simpler binary and ternary systems can be used to describe the more complex systems. This objective requires, however, detailed characterization of silicate speciation in fluids and melts as a function of Al/Si ratio, the type of charge balance for tetrahedrally coordinated cations, and the type and proportion of network-modifying metals (alkali metals and alkaline earths). Currently, a combination of structural information and empirical relationships can be applied to describe liquidus phase relations and fluid/mineral/melt element and stable isotope partitioning. Transport properties can be understood and sometimes quantified in terms of configurational properties of individual Q^n species. However, the experimental database used for these purposes dominantly is from alkali silicate and alkali aluminosilicate systems. This permits application to felsic magmatic systems. However, the lack of much information in alkaline earth aluminosilicate system makes quantitative application to hydrous basaltic less quantitative.

Competing interests

The author declares that he has no competing interests.

Acknowledgements

Many of the ideas and conclusions discussed in this review were developed during the preparation of invited lectures at Tohoku University, Japan in 2013. Portions of the reviewed research were supported by grants from the National Science Foundation (US) (EAR-0707861, EAR1250449, EAR12151931 and EAR-1212754).

References

Adam G, Gibbs JH (1965) On the temperature dependence of cooperative relaxation properties in glass-forming liquids. J Chem Phys 43:139–146

Antignano A, Manning CE (2008) Rutile solubility in H_2O, H_2O -SiO_2, and H_2O-$NaAlSi_3O_8$ fluids at 0.7-2.0 GPa and 700–1000 degrees C; implications for mobility of nominally insoluble elements. Chem Geol 255:283–293

Audetat A, Keppler H (2004) Viscosity of fluids in subduction zones. Science 303:513–516

Audetat A, Keppler H (2005) Solubility of rutile in subduction zone fluids, as determined by experiments in the hydrothermal diamond anvil cell. Earth Planet Sci Lett 232:393–402

Ayers JC, Watson EB (1993) Apatite/fluid partitioning of rare-earth elements and strontium; experimental results at 1.0 GPa and 1000 degrees C and application to models of fluid-rock interaction. Chem Geol 110:299–314

Bassett WA, Shen AH, Bucknum M, Chou IM (1994) A new diamond cell for hydrothermal studies to 2.5 GPa and from −190°C to 1200°C. Rev Sci Instrum 64:2340–2345

Behrens H, Jantos N (2001) The effect of anhydrous composition on water solubility in granitic melts. Amer Miner 86:14–20

Behrens H, Nowak M (1997) The mechanisms of water diffusion in polymerized silicate melts. Contrib Miner Petrol 126:377–385

Behrens H, Meyer M, Holtz F, Nowak M (2001) The effect of alkali ionic radius, temperature, and pressure on the solubility of water in $MAlSi_3O_8$ melts (M=Li, Na, K, Rb). Chem Geol 174:275–289

Bernini D, Audetat A, Dolejs A, Keppler H (2013) Zircon solubility in aqueous fluids at high temperatures and pressures. Geochim Cosmochim Acta 119:178–187

Boettcher AL (1984) The system SiO_2-H_2O-CO_2: Melting solubility mechanisms of carbon and liquid structure to high pressures. Amer Miner 69:823–834

Bottinga Y, Richet P (1995) Silicate melts: The "anomalous" pressure dependence of the viscosity. Geochim Cosmochim Acta 59:2725–2732

Bouhifd MA, Whittington A, Roux J, Richet P (2006) Effect of water on the heat capacity of polymerized aluminosilicate glasses and melts. Geochem Cosmochim Acta 70:711–722

Boyd FR, England JL, Davis BCT (1964) Effect of pressure on the melting and polymorphism of enstatite, $MgSiO_3$. J Geophys Res 69:2101–2109

Buckermann W-A, Muller-Warmuth W, Frischat GH (1992) A further ^{29}Si MAS NMR study on binary alkali silicate glasses. Glasstechnische Bericht 65:18–21

Bureau H, Keppler H (1999) Complete miscibility between silicate melts and hydrous fluids in the upper mantle; experimental evidence and geochemical implications. Earth Planet Sci Lett 165:187–196

Burnham CW (1975) Thermodynamics of melting in experimental silicate-volatile systems. Geochim Cosmochim Acta 39:1077–1084

Cody GD, Mysen BO, Lee SK (2005) Structure vs. composition: A solid state ^1H and ^{29}Si NMR study of quenched glasses along the Na_2O-SiO_2-H_2O join. Geochim Cosmochim Acta 69:2373–2384

Dingwell DB, Webb SL (1990) Relaxation in silicate melts. Eur J Mineral 2:427–449

Dixon JE, Stolper EM (1995) An experimental study of water and carbon dioxide solubilities in mid-ocean ridge basaltic liquids; Part II, Applications to degassing. J Petrol 36:1633–1646

Einstein A (1905) Über die von der molekular-kinetischen Theorie der Wärme geforderte Bewegung von in ruhenden Flüssigkeiten suspendierten Teilchen. Ann Phys 17:549–560

Eyring H (1935a) The activated complex in chemical reactions. J Chem Phys 3:107–115

Eyring H (1935b) The activated complex and the absolute rate of chemical reactions. Chem Rev 17:65–77

Farnan I, Kohn SC, Dupree R (1987) A study of the structural role of water in hydrous silica using cross-polarisation magic angle NMR. Geochim Cosmochim Acta 51:2869–2874

Foustoukos DI, Mysen BO (2012) D/H isotopic fractionation in the H_2-H_2O system at supercritical water conditions: Composition and hydrogen bonding effects. Geochim Cosmochim Acta 86:88–102

Foustoukos DI, Mysen BO (2013) H/D methane isotopologues dissolved in magmatic fluids: Stable hydrogen isotope fractionations in the Earth's interior. Amer Miner 98:946–954

Grove TL, Elkins-Stanton LT, Parman SW, Chatterjee N, Muntner O, Gaetani GA (2003) Fractional crystallization and mantle-melting controls on calc-alkaline differentiation trends. Contrib Miner Petrol 145:515–543

Grove TL, Till CB, Krawczynski MJ (2012) The roleof H_2O in subduction zone magmatism. Ann Rev Earth Planet Sci 40:413–439

Hack A, Thompson AB (2012) Density and viscosity of hydrous magmas and related fluids and their role in subduction zone processes. J Petrol 52:1333–1362

Hamilton DL, Burnham CW, Osborn EF (1964) The solubility of water and the effects of oxygen fugacity and water content on crystallization of mafic magmas. J Petrol 5:21–39

Hess K-U, Dingwell DB (1996) Viscosities of hydrous leucogranitic melts: A non-Ahrrenian model. Amer Miner 81:1297–1300

Holloway JR, Jakobsson S (1986) Volatile solubilities in magmas: Transport of volatiles from mantles to planet surfaces. J Geophys Res 91:505–508

Holtz F, Behrens H, Dingwell DB, Johannes W (1995) H_2O solubility in haplogranitic melts: Compositional, pressure, and temperature dependence. Amer Miner 80:94–108

Hunt JD, Manning CE (2012) A thermodynamic model for the system SiO_2 -H_2O near the upper critical end point based on quartz solubility experiments at 500–1100 degrees C and 5–20 kbar. Geochim Cosmochim Acta 86:196–213

Jackson I (1976) Melting of the silica isotypes SiO_2, BeF_2 and GeO_2 at elevated pressures. Phys Earth Planet Int 13:218–231

Jaeger WL, Drake MJ (2000) Metal-silicate partitioning of Co, Ga, and W; dependence on silicate melt composition. Geochim Cosmochim Acta 64(22):3887–3895

Kawamoto T, Ochiai S, Kagi H (2004) Changes in the structure of water deduced from the pressure dependence of the Raman OH frequency. J Chem Phys 120:5867–5870

Kennedy GC, Wasserburg GJ, Heard HC, Newton RC (1962) The upper three-phase region in the system SiO_2-H_2O. Amer J Sci 260:501–521

Kohn SC, Schofield PF (1994) The importance of melt composition in controlling trace-element behaviour: An exprimental study of Mn and Zn partitioning between forsterite and silicate melt. Chem Geol 117:73–87

Kurkjian CR, Russell LE (1958) Solubility of water in molten alkali silicates. J Soc Glass Techn 42:130T–144T

Kushiro I (1969) The system forsterite-diopside-silica with and without water at high pressures. Amer J Sci 267-A:269–294

Kushiro I (1972) Effect of water on the composition of magmas formed at high pressures. J Petrol 13:311–334

Kushiro I (1990) Partial melting of mantle wedge and evolution of island arc crust. J Geophys Res 95:15929–15939

Kushiro I, Mysen BO (2002) A possible effect of melt structure on the Mg-Fe^{2+} partitioning between olivine and melt. Geochim Cosmochim Acta 66:2267–2273

Kushiro I, Yoder HS, Nishikawa M (1968a) Effect of water on the melting of enstatite. Geol Soc Amer Bull 79:1685–1692

Kushiro I, Syono Y, Akimoto SI (1968b) Melting of a peridotite nodule at high pressures and high water pressures. J Geophys Res 73:6023–6029

Lee SK, Stebbins JF (1999) The degree of aluminum avoidance in aluminum silicate glasses. Amer Miner 84:937–945

Lee SK, Cody GD, Fei Y, Mysen BO (2004) Nature of polymerization and properties of silicate melts at high pressure. Geochem Cosmochim Acta 68:4189–4200

Luth WC, Jahns RH, Tuttle OF (1964) The granite system at pressures of 4 to 10 kilobars. J Geophys Res 69:759–773

Maekawa H, Maekawa T, Kawamura K, Yokokawa T (1991) The structural groups of alkali silicate glasses determined from ^{29}Si MAS-NMR. J Non-Cryst Solids 127(1):53–64

Manning CE (1994) The solubility of quartz in H2O in the lower crust and upper mantle. Geochim Cosmochim Acta 58:4831–4840

Manning CE (2004) The chemistry of subduction-zone fluids. Earth Planet Sci Lett 223:1–16

McMillan PF (1994) Water solubility and speciation models. In: Carroll MR, Holloway JR (ed) Volatiles in magmas. Mineralogical Society of America, Washington DC, pp 131–156

McMillan PF, Holloway JR (1987) Water solubility in aluminosilicate melts. Contr Miner Petrol 97:320–332

Merzbacher CI, White WB (1991) The structure of alkaline earth aluminosilicate glasses as determined by vibrational spectroscopy. J Non-Cryst Solids 130:18–34

Mibe K, Kanzaki M, Kawamoto T, Matsukage KN, Fei Y, Ono S (2007) Second critical endpoint in the peridotite-H2O system. J Geophys Res 112. doi: 10.1029/2005JB004125

Moulson AJ, Roberts JP (1961) Water in silica glass. Trans Faraday Soc 57:1208–1216

Mysen BO (2002) Water in peralkaline aluminosilicate melts to 2 GPa and 1400°C. Gechim Cosmochim Acta 66:2915–2928

Mysen BO (2007) The solution behavior of H2O in peralkaline aluminosilicate melts at high pressure with implications for properties of hydrous melts. Geochim. Cosmochim. Acta, 71, 1820–1834. Geochim Cosmochim Acta 71:1820–1834

Mysen BO (2009) Solution mechanisms of silicate in aqueous fluid and H_2O in coexisting silicate melts determined in-situ at high pressure and high temperature. Geochim Cosmochim Acta 73:5748–5763

Mysen BO (2010a) Structure of H_2O-saturated peralkaline aluminosilicate melt and coexisting aluminosilicate-saturated aqueous fluid determined in-situ to 800°C and ~800 MPa. Geochim Cosmochim Acta 74:4123–4139

Mysen BO (2010b) Speciation and mixing behavor of silica-saturated aqueous fluid at high temperature and pressure. Amer Miner 95:1807–1816

Mysen BO (2011) An experimental study of phosphorous and aluminosilicate speciation in and partitioning between aqueous fluids and silicate melts determined in-situ at high temperature and pressure. Amer Miner 96:1636–1649

Mysen BO (2012a) High-pressure/-temperature titanium solution mechanisms in silicate-saturated aqueous fluids and hydrous silicate melts. Amer Miner 97:1241–1251

Mysen BO (2012b) Silicate-COH melt and fluid structure, their physicochemical properties, and partitioning of nominally refractory oxides between melts and fluids. Lithos 148:228–246

Mysen BO (2013) Hydrogen isotope fractionation between coexisting hydrous melt and coexisting silicate-saturated aqueous fluid: An experimental study in-situ at high pressure and temperature. Amer Miner 98:376–386

Mysen BO, Virgo D (1980) Trace element partitioning and melt structure; an experimental study at 1 atm pressure. Geochim Cosmochim Acta 44(12):1917–1930

Mysen BO, Virgo D (1986) Volatiles in silicate melts at high pressure and temperature. 1. Interaction between OH groups and Si^{4+}, Al^{3+}, Ca^{2+}, Na^+ and H^+. Chem Geol 57:303–331

Mysen BO, Wheeler K (2000) Solubility Behavior of Water in Haploandesitic Melts at high Pressure and high Temperature. Amer Miner 85:1128–1142

Mysen BO, Cody GD (2004) Solubility and solution mechanism of H_2O in alkali silicate melts and glasses at high pressure and temperature. Geochem Cosmochim Acta 68:5113–5126

Mysen BO, Richet P (2005) Silicate glasses and melts: properties and structure. Developments in geochemistry. Elsevier, New York

Mysen BO, Mibe K, Chou I-M, Bassett WA (2013) Structure and equilibria among silicate species in aqueous fluids in the upper mantle: Experimental SiO_2-H_2O and MgO-SiO_2-H_2O data recorded in-situ to 900°C and 5.4 GPa. J Geophys Res 118. doi:10.1002/2013JB010537

Nernst W (1888) Zur Kinetik der in lösung befindtlicher Körper. Erste Abhandlung, Theorie der Diffusion. Z Phys Chem 2:613–637

Newton RC, Manning CE (2002) Solubility of enstatite+forsterite in H_2O in deep crust/upper mantle conditions: 4 to 15 kbar and 700 to 900°C. Geochim Cosmochim Acta 66:4165–4176

Newton RC, Manning CE (2008) Solubility of corundum in the system Al_2O_3–SiO_2–H_2O–NaCl at 800 °C and 10 kbar. Chem Geol 249:250–261

Nowak M, Behrens H (1995) The speciation of water in haplogranitic glasses and melts by in-situ, near-infrared spectroscopy. Geochim Cosmochim Acta 59:3445–3450

Ochs FA, Lange RA (1999) The density of hydrous magmatic liquids. Science 283:1314–1317

Ohtani E, Taulelle F, Angell CA (1985) Al^{3+} coordination changes in liquid silicates under pressure. Nature 314:78–81

Pascal ML, Anderson GM (1989) Speciation of Al, Si, and K in supercritical solutions: Experimental study and interpretation. Geochim Cosmochim Acta 53:1843–1856

Poli S, Schmidt MW (2002) Petrology of subducted slabs. Ann Rev Earth Planet Sci 30:207–235

Richet P (1984) Viscosity and configurational entropy of silicate melts. Geochim Cosmochim Acta 48:471–483

Richet P, Neuville DR (1992) Thermodynamics of silicate melts: configurational properties. In: Saxena SK (ed) Thermodynamic data: systematics and estimation. Advances in physical geochemistry, vol 10. Springer, New York, pp 132–161

Richet P, Lejeune A-M, Holtz F, Roux J (1996) Water and the viscosity of andesite melts. Chem Geol 128:185–197

Richet P, Neuville DR, Holtz F (1997) Viscosity of water-bearing melts. Terra Nova Abstr Supple 9(480)

Roy BN, Navrotsky A (1984) Thermochemistry of charge-coupled substitutions in silicate glasses:-The systems $M_n^{n+}AlO_2$ - SiO_2 (M=LI, Na, K, Rb, Cs, Mg, Ca, Sr, Ba, Pb). J Amer Ceram Soc 67:606–610

Satherley K, Smedley SI (1985) The electrical conductivity of some hydrous and anhydrous molten silicates-as a function of temperature and pressure. Geochim Cosmochim Acta 49:769–777

Schmidt BC, Riemer T, Kohn SC, Holtz F, Dupree R (2001) Structural implications of water dissolution in haplogranitic glasses from NMR spectroscopy: Influence of total water content and mixed alkali effect. Geochim Cosmochim Acta 65:2949–2964

Spallanzani L (1798) *Viaggi alle Due Sicilie e in alcune parti dell' Appennino*, transl. as *Travels in the Two Sicilies and Some Parts of the Appenines*, J. Robinson, London. Stamperia di B, Comini, Pavia

Spera FJ (1974) A thermodynamic basis for predicting water solubilties in silicatemelts and implications for the low velocity zone. Contr Mineral Petrol 45:175–186

Stebbins JF (1987) Identification of multiple structural species in silicate glasses by ^{29}Si NMR. Nature 330:465–467

Stern CR, Huang W-L, Wyllie PJ (1975) Basalt-andesite-rhyolite-H_2O: Crystallization intervals with excess H_2O and H_2O-undersaturated liquidus surfaces to 35 kilobars, with implications for magma genesis. Earth Planet Sci Lett 28:189–196

Stolper E (1982) The speciation of water in silicate melts. Geochim Cosmochim Acta 46:2609–2620

Takata M, Acocella J, Tomozawa M, Watson EB (1981) Effect of water content on the electrical conductivity of $Na_2O{\cdot}3SiO_2$ glass. J Amer Ceram Soc 64:719–724

Toplis MJ, Corgne A (2002) An experimental study of element partitioning between magnetite, clinopyroxene and iron-bearing silicate liquids with particular emphasis on vanadium. Contrib Mineral Petrol 144:22–37

Tuttle OF, Bowen NL (1958) Origin of granite in light of experimental studies in the system $NaAlSi_3O_8$-$KAlSi_3O_8$-SiO_2-H_2O. Geol Soc Amer Mem 74:1–153

Virgo D, Mysen BO, Kushiro I (1980) Anionic constitution of 1-atmosphere silicate melts: implications of the structure of igneous melts. Science 208:1371–1373
Walrafen GE, Yang WH, Chu YC (1996) Raman OD-stretching overtone spectra from liquid D_2O between 22 and 152 decrees C. C. J Phys Chem 100:1381–1391
Wang YB, Cody SX, Cody GD, Mysen BO (2011) ^2H and ^1H NMR study on hydrogen isotope effects in silicate glasses. Paper presented at the 2011 AGU Fall Meeting, San Francisco, CA
Wasserburg GJ (1957) The effects of H_2O in silicate systems. J Geol 65:15–23
Watson EB (1994) Diffusion in volatile-bearing magmas. In: Carroll MR, Holloway JR (ed) Volatiles in magmas: reviews in mineralogy, 30th edition. Mineralogical Society of America, Washington, DC, pp 371–411
Whittington APR, Holtz F (2000) Water and the viscosity of depolymerized silicate melts. Geochim Cosmochim Acta 64:3725–3736
Xue Y, Kanzaki M (2004) Dissolution mechanisms of water in depolymerized silicate melts: Constraints from ^1H and ^{29}Si NMR spectroscopy and ab initio calculations. Geochim Cosmochim Acta 68:5027–5057
Zhang Y-G, Frantz JD (2000) Enstatite-forsterite-water equilibria at elevated temperatures and pressures. Amer Miner 85:918–925
Zhang Y (1999) H_2O in rhyolitic glasses and melts: measurements, speciation, solubility, and diffusion. Rev Geophys Space Phys 37:493–516
Zotov N, Keppler H (1998) The influence of water on the structure of hydrous sodium tetrasilicate glasses. Amer Miner 83:823–834

2

Nitrogen dynamics in a highly urbanized coastal area of Western Japan: impact of sewage-derived loads

Mitsuyo Saito[1*], Shin-ichi Onodera[2], Guangzhe Jin[2,3], Yuta Shimizu[4] and Masanobu Taniguchi[5]

Abstract

In this study, we examined the nitrogen dynamics of a highly urbanized coastal area, focusing on the impacts of sewage-derived nitrogen. High levels of dissolved inorganic nitrogen were detected in seawater near treated sewage effluent (TSE) discharge points before decreasing in the offshore direction, suggesting that the impact zone of sewage effluent is about 1–2 km from the discharge point. The stable isotope ratios of nitrate and particulate organic nitrogen suggest nitrogen uptake by phytoplankton as well as dilution by offshore seawater, which contributed to a decrease in sewage-derived nitrogen levels. However, the extent of the impact zone was controlled by tidal variations and differences in temperature between the TSE and seawater. Our results also identify nitrogen transport processes, through exchange between seawater and sediment pore water, as an additional important source of nitrogen in the study area.

Keywords: Nitrogen dynamics, Coastal area, Urbanization, Sewage, Sediment

Introduction

Human activities significantly influence the global nitrogen cycle and natural water environments such as rivers and coastal seas (Vitousek et al. 1997; Gruber and Galloway 2008). Marine eutrophication in nearshore environments has been attributed to anthropogenic inputs of nitrogen (Nixon 1995; Howarth and Marino 2006). Water pollution from domestic sewage and industrial waste, in particular, is a critical environmental problem related to the early stages of urbanization, where there is significant growth in population (Onodera 2011). For the Osaka metropolitan area in Japan, pollution of the river and coastal areas was significant until the 1970s. Although this has been improved via the development of sewage treatment systems, the removal of nitrogen requires advanced processing, and relatively large proportions of nitrogen are still discharged in treated sewage effluent to coastal areas due to the absence of this treatment. Powley et al. (2016) estimated that nitrogen inputs from the direct discharge of treated and untreated wastewater from coastal cities to the Mediterranean Sea are of the same order of magnitude as riverine inputs. The effects of excess nitrogen supply from wastewater treatment effluent influence ecosystem structure, as well as nitrogen cycling in receiving waters (Waiser et al. 2011; Drury et al. 2013).

Garside et al. (1976) confirmed the seasonal and spatial distribution of sewage-derived nitrogen and its contribution to much of the nitrogen in the Hudson Estuary, New York, and showed that primary production was not limited by nitrogen availability. Costanzo et al. (2001) applied the nitrogen stable isotope ratios ($\delta^{15}N$) of marine plants (seagrass, attached macroalgae, and mangroves) to detect and map the impacts of sewage-derived nitrogen and confirmed $\delta^{15}N$ was elevated near sewage outfalls in Moreton Bay, Australia. Savage (2005) also applied the $\delta^{15}N$ of macroalgae (*Fucus vesiculosus*) and coastal sediments to trace the influence of sewage-derived nitrogen. She confirmed this influence was most pronounced within 10 km of the sewage outfall of the Himmerfjärden embayment of the Baltic Sea.

Eyre (2000) demonstrated the seasonal cycle of nutrient transformation and phytoplankton growth in river-dominated sub-tropical east Australian estuaries. He confirmed that wastewater-derived dissolved inorganic

* Correspondence: misaito@okayama-u.ac.jp
[1]Graduate School of Environmental and Life Science, Okayama University, 3-1-1 Tsushima-naka, Kita-ku, Okayama 700-8530, Japan
Full list of author information is available at the end of the article

nitrogen (DIN) significantly influenced phytoplankton biomass in the estuaries. Pennino et al. (2016) examined the transformations of nitrogen derived from a wastewater treatment facility, based on the nitrogen mass balance and the $\delta^{15}N$ and oxygen stable isotope ratios ($\delta^{18}O$) in the tidal Potomac River estuary from Washington, D.C., to the Chesapeake Bay. Estuaries and shallow bays are dynamic environments, influenced by tidal variations and inflow of freshwater from rivers and channels. Therefore, examination of the temporal dynamics of nitrogen in the sewage impact zone is important. However, the detailed transport process and dynamics of sewage-derived nitrogen were not examined in these previous studies. In this study, we examined the nitrogen dynamics of a highly urbanized coastal area of western Japan that is influenced by sewage-derived nitrogen loads.

Methods/Experimental

Study area

The study area is located in the northeastern part of Osaka Bay in the Seto Inland Sea in western Japan (Fig. 1). The Seto Inland Sea is the largest semi-enclosed coastal sea in Japan. The area is characterized by a temperate climate, with a mean annual precipitation of 1270 mm and a mean temperature of 15.1 °C. Osaka Bay is one of the most enclosed bays in the inland sea and has a water surface area of 1447 km^2 and mean depth of 30.4 m (International EMECS Center 2008). The watershed population of the bay is approximately 16 million, including residents of the major cities of Osaka, Kobe, Kyoto, and Sakai (Nishida et al. 2012). Rapid economic growth progressed urbanization significantly in the Osaka metropolitan area from the 1950s to the 1970s. During this period, increased nutrient loads from domestic and industrial activities induced serious eutrophication and an oxygen deficit in the seawater of the bay (ANSIOB 1996; Nakatsuji 1998). Since the 1970s, the regulation of chemical oxygen demand (COD), phosphorus, and nitrogen loads and the development of sewage treatment systems have decreased nutrient loads and the presence of organic matter in the Bay (International EMECS Center 2008; GESAP 2017).

The target area for this study is the inner bay, near the cities of Osaka and Sakai (Fig. 1). Almost 100% of these

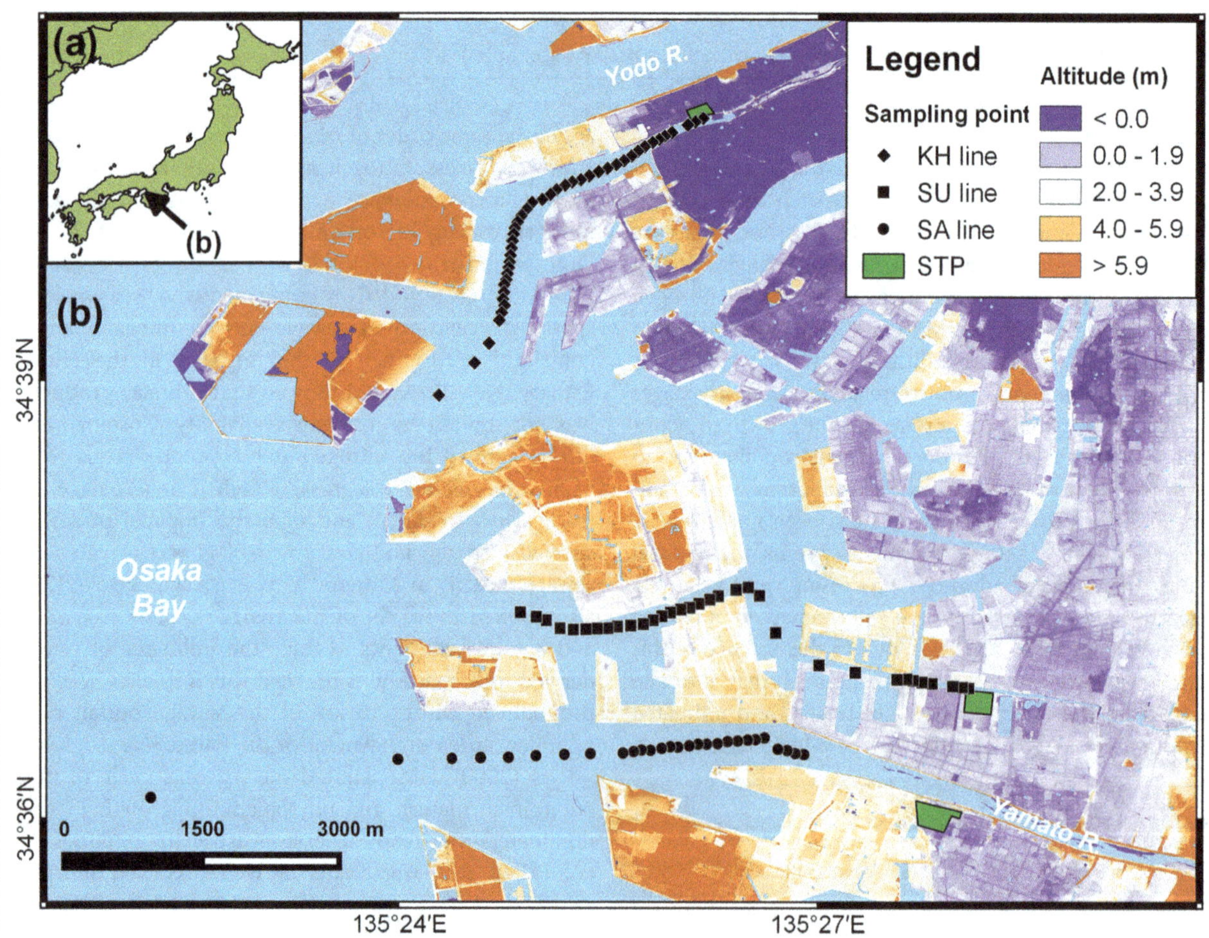

Fig. 1 Map of the study area (**a** location of the study area, **b** locations of seawater sampling points on the KH, SU and SA lines and the STPs)

cities are covered by sewage systems, with 12 public sewage treatment plants (STPs) in Osaka and 3 in Sakai (OPG 2016). However, the removal of nitrogen from this wastewater remains insufficient for preventing nitrogen loading, as most STPs have not yet installed advanced treatment facilities for dissolved nitrogen, such as nitrification–denitrification processes. In Osaka, the average concentration of total nitrogen (TN) and TN load of the influent to these plants are 26 mg L^{-1} and 44.0 t day^{-1}, respectively, which are reduced to 11 mg L^{-1} and 18.6 t day^{-1} in the treated sewage effluent (Osaka City 2016). In Sakai, the average concentration of TN and TN load of the influent to the Sambo STP are 30 mg L^{-1} and 2.3 t day^{-1}, respectively, which are reduced to 3.4 mg L^{-1} and 0.3 t day^{-1} in the treated sewage effluent (SCWSB 2015). In addition, DIN, such as nitrate (NO_3^-) and ammonium (NH_4^+), is a major component of the TN in the treated sewage effluent (TSE) (Osaka City 2017). This indicates that a relatively large proportion of the DIN is discharged in the effluent of these STPs to rivers and coastal areas. Most of the sewage systems in Osaka are "combined systems," which result in increased discharge after high rainfall events because of an inflow of storm water (Broadhead et al. 2013). Therefore, an increase in sewage-derived nitrogen loads from heavy rain has been recognized as a new problem in recent years. We focused on three STPs: Konohana (KH) and Suminoe (SU) in Osaka and Sambo (SA) in Sakai, which are all located in areas of reclaimed land near the coastal bay (STP in Fig. 1). The designed treatment populations for the STPs are 64,302 for KH, 386,431 for SU, and 175,500 for SA (OPG 2017). The treated effluents from KH and SU are discharged directly to the bay, whereas the effluent from SA is discharged to the mouth of the Yamato River.

Field campaigns and sample collection

Field campaigns were conducted in August and October 2014, February 2015, and September 2016 using a fishing boat. Three observation lines (KH line, SU line, and SA line; Fig. 1) were defined from the discharge points of the effluent from the STPs to approximately 5 km offshore. However, we could not access the discharge point of SA by boat because it is located in the Yamato River channel and the water depth is very shallow (< 2 m). There are no freshwater discharge flows of natural rivers and streams from land to the KH line except for its TSE; however, there is a small channel between the Yodo River and the line around 2.5 km from the discharge point (Fig. 1). The KH and SU lines are narrow channels, with bottom materials mainly composed of sludge with high organic matter content. The tidal range, tidal condition, air temperature, 24 h antecedent rainfall, observed lines, and average concentration of dissolved total nitrogen (DTN) in the TSE for the four sampling periods (JMA 2014–2016; Osaka City 2017) are listed in Table 1. The tidal range was larger in February 2015 and September 2016 than in August and October 2014. The air temperature was significantly lower in February 2015 than in the other periods. A 24 h antecedent rainfall was only observed in September 2016. The KH line was observed during three different periods to confirm the seasonal and temporal changes of nitrogen dynamics in the study area. The average DTN concentration in the TSE was highest for KH (14–18 mg L^{-1}) and lowest for SA (approximately 4 mg L^{-1}) (SCWSB 2015; Osaka City 2017).

Vertical profiles of water temperature and electrical conductivity (EC) were measured about every 300 m along the observation lines using a conductivity temperature depth (CTD) sensor (CTD Diver, Schlumberger) and that of chlorophyll-a (Chl-a) concentration via a fluorescent chlorophyll and turbidity sensor (INFINITY-CLW, JFE Advantech Co., Ltd.).

Seawater samples were collected at the surface at approximately 100 m intervals and from less than 1 m above the bottom of the coastal seabed at 300 to 1000 m intervals along the observation lines. At 600 to 1000 m intervals along the observation lines, sediment samples from the surface of the seabed were collected using an Ekman grab sampler. TSE, river water, sediment pore water, and offshore seawater were collected as endmembers of coastal seawater in the study area. TSE samples were collected from land at the effluent discharge point

Table 1 Tide and weather conditions and average DTN concentration in the TSE during the field campaigns

Year	Date	Tidal range (m)	Tidal condition	Air temperature (°C)	24 h antecedent rainfall (mm)	Observed lines	Average DTN concentration in TSE (mg L^{-1})
2014	31 Aug	0.8	Ebb-Rising	27–29	0	SU Line	8
2014	18 Oct	0.6	Rising	15–22	0	KH Line	14
						SA Line	4[a]
2015	7 Feb	1.3	Ebb	5–10	0	KH Line	18
2016	18 Sep	1.4	Ebb Rising	25–28	6.5	KH Line	14[b]

[a]Average annual concentration
[b]Average concentration in September 2014 and 2015

from SA in August 2013. River water samples were collected 2 km upstream from the mouth of the Yamato River in August 2013 and September 2014. Sediment pore water was collected using piezometers installed 50 cm deep using a PVC pipe at the same location for the river water sampling in the Yamato River in August 2013. Offshore seawater samples were collected approximately 7 km from the STP on the SA line in September and October 2014. Water samples were filtered in the field using 0.2 μm cellulose ester filters for analysis of nutrients and stable isotopes and 0.7 μm glass microfiber filters for analysis of particulate organic matter (POM). The samples of water and filtrates were stored in a freezer until analysis.

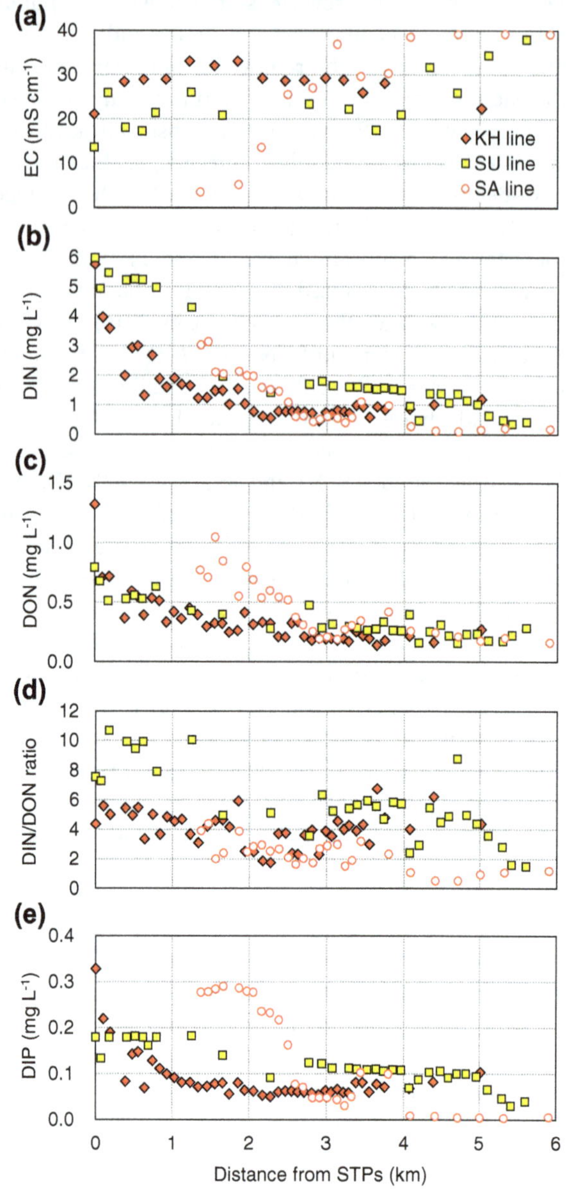

Fig. 2 Spatial variation of **a** EC, **b** DIN, **c** DON, **d** the molar ratio between DIN and DON (DIN/DON), and **e** DIP in the surface seawater on the SU line in August and the KH and SA lines in October 2014

Chemical and isotopic analyses

DIN as a combination of NO_3^-, nitrite (NO_2^-), and NH_4^+; DTN; and dissolved inorganic phosphorus (DIP) measured as phosphate (PO_4^{3-}) in water samples were determined using a continuous-flow automated analyzer (SWAAT; BL TEC K. K.). Concentrations of dissolved organic nitrogen (DON) were calculated from the difference between DTN and DIN. The organic nitrogen content and $\delta^{15}N$ of POM and surface sediment samples were determined using an isotope ratio mass spectrometer (Delta V Advantage and Flash EA 1112; Thermo Fisher Scientific, Inc.). The $\delta^{15}N$ in NO_3^- was determined using the bacterial denitrifier method (Sigman et al. 2001) with an isotope ratio mass spectrometer (Delta V Plus; Thermo Fisher Scientific, Inc.).

Results and discussion

Impact zone of sewage-derived nitrogen

Figure 2 shows the spatial variations in 2014 of EC, DIN, DON, the molar ratio between DIN and DON (DIN/DON), and DIP in the surface seawater along the three observation lines (KH, SU, and SA), starting from the discharge point of the treated sewage and ending approximately 5 km offshore. The KH and SA lines were observed in October, whereas the SU line was observed in August. The tidal and weather conditions, excluding the air temperature, were similar between these two periods (Table 1). The EC was lowest (< 21 mS cm^{-1}) and the concentration of nutrients and DON was highest (DIN > 3.0 mg L^{-1}, DIP > 0.17 mg L^{-1}, and DON > 0.7 mg L^{-1}) at the STPs. DIN, mostly NO_3^- and NH_4^+, was a few times higher than DON, because DIN is a major form of nitrogen derived from TSE, as reported previously (Osaka City 2017). The DIN concentration for the SU line was higher than that for the KH line within 1–2 km of the STPs (Fig. 2b). As already mentioned, the DTN concentration in the TSE was lower for SU (8 mg L^{-1}) than for KH (14 mg L^{-1}) (Table 1); however, the designed treatment population for SU is six times larger than that for KH (OPG 2017). This indicates that the impact of TSE-derived nitrogen is more significant for the SU line. In addition, the higher temperature conditions in August may have accelerated the decomposition of organic matter and the nitrification process, resulting in elevated DIN concentrations. EC increased and nutrients and DON decreased significantly within 1–2 km of the STPs through mixing with seawater of the KH and SU lines. These results indicate that the impact zone of TSE-derived nitrogen is about 1–2 km offshore from the

STPs. The significantly low EC (< 10 mS cm^{-1}) and relatively high DON and DIP in this zone of the SA line may have been influenced by the Yamato River (Fig. 2a–e). The DTN and dissolved total phosphorus (DTP) concentrations in the TSE from SA were approximately 4 and 0.1 mg L^{-1}, respectively (SCWSB 2015). This suggests that the DIP increased (> 0.2 mg L^{-1}) as a result of phosphorus desorption via mixing of seawater (Fox et al. 1986). However, the DTN (DIN+DON) concentration in this zone was approximately 4 mg L^{-1}, which was a similar level to that of the TSE. In addition, DIN/DON was similar among the lines. These results suggest that TSE-derived nitrogen as well as the Yamato River may have influenced the SA line.

Seasonal and temporal changes in the sewage impact zone

The spatial variations of water temperature, EC, Chl-a, NO_3^--N, and NH_4^+-N in the surface seawater of the KH line between October 2014 and February 2015 are shown in Fig. 3. Water temperature was lower in February than in October because of the difference in air temperature (Table 1). The EC was lower and the NO_3^--N and NH_4^+-N concentrations were significantly higher in February than in October from the STP to 2 km offshore. The relatively high Chl-a concentration suggests that phytoplankton were still actively taking up nutrients in February (Fig. 3c). These results suggest that the sewage impact zone expanded in February. As noted previously, the flow of the Yodo River may affect the results for the KH line through the small channel approximately 2.5 km from the STP (Fig. 1). This indicates the effects of the TSE caused the expansion of the impact zone.

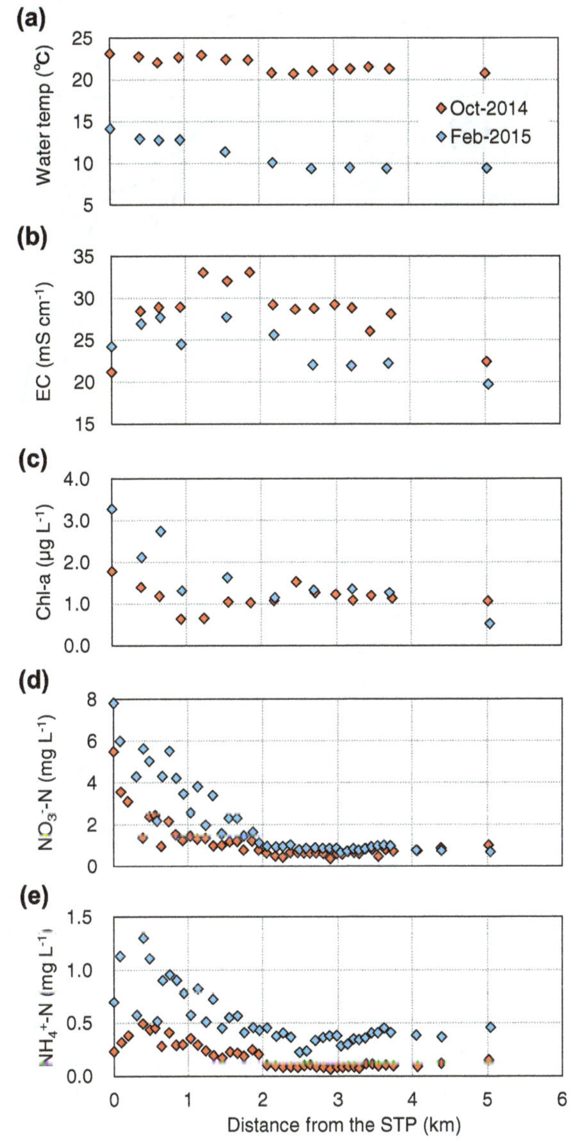

Fig. 3 Spatial variation of **a** water temperature, **b** EC, **c** Chl-a, **d** NO_3^--N, and **e** NH_4^+-N in the surface seawater on the KH line in October 2014 and February 2015

A slightly higher DTN concentration in the TSE in February (18 mg L^{-1}) than in October (14 mg L^{-1}) may have increased the NO_3^--N and NH_4^+-N levels near the STP (Table 1). The tidal range was greater in February than in October, and the February observation was conducted during ebb tide (Table 1). This suggests that the plume of TSE-derived nitrogen was transported offshore with less mixing and dilution by seawater. The water temperature was nearly equal to the air temperature in October; however, it was higher than that of air temperature (> 10 °C) from the STP to approximately 2 km offshore in February (Fig. 3a, Table 1). This suggests that another possible cause for the expansion of the impact zone is the difference in water temperature between the TSE and the surface seawater. TSE is usually discharged at more than 20 °C throughout the year because of the operating parameters of the biological activated sludge treatment process (Hashimoto and Sudo 1986). This would intensify the buoyancy of the effluent plume and allow the plume to extend further offshore during periods of low seawater temperature. The effect of thermal effluent from STPs on water bodies has previously been reported for urban river systems (Nakamuro et al. 2006; Kinouchi et al. 2007).

Next, we discuss the nitrogen dynamics at approximately 2.5 km from the STP, to examine the effect of TSE-derived nitrogen. As noted above, the influence of the Yodo River inflow should be considered for more than 2.5 km offshore from the STP on the KH line. Vertical profiles of EC and Chl-a from the STP to 2.5 km offshore on the KH line during the ebb and rising tide periods in September 2016 are shown in Fig. 4. The surface EC at 2.5 km from the STP was relatively high compared to that of the other sites (Fig. 4a). This

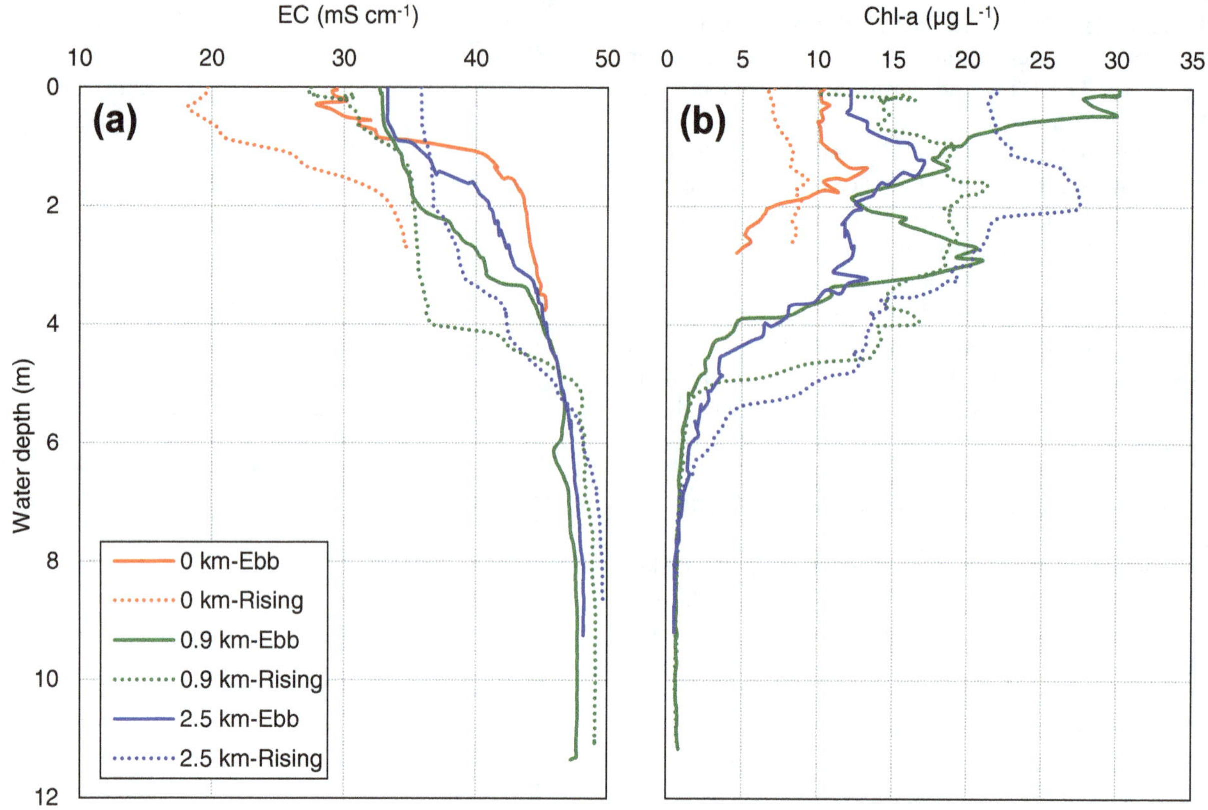

Fig. 4 Vertical profiles of **a** EC and **b** Chl-a from the STP to 2.5 km offshore on the KH line during the ebb and rising tide periods in September 2016

suggests that the influence of the Yodo River inflow was small within 2.5 km of the STP. The EC at depths of less than 5 m was relatively high during the ebb tide because the tidal levels in the observation period were higher during the ebb tide than during the rising tide. Both the EC and the Chl-a concentrations varied significantly between the surface and bottom for both periods. The EC was relatively low and the Chl-a concentration was high at the surface compared to those at the bottom. These results suggest that the surface seawater is significantly affected by TSE-derived nitrogen. The vertical profiles in Fig. 4 suggest that the spatial and temporal variations in DIN should be compared between these two depths.

Figure 5 shows the spatial variations of EC, NO_3^--N, and NH_4^+-N at the surface and less than 1 m above the bottom of the seabed for seawater on the KH line in October 2014 and September 2016. The EC was lower at the surface than at the bottom in both periods (Fig. 5a, b). The NO_3^--N concentration was relatively higher at the surface than that at the bottom; however, it was significantly lower in September than in October at the surface (Fig. 5c, d). These EC and NO_3^--N results suggest that the TSE-derived nitrogen was diluted by rainwater in September because of the antecedent rainfall (Table 1). The NH_4^+-N concentration was at a similar level (< 0.6 mg L^{-1}) between the two depths in October; however, it was significantly higher at the bottom than at the surface in September (Fig. 5e, f).

The results during the ebb and rising tides were compared in September 2016. The EC was lower during the rising tide than during the ebb tide from the STP to 1.5 km offshore at the surface (Fig. 5a) but only near the STP at the bottom (Fig. 5b). Moreover, the NO_3^--N and NH_4^+-N concentrations were relatively high during the rising tide in the lower EC zone at the surface (Fig. 5c, e). These results suggest that the TSE-derived nitrogen was transported about 1.5 km offshore from the STP. In contrast, both the NO_3^--N and NH_4^+-N concentrations were relatively high during the ebb tide at the bottom. The NH_4^+-N concentration, in particular, was detected at a level more than 10 times higher than the concentration at the surface (Fig. 5d, f). These results suggest that the sources of nitrogen were different between the surface and the bottom, especially in September.

Previous research confirmed that estuarine circulation flow is dominant in the inner part of Osaka Bay (Nakajima and Fujiwara 2007; Kobayashi et al. 2017). They imply that nitrogen at the bottom layer of the KH line originated from the bottom seawater of Osaka Bay. However, the reported NH_4^+-N concentrations of the bottom seawater in

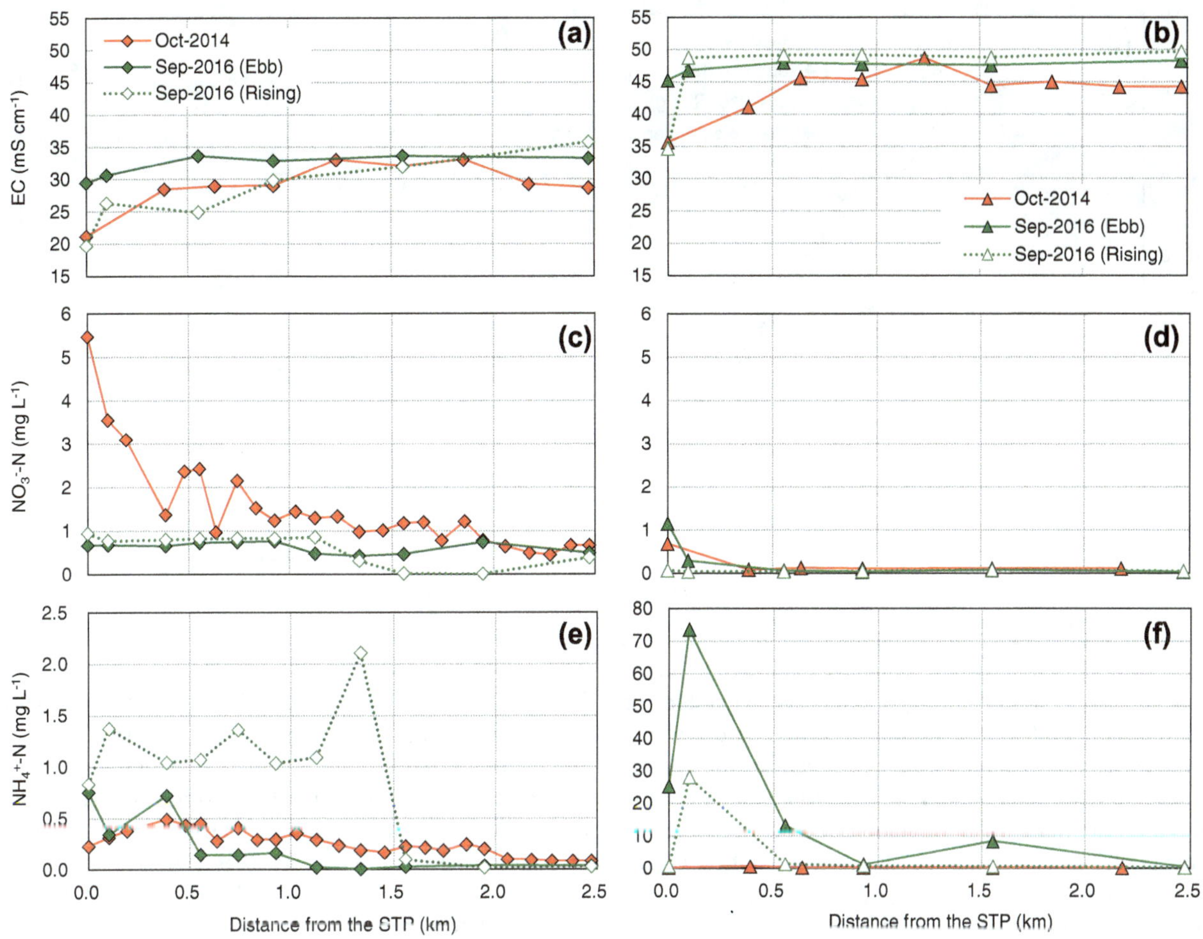

Fig. 5 Spatial variation of **a**, **b** EC, **c**, **d** NO_3^--N, and **e**, **f** NH_4^+-N in the surface (**a**, **c**, **e**) and bottom (**b**, **d**, **f**) seawater of the KH line in October 2014 and September 2016

the bay were less than 0.3 mg L^{-1} (RIEAFOP 2006-2015), which is much lower than the NH_4^+-N concentrations detected in the bottom layer of the KH line in September (Fig. 5f). On the other hand, Onodera et al. (2013) confirmed exchange between surface water and sediment pore water caused by tidal variation in the study area. These results suggest that nitrogen supply via exchange between seawater and sediment pore water occurs in the study area.

Nitrogen dynamics in the sewage impact zone

TSE-derived nitrogen was confirmed to decrease significantly within 1–2 km of the STP (Figs. 2, 3, and 5). Figure 6 shows the relationship between the EC and DIN concentrations and between the Chl-a and particulate organic nitrogen (PON) concentrations in the surface seawater on the KH line in October 2014 and September 2016. Most of the observed DIN concentrations were lower than the concentration derived via the mixing of TSE and offshore seawater (Fig. 6a). This suggests that the dilution effect via mixing with offshore seawater is a major factor in the reduction of DIN concentrations; however, the effect of nitrogen uptake by phytoplankton should also be considered. Figure 6b indicates that the Chl-a and PON concentrations correspond relatively closely to each other and that most of the PON originates from phytoplankton. This result also suggests that the biomass of phytoplankton was about 10 times higher in September 2016 than in October 2014. The Chl-a in September 2016 was also significantly higher than in February 2015 (Fig. 3c). As noted above, the bottom materials in the observation lines are mainly composed of sludge with a high content of organic matter. These results suggest that the sediment in these observation lines originated from POM as phytoplankton.

Figure 7 shows a comparison of spatial variation in PON concentrations and in $\delta^{15}N$ values for NO_3^-, PON, and surface sediment of the KH line. The PON results for October 2014 and the rising tide period in September 2016 are shown (Fig. 7a, b). The $\delta^{15}N$ values forNO_3^- and sediment for October 2014 are shown (Fig. 7b). As also shown in Fig. 6, the PON concentration was at a high level

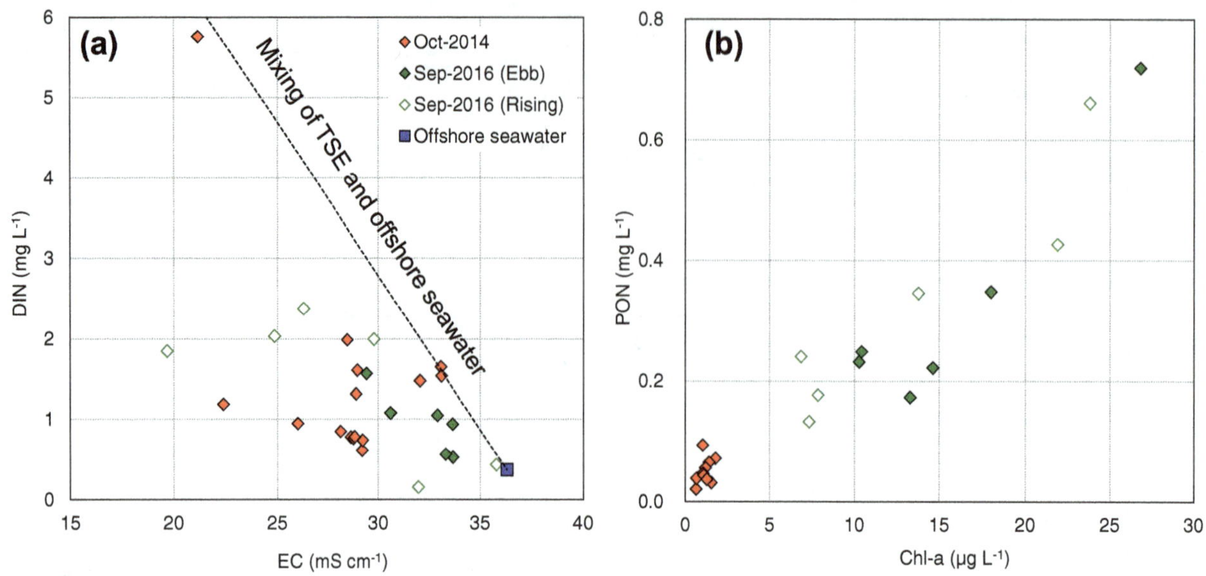

Fig. 6 Relationship between **a** the EC and DIN concentrations and **b** the Chl-a and PON concentrations in the surface seawater of the KH line in October 2014 and September 2016

in September (Fig. 7a). In Fig. 7b, the average values of $\delta^{15}N(NO_3^-)$ for offshore seawater, river water, pore water, and TSE were also shown as endmembers of the coastal seawater in the study area. The $\delta^{15}N(NO_3^-)$ value for the KH line was + 8.8 to + 10.8‰; however, the offshore seawater, river water, and TSE showed higher values (+ 12 to + 14‰), which suggests the $\delta^{15}N(NO_3^-)$ for the KH line may have been influenced by the isotope fractionation via nitrification (Kendall et al. 2007). On the other hand, the lower $\delta^{15}N(NO_3^-)$ value for pore water (+ 5‰) also suggests a contribution of sediment-derived nitrogen to the NO_3^- in the KH line.

Patterns of spatial variation of $\delta^{15}N(PON)$ were similar between the two periods (Fig. 7b). It was high at the STP owing to the influence of the TSE, and then decreased before increasing again in the offshore direction, influenced by the mixing with offshore seawater and river water. $\delta^{15}N$ was higher in NO_3^- than in PON, excluding at the STP in October. The result that NO_3^--N was a major form of DIN at the surface (Fig. 5c) suggests that $\delta^{15}N(PON)$ reflected $\delta^{15}N(NO_3^-)$ and that $\delta^{15}N(NO_3^-)$ was enriched by the nitrogen uptake of phytoplankton in October (Waser et al. 1998; Sigman et al. 1999). Waser et al. (1998) estimated through a batch culture method in a laboratory that $\delta^{15}N$ fractionation is about 5.2‰ during uptake by a coastal diatom (*Thalassiosira pseudonana*). *Thalassiosira* spp. are major diatom species observed in the study area (Oshima et al. 2009). Sigman et al. (1999) estimated that $\delta^{15}N$ fractionation by phytoplankton uptake is 4–6 ‰, based on field observations in the Southern Ocean. For the results of the current study, the average $\delta^{15}N(NO_3^-)$ value was 9.6‰ and $\delta^{15}N(PON)$ was 6.9‰ in October. The enrichment ratio was smaller than the isotope fractionation reported previously.

The $\delta^{15}N(PON)$ in September was lower within 1 km of the STP and higher than in October in the offshore direction (Fig. 7b). The fact that significantly higher NH_4^+-N concentrations were detected in September than in October (Fig. 5e, f) suggests two possibilities: $\delta^{15}N(PON)$ may have been influenced more by NH_4^+ than by NO_3^- and the $\delta^{15}N(PON)$ fractionation by phytoplankton uptake of lower $\delta^{15}N(NO_3^-)$ was derived through nitrification. On the other hand, the tidal range was more than two times greater in September than in October (Table 1). This suggests that a greater tidal range would have increased the exchange between seawater and pore water through the coastal sediment (intrusion and discharge) and increased the contribution of pore water with a low $\delta^{15}N(NO_3^-)$ in September. $\delta^{15}N$ was lower than PON for the surface sediment samples in the same period and had a similar value with the $\delta^{15}N(NO_3^-)$ for pore water (Fig. 7b). The discharge volume of TSE would have been larger in September because of the antecedent rainfall (Table 1). This condition may have intensified the estuarine circulation flow, and exchange between seawater and pore water may have increased in September. It also would have been supported by the high concentration of NH_4^+-N observed in both the bottom and surface seawater samples in September (Fig. 5e, f). The increase in nitrogen supply through sediment during this period would have caused an increase in phytoplankton and PON. However, we require data on

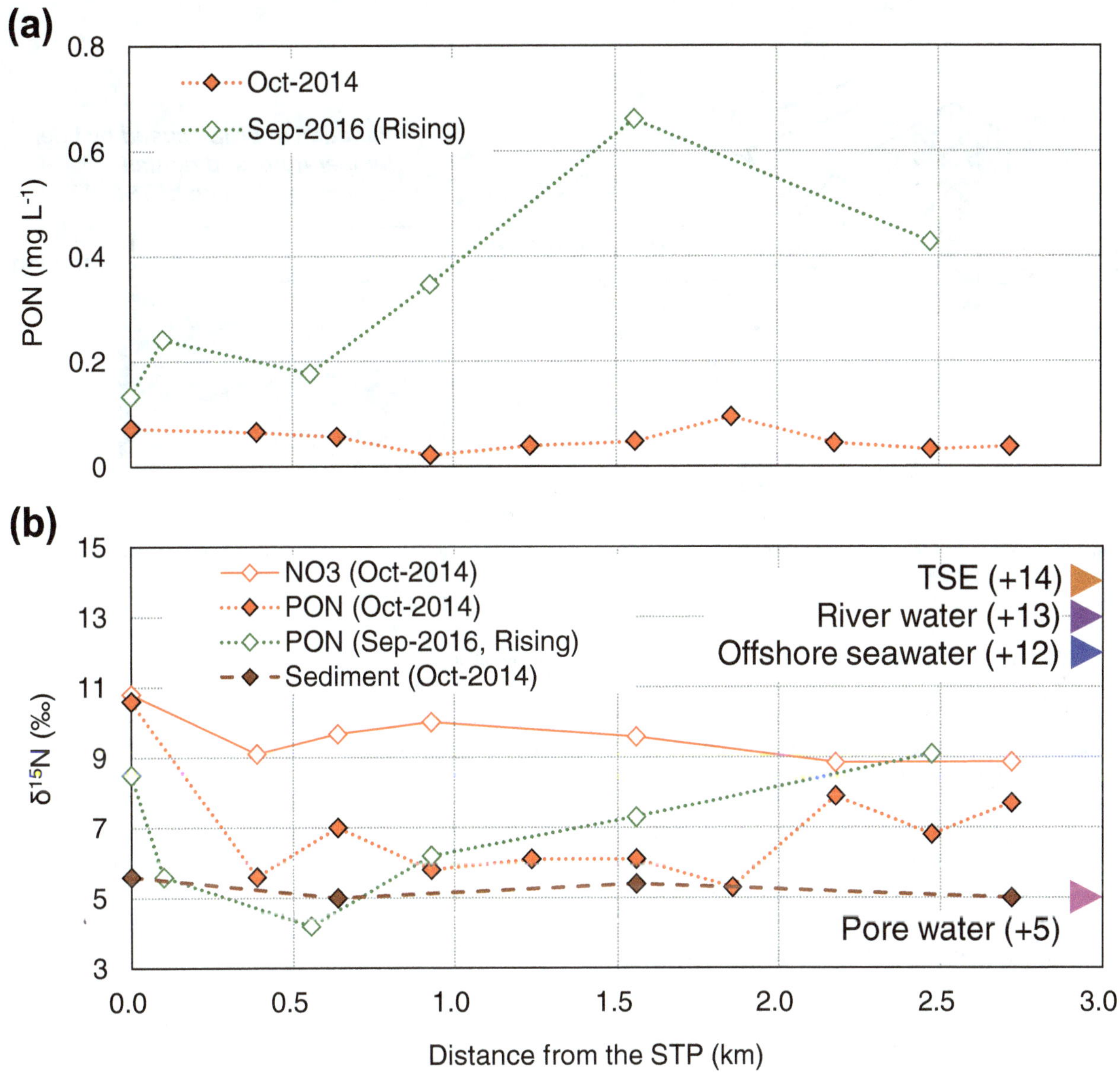

Fig. 7 Spatial variation in **a** PON concentrations and **b** $\delta^{15}N$ values for NO_3^-, PON, and surface sediment on the KH line. The values for PON were measured in October 2014 and during the rising tide period in September 2016. The $\delta^{15}N$ values for NO_3^- and sediment were measured in October 2014. The average $\delta^{15}N(NO_3^-)$ for TSE, river water, offshore seawater, and pore water from 2013, 2013 and 2014, 2014, and 2013, respectively, served as reference values

$\delta^{15}N(NH_4^-)$ to understand these processes in greater detail. Also, seasonal variation in nitrogen concentration and $\delta^{15}N(NO_3^-)$ for the endmembers, such as the Yodo River water, will need to be confirmed because previous research pointed out that $\delta^{15}N(NO_3^-)$ for river water may change according to river discharge (Takagi et al. 2016).

Based on these results, a schematic diagram of nitrogen dynamics in the coastal area, as influenced by sewage-derived loads, is shown in Fig. 8. The sewage-derived nitrogen impact zone (SNIZ) via direct discharge of TSE was estimated to be 1–2 km from the STPs on the KH and SU lines. This result also suggests that the spatial variation in SNIZ is controlled by tidal variation and the water temperature difference between the TSE and seawater. On the other hand, the SA line was confirmed to be significantly influenced by the Yamato River as well as the TSE. However, similar values of $\delta^{15}N(NO_3^-)$ for the TSE and Yamato River water suggest the Yamato River is influenced by the STPs located upstream. Conclusively, this study indicates that coastal sediment is another potential source of nitrogen and is transported via exchange between seawater and pore water. More specifically,

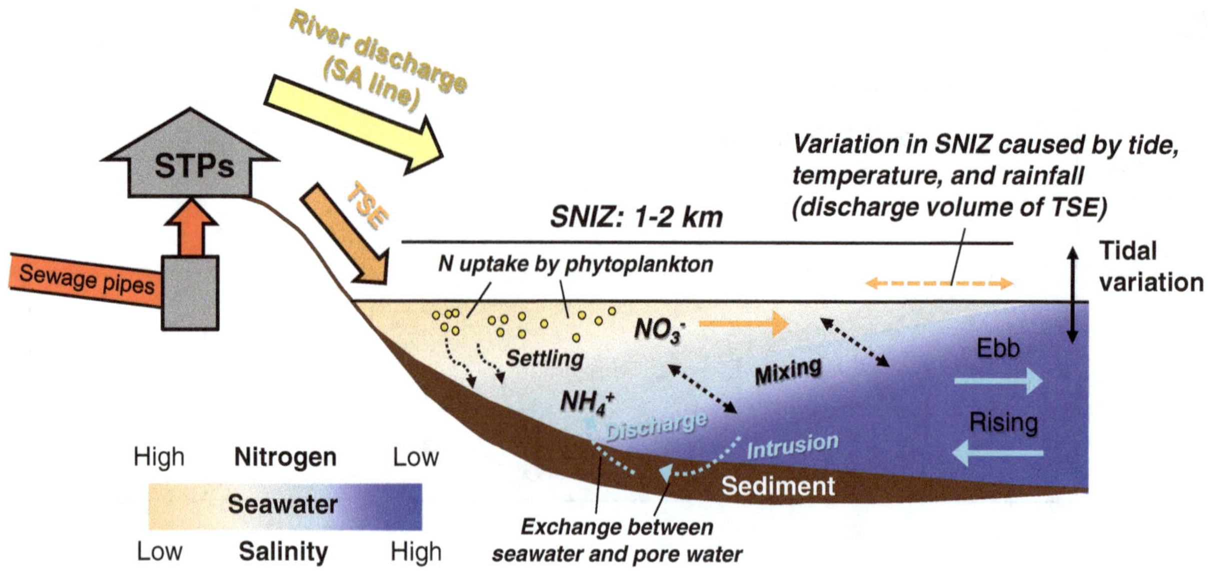

Fig. 8 Schematic diagram of nitrogen dynamics in the coastal sewage-derived nitrogen impact zone (SNIZ). Solid and broken arrows show water movement and suggested processes from the present research, respectively

the sediment may have originated from POM as phytoplankton, which is a secondary product of TSE-derived nitrogen. This suggests that the TSE-derived nitrogen in the coastal environment is not only from direct discharge but also from resupply through the sediment.

However, as the study area is influenced by tidal variation and mixing of several endmembers, the nitrogen dynamics should be quite complex compared with closed systems and the open ocean. Our results indicate that temporal changes in mixing ratios of endmembers and nitrogen uptake by phytoplankton influence both the nitrogen concentration and $\delta^{15}N$ in this study area, although further research is needed for quantitative evaluation of nitrogen transport processes.

Conclusions

In this study, nitrogen dynamics were examined for a sewage-influenced coastal area in western Japan. The impact zone of sewage-derived nitrogen by direct discharge of TSE was estimated to be 1–2 km from STPs in the study area. DIN uptake by phytoplankton as well as dilution by offshore seawater contributed to the decrease in sewage-derived nitrogen. However, the extent of the impact zone was shown to be controlled by tidal variation temporally and water temperature differences between TSE and seawater seasonally. These results also indicate that nitrogen supply via exchange between seawater and sediment pore water is another important source of nitrogen beyond the direct discharge of TSE.

Abbreviations

SNIZ: Sewage-derived nitrogen impact zone; STPs: Sewage treatment plants; TSE: Treated sewage effluent

Acknowledgements

We thank Mr. Yutaka Maruyama and Mr. Yusuke Tomozawa for their assistance in our field campaigns and analysis of the samples. MS gratefully acknowledges Dr. Seiko Yoshikawa and Dr. Masahiro Kobayashi, who were session conveners of A-HW16 in Japan Geoscience Union Meeting 2016, for their recommendation of the present study to PEPS. Finally, the authors would like to thank the two reviewers for their helpful comments and suggestions.

Funding

This work was supported by Grant for Environmental Rehabilitation and Creation of the Osaka Bay Area (Head; SO, 2014–2016, Grant Number 280007) and JSPS bilateral joint research projects (Head; SO, 2016–2018). MS was partly funded by the tenure track program of JST for Okayama University (2013–2014).

Authors' contributions

MS proposed the topic and conceived and designed the study. SO collaborated with the corresponding author in the construction of the manuscript. GJ, YS, and MT analyzed the data and helped in their interpretation. All authors read and approved the final manuscript.

Authors' information

Not applicable

Competing interests

The authors declare that they have no competing interest.

Author details

[1]Graduate School of Environmental and Life Science, Okayama University, 3-1-1 Tsushima-naka, Kita-ku, Okayama 700-8530, Japan. [2]Graduate School of Integrated Arts and Sciences, Hiroshima University, 1-7-1 Kagamiyama, Higashi-Hiroshima 739-8521, Japan. [3]Guangdong Ocean University, Huguangyan, Zhanjiang, Guandong Province 524-088, China. [4]Western Region Agricultural Research Center, National Agriculture and Food Research Organization, 6-12-1 Nishi-Fukatsu-cho, Fukuyama 721-8514, Japan. [5]Faculty of Systems Engineering, Wakayama University, Sakaedani 930, Wakayama 640-8510, Japan.

References

ANSIOB (Association for New Social Infrastructure of Osaka Bay) (1996) Communication tool for sustainable development. Institute of Coastal Environment Inc. (in Japanese)

Broadhead AT, Horn R, Lerner DN (2013) Captured streams and springs in combined sewers: a review of the evidence, consequences and opportunities. Water Res 47:4752–4766

Costanzo SD, O'donohue MJ, Dennison WC, Loneragan NR, Thomas M (2001) A new approach for detecting and mapping sewage impacts. Mar Pollut Bull 42:149–156

Drury B, Rosi-Marshall E, Kelly JJ (2013) Wastewater treatment effluent reduces the abundance and diversity of benthic bacterial communities in urban and suburban rivers. Appl Environ Microbiol 79:1897–1905

Eyre BD (2000) Regional evaluation of nutrient transformation and phytoplankton growth in nine river-dominated sub-tropical east Australian estuaries. Mar Ecol Prog Ser 205:61–83

Fox LE, Sager SL, Wofsy SC (1986) The chemical control of soluble phosphorus in the Amazon estuary. Geochim Cosmochim AC 50:783–794

Garside C, Malone TC, Roels OA, Sharfstein BA (1976) An evaluation of sewage-derived nutrients and their influence on the Hudson Estuary and New York Bight. Estuar Coast Mar Sci 4:281–289

GESAP (GEC Sanitation Programme) (2017) Sewage works in Osaka, Japan. http://nett21.gec.jp/GESAP/themes/index.html. Accessed 27 May 2017

Gruber N, Galloway JN (2008) An Earth-system perspective of the global nitrogen cycle. Nature 451:293–296

Hashimoto S, Sudo R (1986) New method of activated sludge. The Industrial Water Institute, Tokyo (in Japanese)

Howarth RW, Marino R (2006) Nitrogen as the limiting nutrient for eutrophication in coastal marine ecosystems: evolving views over three decades. Limnol Oceanogr 51:364–376

International EMECS (Environmental Management of Enclosed Coastal Seas) Center (2008) Environmental conservation of the Seto Inland Sea. https://www.emecs.or.jp/upload/publish/seto_inland_sea_en.pdf. Accessed 27 May 2017

JMA (Japan Meteorological Agency) (2014–2016) Past metrological data and tidal observation data. http://www.jma.go.jp/jma/menu/menureport.html. Accessed 14 June 2017

Kendall C, Elliott MM, Wankel SD (2007) Tracing anthropogenic inputs of nitrogen to ecosystems. In: Michener R, Lajtha K (eds) Stable isotopes in ecology and environmental science. Blackwell Publishing, Oxford

Kinouchi T, Yagi H, Miyamoto M (2007) Increase in stream temperature related to anthropogenic heat input from urban wastewater. J Hydrol 335:78–88

Kobayashi S, Matsumura Y, Kawamura K, Nakajima M, Yamamoto K, Akiyama S, Ueta Y (2017) Estimation of the origin of dissolved organic matter and biogeochemical cycles of nutrients in Osaka Bay, Japan. J Jpn Soc Water Environ 40:97–105 (in Japanese)

Nakajima M, Fujiwara T (2007) Estuarine circulation and hypoxic water mass in Osaka Bay. Bull Coast Oceanogr 44:157–163 (in Japanese)

Nakamuro K, Okuno T, Takata S, Taguchi H, Inoue T, Ueda A, Fukunaga I, Kagajyo N, Hattori Y, Koi H (2006) Influence of sewage effluent on temperature of urban river water. J Jpn Soc Water Environ 29:177–181 (in Japanese)

Nakatsuji K (1998) Water environment in coastal area. In: Takahashi Y, Kawada K (eds) Water cycle and catchment environment. Iwanami Shoten, Tokyo (in Japanese)

Nishida S, Kawasumi R, Nakatani Y, Murakami Y (2012) Changes and current state of nutrient loads in the Osaka Bay basin. J Jpn Soc Civ Eng Ser G (Environmental Research) 68:III751–III760 (in Japanese)

Nixon SW (1995) Coastal marine eutrophication: a definition, social causes, and future concerns. Ophelia 41:199–219

Onodera S (2011) Subsurface pollution in Asian megacities. In: Taniguchi M (ed) Groundwater and subsurface environments: human impacts in Asian coastal cities. Springer, Tokyo

Onodera S, Saito M, Yoshikawa M, Onishi K, Shimizu Y, Ito H (2013) Nutrient transport and surface water groundwater interactions in the tidal zone of the Yamato River, Japan. IAHS Publ 361:204–211

OPG (Osaka Prefectural Government) (2016) Municipal coverage of sewage treatment system in 2016 (in Japanese). http://www.pref.osaka.lg.jp/gesui_jigyo/tokei/fukyuh20.html. Accessed 27 May 2017

OPG (Osaka Prefectural Government) (2017) Statistics of sewage treatment system in Osaka prefecture (in Japanese). http://www.pref.osaka.lg.jp/gesui_jigyo/tokei/index.html. Accessed 23 Oct 2017

Osaka City (2016) Osaka city environmental white paper (in Japanese). http://www.city.osaka.lg.jp/kankyo/page/0000370176.html. Accessed 27 May 2017

Osaka City (2017) Results of water quality test of sewage treatment plants (in Japanese). http://www.city.osaka.lg.jp/kensetsu/page/0000009603.html. Accessed 23 June 2017

Oshima A, Nishio T, Fujiwara Y, Shinya M, Kitano M (2009) Dynamics of a red tide of *Skeletonema costatum* and *Thalassiosira* spp. (Bacillariophyceae) around the Yume-Shima landfill site in Osaka Bay. Annu Rep Osaka City Inst Public Health Environ Sci 71:49–55

Pennino MJ, Kaushal SS, Murthy SN, Blomquist JD, Cornwell JC, Harris LA (2016) Sources and transformations of anthropogenic nitrogen along an urban river–estuarine continuum. Biogeosciences 13:6211–6228

Powley HR, Dürr HH, Lima AT, Krom MD, Van Cappellen P (2016) Direct discharges of domestic wastewater are a major source of phosphorus and nitrogen to the Mediterranean Sea. Environ Sci Technol 50:8722–8730

RIEAFOP (Research Institute of Environment, Agriculture and Fisheries, Osaka Prefecture) (2006-2015) Annual project reports (in Japanese). http://www.kannousuiken-osaka.or.jp/publication/suisan_shiryo/. Accessed 27 Oct 2017

Savage C (2005) Tracing the influence of sewage nitrogen in a coastal ecosystem using stable nitrogen isotopes. Ambio 34:145–150

SCWSB (Sakai City Waterworks and Sewerage Bureau) (2015) Annual report of sewerage quality in 2015 (in Japanese). http://water.city.sakai.lg.jp/customer/suisitujhoho/gesuidou/1444721746786.html. Accessed 23 Oct 2017

Sigman DM, Altabet MA, McCorkle DC, Francois R, Fischer G (1999) The $\delta^{15}N$ of nitrate in the Southern Ocean: consumption of nitrate in surface waters. Glob Biogeochem CY 13:1149–1166

Sigman DM, Casciotti KL, Andreani M, Barford C, Galanter M, Böhlke JK (2001) A bacterial method for the nitrogen isotopic analysis of nitrate in seawater and freshwater. Anal Chem 73:4145–4153

Takagi S, Kondo M, Kobayashi S, Abo K (2016) Influence on the changes of nutrient conditions and nitrogen stable isotope to the evaluation in nutrient load from land to the sea. J Jpn Soc Civ Eng, Ser B2 (Coastal Engineering) 72: I_1255–I_1260 (in Japanese)

Vitousek PM, Aber JD, Howarth RW, Likens GE, Matson PA, Schindler DW, Schlesinger WH, Tilman DG (1997) Human alteration of the global nitrogen cycle: sources and consequences. Ecol Appl 7:737–750

Waiser MJ, Tumber V, Holm J (2011) Effluent-dominated streams. Part 1: presence and effects of excess nitrogen and phosphorus in Wascana Creek, Saskatchewan, Canada. Environ Toxicol Chem 30:496–507

Waser NAD, Harrison PJ, Nielsen B, Calvert SE (1998) Nitrogen isotope fractionation during the uptake and assimilation of nitrate, nitrite, ammonium, and urea by a marine diatom. Limnol Oceanogr 43:215–224

3

An experimental study on the rate and mechanism of capillary rise in sandstone

Yuya Tsunazawa[1*], Tadashi Yokoyama[2] and Naoki Nishiyama[3]

Abstract

The Lucas-Washburn equation is a fundamental expression used to describe capillary rise in geologic media on the basis of pore radius, liquid viscosity, surface tension, contact angle, and time. It is known that a radius value significantly smaller than the main pore radius must be used in the equation in order for the predictions to fit the experimentally measured values. To evaluate this gap between theoretical predictions and experimental data, we conducted several capillary rise experiments using Berea sandstone. First, to investigate conditions in which pores of any size are available for capillary rise, an experiment was conducted using a dried core. Next, to adjust the size distribution of pore water before the capillary rise, gas pressure was applied to a water-saturated core and water was expelled from pores of $r > 10$ μm; then, capillary rise was initiated. Under this condition, capillary rise occurred only in the pores of $r > 10$ μm. The same experiment was conducted for $r = 3$, 1, and 0.36 μm. When narrower pores were made available for capillary rise, the overall rate of rise decreased and approached the rate observed when the sample was dry initially. This observation suggests that the capillary rise in narrow pores plays a significant role in the overall rate. Based on these results, we propose a conceptual capillary rise model that considers differing radii in branched pores and provide an example of a quantitative description of capillary rise.

Keywords: Capillary rise, Capillary pressure, Lucas-Washburn equation, Water expulsion method, Sandstone

Background

Geologic media commonly have pores of various sizes and shapes. When water comes in contact with a geologic medium, it is drawn into the pores by capillary force. At equilibrium, the height of capillary rise in a vertical pipe with a radius of r at 20 °C is estimated to be 15, 1.5, and 0.15 m for $r = 1$, 10, and 100 μm, respectively. An understanding of the mechanism and rate of capillary rise is important when considering the transport of water in geologic media near the ground surface (i.e., imbibition and drying), associated processes such as rock weathering and soil formation, water availability to plants, and contaminant migration.

Capillary rise is driven by capillary pressure, ΔP (Pa), and ΔP is correlated to pore radius r (m) as follows:

$$\Delta P = \frac{2\gamma \cos\theta}{r}, \tag{1}$$

where γ is the surface tension of the water-air interface (N/m) and θ is the contact angle. This equation shows that capillary pressure increases with decreasing pore radius. The volumetric flow rate q in a pipe (m^3/s) is given by the following Hagen-Poiseuille equation:

$$q = \pi r^2 v = \frac{\pi r^4}{8\mu}\frac{\Delta P}{l}, \tag{2}$$

where l (m) is the length of the pipe, ΔP is the pressure differential between the inlet and outlet of the pipe (Pa), and μ is the viscosity of water (Pa s). In addition, if capillary rise in a pipe with a tortuosity of τ (dimensionless) occurs at a rate of v (m/s), we have

$$\frac{dx}{dt} = \frac{v}{\tau}, \tag{3}$$

where x is the height of capillary rise (m) ($\tau x = l$) and t is the elapsed time after the onset of capillary rise (s). By combining Eqs. 1–3 with $\tau = 1$, the Lucas-Washburn (L-W) equation is obtained:

* Correspondence: ytsunazawa@eps.s.u-tokyo.ac.jp
[1]Department of Earth and Planetary Science, Graduate School of Science, The University of Tokyo, 7-3-1 Hongo, Bunkyo-ku, Tokyo 113-0033, Japan
Full list of author information is available at the end of the article

$$x^2 = \frac{r\gamma \cos\theta}{2\mu} t. \tag{4}$$

This is a fundamental equation for understanding capillary rise in pores. However, it has been found that the *x*-*t* relationship predicted by the L-W equation with simple assumptions (e.g., a constant, realistic r and a constant θ) does not agree well with measured, experimental data (Dullien et al. 1977; Hammecker et al. 1993). Methods to improve the L-W equation involving change in contact angle (Einset 1996; Siebold et al. 2000; Heshmati and Piri 2014), non-uniformity of pore radius (Dullien et al. 1977; Erickson et al. 2002), tortuosity of pores (Benavente et al. 2002; Cai et al. 2014), non-circularity of the cross section of pores (Benavente et al. 2002; Cai et al. 2014), and inertial effect and pore wall roughness especially at the earliest stage of capillary rise (Szekely et al. 1971) have been discussed. A geologic medium usually has a complex pore structure, and any of the variables τ, r, and θ can change depending on the position in the medium and the elapsed time of capillary rise. Therefore, it is often not easy to determine which factor most affects the overall rate of capillary rise.

For this study, several different experiments were performed using sandstone, each designed to evaluate the effects of pore size on the rate of capillary rise. We compare experimental results with theoretical predictions based on the L-W equation, discuss the mechanism of capillary rise, and propose a conceptual model that can account for the experimental results.

Methods

Sample description

Berea sandstone from Ohio, USA, was used in the experiments. The sandstone is composed of 88.9 % quartz, 3.4 % feldspar, 2.2 % carbonate, 3.9 % clay, 0.5 % evaporite, and 1.1 % other (Zhan et al. 2010). Figure 1 shows a backscattered electron image of the polished surface of a sample. Typical grain sizes were 150–250 μm. The open-pore porosity of the sample (ϕ_{open}) was 18.7–21.9 %. The value of ϕ_{open} was determined using "$\phi_{open} = (W_{wet,sat} - W_{dry}) / V$", where $W_{wet,sat}$ is the water-saturated sample weight, W_{dry} is the dry sample weight, and V is the sample volume (water density = 0.998 g/cm^3 at 20 °C). The pores were saturated with water under a vacuum following the procedure of Yokoyama (2013). The transport-pore porosity ϕ_{tra}, calculated by subtracting the fraction of dead-end pores from the fraction of open pores following the method of Yokoyama and Takeuchi (2009), was 17.6 %. Figure 2 shows pore size distribution determined by the water expulsion method (Nishiyama et al. 2012) using an initially water-saturated sample of Berea sandstone. The majority of the pores (76 %) had radii ≥ 10 μm. The fractions of pores with radii of 1–9 μm and < 1 μm were 22 and 2 %, respectively.

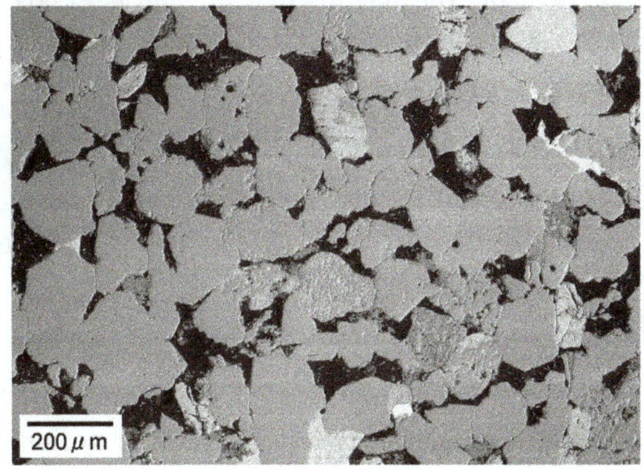

Fig. 1 Backscattered electron image of Berea sandstone. *Gray*, *white*, and *black* correspond to quartz (with minor feldspar and clay), Ca-Mg carbonates, and pores (filled with resin), respectively

In the analysis of capillary rise, we incorporated the effect of pore tortuosity (τ) into the L-W equation. The value of τ was determined as follows. First, permeability (k) of the Berea sandstone under saturated conditions was determined by a constant-head permeability test (Yokoyama 2013). The value of k was 2.16×10^{-13} m^2 at 20.2–20.6 °C, which was determined by Darcy's law, as follows:

$$\frac{Q}{A} = \frac{k}{\mu} \frac{\Delta P}{L}, \tag{5}$$

where Q is the flow rate (1.13×10^{-8} m^3/s), A is the cross-sectional area of the rock (5.15×10^{-4} m^2), ΔP is the differential pressure of water (1.16×10^4 Pa), L is the

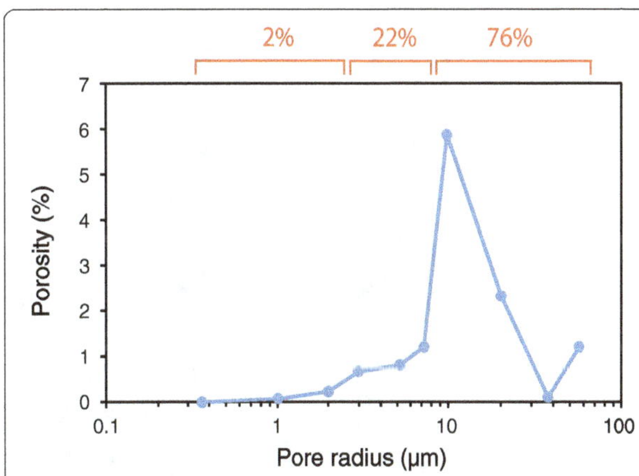

Fig. 2 Pore size distribution of sandstone measured by the water expulsion method. The percentages and ranges at the top of the graph show the volumetric proportion of pores for each pore radius range

thickness of the rock (1.14×10^{-2} m), and $\mu = 1.002 \times 10^{-3}$ Pa·s. For the case of a circular tube of radius r, permeability k and pore radii (r) can be correlated by the following Kozeny-Carman relation (Carman 1956; Paterson 1983; Walsh and Brace 1984):

$$k = \frac{\phi_{\rm tra} r^2}{8\tau^2}. \tag{6}$$

By inserting measured values of k, $\phi_{\rm tra}$, and r (1×10^{-5} m from Fig. 2) into Eq. 6, τ was calculated to be 3.19.

Experiment 1: measurement of capillary rise using a dried core

Figure 3a shows a photograph of the first experiment, which investigated capillary rise using a dried rock core. The rock core was 3.44 cm in diameter, 15.2 cm high, and 298.0 g in dry mass. At the onset, the bottom of the core was dipped in pure water (milliQ) to a depth of approximately 3 mm. The wet and dry portions were visually distinct, and the variation of the height of the wet front over time was monitored. The temperature and relative humidity were 19.4–20.3 °C and 40.6–47.3 %, respectively, measured every 10 min for the duration of the experiment.

Experiment 2: measurement of capillary rise after controlling for the size distribution of pore water

In experiment 1, water could infiltrate pores of any size because the sample was initially dry. Also of interest is the rate of capillary rise when pores smaller than a given radius are filled with water before the onset of capillary rise. To achieve this condition, we applied the following water expulsion method. First, the lateral side of a rock core sample (2.46 cm in diameter and 4.94 cm high) was sealed with resin (Technovit, 4004; Fig. 3b), and the sample's pores were filled with water. The sample was installed on an apparatus for water expulsion (Fig. 4a), and gas pressure was applied to the bottom of the sample. When the difference in gas pressure between the top and bottom of the sample ($\Delta P_{\rm gas}$) overcomes the capillary pressure in a pore, water in the pore is expelled to the top of the sample. This occurs when the following condition is satisfied:

$$r > \frac{2\gamma \cos\theta}{\Delta P_{\rm gas}}, \tag{7}$$

where r is the radius of pore from which water is expelled (Yokoyama and Takeuchi 2009; Nishiyama et al. 2012; Nishiyama and Yokoyama 2014). Equation 7 shows that the minimum radius of empty pores after water expulsion decreases as $\Delta P_{\rm gas}$ increases (Fig. 4b). Expelled water was wiped away with tissue, and pores larger than a given radius emptied. After this water expulsion treatment, the sample was weighed, and the capillary rise experiment was initiated (Fig. 3b). The water expulsion treatment was conducted for four pore radii using the same sample: the values of $\Delta P_{\rm gas}$ applied were 146, 485, 1441, and 4000 hPa, corresponding to radii of 10, 3, 1, and 0.36 μm, respectively. For comparison, the capillary rise experiment was also carried out with the fully dried sample. In experiment 2, the position of the wet front could not be seen because the sample was sealed with resin. Therefore, the sample was removed from the apparatus intermittently and weighed to determine the amount of water absorbed. The temperature and relative humidity were 17.6–22.0 °C and 33.1–44.2 %,

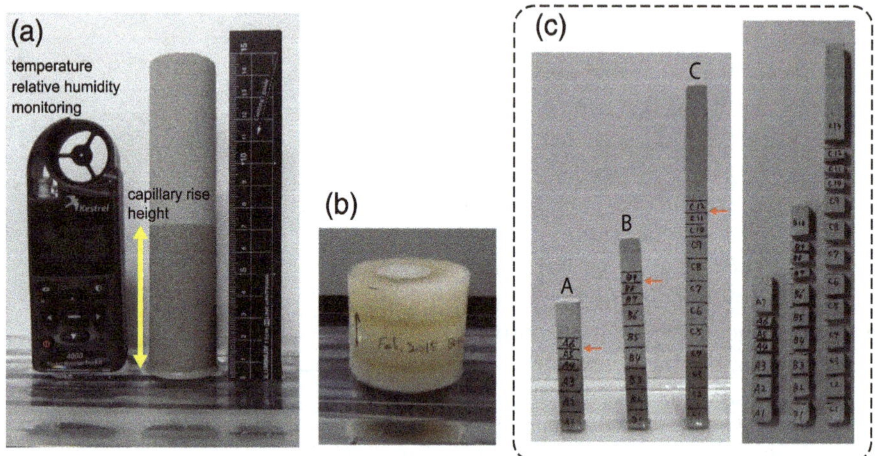

Fig. 3 Photographs of the three types of capillary rise experiments. **a** Experiment using a dried core without a lateral seal. **b** Experiment using a core with a lateral seal. Capillary rise was initiated after water in pores with a radius >10, >3, >1, or >0.36 μm was expelled. **c** Experiment for evaluating the height profile of water saturation. *Left*: during capillary rise; *right*: after samples are cut

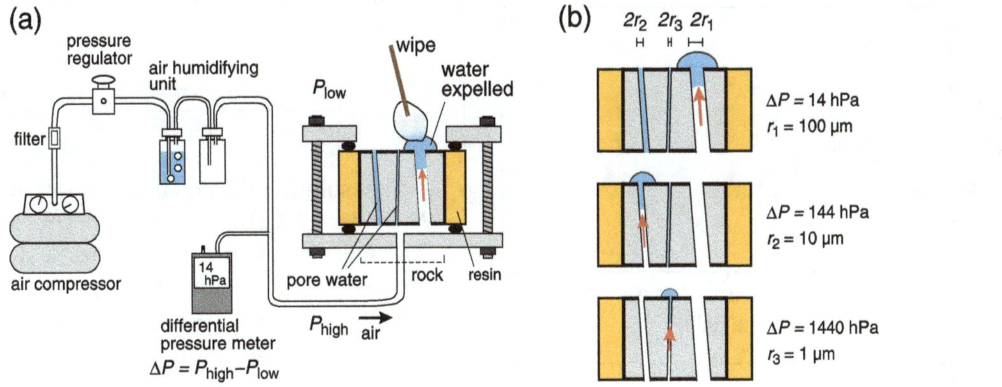

Fig. 4 Schematic illustration of the water expulsion method. **a** Experimental apparatus. **b** Diagram illustrating the process of water expulsion through increasing the differential pressure applied to the sample

respectively (humidity was not measured during the experiments for $r = 1$ and 3 μm).

Experiment 3: measurement of the height profile of water saturation

This experiment aimed to evaluate how the height profile of water saturation (S) in the sample changes as capillary rise proceeds. First, stick-shaped samples (A, B, and C) approximately 1.0 cm wide, 0.7 cm thick, and 6–15 cm high were cut from Berea sandstone. The bottom of each sample was dipped in pure water to a depth of approximately 3 mm to initiate capillary rise (Fig. 3c, left). After time had passed such that the height of the wet front rose to 3.7 cm (sample A), 6.7 cm (sample B), and 9.7 cm (sample C), the samples were cut into 7–16 pieces (Fig. 3c, right) using a circular saw (HOZAN K-210) without adding water, and each piece was weighed. Then, the pieces were dried, weighed again, and S was determined by

$$S = (W_{\text{wet},S} - W_{\text{dry}})/\phi_{\text{open}}V, \tag{8}$$

where $W_{\text{wet},S}$ is the sample weight at S (water density = 0.998 g/cm^3). Samples were cut in intervals of approximately 5 mm for portions near the wet front and 1 cm for portions distal to the wet front, except for the dry portion, which was left intact. Times required for the completion of sample cutting ranged from approximately 1 min (sample A) to approximately 2.7 min (sample C). To minimize migration of water during cutting, the portions near the wet front were cut first. The temperature and relative humidity during the experiment were 19.3–19.6 °C and 38.6–40.1 %, respectively.

Results and discussion

Height profile of water saturation

Figure 5 shows height profiles of S during the capillary rise experiments. Measurements of S at various heights in samples A, B, and C are plotted to show vertical trends in capillary rise; arrows show the position of the visible wet front in each sample, i.e., 3.7, 6.7, and 9.7 cm, respectively. From these results, it is observed that S was relatively constant at 0.52 ± 0.05 from the bottom of the sample to approximately 70 % of the height of wet front at each stage, above which S decreased rapidly upward toward the wet front. The value of S decreased almost to zero just above the wet front, which indicates that the difference in capillary rise height between the interior and exterior of the samples was small.

Time variation in capillary rise height in initially dry sandstone

The results of the capillary rise experiment conducted using the dried core are shown in Fig. 6a. The rise of the

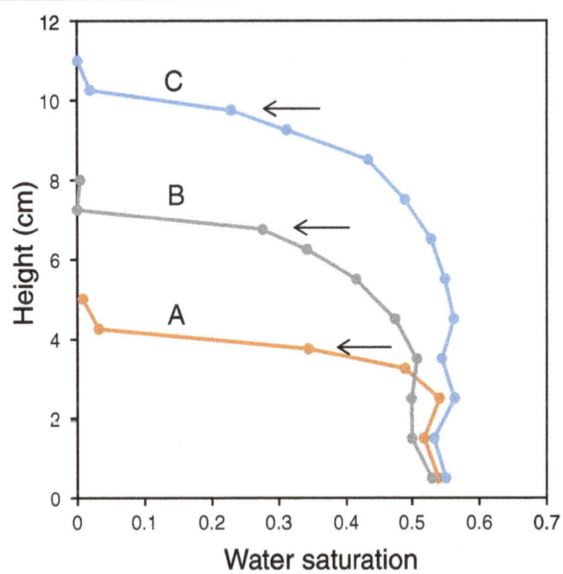

Fig. 5 Time variation in the height profile of water saturation in experiment 3. *Arrows* show the positions of the wet front in each sample

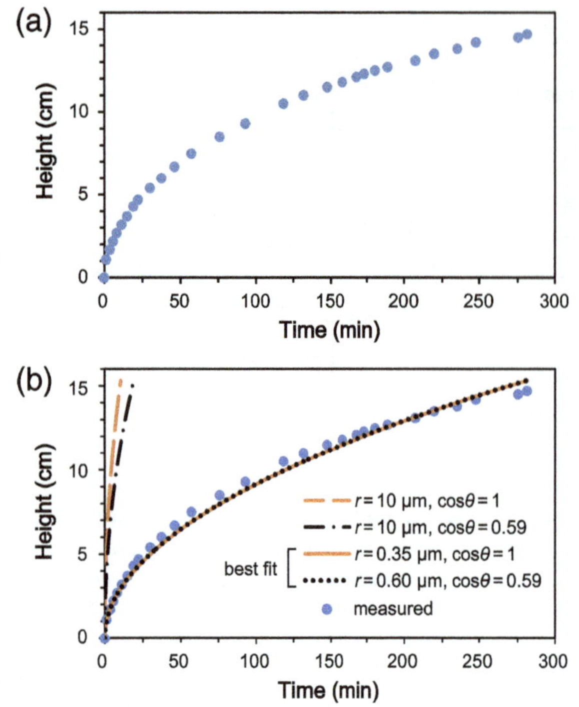

Fig. 6 Measured and calculated values of capillary rise over time. Examples of capillary rise curves calculated using the L-W equation under the conditions of $r = 10$ μm and $\cos\theta = 1$ or 0.59 and best fit curves following the L-W equation at $\cos\theta = 1$ or 0.59

wet front was rapid at the start of the experiment and gradually slowed over time. As observed in the case of the stick-shaped samples, the difference in capillary rise height between the interior and exterior of the core was small, which was confirmed by observing how the wetting at the top of the core spread at the completion of the capillary rise. The measured capillary rise height can be compared with values calculated using the L-W equation (Eq. 4). The L-W equation is often used in a way that neglects the effects of gravity on the water and the tortuosity of pores, but we considered these factors. Because capillary rise is driven by capillary pressure from which the effect of gravity is subtracted, we have

$$\Delta P = \frac{2\gamma \cos\theta}{r} - x\rho g, \tag{9}$$

where ρ is the density of water at 20 °C (998 kg/m^3) and g is gravitational acceleration (9.80 m^2/s). By inserting Eq. 9 into Eq. 2, the flow rate q becomes

$$q = \frac{\pi r^4}{8\mu\tau l}\left(\frac{2\gamma \cos\theta}{r} - x\rho g\right). \tag{10}$$

Thus, ΔP and q decrease with increasing x and eventually become zero at the equilibrium height. The analytical solution of Eq. 10 with $q = \tau\pi r^2(\mathrm{d}x/\mathrm{d}t)$ is computed as follows:

$$t = -\frac{8\mu\tau^2}{r^2\rho g}x - \frac{16\gamma\mu\tau^2 \cos\theta}{r^3\rho^2 g^2}\log\left(1 - \frac{r\rho g}{2\gamma\cos\theta}x\right). \tag{11}$$

The solution is equal to those of Hamraoui and Nylander (2002) and Fries and Dreyer (2008) if $\tau = 1$. The most dominant pore radius in the rock sample is approximately 10 μm (Fig. 2); therefore, we initially assume $r = 10$ μm. As for θ, the assumption of $\cos\theta = 1$ ($\theta = 0°$) has been used previously to analyze capillary rise in sandstone (Dullien et al. 1977; Hammecker and Jeannette 1994) and granitic rocks (Mosquera et al. 2000). In addition, Heshmati and Piri (2014) reported that if capillary numbers ($\mu v/\gamma$) are <~0.001, θ becomes ~10° ($\cos\theta = 0.98$) for the case of capillary rise in a glass tube. The capillary number of our sample was calculated to be <0.001 for $x > 1$ mm based on the measured value of v. The assumption of $\cos\theta = 1$, therefore, seems to be reasonable. However, the contact angle of quartz, the predominant mineral in Berea sandstone, has been reported to range between 0° and 54° (Jańczuk et al. 1986). Therefore, we also considered the case of $\cos\theta = 0.59$ ($\theta = 54°$) as an extreme case. Figure 6b shows the variation of water height with time, calculated by Eq. 11 with $\cos\theta = 1$ and 0.59 ($\gamma = 0.0727$ N/m at 20 °C) and plotted with the measured data. Calculated water heights were significantly higher than the measured values, both in the case of $\cos\theta = 1$ and 0.59.

Table 1 shows the combinations of r and $\cos\theta$ with the measured and calculated values that match best, and the results of $\cos\theta = 1$ ($r = 0.35$ μm) and $\cos\theta = 0.59$ ($r = 0.60$ μm) are shown in Fig. 6b. A higher $\cos\theta$ is associated with a smaller r. Even considering wide variation of the contact angle, the calculated values of $r = 0.35$ and 0.60 μm are significantly smaller than the main pore radii of the rock, which is consistent with the findings of Dullien et al. (1977). This result suggests that the overall rate of capillary rise is related to some process occurring in narrow pores.

Capillary rise after controlling for the size distribution of pore water

Figure 7a–e shows the variation of sample weight over time (dry weight + initial water + newly absorbed water) during the capillary rise experiments beginning after the

Table 1 Pore radius and $\cos\theta$ ($\theta = 0°, 24°, 40°, 54°$) values at which the measured and calculated values match best

$\cos\theta$	1	0.91	0.77	0.59
r (μm)	0.35	0.39	0.46	0.60

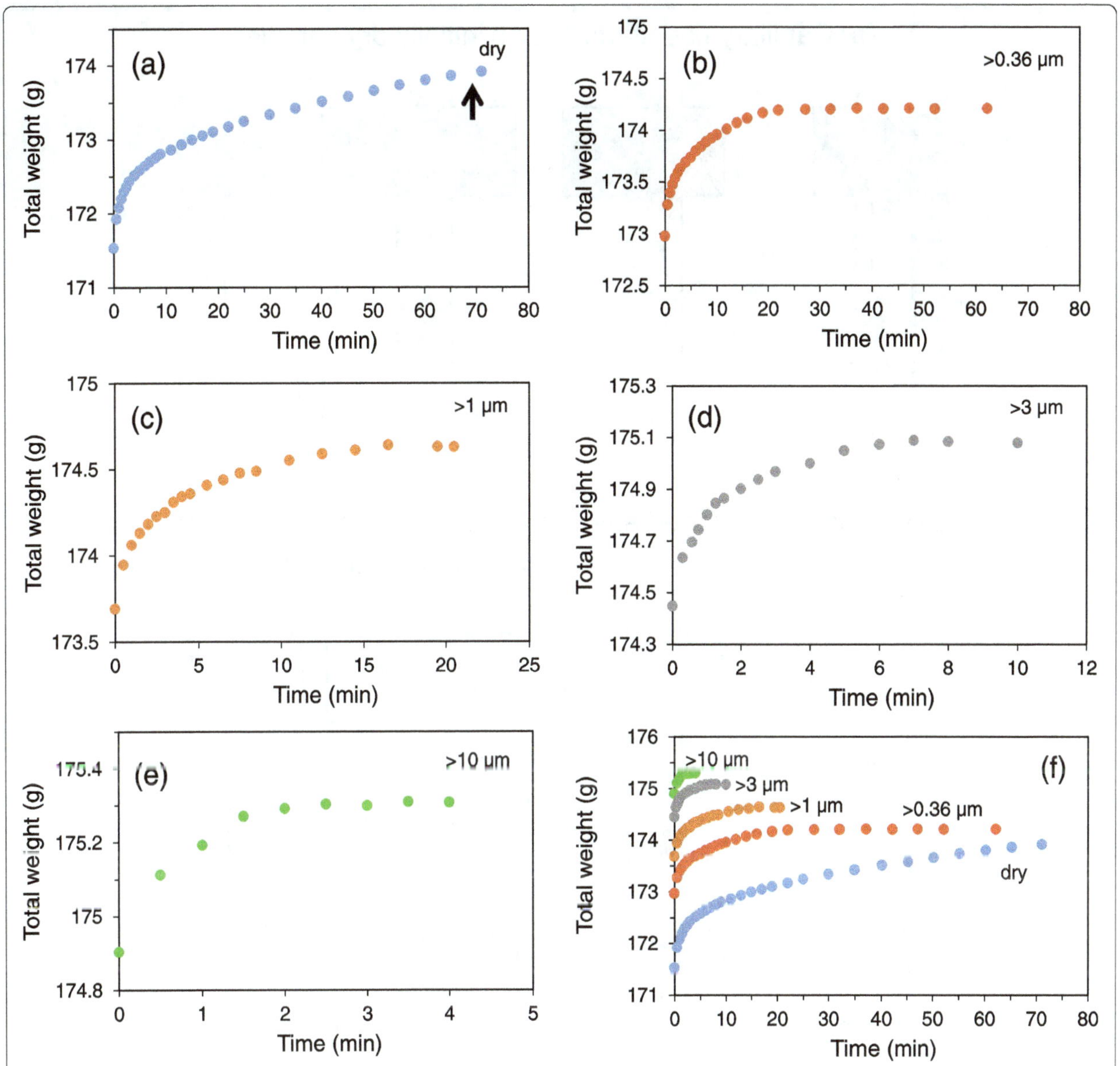

Fig. 7 Sample weight over time during capillary rise after the water expulsion method. The arrow in **a** indicates the time at which wetting was observed at the top of the core. The experiment was started with a dry sample (**a**) or a sample in which the water in pores with a radius > 0.36, > 1, > 3, or > 10 μm was expelled (**b**–**e**, respectively). The results are summarized in **f**

water expulsion treatments, and Fig. 7f summarizes these data. As described above, water infiltrates the "empty pores" formed by water expulsion. The position of the wet front cannot be seen in the initially wet samples used for these experiments (Fig. 7b–e), but for the experiment initiated with a fully dried sample (Fig. 7a), the time at which the wet front reached the top of the sample could be visually confirmed (70 min, arrow in Fig. 7a). The weight was still increasing after 70 min in the trial, as depicted in Fig. 7a, indicating that water absorption was continuing. Conversely, for the case in which pores <0.36 μm were initially filled with water (Fig. 7b), the weight did not increase after approximately 20 min. Based on the results reported in Fig. 7a, the height of capillary rise in the experiment represented by Fig. 7b likely reached the top of the sample before a stable weight was attained. Therefore, capillary rise after expelling water in pores >0.36 μm occurred much faster than under dry conditions. Likewise, as the size of empty pores increased (Fig. 7c–e), the rate of capillary rise increased, as demonstrated in Fig. 7f.

A conceptual model for capillary rise

Dullien et al. (1977) and Einset (1996) have proposed that non-uniformity of pore radius in a single pore (see Fig. 8a for an example) is a major reason for slower than

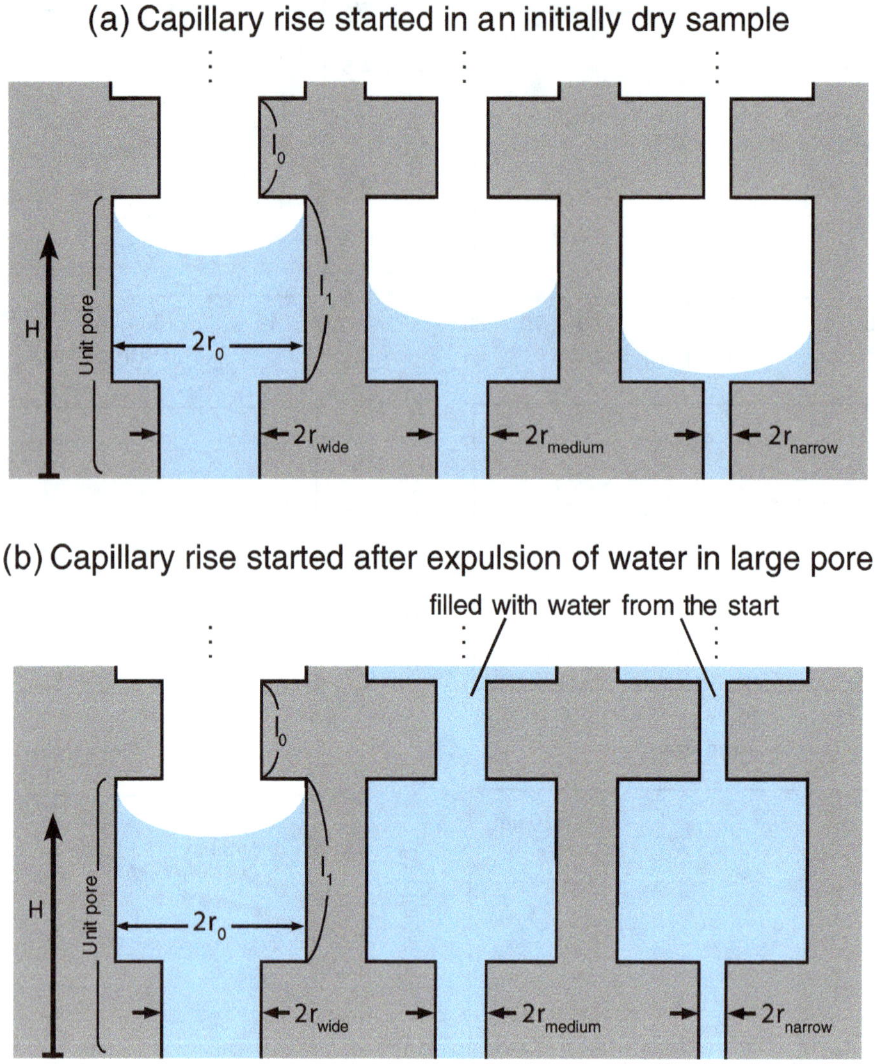

Fig. 8 Effect of non-uniformity of the pore radius, as predicted by Dullien (1977)'s model. No difference of capillary rise height H is predicted whether the experiment was started with a dry sample (**a**) or a sample that has undergone water expulsion treatment (**b**)

predicted capillary rise when using the L-W equation with the dominant pore size of a sample. When a pore consists of a narrow part followed by a wider part, the driving force of water at the wide part is relatively weak because of low capillary pressure, while flow resistance in the narrow part is high. Consequently, the rate of capillary rise in such a non-uniform pore is slower than that in a consistently narrow pore. If the porosity of a rock is formed by repeated "unit pores" such as the pores illustrated in Fig. 8a, the effective pore radius (r_{eff}) for capillary rise in the rock is given by

$$r_{\text{eff}} = \left(\sum_k l_k\right)^2 \left[\sum_k l_k \sum_j \left(\frac{l_j}{2r_j}\right)\left(\frac{r_k}{r_j}\right)^3\right]^{-1}, \tag{12}$$

where r and l are the radius and the length of a component of the unit pore, respectively (Dullien et al. 1977). The summations represent overall components of the unit pore. In Eq. 12, the factor of 1/3 originally included by Dullien et al. (1977) is excluded because the factor corresponds to the effect of tortuosity, which is already taken into account in this study (Eq. 4 or Eq. 11). If $r_{\text{wide}} = 10$ μm, $r_0 = 30$ μm, and $l_0 = l_1 = 200$ μm are assumed on the basis of grain size (Fig. 1) and pore size distribution (Fig. 2), r_{eff} is calculated to be 1.4 μm, which means that the rate of capillary rise in the rock is equivalent to that in a pore with a 1.4 μm radius, even though the pore radii of the rock are 10 and 30 μm. Therefore, this effect may partially explain the slow capillary rise indicated by our results. However, this effect is unlikely to be the sole factor; according to Dullien's model (Dullien et al. 1977), the height of capillary rise is predicted to be similar between experiments started with dry samples and with samples

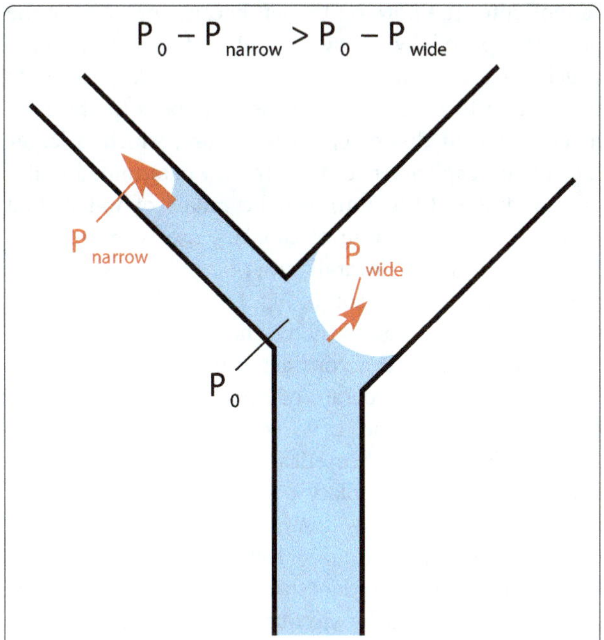

Fig. 9 Capillary rise in Y-shaped piping with different radii, as proposed by Mehrabian et al. (2011) and Sadjadi et al. (2015). Because capillary pressure in narrower pores is greater than that in wider pores, water advances first into narrower pores

subjected to water expulsion treatment prior to the experiment, as shown in Fig. 8a, b. This prediction is inconsistent with our results (Fig. 7f).

A theoretical calculation of capillary rise in Y-shaped piping with different pipe radii (Mehrabian et al. 2011) indicates that water selectively advances in the narrower pipe because the capillary pressure in the narrower pipe is greater than that in the wider pipe (Fig. 9). Sadjadi et al. (2015) also reported a similar phenomenon based on theoretical calculations and experimental observations of capillary rise in a simple Y-shaped junction. We extend these concepts to rocks with complex pore structure and propose a conceptual model that accounts for the effect of branched pores with different pore radii (Fig. 10a). When capillary rise occurs in dry conditions (Fig. 10a), water advances through a narrower pore (Fig. 10b), while capillary rise in an adjacent wider pore almost stops near the junction (light backward movement of water front in a wider pore toward the junction may even occur). When water in the narrow pore rises to a certain height (l_{resume} in Fig. 10b), the pressure drop in the narrow pore overcomes the difference in capillary pressure between the narrow and wide pores, and water begins to advance up the wide pore. By extending the model presented by Sadjadi et al. (2015), l_{resume} can be calculated by

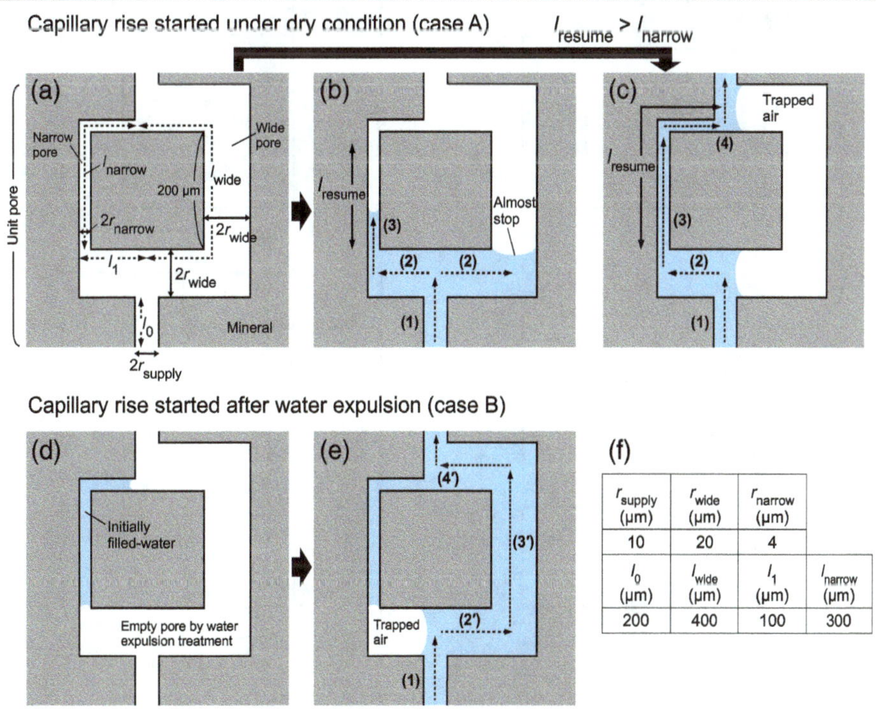

r_{supply} (μm)	r_{wide} (μm)	r_{narrow} (μm)	
10	20	4	
l_0 (μm)	l_{wide} (μm)	l_1 (μm)	l_{narrow} (μm)
200	400	100	300

Fig. 10 Conceptual models of capillary rise in rock pores with a branched structure. **a–c** Capillary rise in dried rock. **d**, **e** Capillary rise after emptying wider pores by water expulsion. **f** Values of the radii and lengths of each pore used to calculate r_{eff} for the sample. These values were chosen based on the grain size and pore size distributions of the sample

$$l_{\mathrm{resume}} = l_{\mathrm{supply}}\left(\frac{r_{\mathrm{narrow}}}{r_{\mathrm{supply}}}\right)^4\left(\frac{r_{\mathrm{wide}}}{r_{\mathrm{narrow}}}-1\right)-l_1\left(\frac{r_{\mathrm{narrow}}}{r_{\mathrm{wide}}}\right)^4, \quad (13)$$

where l_{supply} is the distance from the bottom of the rock to the junction. Definitions of the other variables are given in Fig. 10a. In the case that $l_{resume} > l_{narrow}$ (Fig. 10c), water in the narrow pore arrives at the next junction faster than the water in the wide pore (path 4 in Fig. 10c) and is likely to trap air in the wide pore. Therefore, the wide pore cannot be used as a path for capillary rise, and, consequently, capillary rise through the narrow pore occurs slowly (path 1–4 in Fig. 10c). If the parameters listed in Fig. 10f are used as an example, and the rock porosity is assumed to be formed by repeated unit pores as in Fig. 10a, then l_{resume} exceeds 300 μm (l_{narrow}) when the height of capillary rise (l_{supply}) is above 6 mm. Therefore, paths 1–4 in Fig. 10c are the possible paths of capillary rise under dry conditions, and the corresponding r_{eff} is determined to be 0.30 μm. When capillary rise occurs after the water expulsion treatment (Fig. 10d), water can enter the pore on the right (path 2′ in Fig. 10e), whereas water probably cannot enter the pore on the left because one end of the pore is blocked by the water filling the narrow pore, which traps air beneath it. Therefore, capillary rise here is expected to progress along the wide pore (paths 1–4′ in Fig. 10e). In this case, r_{eff} is 4.7 μm, which is larger (i.e., higher capillary rise) than the r_{eff} determined under dry conditions (0.30 μm). As the radius of water-filled pores increases, the rate of capillary rise is expected to increase because the overall path of capillary rise widens, as seen in the comparison of Fig. 11a, b. The result illustrated in Fig. 7f supports the validity of this model.

Inertia has also been considered to explain deviations between the measured rate of capillary rise and predictions made using the L-W equation (e.g., Bosanquet 1923; Quéré 1997). Theoretical studies and experiments conducted using a capillary pipe demonstrated that the height of capillary rise dominated by inertial force is proportional to time (Quéré 1997; Siebold et al. 2000), whereas the L-W equation shows that the height of capillary rise governed by viscous force is proportional to the square root of time. Because the height of capillary rise in our sample did not increase linearly with time (Fig. 6), the effect of inertia is likely to be negligible, at least for the time span considered in this study.

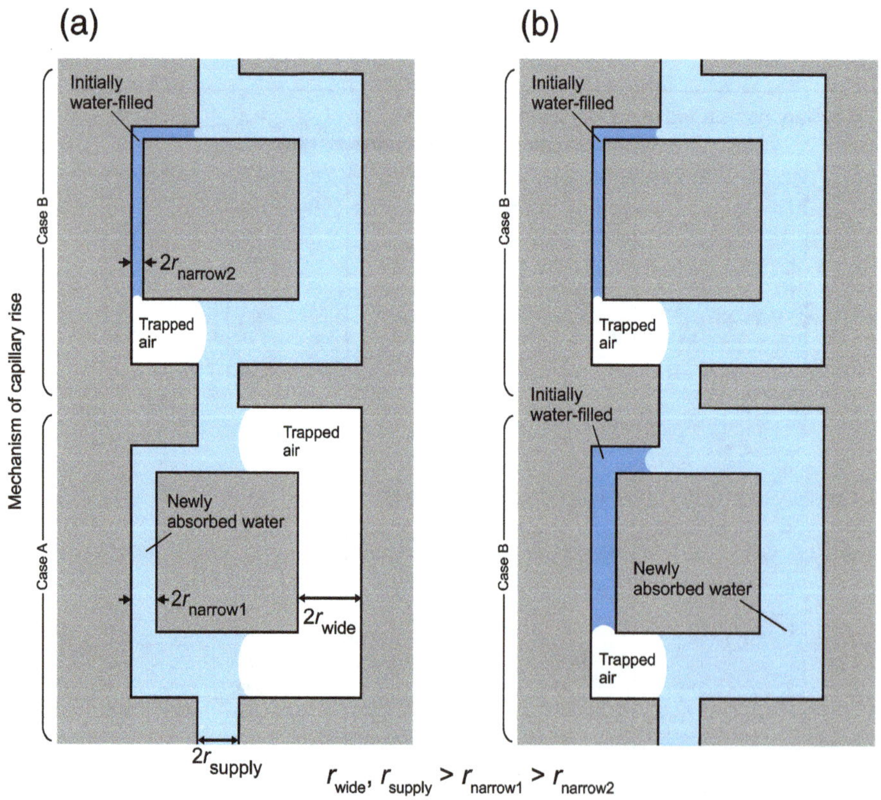

Fig. 11 Conceptual models of capillary rise after the water expulsion treatment with different ΔP in Eq. 7. **a** Only the pores with a radius of $r_{narrow2}$ are initially water-filled. (**b**) The pores with radii of $r_{narrrow1}$ and $r_{narrow2}$ become water-filled. Water advances depending on whether the narrower pore is filled with water (case a or b in Fig. 10)

Conclusions

To evaluate the relationship between the rate of capillary rise and pore size, capillary rise experiments using Berea sandstone were conducted using controlled size distribution of pore water, and the results were analyzed using the L-W equation. The key results and interpretations are as follows:

(1) A pore radius value significantly smaller than dominant pore radius of the rock had to be used to reproduce the experimental results using the L-W equation, assuming $\cos\theta = 1$ and a uniform pore radius.
(2) In the experiment that was initiated with a dry sample, water saturation in the sample was relatively constant below approximately 70 % of the height of the wet front and decreased rapidly with increasing height toward the wet front.
(3) As the radius of the pores available for capillary rise decreased, the rate of capillary rise decreased and approached that of the initially dry sample.
(4) The effect of the nonuniformity of pore radii in a single flow path can explain the result of (1) at least partly but is unlikely to account for the result of (3). These results can be better explained by considering the capillary rise in branched pores with different pore radii.

Abbreviations

L-W: Lucas-Washburn.

Competing interests

The authors declare that they have no competing interests.

Authors' contributions

YT conducted the experimental study and analyzed the data. TY proposed the topic and designed the study. NN helped interpret the experimental results. All authors collaborated to write the manuscript and approved the final manuscript.

Acknowledgements

We thank two anonymous reviewers for their helpful comments.

Author details

[1]Department of Earth and Planetary Science, Graduate School of Science, The University of Tokyo, 7-3-1 Hongo, Bunkyo-ku, Tokyo 113-0033, Japan. [2]Graduate School of Integrated Arts and Sciences, Hiroshima University, 1-7-1 Kagamiyama, Higashi-Hiroshima, Hiroshima 739-8521, Japan. [3]Research Center for Functional Materials, National Institute for Materials Science, 1-1 Namiki, Tsukuba, Ibaraki 305-0044, Japan.

References

Benavente D, Lock P, García Del Cura MÁ, Ordóñez S (2002) Predicting the capillary imbibition of porous rocks from microstructure. Transport Porous Med 49:59–76

Bosanquet H (1923) On the flow of liquids into capillary tubes. Philos Mag Ser 6: 525–531

Cai J, Perfect E, Cheng C-L, Hu X (2014) Generalized modeling of spontaneous imbibition based on Hagen-Poiseuille flow in tortuous capillaries with variably shaped apertures. Langmuir 30:5142–5151

Carman PC (1956) Flow of gases through porous media. Elsevier, New York

Dullien FAL, El-Sayed MS, Batra VK (1977) Rate of capillary rise in porous media with nonuniform pores. J Colloid Interface Sci 60:497–506

Erickson D, Li D, Park CB (2002) Numerical simulations of capillary-driven flows in nonuniform cross-sectional capillaries. J Colloid Interface Sci 250:422–430

Einset EO (1996) Capillary infiltration rates into porous media with applications to silcomp processing. J Am Ceram Soc 79:333–338

Fries N, Dreyer M (2008) The transition from inertial to viscous flow in capillary rise. J Colloid Interface Sci 327:125–128

Hammecker C, Mertz JD, Fischer C, Jeannette D (1993) A geometrical model for numerical simulation of capillary imbibition in sedimentary rocks. Transport Porous Med 12:125–141

Hammecker C, Jeannette D (1994) Modelling the capillary imbibition kinetics in sedimentary rocks: role of petrographical features. Transport Porous Med 17:285–303

Hamraoui A, Nylander T (2002) Analytical approach for the Lucas-Washburn equation. J Colloid Interface Sci 250:415–421

Heshmati M, Piri M (2014) Experimental investigation of dynamic contact angle capillary rise in tubes with circular and noncircular cross sections. Langmuir 30:14151–14162

Jańczuk B, Chibowski E, Białopiotrowicz T (1986) Time dependence wettability of quartz with water. Chem Papers 40:349–356

Mehrabian H, Gao P, Feng JJ (2011) Wicking flow through microchannels. Phys Fluids 23:122108

Mosquera MJ, Rivas T, Prieto B, Silva B (2000) Capillary rise in granitic rocks: interpretation of kinetics on the basis of pore structure. J Colloid Interface Sci 222:41–45

Nishiyama N, Yokoyama T, Takeuchi S (2012) Size distributions of pore water and entrapped air during drying-infiltration processes of sandstone characterized by water-expulsion porosimetry. Water Resour Res 48, W09556

Nishiyama N, Yokoyama T (2014) Estimation of permeability of sedimentary rocks by applying water-expulsion porosimetry to Katz and Thompson model. Eng Geol 177:75–82

Paterson MS (1983) The equivalent channel model for permeability and resistivity in fluid-saturated rock—a re-appraisal. Mech Mater 2:345–352

Quéré D (1997) Inertial capillary. Europhys Lett 39:533–538

Sadjadi Z, Jung M, Seemann R, Rieger H (2015) Meniscus arrest during capillary rise in asymmetric microfluidic pore junctions. Langmuir 31:2600–2608

Siebold A, Nardin M, Schultz J, Walliser A, Oppliger M (2000) Effect of dynamic contact angle on capillary rise phenomena. Colloid Surface A 161:81–87

Szekely J, Neumann AW, Chuang YK (1971) The rate of capillary penetration and the applicability of the Washburn equation. J Colloid Interface Sci 35:273–278

Walsh JB, Brace WF (1984) The effect of pressure on porosity and the transport properties of rock. J Geophys Res 89:9425–9431

Yokoyama T, Takeuchi S (2009) Porosimetry of vesicular volcanic products by a water-expulsion method and the relationship of pore characteristics to permeability. J Geophys Res 114, B02201

Yokoyama T (2013) Characterization of the reaction and transport properties of porous rhyolite and its application to the quantitative understanding of the chemical weathering rate. Geochim Cosmochim Acta 118:295–311

Zhan X, Schwartz LM, Toksoz MN, Smith WC, Morgan FD (2010) Pore-scale modeling of electrical and fluid transport in Berea sandstone. Geophysics 75:135–142

4

Hydrogen mobility in transition zone silicates

Razvan Caracas[1*] and Wendy R. Panero[2]

Abstract

We study the hydrogen mobility in ringwoodite and wadsleyite considering multiple charge-balanced defects, including Mg < = > 2H, Si < = > Mg + 2H, and the hydrogarnet defect, Si < = > 4H, using molecular dynamics simulations based on the density functional theory at transition zone pressures and temperatures between 1500 and 2500 K. We determine the diffusion coefficients and study in detail the mechanism of hydrogen mobility during lengthy simulations. Our results show that temperature, water concentration, and defect mechanism have a significant effect on mobility. We find that the fastest diffusion is for the Mg < = > 2H defect, while H is more mobile when incorporated as Si < = > Mg + 2H than as hydrogarnet defects. The computed diffusivities for ringwoodite are larger than for wadsleyite: at 2000 K, diffusivity is 1.13×10^{-09} m^2/s for ringwoodite compared to 0.93×10^{-09} m^2/s for wadsleyite. In general, the hydrogen atoms spend on the order of tens of picoseconds or more trapped in or around the vacancy sites with net migration between sites over timescales of tens of femtoseconds. At 2500 K, some of these hydrogen excursions take place over several angstroms, while at 2000 K, they do not always result in net diffusion. At 1500 K, most of the defects fail to make excursions from their defect sites resulting in diffusion.

Keywords: Diffusion, Transition zone, Wadsleyite, Ringwoodite, Transport, Electrical conductivity, Defects, Hydrogen

Introduction

The incorporation of water in major mantle minerals affects the physical properties of these minerals, even with low concentrations of water. The introduction of water as hydrogen defects reduces seismic wave velocities (e.g., Jacobsen and Smyth 2006), enhances deformation (Kavner 2003), and increases electrical conductivity (Huang et al. 2005; Yoshino et al. 2009).

The transition zone is dominated by polymorphs of olivine, in which wadsleyite (*I2/m*) is stable from 410–520 km depth, transforming to ringwoodite (*Fd3m*), stable from 520–660 km depth. As synthesized in the laboratory, each of these polymorphs can store water in excess of 2 wt%, making them the primary focus for detecting water in the deep mantle. Wadsleyite incorporates hydrogen predominantly through Mg < = > 2H defects (Smyth et al. 1997) while ringwoodite accommodates hydrous defects through multiple defect mechanisms, dominated by the same Mg < = > 2H defect, but also with the hydrogarnet defect, Si < = > 4H, and minor Si < = > Mg + 2H (Panero 2010; Panero et al. 2013).

The amount of water actually stored in the transition zone, however, may be significantly less than the total capacity. While an inclusion of ringwoodite in diamond suggests that localized regions of hydrated ringwoodite are possible (Pearson et al. 2014), seismic velocity and phase boundaries (Houser 2016) and electrical conductivity of the transition zone are more consistent with a mantle significantly undersaturated in water.

Electrical conductivity of the Earth's mantle varies significantly with depth and tectonic regime, including in the transition zone, in which the conductivity at the base of the transition zone is greater than the upper part of the transition zone (e.g., Kelbert et al. 2009; Shimizu et al. 2010). This variability is likely a complex function of temperature, composition (particularly H, Fe, and redox), and tectonic history. Indeed, the electrical conductivity of ringwoodite and wadsleyite vary as a function of water and iron content (e.g., Huang et al. 2005; Yoshino et al. 2009), with likely significant contributions from both proton and polaron conductivities. While disagreement remains as to the absolute values of conductivity in these

* Correspondence: razvan.caracas@gmail.com
[1]CNRS, Ecole Normale Supérieure de Lyon, Université Claude Bernard Lyon 1, Laboratoire de Géologie de Lyon, CNRS UMR 5276, Lyon, France
Full list of author information is available at the end of the article

materials, the electrical conductivity of hydrous ringwoodite is consistently found to be greater than hydrous wadsleyite for comparable water contents and experimental conditions but with different dependencies on absolute water content.

Direct measurements of hydrogen diffusivity in wadsleyite and ringwoodite under transition zone pressures and conditions are limited (Hae et al. 2006; Ohtani and Zhao 2009; Sun et al. 2015), in which these measurements include the complicating effects of iron that appears to enhance electrical conductivity through proton-polaron exchange mechanisms. The hydrogen diffusivity data are consistent with the electrical conductivity in which the H conductivity in ringwoodite exceeds that of wadsleyite.

We therefore calculate the diffusivity of hydrogen in ringwoodite at mantle-relevant conditions from ab initio molecular dynamics. While this is a well-established standard technique for melts and fluids (e.g., Tse 2002; Schweiger et al. 2008; Mookherjee et al. 2008), it is an unusual approach for solids due to the length of required calculations for appropriate statistics. For the latter, the large majority of the previous calculations (e.g., Bunson et al. 1999; Ammann et al. 2011) instead resolved to determine the activation energy along an assumed diffusion path, which combined with the concentration of vacancies, allowed to estimate the diffusion coefficients. Our approach is more daring: we insert hydrogen in the structures as defects in exchange for cations and obtain the self-diffusion coefficients directly from long molecular dynamics simulations; consequently, we also capture the diffusion path.

With these considerations, we investigate in detail the mobility of hydrogen in ringwoodite as a function of temperature, defect mechanism, and defect geometry. We compare our results to calculations in wadsleyite and further consider the effects of shear and strain on wadsleyite. We also examine the pathway through which hydrogen diffuses to suggest the consequences of interactions with point defects and dislocations in hydrogen diffusivity. We estimate the effect of hydrogen on the electrical conductivity of the major minerals of the Earth's transition zone through the Nernst-Einstein relation and assess the contribution of hydrogen to the measured mantle conductivity.

Methods/experimental

We perform first-principles molecular dynamics simulations. We allow the atoms to move in a crystal according to the Newtonian dynamics (*molecular dynamics*) under the action of interatomic forces computed within the density functional theory (DFT) (*first principles*). We use the planar-augmented wavefunction (PAW) formulation (Blochl 1994) of the DFT the VASP implementation (Kresse and Hafner 1993; Kresse and Joubert 1999) to compute the forces. The exchange-correlation energy is expressed in the Perdew-Burke-Ernzerhof form (Perdew et al. 1996) of the generalized gradient approximation. We employ a 500 eV cutoff for the kinetic energy, 700 eV for the augmented cutoff, and the Γ point to sample the reciprocal space.

We use isokinetic NVT ensembles, where the velocity of the atoms is rescaled at each time step to fit the temperature, T, at fixed volume, V, and number of atoms, N, in the simulation box. The isokinetic NVT can be seen as an extreme version of the Nose-Hoover thermostat where the oscillation period of the temperature is set to the time step. We perform the simulations at volumes corresponding to transition zone pressures.

In all our calculations, we allow for long simulation times, between 70 and 200 ps at each condition with a time step of 1 fs. A 70 ps simulation requires the equivalent of about 500 processor days of computing time.

For ringwoodite, we consider several hydrogen incorporation mechanisms (Panero 2010): $Mg^{2+} <=> 2H^+$, $Si^{4+} <=> Mg^{2+} + 2H^+$, and the hydrogarnet substitution, $Si^{4+} <=> 4H^+$. We consider two different H/O atomic ratios per simulation cell: 1/32 and 2/32 each in two different geometries: a $2 \times 1 \times 1$ supercell of the conventional cell and a $2 \times 2 \times 2$ supercell of the primitive cell, with 16 Mg_2SiO_4 formula units. The simulations are run at 1500, 2000, and 2500 K with the same initial specific volume of ringwoodite, corresponding to about 25 GPa. Temperature and the substitution mechanisms affect the actual pressure of the simulations, but in all cases, the pressure remains in an interval of about 23 to 33 GPa. Over this pressure range, we do not expect the values of the diffusivity to change significantly.

We compare the ringwoodite diffusivities to the $Mg^{2+} <=> 2H^+$ defect on the M3 site in wadsleyite, the only active defect mechanism (Tsuchiya and Tsuchiya 2009), in a simulation box of one unit cell, or eight molecular units. To accelerate the diffusion, we also study the effect of strain. In order to compensate for the smaller simulation sizes, the simulation times of wadsleyite are longer, on the order of 100 ps.

The diffusion coefficient of hydrogen, D_O, in each calculation is obtained as the slope of the mean square displacements, MSD, as a function of time:

$$D = \mathrm{MSD}/6t \quad (1)$$

where, t is time, and 6 is a geometric factor describing motion in three dimensions. MSD is the mean square displacement defined as:

$$\mathrm{MSD} = < [r(t) - r(t_O)]^\wedge 2 > \quad (2)$$

Where $r(t)$ is the position in three-dimensional space at time, t. MSD is computed after an initial variable thermalization type of at least 2 ps, and t-t_0 window glides over the entire remaining simulation time. Values less than ~1.0×10^{-10} m^2/s are an indication that the hydrogen atoms jump on a timescale similar to or greater than the total simulation time, and we include diffusivities for calculations only when the MSD is a linear function of time. Otherwise, we interpret that there are insufficient statistics to obtain reliable diffusion coefficients.

The diffusivity is a function temperature and can be modeled with D_0 as prefactor and the activation enthalpy H,

$$D = D_0 \exp(-H/RT) \tag{3}$$

Where R is the gas constant and T is temperature in Kelvin.

We estimate the contribution proton conductivity makes to the apparent electrical conductivity, σ, with the Nernst-Einstein relation,

$$\sigma = \left(F^2/R\right)\left(Dq^2c/T\right) \tag{4}$$

where F is Faraday's constant, q is the charge on the ion (+1), and c is concentration.

Results and discussion

Typical mean square displacement plots as a function of time are shown in Fig. 1 at both 2000 and 2500 K. Other MSD plots at 2000 K show larger irregularities; in those cases only the slope in the second half of the diagram was considered, which is linear. If the MSD is unchanging with time (e.g., for Mg, Si, and O in Fig. 1), this was a sign for no diffusion during the duration of our simulation and those points were discarded.

Table 1 summarizes the diffusion results of all the simulations performed for the $Mg^{2+} <=> 2H^+$, $Si^{4+} <=> Mg^{2+} + 2H^+$, and $Si^{4+} <=> 4H^+$ substitutions in ringwoodite and different H concentrations, and 2000 and 2500 K, and for the $Mg^{2+} <=> 2H^+$ substitution in wadsleyite, at 1500 and 2000 K. There is no net diffusion in ringwoodite at 1500 K. The calculations span a range of temperatures greater than the entire transition zone. For each simulation, we also indicate the average pressure of the constant volume calculations.

Based on the computed diffusion coefficients, we fit the prefactor D_0 and the activation enthalpy, H, for each defect mechanism, concentration and phase according to Eq. (3). The values are listed in Table 2. They should be taken with care, as pressure increases with temperature in these isochoric calculations.

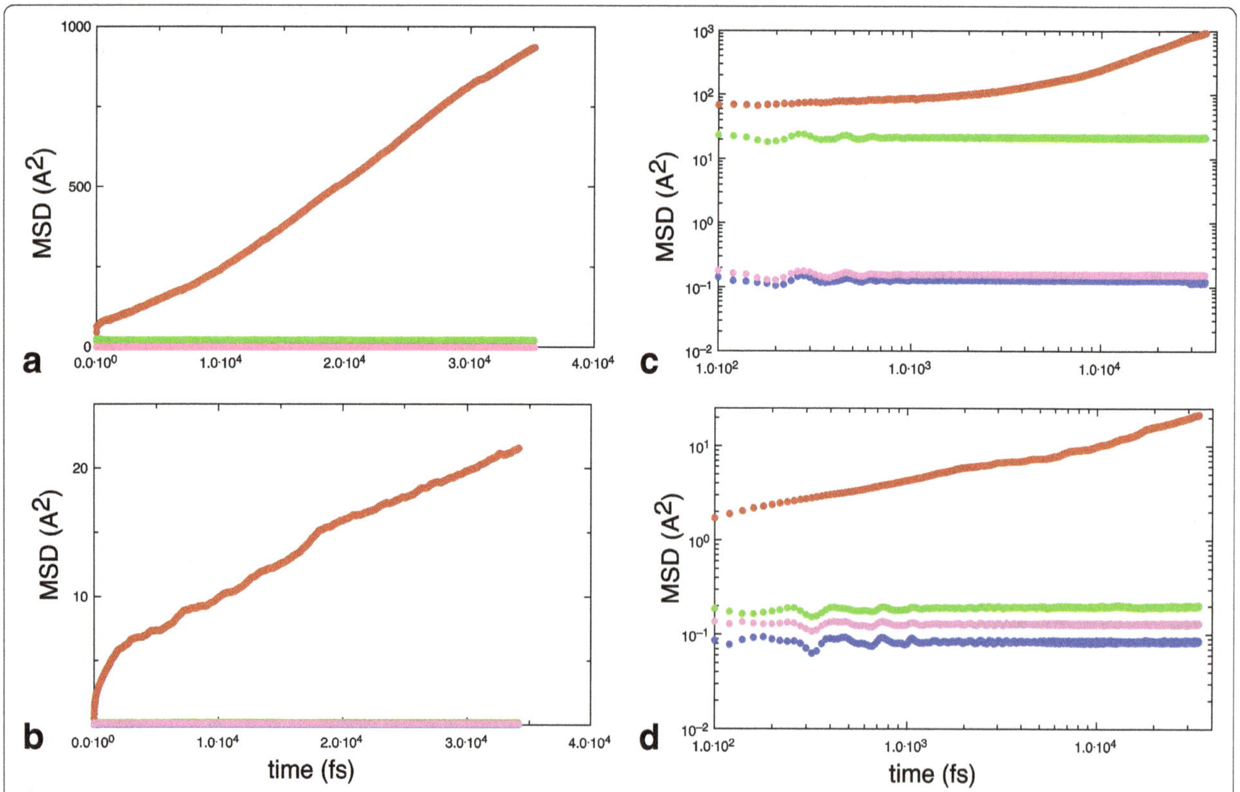

Fig. 1 Examples of the mean square displacements vs time. **a**, **c** Mg < − > 2H T = 2500 K, 1.98 wt% H_2O, 2 × 2 × 2 primitive supercell. **b**, **d** Si < − > Mg + 2H T = 2000K, 1.98 wt% H_2O, 2 × 1 × 1 conventional supercell. The MSD for Mg, Si, O, and H are respectively in *green*, *blue*, *pink*, and *red*. The only diffusive species is H. **a** and (**b**) are linear plots, while (**c**) and (**d**) are the log-log plots

Table 1 Calculations on H-bearing ringwoodite and wadsleyite

Defect	Mineral	H/O ratio	H_2O wt%	Geometry	P (GPa)	T (K)	D_H (10^{-9} m^2/s)	log(σ) (S/m)
Mg < = > 2H	Rw	1/32	0.81	2 × 1 × 1 C	23.6	2000	0.569	0.03
		1/16	1.63	2 × 2 × 2 P	25.0	2000	1.132	−1.12
		1/16	1.63	2 × 1 × 1 C	24.7	2000	[a]	
		1/32	0.81	2 × 2 × 2 P	25.1	2000	[a]	
		1/32	0.81	2 × 2 × 2 P	25.1	2000	[a]	
		1/32	0.81	2 × 1 × 1 C	27.1	2500	8.215	1.09
		1/16	1.63	2 × 1 × 1 C	26.8	2500	1.077	0.51
		1/32	0.81	2 × 2 × 2 P[b]	27.0	2500	8.215	1.09
		1/32	0.81	2 × 2 × 2 P[b]	27.0	2500	10.005	1.18
		1/16	1.63	2 × 2 × 2 P	26.7	2500	43.727	2.12
Si < = > Mg + 2H	Rw	1/32	0.80	2 × 1 × 1 C	22.1	2000	0.217	−0.40
		1/16	1.60	2 × 1 × 1 C	24.6	2000	0.796	0.47
		1/32	0.80	2 × 1 × 1 C	30.3	2500	2.698	0.60
		1/16	1.60	2 × 1 × 1 C	26.6	2500	5.448	1.20
Si < = > 4H	Rw	1/16	1.62	2 × 1 × 1 C	24.0	2000	0.077	−0.54
		1/8	3.27	2 × 1 × 1 C	25.2	2000	0.027	−0.70
		1/16	1.62	2 × 1 × 1 C	32.6	2500	0.263	−0.11
		1/8	3.27	2 × 1 × 1 C	27.0	2500	0.263	0.20
Mg < = > 2H and Si < = > Mg + 2H	Rw	1/16	1.62	2 × 2 × 2 P	28.0	2000	0.452	0.23
		1/16	1.62	2 × 2 × 2 P	29.8	2500	0.74	0.35
Mg < = > 2H	Wd	1/16	1.63	1 × 1 × 1	21.5	1500	0.024	−0.94
		1/16	1.63	1 × 1 × 1 5% strain	36.9	1500	0.003	−1.80
		1/16	1.63	1 × 1 × 1 7% shear	21.6	1500	0.107	−0.29
		1/16	1.63	1 × 1 × 1 11% shear	20.2	1500	0.617	0.47
		1/16	1.63	1 × 1 × 1	21.8	2000	0.926	0.52
		1/16	1.63	1 × 1 × 1	23.6	2500	88.43[c]	2.41

Summary of all the simulations performed on ringwoodite (R) and wadsleyite (W). Temperatures (T) and pressures (P) are also indicated. The 2 × 1 × 1 C and 2 × 2 × 2 P geometries represents supercell of the respectively Conventional and Primitive unit cells of ringwoodite

Rw ringwoodite structure, *Wd* wadsleyite structure

[a]There is no net diffusion observed during the duration of the simulation

[b]The two values are for two different simulations

[c]Wadsleyite at 2500 K is molten; the value is given here only as reference

Table 2 Hydrogen diffusivity

Defect	Mineral	H/O ratio	H_2O wt%	Geometry	D_0(m^2/s)	H (kJ/mol)
Mg < = > 2H	R	1/16	1.63	2 × 2 × 2 P	9.73E-02	303.8
		1/32	0.81	2 × 1 × 1 C	4.16E-04	224.5
Si < = > Mg + 2H	R	1/16	1.60	2 × 1 × 1 C	1.19E-05	159.9
		1/32	0.80	2 × 1 × 1 C	6.44E-05	209.5
Si < = > 4H	R	1/8	1.62	2 × 1 × 1 C	5.05E-04	260.9
		1/16	3.27	2 × 1 × 1 C	2.26E-06	188.3
Mg < = > 2H and Si < = > Mg + 2H	R	1/16	1.62	2 × 2 × 2 P	5.32E-04	232.4
Mg < = > 2H	W	1/16	1.63	1 × 1 × 1	5.62E-05	183.2

Diffusivity fits to the computed diffusion coefficients

The hydrogen atoms typically spend several tens of picoseconds in one given cation vacancy site in between the diffusion jumps. The shape of their local trajectory depends on this site. Diffusion is realized in rapid (<100 fs) transfers between stable defect sites. During a time that is on the order of up to a tenth of a picosecond, the hydrogen atoms may undergo an excursion that can be up to several angstroms long. There are cases, especially at low temperature (1500 K) when the outcome of such an excursion is cut short and the atoms return to the initial cage. But at 2000 and 2500 K, most of these excursions result in a net displacement. This is the mechanism of the diffusion. The diffusion trajectory is a function of defect.

H migration in ringwoodite

In ringwoodite, the hydrogen atoms that are part of the Si < = > 4H defects and the Si < = > Mg + 2H defects rapidly switch between the predicted lowest energy positions on the outside face of each polyhedron, generally maintaining maximum distance between the hydrogen atoms associated with each defect. Their trajectory results in apparent crosses on the faces of the tetrahedra (Figs. 2 and 3).

Figure 2 shows a part of the trajectories of the four H atoms that substitute one Si atom for the Si < = > 4H defect. The H atoms remain around the initial vacancy. After 11 ps in the simulation one of them exits the vacancy, moves away around an O atom (Fig. 2) to be trapped in an empty adjacent octahedron, inside which it moves around describing a donut shape. Finally, the H atom undergoing the excursion falls back into a Si vacancy close to the other three hydrogen atoms but situated at a distance of one unit cell from the initial defect. During this time, it covers the equivalent net distance of one lattice parameter. The case analyzed here corresponds to one substituted Si in a $2 \times 1 \times 1$ conventional supercell. As expected, the H diffuses along the shortest path, the *y*- or *z*-axis. In Fig. 2b, c, it is clearly seen a diffusion pattern similar to a channel. An increase in temperature by 500 K leads to an increase in diffusion by more than one order of magnitude.

In case of the Si < = > Mg + 2H substitution, an excess Mg lies on the Si vacancy site; the H atoms are on the outer parts of the SiO_6 tetrahedra. The pathway to initiate diffusion passes through an empty interoctahedral space. Not all H excursions result in net diffusion (Fig. 3) at 2000 K.

For the Mg < = > 2H defect mechanism, the hydrogen atoms migrate rapidly within a plane of the vacant Mg octahedron defined by the direction perpendicular to the edge-sharing Mg octahedra and sub-parallel to the plane defined by corner-sharing Si tetrahedra (Fig. 4) mapping out a "donut" shape about the octahedral vacancy. This plane reflects the crystallographic equivalent sites with the lowest energy positions as predicted by Panero (2010). The pair of hydrogen atoms is oriented such that their distance to the neighboring Mg octahedra is maximized, which also carry a positive charge. The actual trajectory is easily identified due to its donut shape.

All Mg < = > 2H defects exchange H atoms during the course of the simulations; their diffusion paths follow not only the directions of the unit cell axis, as in the cases discussed above, but also along a diagonal.

Finally, the simulation of two adjacent defects of different types (Mg < = > 2H and Si < = > Mg + 2H) in the $2 \times 2 \times 2$ primitive supercell shows how these defects interact. About 34.7 ps into the simulation, three hydrogen atoms are found around the Mg vacancy, and one H atom around the Si vacancy. Figure 5a shows the trajectory of the Mg atoms over about 0.8 ps and a snapshot of the position of the other atoms before the diffusion jump. It is clearly seen that the defect Mg stays inside the Si vacancy. The same snapshot with the H trajectory over 2 ps is

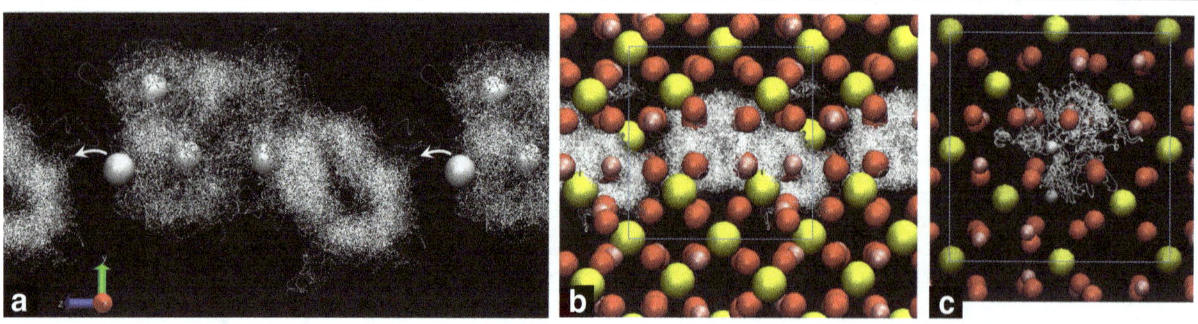

Fig. 2 Diffusion path of H for the Si < = > 4H defect at 2500 K. Mg, Si, O, and H are represented respectively by *pink*, *yellow*, *red*, and *white* spheres. Sphere radii not to scale. The trajectory of the H atoms is shown with the dotted line over several picoseconds. **a** shows only the H trajectory for clarity. The diffusion path is shown in the context of the entire structure in (**b**) and (**c**). The hydrogen atoms occupy the Si vacancy. After escape, indicated by the *white arrow*, one of them described donut shapes while trapped in an empty octahedron, before falling into the neighboring Si vacancy and thus realizing a net proton transfer (**a**). The diffusion path describes channel-like patterns along the *z*-axis, as viewed in the image along *x*-axis (**b**) and *z*-axis (**c**). The unit cell is outlined with *dashed line*

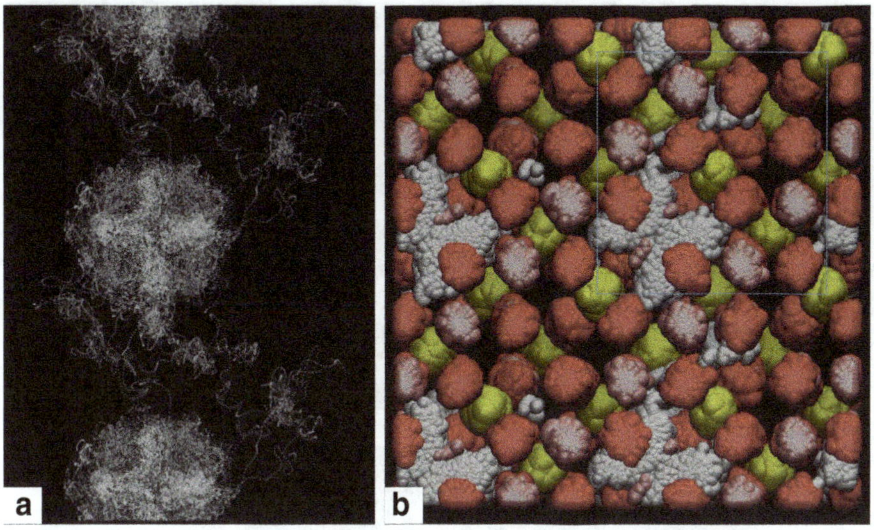

Fig. 3 H diffusion for the Si ⇔ Mg + 2H defects at 2500K. They are computed in a 2 × 1 × 1 convention supercell with one defect (**a**) and in a 2× 2× 2 primitive supercell containing two defects (**b**). **a** Only the trajectory of H is shown, with all the other atoms omitted for clarity. H atoms describe crosses-like patterns on the outside parts of the SiO_4 tetrahedra. H atoms undergo lengthy excursions before falling back into the vacancy site. **b** The trajectory described by all the atoms during 35 ps is represented here. *Pink*, *yellow*, *red* and *white* represent respectively the Mg, Si, O, and H atoms. Not all excursions of H result in net diffusion, as seen from the defect in the *lower left* part of the image: H falls back into the initial site after escaping into the adjacent empty octahedron for a short time. The unit cell is outlined with *dashed line*

shown in Fig. 5b: the characteristic donut and cross shapes are clearly identifiable. One hydrogen atom is shown during its jump in snapshot in Fig. 5c. The jump requires just a few tens of femtoseconds. Eventually, the H falls into the neighboring vacancy, each of which containing now two H atoms (Fig. 5d). A similar but reverse process was responsible for obtaining the three H atoms in the Mg vacancy site. Figure 4e shows a 20 ps duration of trajectories of all the atoms from the same structural fragment.

H migration in wadsleyite

In wadsleyite, static calculations of H defects show that pairs of H in the Mg < = > 2H defect tend to align with the OH dipoles parallel to the *z*-axis (Tsuchiya and Tsuchiya 2009). In our dynamical simulations at the lowest temperatures in our simulations (1500 K), the H atoms explore the vacancy site, in *trans* positions. At 2000 K, they escape easily from their initial sites and wander around to realize a net diffusion.

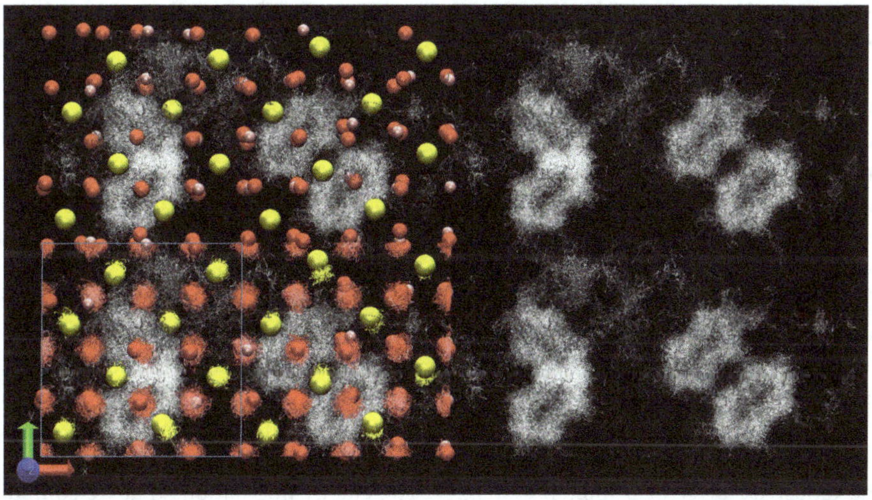

Fig. 4 Trajectory of the H atoms for the Mg < = > 2H defect during about 35 ps. A structural fragment of 4 × 2 × 2 unit cells is represented; on the right half all the atoms except for H are omitted for clarity. The example shown here is for a simulation at 2500 K with two defects in a 2 × 2 × 2 primitive supercell. *Pink*, *yellow*, and *red* represent respectively the Mg, Si, and O atoms. The unit cell is outlined with *dashed line*

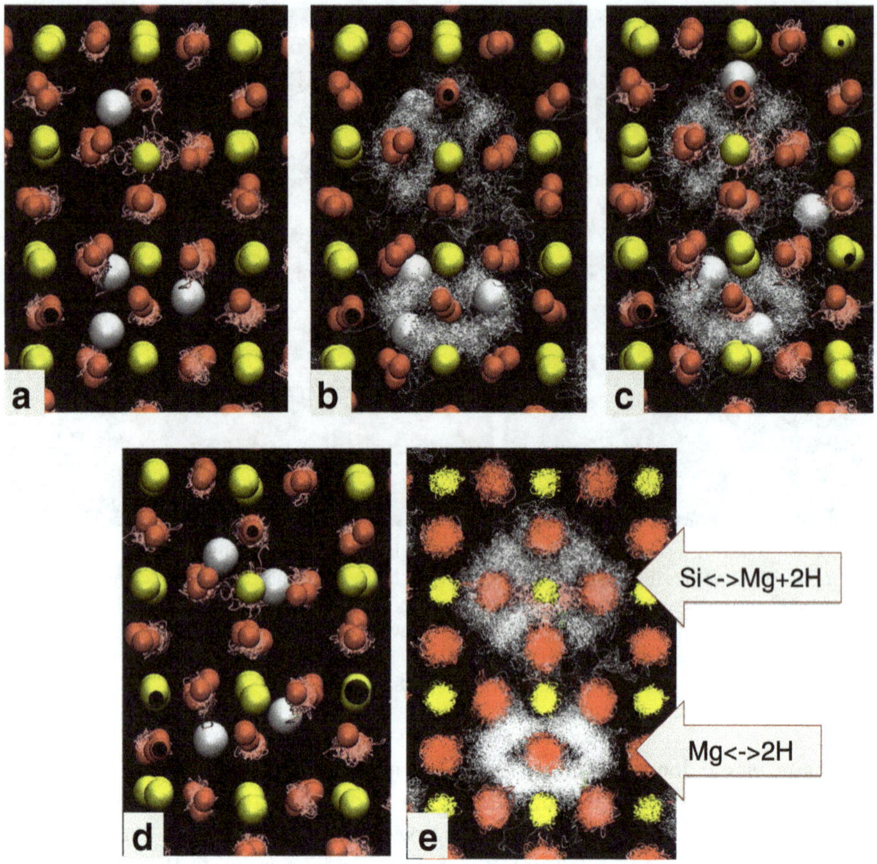

Fig. 5 Snapshots of the diffusion of hydrogen in a 2 × 2× 2 primitive supercell containing two defects at 2000 K. Mg < = > 2H lies in the *lower* half of the image above and Si < = > Mg + 2H lies in the *upper* half. **a** A snapshot of the structure when 3 H atoms occupy the Mg vacancy, 1 Mg the Si vacancy, and one last H gravitates around the Si vacancy. The Mg atoms are represented with their trajectory (*pink dotted lines*) during half a picosecond. There is no diffusion of Mg. **b** The H trajectory is donut- and cross-shaped in the respective two defects as represented over 2 ps simulation time. **c** One of the three H atoms escape the Mg vacancy. The excursion lasts several tens of femtoseconds. **d** The escaped H atoms falls around the Si vacancy. **e** Trajectory of all atoms during 20 ps of simulation time. The only diffusing species is H

Comparable simulations on wadsleyite were performed in one unit cell, with eight formula units, so that water concentrations are comparable between the wadsleyite and ringwoodite simulations. The diffusion of hydrogen is realized in channel-like trajectories continuous along the *z*-axis and interconnected along the *y*-axis (Fig. 6). The H atoms, as in ringwoodite, diffuse by hopping between the O atoms and avoiding the Mg sites. They move around the SiO_4 tetrahedra but prefer the vacant inter-tetrahedral space.

At 2500 K and 23.6 GPa in wadsleyite (Table 1), MSD shows net displacements for all oxygen, silicon, and magnesium atoms in the structure, and therefore, we interpret that this calculation reflects a melt, consistent with the melting temperature of wadsleyite. The hydrogen diffusivity calculated in this simulation, 88.10^{-9} m^2/s, then reflects the diffusivity of H in a molten system. Our hydrogen-bearing calculations show the presence of melt of lower temperatures than both previous freezing simulations on anhydrous liquid Mg_2SiO_4 (De Koker et al. 2008) and recent static experimental data (Liebske and Frost 2012) consistent with melting-point depression for hydrated materials.

Energy considerations

There is no obvious relation between the position of the hydrogen atoms along their diffusion path and the fluctuation in total energy of the system. There are examples when the local energy maximum corresponds to migration of one H atom between sites, but in these cases, the energy profile does not resemble that of a saddle point. Furthermore, the next local energy maximum corresponds to H inside or around its new host site. This behavior reflects the complex relationship between the position of all the atoms in the lattice and the total energy of the lattice, not just the hydrogen positions. The energy fluctuations that we compute include contributions from not only the position of the diffusing atoms but also the vibrational contributions. For example, Fig. 7a shows the fluctuations in total energy computed

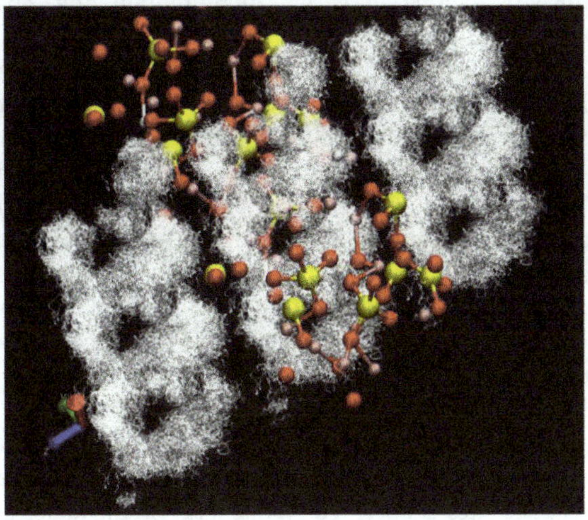

Fig. 6 Diffusion of H in the wadsleyite structure at 2000 K. The structural fragment shown here represents the H trajectories during 60 picoseconds after thermalization at 2000 K spanning several unit cells. The diffusion is realized in the (010) plane, containing the shortest lattice parameters

during several picoseconds for the Si < = > 4H substitution in ringwoodite. These fluctuations are regularly observed during NVT simulations. Details in Fig. 7b are shown prior, during, and after the escape of the H atom in the Si < = > 4H substitution depicted in Fig. 2 and discussed in detail above. There is no clear relation between the position of the hydrogen atoms along the diffusion path and the total energy of the system. This might also raise questions about the validity of the activation energy approach within a partially fixed lattice in static (i.e., 0 K) or nudged-elastic band simulations.

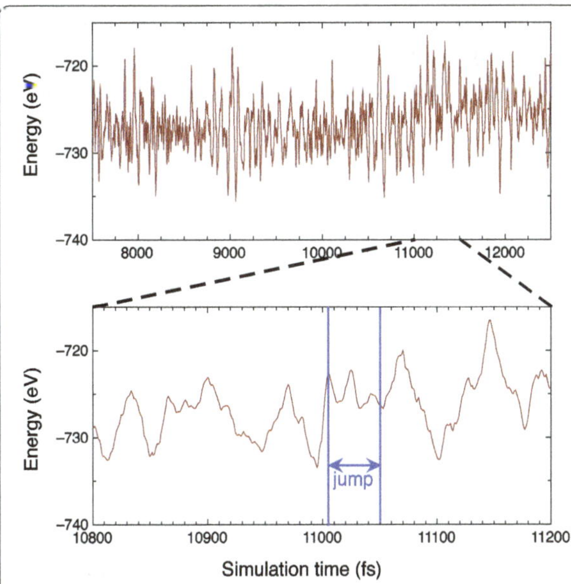

Fig. 7 Energy variation during parts of the Si < = > 4H simulation at 2500 K. **a** The variations are typical for an NVT simulation. **b** The detail around the H jump from one vacancy site to the neighboring one does not show any relation between the position of H in the structure and the energy

Factors affecting diffusion rate: phase, substitution mechanism, concentration, and temperature

For ringwoodite, we find no statistically significant net migration of hydrogen at 1500 K in any of the calculations. Figure 8 shows the diffusion rates we obtained for each calculation in Table 1, with fits to Eq. (3) in Table 2. In ringwoodite, the Mg < = > 2H defect leads to the greatest H diffusivities at both 2000 and 2500 K significantly increasing with increased water concentration. We find that the apparent diffusivity is greater in the $2 \times 2 \times 2$ primitive cell than in the $2 \times 1 \times 1$ conventional cell, and that for constant temperature and geometry, the diffusivity increases with increasing defect density. The largest values of diffusion are greater than 10^{-9} m^2/s at 2000 K and increase to 4.4×10^{-8} at 2500 K. Such high values of the H diffusion are comparable to the ones obtained in cold hydrous $MgSiO_3$ silicate melts (Karki et al. 2010): on the order of 2×10^{-8} m^2/s at

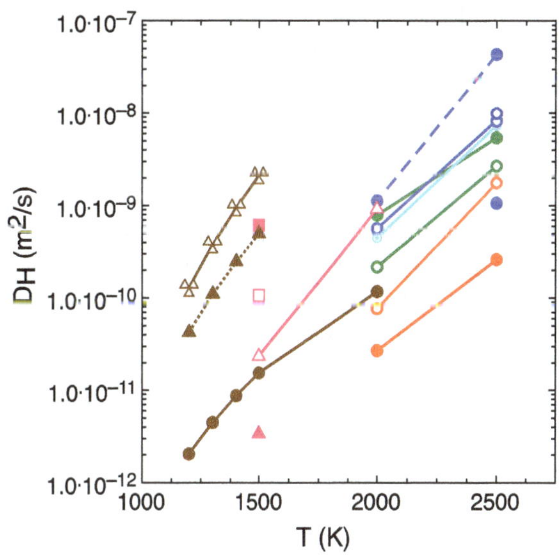

Fig. 8 Hydrogen diffusion coefficients in ringwoodite and wadsleyite as a function of temperature. Mg < = > 2H, Si < = > Mg + 2H, and Si < = > 4H defect mechanisms are respectively plotted in *blue*, *green*, and *red*. Ringwoodite results are represented with one (*open circles*) and two (*solid circles*) defects per simulation supercell in the $2 \times 2 \times 2$ primitive (*dashed lines*) and $2 \times 1 \times 1$ conventional (*solid lines*) supercells. The combined Mg < = > 2H and Si < = > Mg + 2H defects in the same $2 \times 2 \times 2$ supercell is shown with *dotted circles*. Wadsleyite is in *purple*: *empty triangles* for Mg < = > 2H defect, *empty squares* with 7% shear, *solid squares* with 15% shear, *solid triangle* with 5% uniaxial strain. *Brown circles* are extrapolated experimental data from Sun et al. (2015) for ringwoodite, *brown triangles* from Hae et al. (2006) for wadsleyite. Triple empty *triangles* are for experiments in polycrystalline wadsleyite (Ohtani and Zhao 2009)

2500 K and 0 GPa, and 3×10^{-8} m^2/s at 3000 K and 20 GPa. This suggests that either solid hydrous ringwoodite at 2500 K is superheated or that the hydrous melts at 3000 K are supercooled.

Consistently, lower hydrogen diffusivities are found in the other two defects. Hydrogen in the Si < = > Mg + 2H defect diffuse faster at each temperature than hydrogen in Si < = > 4H defects for all concentrations and supercell geometries calculated. Doubling the water content in each calculation changes the inferred hydrogen diffusivity by a factor of two but with opposite effect: it increases mobility of hydrogen associated with the Si < = > Mg + 2H defect yet decreases when associated with the Si < = > 4H defect.

The simulations with the combined Mg < = > 2H and Si < = > Mg + 2H defects yield diffusion coefficients that fall, as expected, between the equivalent concentration Mg < = > 2H calculations and the Si < = > Mg + 2H calculations. While each defect is active in ringwoodite, the Mg < = > 2H defect is dominant, particularly at concentrations below 1 wt% (Panero et al. 2013). Therefore, as the mechanism with both the greatest hydrogen diffusivity and greatest concentration, we consider the diffusivity of hydrogen in the Mg < = > 2H defect to control the diffusivity of hydrogen in ringwoodite.

The experimental hydrogen diffusion data for ringwoodite extrapolated to 2000 K falls in the middle of the range of values we obtained in our simulations (Ohtani and Zhao 2009; Sun et al. 2015). The inferred activation enthalpy in 2000–2500 K for ringwoodite in the fastest mechanism is 300 kJ/mol, significantly greater than the experimental value for iron-bearing polycrystalline and single-crystal ringwoodite, 140 and 101 kJ/mol, respectively. In contrast, the activation enthalpy in wadsleyite, 180 kJ/mol, is more similar to the polycrystalline, iron-bearing experimental measurements in wadsleyite of 123 kJ/mol. The presence of iron, additional defects, or experimental strain may explain the difference. Our computed activation enthalpies are larger than the ones obtained from static simulations (Verma and Karki 2009), by almost one order of magnitude in case of ringwoodite. Table 2 lists the diffusivity fits to our computed diffusion coefficients.

For wadsleyite however, the extrapolated experimental data (Hae et al. 2006) are greater than our theoretical results. As with the Ohtani and Zhao (2009) data, these experiments are measured in an iron-bearing polycrystalline sample, which may explain their faster diffusivities values.

At 2000 K, the H diffusion in wadsleyite for the Mg < = > 2H defect is 0.93×10^{-9} m^2/s. This is 16% less than the fastest value of 1.13×10^{-9} m^2/s computed for ringwoodite for the same defect and at the same temperature, consistent to the experimental hydrogen diffusivity data (Hae et al. 2006; Ohtani and Zhao 2009) (Table 1).

To explore if the discrepancies with previously published diffusivities as discussed above are due to lattice strain, we calculated the diffusion under various systems of strains applied to the wadsleyite structure at 1500 K. We found that a 5% uniaxial strain along the x-axis reduces the diffusivity by one order of magnitude according to the mean square-displacement analysis. However, as the values are already below 10^{-10} m^2/s, they do not show diffusion; rather, they reflect an enhancement of the H localization. When pure sheared is applied, the values increase by about one order of magnitude for 7% shear and almost two orders for 10% shear. It is only this last case where actual net hydrogen transfers between the lattice sites is observed. While large strain values are highly unlikely to exist in a natural setting, we show that residual strains in experiments may lead to an underestimate of diffusivity due to H localization.

Electrical conductivity of the transition zone

Finally, we use the Nernst-Einstein relation to obtain the electrical conductivity from the diffusion coefficients. The values are reported on the last column of Table 1. They correspond to diffusivity inside single crystals.

The last column of Table 1 lists the computed conductivities for each simulation. For ringwoodite, the highest values computed at 2500 K are on the order of 1.1 to 2.1 log(σ) (S/m). They correspond to the Mg < = > 2H defect that dominate the transport properties. Values at 2000 K are one order of magnitude lower. For wadsleyite, the computed log values of the conductivity are on the order of 0.5 log(σ) (S/m) at 2000 K.

Our values fall between the same order of magnitude down to two orders of magnitude below experimental values (Huang et al. 2005; Sun et al. 2015). This is most probably due to crystallinity: the diffusion along the grain boundaries plays a major role in the total transport. Measurements on real crystals show values order of magnitude higher as obtained for single crystal (Sun et al. 2015) than polycrystalline samples (Ohtani and Zhao 2009).

We conclude from our suite of calculations in ringwoodite that increased concentration of defects significantly enhances proton migration in the crystal, consistent with the concentration-dependent electrical conductivity (Huang et al. 2005). We find that hydrogen clearly rests in charge-neutral defect sites within the crystal for the majority of the simulation time, and therefore, it is the increased density of defects that leads to an increased conductivity through shortened hopping distance for the hydrogen. As with any diffusion-limited processes, the proton conductivity will be dominated by the fastest diffusing species, which in the case of ringwoodite, is the hydrogen atoms associated with the dominant Mg < = > 2H defect.

We find that the associated inferred conductivity of ringwoodite is greater than that of wadsleyite at realistic mantle conditions. The computed values of the conductivity and the difference between ringwoodite and wadsleyite are fully consistent with an increase in conductivity at the base of the Earth's mantle (Kelbert et al. 2009; Shimizu et al. 2010) and may help signal the presence of significant (~1 wt%) water in the transition zone.

Conclusions

We study the main three mechanisms of incorporating hydrogen into the ringwoodite structure via chemical substitutions: Mg < = > 2H, Si < = > Mg + 2H, and the hydrogarnet defect, Si < = > 4H. We perform first-principles molecular dynamics simulations lasting up to a couple hundred picoseconds. Three Mg < = > 2H calculations at 2000 K did not yield sufficient migration statistics to results in a reliable diffusivity in 200 ps of simulation time, which highlights that the starting conditions and the thermal history of the calculations may play a factor in the quantitative results.

We find that temperature and defect concentration increase the rate of the H diffusion. More isometric geometries, like the $2 \times 2 \times 2$ primitive supercell also show faster diffusion than more anisotropic, like the $2 \times 1 \times 1$ conventional supercell, similar to the reduction of diffusivity found in wadsleyite for 5% strain. We find that the Mg < = > 2H defect mechanism yields the greatest H diffusion rates, with the hydrogarnet substitution, Si < = > 4H, showing the slowest diffusion.

During the simulation time of several tens of picoseconds, the H atoms travel over several lattice sites. However, during their path, they spend a considerable amount of time pinned in the defect sites. The diffusion is realized via jumps between the vacancy sites; during these jumps, the hydrogen atoms move around the neighboring O atoms, such as to continuously form OH bonds. They try to avoid at all cost the Mg cations, which are also positively charged; usually they stay in the void octahedral space. There is no obvious relation in the energy fluctuations during the simulations and the positions of the hydrogen atoms in the lattice—be they in the vacancy sites or during the diffusive jumps.

The Mg < = > 2H defects are more mobile in ringwoodite than wadsleyite. This result agrees with the proton conductivity inferences of Yoshino et al. (2009) and Huang et al. (2005), as well as the observed increase in electrical conductivity with depth through the transition zone (e.g., Khan et al. 2011; Olsen 1998).

Abbreviations

MSD: Mean square displacements

Acknowledgements

The calculations were performed on the ADA machine at the IDRIS Supercomputing Center, France, under eDARI grant x2016106368.

Funding

RC was funded by the Extreme Physics and Chemistry directorate of the Deep Carbon Observatory. WRP was funded by NSF via grant EAR 09-55647.

Authors' contributions

RC and WRP proposed the topic, conceived and designed the study, analyzed the data, and constructed the manuscript. RC ran the calculations. Both authors read and approved the final manuscript.

Competing interests

The authors declare that they have no competing interests.

Author details

[1]CNRS, Ecole Normale Supérieure de Lyon, Université Claude Bernard Lyon 1, Laboratoire de Géologie de Lyon, CNRS UMR 5276, Lyon, France. [2]School of Earth Sciences, Ohio State University, Columbus, OH 43210, USA.

References

Ammann MW, Brodholt JP, Dobson DP (2011) Ferrous iron diffusion in ferro-periclase across the spin transition. Earth Planet Sci Lett 302:393–402. doi:10.1016/j.epsl.2010.12.031

Blochl PE (1994) Projector augmented-wave method. Phys Rev B 50:17953–17979

Bunson PE, Di Ventra M, Pantelides ST, Schrimpf RD, Galloway KF (1999) Ab initio calculations of H^+ energetics in SiO_2: implications for transport. IEEE Trans Nuclear Sci 46:1568–1573

De Koker NP, Stixrude L, Karki BB (2008) Thermodynamics, structure, dynamics, and freezing of Mg_2SiO_4 liquid at high pressure. Geoch Cosmochimica Acta 72:1427–1441. doi:10.1016/j.gca.2007.12.019

Hae R, Ohtani E, Kubo T, Koyama T, Utada H (2006) Hydrogen diffusivity in wadsleyite and water distribution in the mantle transition zone. Earth Planet Sci Lett 243:141–148. doi:10.1016/j.epsl.2005.12.035

Houser C (2016) Global seismic data reveal little water in the mantle transition zone. Earth Planet Sci Lett 448:94–101. doi:10.1016/j.epsl.2016.04.018

Huang X, Xu Y, Karato SI (2005) Water content in the transition zone from electrical conductivity of wadsleyite and ringwoodite. Nature 434:746–749. doi:10.1038/nature03426

Jacobsen SD, Smyth JR (2006). Effect of water on the sound velocities of ringwoodite in the transition zone. Earth's Deep Water Cycle. In: Jacobsen SD, van der Lee S. Geophys Monogr Ser, vol. 168. pp. 131-145, AGU, Washington, DC

Karki BB, Bhattarai D, Mookherjee M, Stixrude L (2010) Visualization-based analysis of structural and dynamical properties of simulated hydrous silicate melt. Phys Chem Minerals 37:103–117

Kavner A (2003) Elasticity and strength of hydrous ringwoodite at high pressure. Earth Planet Sci Lett 214:645–654. doi:10.1016/S0012-821X(03)00402-3

Kelbert A, Schultz A, Egbert G (2009) Global electromagnetic induction constraints on transition-zone water content variations. Nature 460:1003–1006

Khan A, Kuvshinov A, Semenov A (2011) On the heterogeneous electrical conductivity structure of the Earth's mantle with implications for transition zone water content. J Geophys Res Solid Earth 116:B01103. doi:10.1029/2010JB007458

Kresse G, Hafner J (1993) Ab initio molecular dynamics for liquid metals. Phys Rev B 47:558

Kresse G, Joubert D (1999) From ultrasoft pseudopotentials to the projector augmented-wave method. Phys Rev B 59:1758

Liebske C, Frost DJ (2012) Melting phase relations in the MgO–MgSiO3 system between 16 and 26GPa: implications for melting in Earth's deep interior. Earth Planet Sci Lett 345–348(C):159–170. doi:10.1016/j.epsl.2012.06.038

Mookherjee M, Stixrude L, Karki BB (2008) Hydrous silicate melt at high pressure. Nature 452:983–986

Ohtani E, Zhao D (2009) The role of water in the deep upper mantle and transition zone: dehydration of stagnant slabs and its effects on the big mantle wedge. Russian Geol Geophys 50:1073–1078. doi:10.1016/j.rgg.2009.11.006

Olsen N (1998) The electrical conductivity of the mantle beneath Europe derived from C-responses from 3 to 720 hr. Geophys J Int 133:298–308

Panero WR (2010) First principles determination of the structure and elasticity of hydrous ringwoodite. J Geophys Res 115:B03203. doi:10.1029/2008JB006282

Panero WR, Smyth JR, Pigott JS, Liu Z, Frost DJ (2013) Hydrous ringwoodite to 5 K and 35 GPa: multiple hydrogen bonding sites resolved with FTIR spectroscopy. Amer Mineral 98:637–642. doi:10.2138/am.2013.3978

Pearson DG, Brenker FE, Nestola F, McNeill J, Nasdala L, Hutchison MT et al (2014) Hydrous mantle transition zone indicated by ringwoodite included within diamond. Nature 507:221–224. doi:10.1038/nature13080

Perdew JP, Burke K, Ernzerhof M (1996) Generalized gradient approximation made simple. Phys Rev Lett 77:3865–3868

Schweiger E, Sharma M, Gygi F, Galli G (2008) Melting of ice under pressure. Proc Nat Acad Sci 105:14779–14783

Shimizu H, Utada H, Baba K, Koyama T, Obayashi M, Fukao Y (2010) Three-dimensional imaging of electrical conductivity in the mantle transition zone beneath the North Pacific Ocean by a semi-global induction study. Phys Earth Planet Inter 183:252–269

Smyth JR, Kawamoto T, Jacobsen SD, Swope RJ, Hervig RL, Holloway JR (1997) Crystal structure of monoclinic hydrous wadsleyite [β-(Mg, Fe)$_2$SiO$_4$]. Amer Mineral 82:270–275

Sun W, Yoshino T, Sakamoto N (2015) Hydrogen self-diffusivity in single crystal ringwoodite: implications for water content and distribution in the mantle transition zone. Geophy Res Lett 42:06448. doi:10.1002/2015GL064486

Tse JS (2002) Ab initio molecular dynamics with density functional theory. Ann Rev Phys Chem 62:249–290. doi:10.1146/annurev.physchem.53.090401.105737

Tsuchiya J, Tsuchiya T (2009) First principles investigation of the structural and elastic properties of hydrous wadsleyite under pressure. J Geophys Res 114: B02206. doi:10.1029/2008JB005841.

Verma, AK, Karki, BB (2009) First-principles simulations of native point defects and ionic diffusion in high-pressure polymorphs of silica. Phys Rev B 79:214115.

Yoshino T, Matsuzaki T, Shatskiy A, Katsura T (2009) Earth and planetary science letters. Earth Planet Sci Lett 288:291–300. doi:10.1016/j.epsl.2009.09.032.

5

Integrated tephrostratigraphy and stable isotope stratigraphy in the Japan Sea and East China Sea using IODP Sites U1426, U1427, and U1429, Expedition 346 Asian Monsoon

Takuya Sagawa[1*], Yoshitaka Nagahashi[2], Yasufumi Satoguchi[3], Ann Holbourn[4], Takuya Itaki[5], Stephen J. Gallagher[6], Mariem Saavedra-Pellitero[7], Ken Ikehara[5], Tomohisa Irino[8] and Ryuji Tada[9]

Abstract

Integrated Ocean Drilling Program Expedition 346 "Asian Monsoon" obtained sediment successions at seven sites in the Japan Sea (Sites U1422–U1427 and U1430) and at two closely located sites in the northern East China Sea (Sites U1428 and U1429). The Quaternary sediments of the Japan Sea are characterized by centimeter- to decimeter-scale dark–light alternations at all sites deeper than 500 m water depth. The sedimentary records from these sites allow an investigation of the regional environmental response to global climate change, including changes in the Asian Monsoon and eustatic sea level. However, the discontinuous occurrence of calcareous microfossils in the deep-sea sediments and their distinct isotope signature that deviates from standard marine $\delta^{18}O$ records do not permit the development of a detailed stable isotope stratigraphy for Japan Sea sediments. Here, we present the tephrostratigraphy for the two southernmost sites drilled in the Japan Sea (Sites U1426 and U1427) and for one site drilled in the East China Sea (Site U1429) along with the benthic $\delta^{18}O$ isotope stratigraphy for the shallower Site U1427 and the East China Sea Site U1429. Eighteen tephra layers can be correlated between sites using the major-element composition and morphology of volcanic glass shards, and the compositions of grains and heavy minerals. Tephra correlations show that negative $\delta^{18}O$ peaks in the Japan Sea correspond to positive glacial maxima peaks in the East China Sea. Using this integrated stratigraphic approach, we establish an orbital-scale age model at Site U1427 for the past 1.1 Myr. The correlation of tephra layers between the shallower Site U1427 (330 m below sea level: mbsl) and the deeper Site U1426 (903 mbsl) in the southern Japan Sea provides the opportunity for further age constraints. Our results show that alternations in sediment color at Sites U1426 and U1427 can be correlated for the past 1.1 Myr with minor exceptions. Thus, the stable isotope stratigraphy established at the shallower Site U1427 can be correlated to Site U1426, and in turn to all sites drilled during Expedition 346, based on correlations of dark–light layering.

Keywords: Japan Sea, East China Sea, Tephrostratigraphy, Stable isotope stratigraphy, Integrated Ocean Drilling Program, Expedition 346, Inter-site correlation

* Correspondence: tsagawa@staff.kanazawa-u.ac.jp
[1]Institute of Science and Engineering, Kanazawa University, Kakuma-machi, Kanazawa, Ishikawa 920-1192, Japan
Full list of author information is available at the end of the article

Introduction

Integrated Ocean Drilling Program (IODP) Expedition 346 "Asian Monsoon" drilled seven sites in the Japan Sea (JS) and two closely located sites in the northern East China Sea (ECS) (Fig. 1). The main scientific goal of this expedition was to unravel the evolution and variability of the East Asian monsoon over multiple timescales. The JS drilling sites spanned both latitudinal and depth transects in order to explore linkages between atmospheric and oceanic processes, along the edge of the Eurasian continent, the Korean Peninsula, and the Japanese islands, and to investigate regional responses to far-field forcing. However, detailed correlation of the suite of cores obtained during Expedition 346 is a pre-requisite for any regional paleoclimatic and paleoceanographic reconstructions.

The late Quaternary deep-sea sediment in the JS is characterized by centimeter- to decimeter-scale alternations of dark and light layers (Tada et al. 1992 1999; Nakajima et al. 1996), with the darker sediment occasionally showing fine laminations. The alternation of color was caused by changes in deep-water oxygenation in sediments recovered from depths > 500 m below sea level (mbsl) (Ikehara et al. 1994). Dark–light color alternations are present at six of the seven sites drilled during Expedition 346 (Sites U1422–U1426 and U1430), and they occupy water depths of 990–3430 mbsl, whereas the succession at Site U1427 on the outer margin of a marginal terrace (~ 330 mbsl) shows gradual and subtle color changes (Fig. 1). Previous studies have demonstrated that the dark–light layers in sediments from deeper sites can be correlated across the JS (e.g., Tada et al. 1992; Nakajima et al. 1996). This is also the case for the six deeper sites of Expedition 346, which are readily correlated over the past 1.5 Myr (Tada et al. 2018). Due to its tectonic setting, the JS and

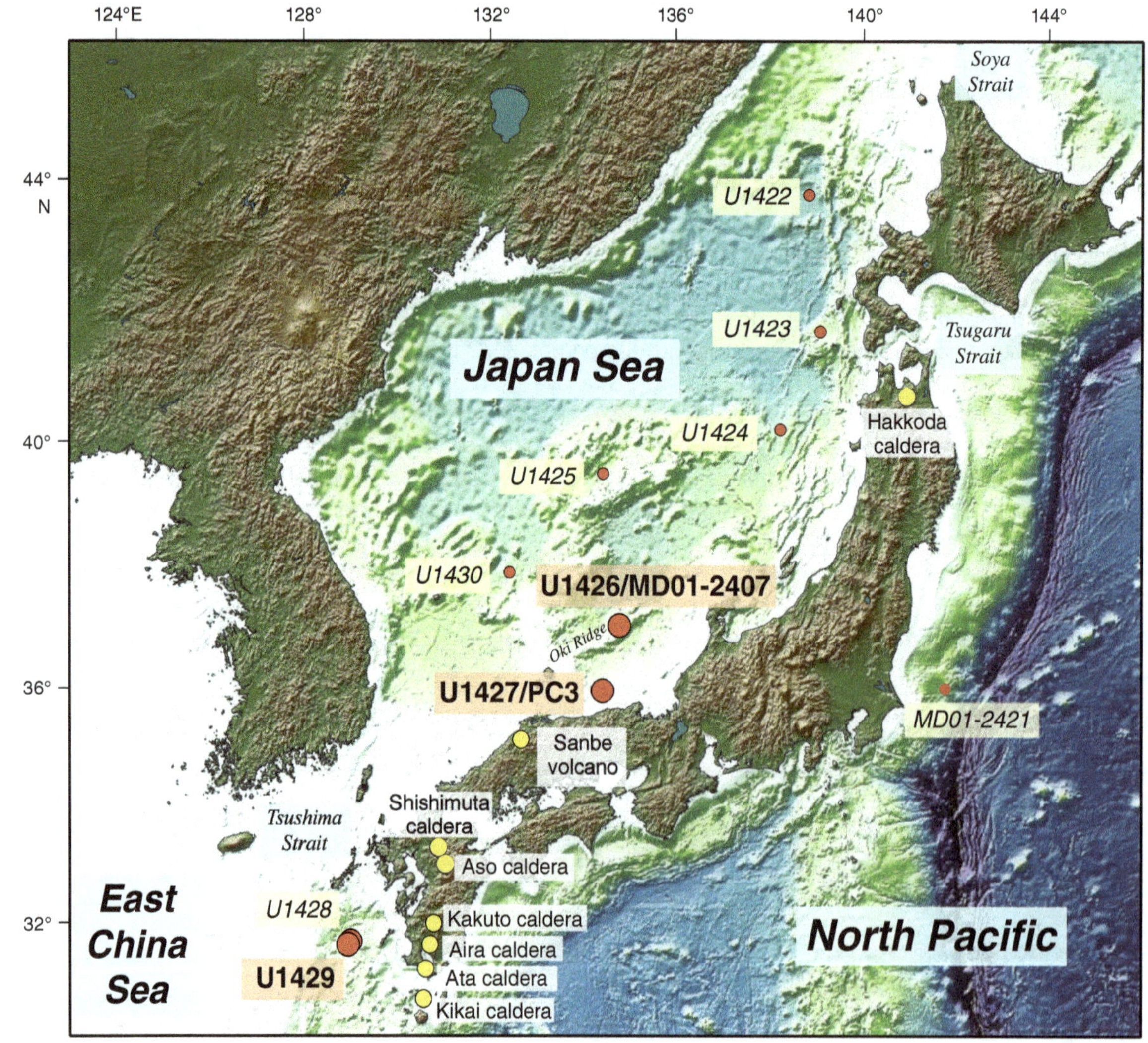

Fig. 1 Locations of sites drilled during IODP Expedition 346 in the Japan Sea and East China Sea. The source volcanoes of some of the regional tephra layers are shown

northern ECS sediments contain a large number of tephra layers that were erupted from volcanoes on the Japanese islands and the Korean Peninsula (Ikehara 2015). Tephra deposition is geologically synchronous; therefore, tephra layers provide an excellent way to precisely correlate sediment successions in the JS and the ECS.

The Pliocene and Pleistocene tephrostratigraphy and tephrochronology of the Japanese islands were summarized by Machida (1999), Tamura and Yamazaki (2010), and Satoguchi and Nagahashi (2012). Major- and trace-element compositions of volcanic glass shards of the more regionally widespread tephra layers were investigated by Nagahashi et al. (2004, 2007, 2015). These element compositions of tephra layers have been used to identify and correlate the layers (e.g., Nagahashi et al. 2004). Tephra layers in JS sediments were originally described by Pouclet and Scott (1992), who summarized their temporal distribution in the sediments of ODP Leg 127/128. These tephra layers were subsequently investigated to understand the nature of volcanic activity (Nakagawa et al. 1994) in order to identify the sources and to establish relationships with oxygen isotope stratigraphy for the past 400 kyr (Shirai et al. 1997). The stratigraphy and petrography of Upper Pleistocene marine tephras from the JS and the North Pacific Ocean were summarized by Ikehara (2015) and Aoki (2008), respectively. However, the long-term tephrostratigraphy of JS sediments, its correlation with known widespread tephra layers, and the tephrochronology for the past several million years have not been established to date. If the tephrostratigraphy of JS sediments is known, comparison of environmental changes between the Japanese islands and the Japan Sea will be possible. Here, we provide correlations between tephra layers that are interbedded in sediment successions at Sites U1426, U1427, and U1429 with regionally occurring tephras in order to determine isochronous horizons for these sites.

The generation of orbital-scale age models for the JS sediments is complicated by the sporadic occurrence and variable preservation of calcareous microfossils, which hinders the construction of a detailed oxygen isotope stratigraphy. Furthermore, the JS carbonate compensation depth (CCD) has varied from < 1000 to > 3000 m over the past 30 kyr (Oba et al. 1991), and the JS oxygen isotope signal is different from that of the open ocean, as it is a semi-enclosed environment (e.g., Oba et al. 1991; Kido et al. 2007; Yokoyama et al. 2007). The JS is connected to the North Pacific Ocean and adjacent marginal seas (ECS and Okhotsk Sea) only by shallow and narrow straits (Fig. 1). Even the sill depths of the deepest connections to the open ocean, the Tsushima and Tsugaru Straits, are only ~ 130 m, while the sill depths in the Soya Strait and the Mamiya Strait are ~ 55 and ~ 15 m, respectively. Therefore, glacial–interglacial sea-level change has had a marked influence on seawater exchange through the straits. For example, during the last glacial maximum (LGM), the JS was almost isolated from the surrounding seas, resulting in planktonic $\delta^{18}O$ minima that are interpreted as a signal of low-salinity surface water (Oba et al. 1991).

The shallow Site U1427 is an important site for Expedition 346 site correlation for two main reasons. First, as it is in the southern part of the JS, the sediment succession at Site U1427 contains tephra layers that are also present at the ECS sites. These tephra layers can be used to constrain the JS $\delta^{18}O$ stratigraphy and to correlate the $\delta^{18}O$ stratigraphy of the JS to the ECS, the latter with an open-ocean $\delta^{18}O$ signal. Second, the sediment at this site contains calcareous microfossils throughout the succession, except for two intervals (Tada et al. 2015a), thereby allowing high-resolution $\delta^{18}O$ analyses. Once inter-site correlation between Site U1427 and the deeper JS sites is achieved, the U1427 $\delta^{18}O$ stratigraphy can be projected to the deeper sites, where the preservation of foraminifera is poor.

Objectives of this study

This study aims to (1) establish a tephrostratigraphy for the past 1.1 Myr based on tephra layers interbedded in the sediment successions of Sites U1426, U1427, and U1429; (2) construct age models for Sites U1427 and U1429; and (3) correlate the sediment succession of the shallow JS Site U1427 with the deeper JS Site U1426. To achieve these objectives, we investigate the petrography and major-element chemistry of tephra layers within the sediment successions at these three sites, analyze benthic foraminiferal oxygen isotopes at Sites U1427 and U1429, and analyze radiolarian, planktonic foraminiferal, and calcareous nannofossil assemblages at Site U1427. By using tephras as the primary correlating horizons, we are also able to correlate sediment color variations at shallow- and deep-water sites in the JS.

Geographical and oceanographical setting

The JS and ECS are marginal seas located along the eastern margin of the Eurasian continent (Fig. 1). The JS is connected with the North Pacific Ocean, Okhotsk Sea, and East China Sea. Although the maximum water depth in the JS is 3700 m, the straits connecting to adjacent seas are shallow (< 130 m), meaning that intermediate and deep waters in the JS are separated from the adjacent seas. The only surface water flowing into the JS is the Tsushima Warm Current (TWC) flowing from the ECS through the Tsushima Strait. Since the TWC originates from the Kuroshio Current in the subtropical North Pacific Ocean and mixes with nutrient-rich coastal water derived from the Yangtze River in the ECS, this surface current supplies heat, salt, and nutrients to the JS.

During the LGM, the sill depths of the two deepest straits were ~10 m (Matsui et al. 1998). Thus, surface water exchange through the straits was limited, and TWC inflow was restricted, resulting in reduced salt supply from the ECS. The surface water freshening is recorded in the planktonic foraminiferal $\delta^{18}O$ data of deep-sea sediments across the JS (Oba et al. 1991). Therefore, the oxygen isotope stratigraphy in the JS is complicated by glacio-eustatic effects on surface salinity. Since carbonate $\delta^{18}O$ values reflect both temperature and seawater $\delta^{18}O$, the JS planktonic foraminifera $\delta^{18}O$ values reflect two kinds of minima on a glacial–interglacial time scale: (1) during interglacial maxima (warm conditions) and (2) during glacial maxima (low seawater $\delta^{18}O$; i.e., low salinity). Therefore, to correlate with marine isotope stages, one should interpret planktonic $\delta^{18}O$ data along with temperature information derived from, for example, planktonic microfossil assemblages (Kido et al. 2007). In this study, we focus on the benthic foraminiferal $\delta^{18}O$ record in the shallow Site U1427 (Fig. 1). The water depth of this site is ~330 mbsl, well below the modern thermocline, which remains at ~100–250 m water depth throughout the year (World Ocean Atlas 2013 (Locarnini et al. 2013)). Thus, the warm signal, attributed to intrusion of the TWC during interglacial maxima, is not expected to be recorded in the benthic $\delta^{18}O$ values, thereby eliminating $\delta^{18}O$ minima induced by warm temperatures during the interglacial maxima.

Glacio-eustatic sea-level change also influenced the ecological environment in the JS. Since the sea-level drop during glacial periods restricted the inflow of the TWC, assemblages of planktonic microfossils during glacial and interglacial periods differed markedly (Oba et al. 1991), and subtropical species were present only during interglacials. Therefore, combining records of benthic $\delta^{18}O$ and TWC planktonic microfossil species at Site U1427 allows glacials to be distinguished from interglacials, which can be used to establish an orbital-scale age model.

Methods/Experimental

Sediment cores

Site U1426 is in the south-central part of the JS, near the top of the Oki Ridge (37°2.00′N, 134°48.00′E, 903 mbsl) (Fig. 1). The sediment succession of the upper 120 m is characterized by decimeter-scale color banding with alternations of dark and lighter brown layers (Fig. 2a). The organic-rich darker layers generally display a sharp base, whereas the upward transition from dark to light layers is gradational and commonly strongly bioturbated (Tada et al. 2015b). Some darker layers are thinly laminated. Calcareous microfossil preservation is relatively good in the upper part of the succession at this site compared with deeper sites, although foraminifera are poorly preserved or absent in some intervals (Tada et al. 2015b). Several tephra layers occur in the succession.

Site U1427 is located in the south-central part of the JS on the outer margin of the marginal terrace off Tottori (35°57.92′N, 134°26.06′E, 330 mbsl), ~110 km south of Site U1426 (Fig. 1). The sediment is dominated by heavily bioturbated and homogeneous clayey silt. Faint laminations are present in the intervals at 162–177, 225–246, and 346–372 m (U1427_patched_CCSF-D_rev20170310, CCSF: core composite depth below seafloor, see below for explanation of the "patched" depth-scale). Centimeter-thick tephra layers occur frequently throughout the succession. Owing to its proximity to Site U1426, it is likely that the same tephra layers are present at both sites. The sediment at this site differs from that of deeper JS sites, including Site U1426, as it lacks dark–light color banding. Instead, the sediment shows gradual meter-scale color changes accompanied by lithological changes (Fig. 2a). The sediment contains well-preserved foraminifera throughout, except for the two laminated intervals in the upper part of the core (Tada et al. 2015a). Sites U1427 and U1426 are close to the first and second branch of the TWC, respectively.

Site U1429 is in the southern part of the Danjo Basin in the northwestern part of the ECS (31°37.04′N, 128° 59.85′E, 732 mbsl). The upper ~180 m of the succession is characterized by calcareous-nannofossil-rich clay with diatoms and foraminifera (Tada et al. 2015c). The sediment is heavily bioturbated throughout and shows subtle meter-scale color variability (Fig. 2a). Tephra layers of various thickness, ranging from centimeter-scale to > 0.5 m, are present throughout the succession.

The shipboard composite sections, so-called "splice sections," of Sites U1426 and U1427 (Tada et al. 2015a, 2015b) were revised post-cruise based on re-examination of close-up section photographs; the splice of Site U1429 is from Tada et al. (2015c). During this re-evaluation, a stratigraphical gap and repeated successions were found in the original shipboard splice (Irino et al. 2018). Detailed inter-hole correlation and revision of the shipboard splices for Sites U1426 and U1427 are presented by Irino et al. (2018). The depth-scale for the revised splice section is expressed as "patched_CCSF-D" in this study. The depth-scale for Sites U1426 and U1427 used in this study are the latest versions, "U1426_patched_CCSF-D_rev20150519" and "U1427_patched_CCSF-D_rev20170310," respectively, unless noted otherwise.

Shipboard measurements

Sediment cores recovered during the expedition were analyzed using a Whole-Round Multisensor Logger (WRMSL; measuring *P* wave velocity, density, and magnetic susceptibility), Natural Gamma Radiation Logger (NGRL), Section

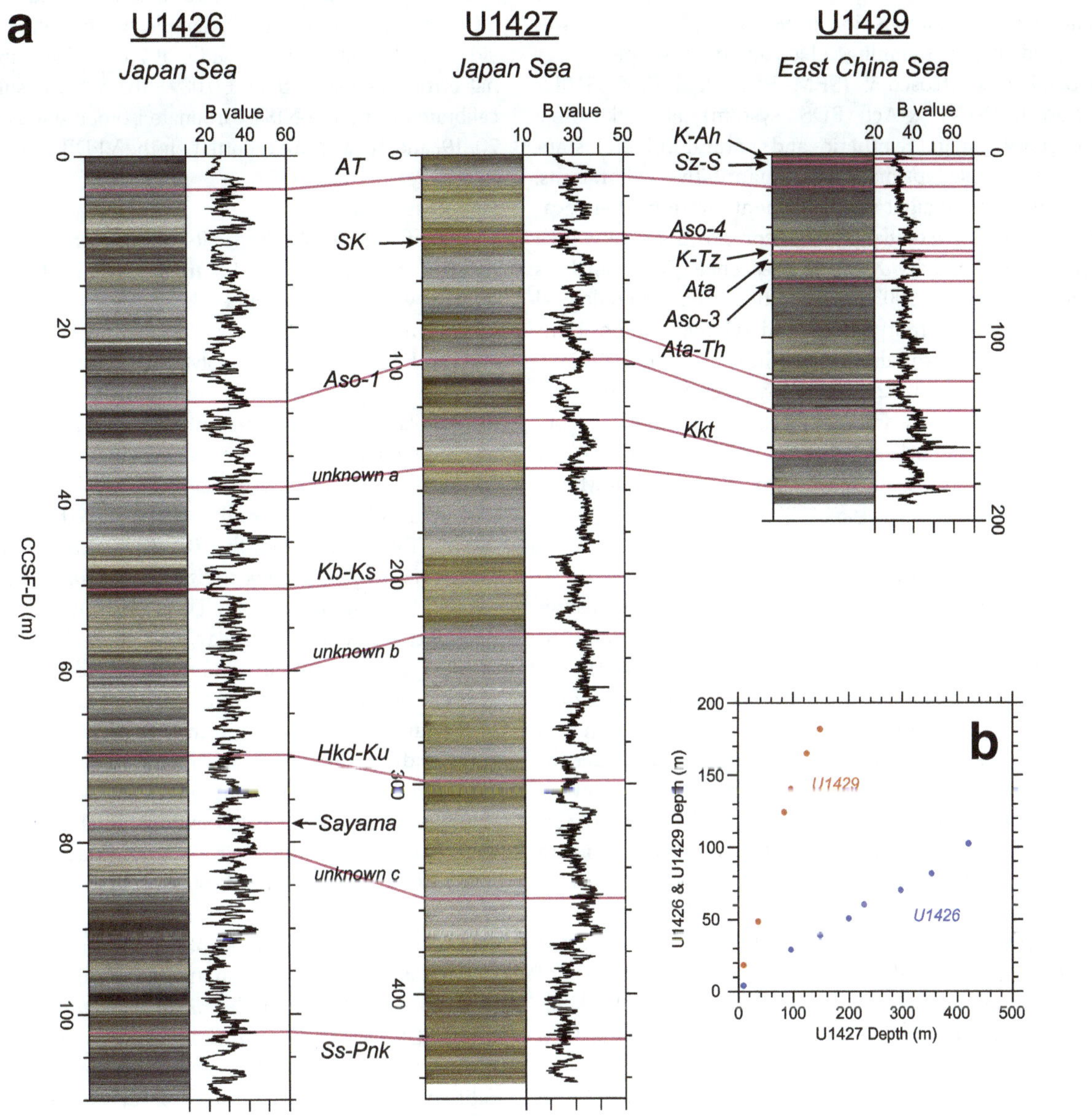

Fig. 2 a Composite images and color profiles of the studied intervals of Sites U1426, U1427, and U1429. Shipboard line-scan images of the surfaces of split sediment cores. Brightness and contrast of the images are enhanced to highlight color changes. Positions of regional tephras and correlative layers in the successions are indicated by magenta lines. **b** Depth relationships of correlative tephra layers between sites

Half Imaging Logger (SHIL), and Section Half Multisensory Logger (SHMSL). For details of each method, please see the Methods chapter of the Proceedings volume (Tada et al. 2015d). The SHIL captures continuous high-resolution images of the archive half-surface. Images were collected shortly after core splitting and surface scraping to avoid color changes resulting from excessive sediment drying and oxidation of the surface. High-resolution color data (red, green, and blue) were acquired by digital color imaging. Reflectance of visible light and magnetic susceptibility (MS) were measured on the surfaces of sediment cores, which were covered with clear plastic wrap, on the SHMSL with an Ocean Optics USB4000 spectrophotometer and a Bartington Instruments MS2E point sensor, respectively. These measurements were made at 1 cm spacing to provide high-resolution records. L*, a*, and b* were calculated from the obtained reflectance spectra.

Tephra

We analyzed 56 samples from Site U1426, 56 samples from Site U1427, and 28 samples from Site U1429. Occurrences of tephra layers were recorded during shipboard visual core descriptions. Recognition of tephra layers was based on color, sedimentary structures, and components. The samples were *ultrasonicated* to remove clay contaminants and then separated into four grain-size fractions using 32, 63, and 250 μm sieves. The major-element composition of volcanic glass shards was analyzed using scanning electron microscope–energy dispersive spectroscopy (SEM–EDS, Jeol JSM-6510LV +Oxford INCA x-Act EDS system) at Fukushima University, Japan. Synthetic and natural mineral standards were used primarily for quantitative calibrations. Compositions of eight major elements were further recalibrated by sub-standards, the compositions of which were determined by X-ray fluorescence (XRF) analysis (Nagahashi et al. 2003). In addition, the mineralogical composition, the morphology and relative compositions of volcanic glass shards, and the heavy mineral composition (specific gravity 2.8) were investigated for grains in the 63–250 μm size fraction of samples collected from key correlative horizons. Images of glass morphology were obtained using a low-vacuum SEM (Hitachi Miniscope TM-1000).

Oxygen isotope analyses

Sediment samples for benthic foraminiferal isotope analysis at Site U1427 were washed using a 63 μm sieve, and the residues were dried at 40 °C in an oven overnight. *Uvigerina* spp. and *Cassidulina* spp. were picked from the 250–355 μm fraction of the dried residues. Picked foraminiferal shells were gently crushed using two glass plates under the microscope. The fragments were ultrasonicated several times with Milli-Q and methanol. The $\delta^{18}O$ values of foraminifera in samples from Site U1427 were measured with a Thermo Finnigan MAT253 mass spectrometer connected to a Kiel III carbonate device at the Kochi Core Center, Japan. The isotope results were calibrated to the Vienna Pee Dee belemnite (VPDB) using the National Institute of Standards and Technology (NIST: Gaithersburg, Maryland) standard NBS 19. The analytical precision was better than ± 0.06‰.

Sediment samples from Site U1429 were dried in an oven at 40 °C, wet-sieved at 63 μm, and dried at 40 °C. Between three and seven well-preserved tests of *Uvigerina* spp. were selected for analysis, except for a few samples in which *Uvigerina* was rare or absent, in which case *Bulimina* spp. or *Cibicidoides* spp. were selected. Tests were broken into large fragments, cleaned in alcohol in an ultrasonic bath, and then dried at 40 °C. Replicate measurements of Site U1429 samples and core tops from several ocean basins (Holbourn et al. unpublished data) indicate no significant offset between the $\delta^{18}O$ values of *Uvigerina* and *Bulimina*. The $\delta^{18}O$ values of *Cibicidoides* were adjusted by adding 0.71‰ to the original measurements to compensate for the offset between *Uvigerina* and *Cibicidoides* (Holbourn et al., unpublished core-top data). Measurements of benthic foraminiferal $\delta^{18}O$ at Site U1429 were made using a Finnigan MAT251 mass spectrometer at the Leibniz Laboratory, Kiel University, Germany. The instrument is coupled on-line to a Carbo-Kiel Device (type I). Samples were reacted by individual acid addition (99% H_3PO_4 at 73 °C). Standard external error was better than ± 0.08‰ for $\delta^{18}O$. Results were calibrated using the NIST carbonate isotope standards NBS 20, 19, and 18, and are reported on the VPDB scale.

Microfossil analysis

A total of 455 samples from Site U1427 were processed for radiolarian analysis (at ~ 1.5 m intervals). Of these, 270 samples were from the upper 350 m, which was the depth range targeted in this study. Detailed analytical methods and results for all samples analyzed are provided by Itaki et al. 2018. Based on previous studies of JS sediment cores, the radiolarian taxon, *Tetrapyle octacantha* group, occurred commonly during interglacial periods and only sporadically during glacial periods over the past 640 kyr (Itaki 2007; Itaki et al. 2007). In this study, we considered the following taxa as TWC indicators: the *Tetrapyle octacantha* group, *Dictyocoryne* spp., *Didymocyrtis tetrathalamus*, *Euchitonia* spp., and *Spongaster tetras*. These taxa are common in the subtropical regions today, and their distribution in the JS surface sediment is closely related to the inflow of the TWC (Motoyama et al. 2016). These TWC indicators are expressed as numbers of individuals per gram of dry sediment.

A total of 231 samples (at ~ 1 m intervals) were processed for foraminiferal paleoenvironmental analyses. Sediment was washed with hydrogen peroxide over a 63 μm wire mesh sieve. Dried residues were divided into fractions using a microsplitter, and a minimum of 100 individuals (typically > 200 tests) were identified and counted in the > 150 μm size fraction. Four subtropical planktonic foraminifera species (*Globigerinoides ruber, Pulleniatina obliquiloculata*, *Globorotalia menardii*, and *Globorotalia tumida*) are indicators of the TWC, along with the percentage of right-coiling *Neogloboquadrina pachyderma* (*Neogloboquadrina incompta*; sensu Darling et al. 2006) (Oba et al. 1991; Domitsu and Oda 2005; Gallagher et al. 2015). The %TWC species and %*N. incompta* are expressed as percentages of the total planktonic foraminifera assemblage.

Sixty samples were processed for coccolith analysis (at ~ 3 m intervals). Sediment samples were freeze-dried and prepared following combined dilution and filtering (Andruleit 1996), which enabled estimates of the absolute numbers of coccoliths per gram of dry sediment. Between 61 and 129 mg of dry bulk sediment was suspended in buffered water and treated ultrasonically for ~ 30 s. Samples were wet-split using a rotary sample divider, and 1/100 of the original solution was filtered through polycarbonate filters (0.45 μm pore size) with a vacuum pump. The filters were dried in an oven at ~ 40 °C for at least 24 h and stored. A small part of each filter was cut out, fixed on an aluminum stub, and sputtered

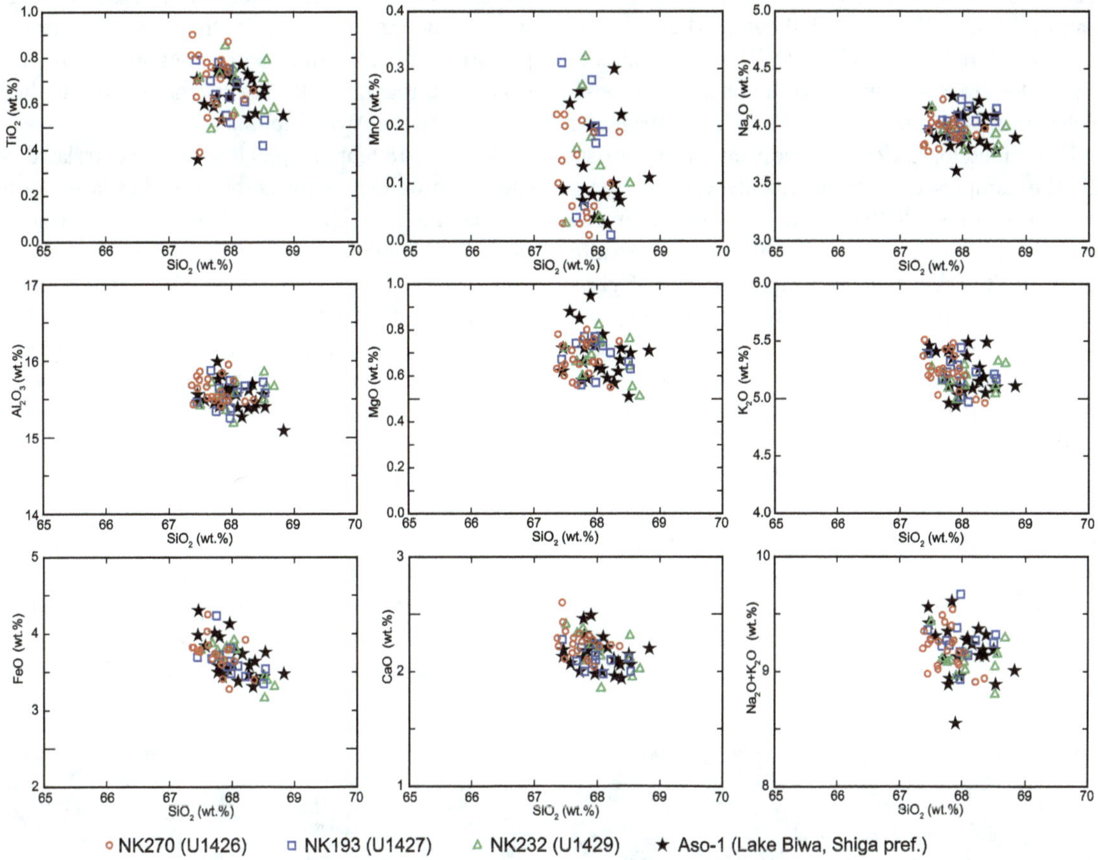

Fig. 3 Example of the Harker diagrams for correlative tephra samples from Sites U1426 (NK270), U1427 (NK193), U1429 (NK232), and the Aso-1 tephra in Lake Biwa sediment

with gold/palladium. Following a counting routine, a minimum of 400 coccoliths per sample were identified and classified under SEM at × 3000 magnification. However, in a few cases, this minimum number was not reached. Identification of coccoliths followed Young et al. (2003) and the electronic guide Nannotax 3 (http://www.mikrotax.org/Nannotax3/). TWC indicators included the taxa *Calcidiscus leptoporus*, *Gephyrocapsa oceanica*, and *Helicosphaera carteri* (Muza 1992).

Results and discussion

Correlation of tephra layers between sites and with widespread tephra layers

The following screening procedure was applied for detecting correlative tephra between sites.

1) The concentrations of four major elements (Al_2O_3, FeO, CaO, and K_2O) in volcanic glass shards were compared between sites, and samples with a consistent composition within two standard deviations (2σ) for each element were selected. We used these four oxides for the first screening as they are more accurately quantified and suitable for the discrimination of tephras than other elements.
2) All major-element compositions were plotted against SiO_2 (i.e., Harker diagrams; Harker 1909), excluding data that differed between sites by more than 1σ.
3) If potentially correlative tephra layers passed the two previous screening steps but occurred in an inconsistent stratigraphic position in the successions between sites, they were eliminated from the list of correlative horizons.
4) Finally, the correlations were carefully assessed on the basis of lithofacies, mineralogical composition, the morphology and composition of volcanic glass shards, and heavy mineral composition.

Only tephras that passed all steps above were identified as correlative tephra layers. Furthermore, all tephra layers investigated in this study were examined to determine whether they are correlated to well-known widespread tephra layers of the Japanese islands, using the procedure noted above.

The well-known and widespread tephra Aso-1, which was deposited at ~ 0.25 Ma, can be correlated between sites. The average major-element compositions of volcanic glass shards for NK270 at Site U1426, NK193 at Site U1427, and NK232 at Site U1429 show good agreement within 1σ (Additional file 1: Table S1). The SiO_2 contents of these samples are 67–69 wt% and the relative compositions of elements plot in similar areas on Harker diagrams (Fig. 3). The mineralogical compositions are also consistent, with the samples comprising mainly volcanic glass shards and heavy minerals that are mainly orthopyroxene and clinopyroxene with minor hornblende and apatite (Additional file 2: Table S2). The morphologies of glass shards are platy and intermediate (platy and pumice) types (Fig. 4 and Additional file 2: Table S2). Based on the data above, NK270, NK193, and NK232 are correlated with each other. The major-element compositions of volcanic glass shards, grain composition, heavy minerals, and glass morphology of these tephra samples are similar to the BT61 tephra in the Takashima-oki sediment core from Lake Biwa (Yoshikawa and Inouchi 1993; Nagahashi et al. 2004) (Figs. 3 and 4; Additional files 1 and 2: Tables S1 and S2). The BT61 tephra is a co-ignimbrite ash (Sparks and Walker 1977) that was associated with the Aso-1 pyroclastic flow deposit erupted from the Aso caldera in the southern part of Kyushu Island (see Fig. 1 for the location of the Aso caldera). Furthermore, the stratigraphic positions of these samples are consistent with the stable isotope stratigraphy. Therefore, these three tephra layers correlate to the Aso-1 tephra.

There are 18 tephra layers that can be correlated between sites or to well-known widespread tephra layers of the past 1.1 Ma (Tables 1, S1, and S2). These are the Kikai-Akahoya (K-Ah), Sakurajima-Satsuma (Sz-S), Aira-Tanzawa (AT), Aso-4, Kikai-Tozurahara (K-Tz), Ata, Sanbe-Kisuki (SK), Aso-3, Ata-Torihama (Ata-Th), Aso-1, Kakuto (Kkt), Kobayashi-Kasamori (Kb-Ks), Hakkoda-Kokumoto (Hkd-Ku), and Shishimuta-Pink (Ss-Pink), from youngest to oldest (Table 1). All of the tephra layers, with the exception of three tephras (Sz-S, SK, and Hkd-Ku), are co-ignimbrite ash associated with large-scale pyroclastic flow eruptions on Kyushu island. Sz-S and SK are Plinian pumice-fall deposits from Sakurajima volcano in southern Kyushu and Sanbe volcano in the San-in area (Fig. 1). Hkd-Ku is a co-ignimbrite ash associated with a large-scale pyroclastic flow eruption of Hakkoda Caldera in northern Honshu Island. Figure 2b shows the depth relationship of tephra layers at

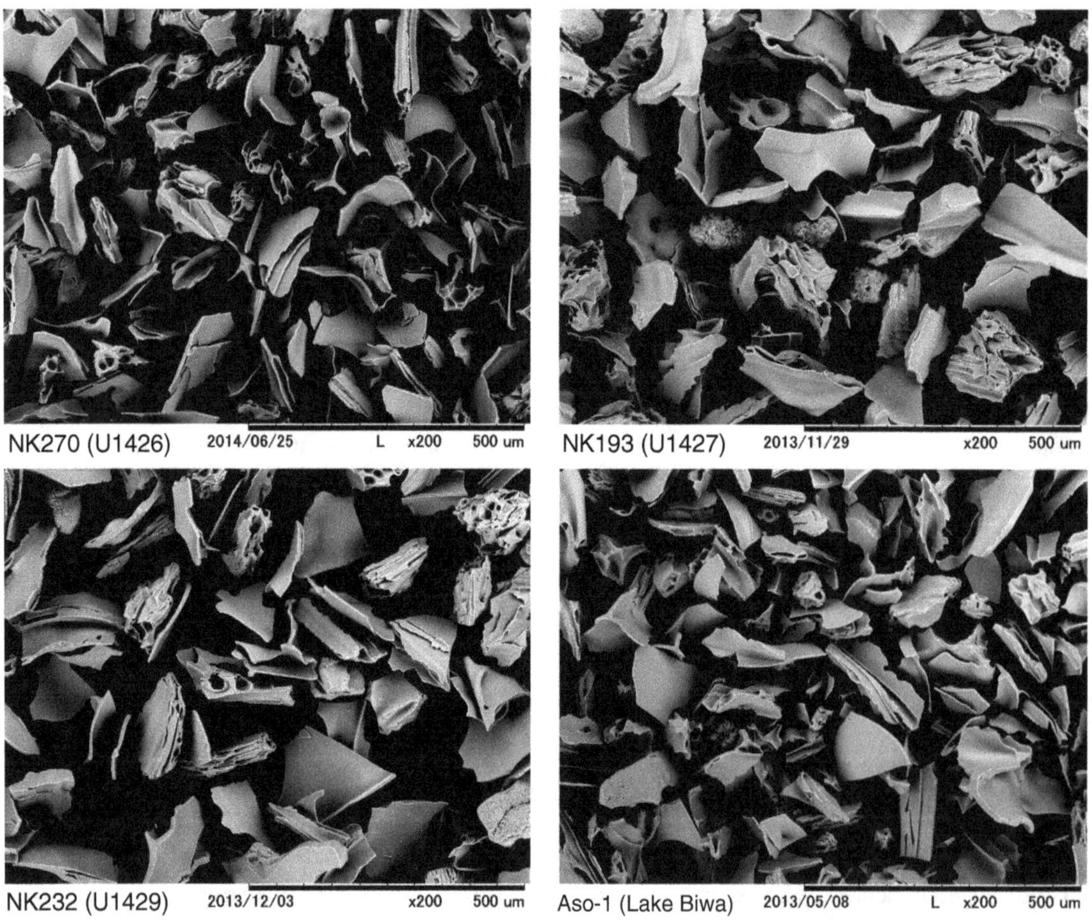

Fig. 4 Scanning electron images of volcanic glass shards of the Aso-1 tephra at Sites U1426, U1427, U1429, and Lake Biwa

Table 1 Tephra layers identified at Sites U1426, U1427, and U1429

U1429		U1427		U1426		Tephra name
Sample no.	Depth (m CCSF-D)	Sample no.	Depth (m, Patched_ CCSF-D_rev20170310)	Sample no.	Depth (m, Patched_CCSF-D_ rev20150519)	
NK454	2.97				(1.51)*	K-Ah
NK457	6.14					Sz-S
NK458	18.06	NK191	10.69	NK268	3.92	AT
NK224	48.26	NK491	37.54		(10.28)*	Aso-4
NK225	52.68					K-Tz
NK226	56.23					Ata
		NK192	41.23			SK
NK227	69.83				(16.39)*	Aso-3
NK341	126.40	NK299	84.67		(26.84)*	Ata-Th
NK232	140.09	NK193	97.70	NK270	28.63	Aso-1
NK234	164.59	NK194	126.74			Kkt
NK235	181.30	NK195	149.77	NK168	38.57	*Unknown-a*
		NK197	196.86	NK169	50.45	Kb-Ks
		NK199	228.70	NK274	60.00	*Unknown-b*
		NK204	298.37	NK280	69.79	Hkd-Ku
				NK174	77.92	Sayama
		NK208	354.53	NK175	81.38	*Unknown-c*
		NK312	421.80	NK176	102.12	Ss-Pink

*Depths are calculated by interpolation of correlation tie points with MD01-2407

the three sites. The data suggest that sedimentation at all three sites was continuous and that the sedimentation rates at Sites U1426 and U1429 were 0.23 and 1.2 times higher than at Site U1427, respectively.

Oxygen isotopes at Sites U1427 and U1429

Figure 5a shows the benthic and planktonic foraminiferal $\delta^{18}O$ records of core KR07-12 PC3, which was retrieved near Site U1427. The age model of this core is based on radiocarbon measurements of planktonic foraminifera from 16 horizons and the location of the AT tephra layer. The amplitude of the negative excursion of planktonic foraminiferal $\delta^{18}O$ at the LGM is similar to that reported from deeper sites in previous studies (e.g., Oba et al. 1991; Yokoyama et al. 2007). There is no significant offset between the $\delta^{18}O$ values of *Uvigerina* spp. and *Cassidulina* spp. The LGM benthic and planktonic foraminiferal isotope records show low $\delta^{18}O$ values, suggesting that the benthic foraminiferal $\delta^{18}O$ curve at Site U1427/PC3 records surface-freshening events at glacial maxima. The peak of freshening occurred when sea level was lowest (Yokoyama et al. 2000; Mix et al. 2001). Benthic foraminiferal $\delta^{18}O$ values in the upper 350 m of core from Site U1427 vary between 1.3 and 5.0‰ (Fig. 5b). Samples are barren of benthic foraminifera at 160–180 and 224–245 m. Several negative peaks are present, five of which (at 6, 50, 132, 180, and 246 m) have relatively high amplitudes of ~ 2‰, whereas other peaks show smaller amplitudes. The benthic isotope and sediment color (b*) records show similar patterns, with $\delta^{18}O$ minima corresponding to b* minima (Fig. 5b).

The $\delta^{18}O$ record of ECS Site U1429 is shown in Fig. 6b. The $\delta^{18}O$ values vary between 2.3 and 4.0‰, representing three glacial–interglacial cycles. The age model for Site U1429 was derived by correlating the benthic $\delta^{18}O$ record with the LR04 global stack $\delta^{18}O$ curve (Lisiecki and Raymo 2005) and by integrating tephra layers for which the eruptive ages are known. The tie points with LR04 are shown in Fig. 6b by red broken lines and are listed in Table 2. We used the ages of the following four tephra layers: K-Ah (7.2 ± 0.1 ka: Smith et al. 2013); Sz-S (~ 13 ka: Okuno 2002); AT (30.0 ± 0.2 ka: Smith et al. 2013); and Aso-4 (87.0 ± 0.3 ka: Aoki 2008). The K-Ah and AT tephras are widely found in marine, lacustrine, and terrestrial sediments. The stratigraphic positions of these tephras relative to the global oxygen isotope curve are well established (e.g., Oba et al. 1991; Ijiri et al. 2005; Domitsu and Oda 2006; Kubota et al. 2010). Smith et al. (2013) estimated the eruption ages of K-Ah and AT using tephra layers preserved in the sediment of Lake Suigetsu, the chronology of which has been established by varve counting with high-resolution accelerator mass spectrometry radiocarbon measurements (Bronk Ramsey et al. 2012). Radiocarbon measurements of soil just

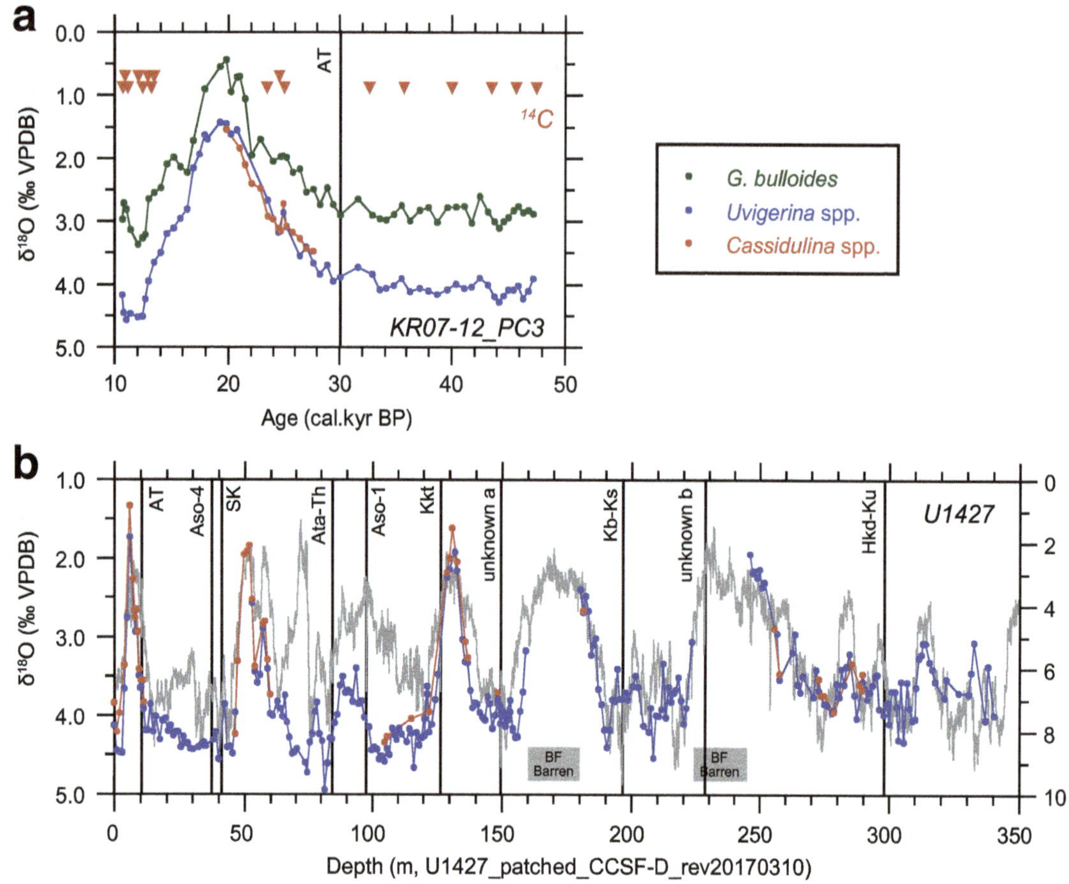

Fig. 5 $\delta^{18}O$ records of **a** KR07-12 PC3 and **b** Site U1427. Red triangles in **a** indicate the positions of planktonic foraminiferal radiocarbon dates. The gray line in **b** shows the sediment color profile b* at Site U1427. The vertical lines in **a** and **b** denote tephra horizons

beneath the Sz-S tephra yielded an age of ~ 13 cal. kyr BP (Okuno 2002). This age is consistent with the $\delta^{18}O$ stratigraphy of marine core MD98-2195 (Moriwaki et al. 2011). Aoki (2008) established the stratigraphic position of the Aso-4 tephra relative to the benthic foraminiferal $\delta^{18}O$ record of marine core MD01-2421 off central Japan in the Pacific Ocean. These results are in line with the $\delta^{18}O$ stratigraphy of Site U1429 (Fig. 6b). The average sedimentation rate at Site U1429 is ~ 48 cm/kyr (Fig. 6c), and the age at the base of the site is ~ 385 ka. The error in the correlation with LR04 is estimated to be ± 2.5 kyr based on the average time resolution of the benthic $\delta^{18}O$ record. The nomenclature of marine isotope stages (MISs) and substages used in this study follows Railsback et al. (2015). The benthic foraminiferal $\delta^{18}O$ data at Sites U1429 and U1427 are available in Additional files 3 and 4: Table S3 and S4, respectively.

Correlation of Sites U1427 and U1429

Six tephra layers are present at Sites U1427 and U1429 (Table 1 and Fig. 2). Five of them are identified as AT, Aso-4, Ata-Th, Aso-1, and Kkt (from youngest to oldest), whereas the source volcano and eruption age of the lowest tephra, tentatively called "unknown-a" in this study, are not known. These tephra layers allow the JS and ECS benthic $\delta^{18}O$ records to be constrained (Fig. 6). The unknown-a tephra at the bottom of Site U1429 is near the MIS 11/10 boundary. The Kkt tephra is near the end of MIS 10 at Site U1429 and at the end of the light $\delta^{18}O$ peak at 132 m depth at Site U1427, suggesting that this $\delta^{18}O$ peak is equivalent to MIS 10a. Two tephra layers, Aso-1 and Ata-Th, are present before and after the MIS 8 maximum at Site U1429, respectively; therefore, the small peak at ~ 90 m at Site U1427 is identified as MIS 8. Two negative peaks at ~ 58 and 50 m at Site U1427, between Aso-4 and Ata-Th, are correlated to MIS 6c and 6a. The AT tephra, which was formed just prior to the LGM, has already been described. Assuming that $\delta^{18}O$ minima correspond to the lowest sea levels, then the $\delta^{18}O$ minima at Site U1427 during glacial periods may be correlated to $\delta^{18}O$ maxima at Site U1429. The tie points for this correlation are shown in Fig. 6 (blue dotted lines) and listed in Table 3.

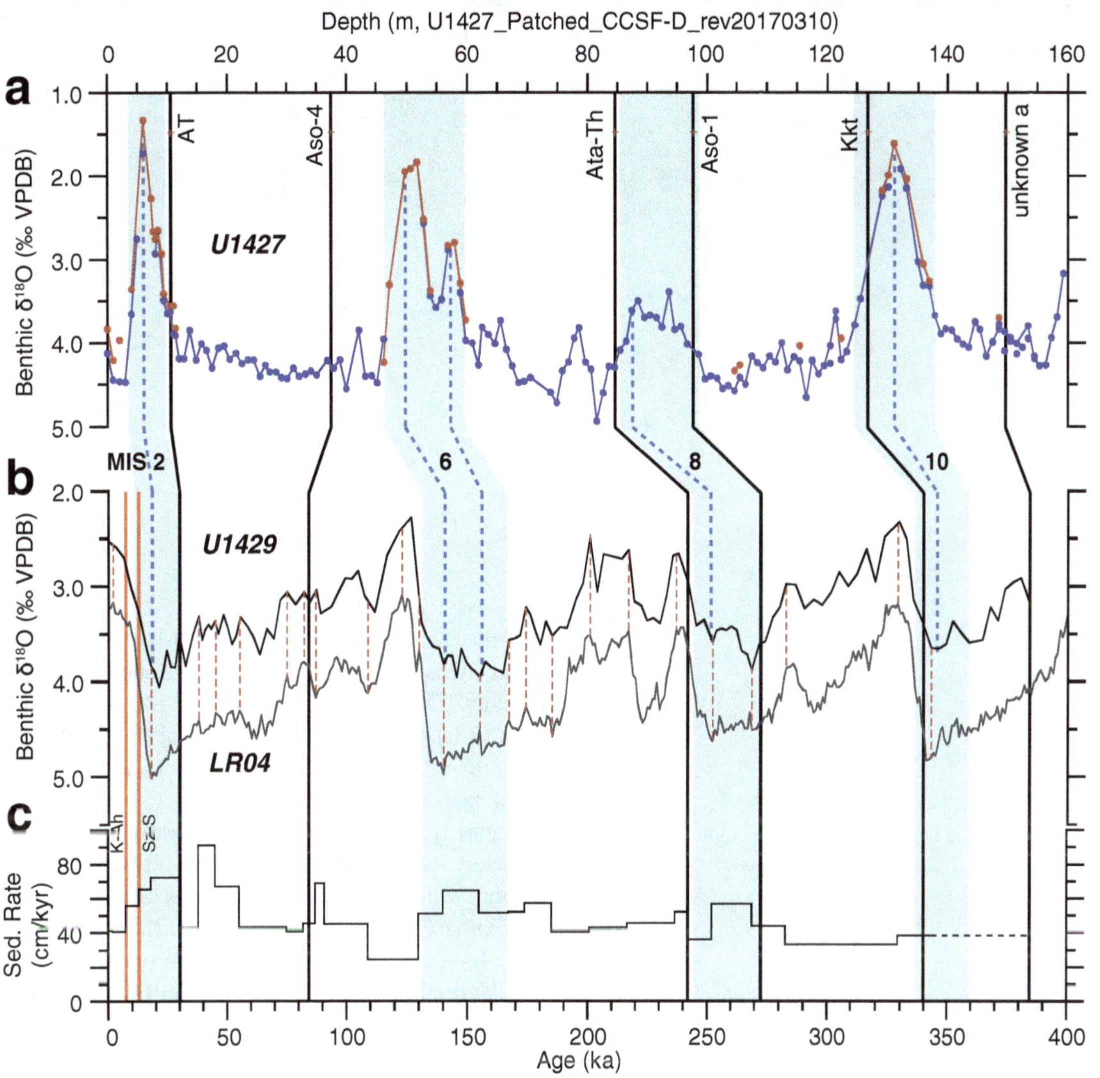

Fig. 6 Comparison of benthic foraminiferal $\delta^{18}O$ records at Sites U1427 (**a**) and U1429 (**b**). Red dotted lines in **b** indicate age control points correlating U1429 $\delta^{18}O$ measurements and the LR04 stack. **c** Sedimentation rate at Site U1429. The positions of two tephra layers, K-Ah and Sz-S, used to constrain the Site U1429 age model are shown by vertical lines (see text for details)

Age model of Site U1427

The age model of the sediment succession at Site U1427 was constructed by combining age constraints from benthic foraminiferal $\delta^{18}O$ measurements and tephrostratigraphy, the relative abundance of subtropical microfossils, and radiolarian biostratigraphy (Figs. 7 and 8). A summary of this approach is provided in Table 4.

As discussed above, the negative peaks in benthic foraminiferal $\delta^{18}O$ values at Site U1427 correlate to glacial maxima in the upper 160 m, equivalent to the past 400 kyr (Fig. 6). Given that eustatic change is a major cause of negative $\delta^{18}O$ excursions (e.g., Oba et al. 1991; Kido et al. 2007), such a correlation is likely to be also applicable to glacial stages older than 400 ka. Two distinct negative excursions, at ~ 180 and ~ 246 m, are present below 160 m (Fig. 7b). The upper excursion occurs prior to MIS 10 and therefore is equivalent to MIS 12. The lower excursion corresponds to MIS 16, not to MIS 14. This interpretation is based on sea-level and ice-volume estimates which suggest that the sea-level fall during MIS 14 was less than that during MIS 16, when the ice volume was as large as during MIS 12 (Bintanja et al. 2005; Elderfield et al. 2012). Therefore, the sea-level drop at MIS 14 is likely to have resulted in less intense surface freshening in the absence of a strong negative $\delta^{18}O$ excursion. The two zones that lack benthic foraminifera at 160–180 and 224–245 m occur just above negative excursions and correspond to intervals of laminated sediment. The shipboard data show that all microfossil groups, except diatoms, are rare or absent in these intervals (Tada et al. 2015a). Taking into account the higher abundance of freshwater diatom species in these intervals (Tada et al. 2015a), we suggest that surface water freshening during

Table 2 Correlation tie points between Sites U1429 and LR04

U1429 depth (m, CCSF-D)	Age (kyr)
0.82	2
2.97	7.3
6.14	13
9.40	18
18.06	30
21.48	38
27.85	45
34.55	55
43.14	75
45.99	82
48.26	87
51.00	91
53.11	91
61.18	109
64.56	123
66.24	130
71.34	140
81.00	155
87.14	167
90.75	174
97.04	185
103.46	201
110.29	217
119.40	237
122.00	242
126.40	242
130.00	252
139.58	269
145.70	283
160.72	329
166.04	343

these two glacial maxima was too extreme for some marine groups.

The trend of benthic foraminiferal $\delta^{18}O$ variations at Site U1427 closely follows the sediment color reflectance b* data (Fig. 5b). All negative excursions in $\delta^{18}O$ values correspond to b* minima. The b* variation shows a negative correlation with natural gamma radiation (NGR) and MS trends and a positive correlation with the relative abundance of biogenic components such as calcareous nannofossils (figure F13 in Tada et al. 2015a). Since NGR and MS values reflect terrigenous clastic content and show a negative correlation with biogenic component abundance, b* is interpreted to reflect the ratio of biogenic to terrigenous input. As Site U1427 is on the outer margin

Table 3 Correlation tie points between Sites U1427 and U1429

U1427 depth (m, U1427_ Patched_CCSF-D_ rev20170310)	U1429 depth (m, CCSF-D)	U1429 Age (ka)	Event
0.00	0.0	0.0	Core top
6.01	9.40	18.0	MIS 2
10.69	18.06	30.0	AT
37.54	48.26	87.0	Aso-4
50.78	71.34	140.0	MIS 6a
57.84	81.00	155.0	MIS 6c
84.67	126.40	242.0	Ata-Th
88.51	130.00	252.0	MIS 8
97.70	140.09	270.2	Aso-1
126.74	164.59	339.2	Kkt
131.00	167.18	346.0	MIS 10
149.77	181.30	383.2	Unknown-a tephra

of the marginal terrace (~ 330 mbsl), terrigenous input at this site is likely to be strongly controlled by sea-level variability. Therefore, $\delta^{18}O$ values and b* should exhibit similar variations on glacial–interglacial time scales. However, two b* minima are present at ~ 30 m and ~ 70 m that are not associated with a negative $\delta^{18}O$ excursion (Fig. 5b). The sediment in these intervals is characterized by black to very dark greenish-gray clayey silt with heavy bioturbation, with an appearance that is not typical of glacial maxima (Tada et al. 2015a). Nevertheless, assuming that sea-level falls at glacial maxima are typically accompanied by b* minima, then the MIS 12 and 16 glacial maxima are at 167.2 and 232.0 m, respectively (Fig. 7 and Table 4). In the absence of a stable-isotope record below 350 m, we have correlated b* minima to glacial maxima, with MIS 24–22 at 345–370 m and MIS 26 at 382 m.

The regional tephra layers also constrain the age of sediment in this section. We used four tephra layers (AT, Aso-4, Hkd-Ku, and Ss-Pink) for which the relative positions in the stable isotope stratigraphy are known. Pickering et al. (1999) established a stable isotope stratigraphy and magnetostratigraphy for the deep-marine successions of the middle Kazusa Group in the Yoro River section in southeast Japan. The stratigraphic position of the Ku1 tephra, which is also known as the Hkd-Ku tephra (Satoguchi and Nagahashi 2012), is at the base of MIS 18, slightly above the Brunhes–Matuyama boundary (Pickering et al. 1999). Based on this, the two minor negative $\delta^{18}O$ peaks at 284.7 and 295.4 m, above the Hkd-Ku tephra layer, correspond to MIS 18, and the $\delta^{18}O$ minimum below Hkd-Ku at 314.1 m is the MIS 20 maximum. The Ss-Pink lies in the Jaramillo Subchron in lacustrine deposits of the Osaka and Kazusa groups (Torii et al. 1974; Hayashida et al.

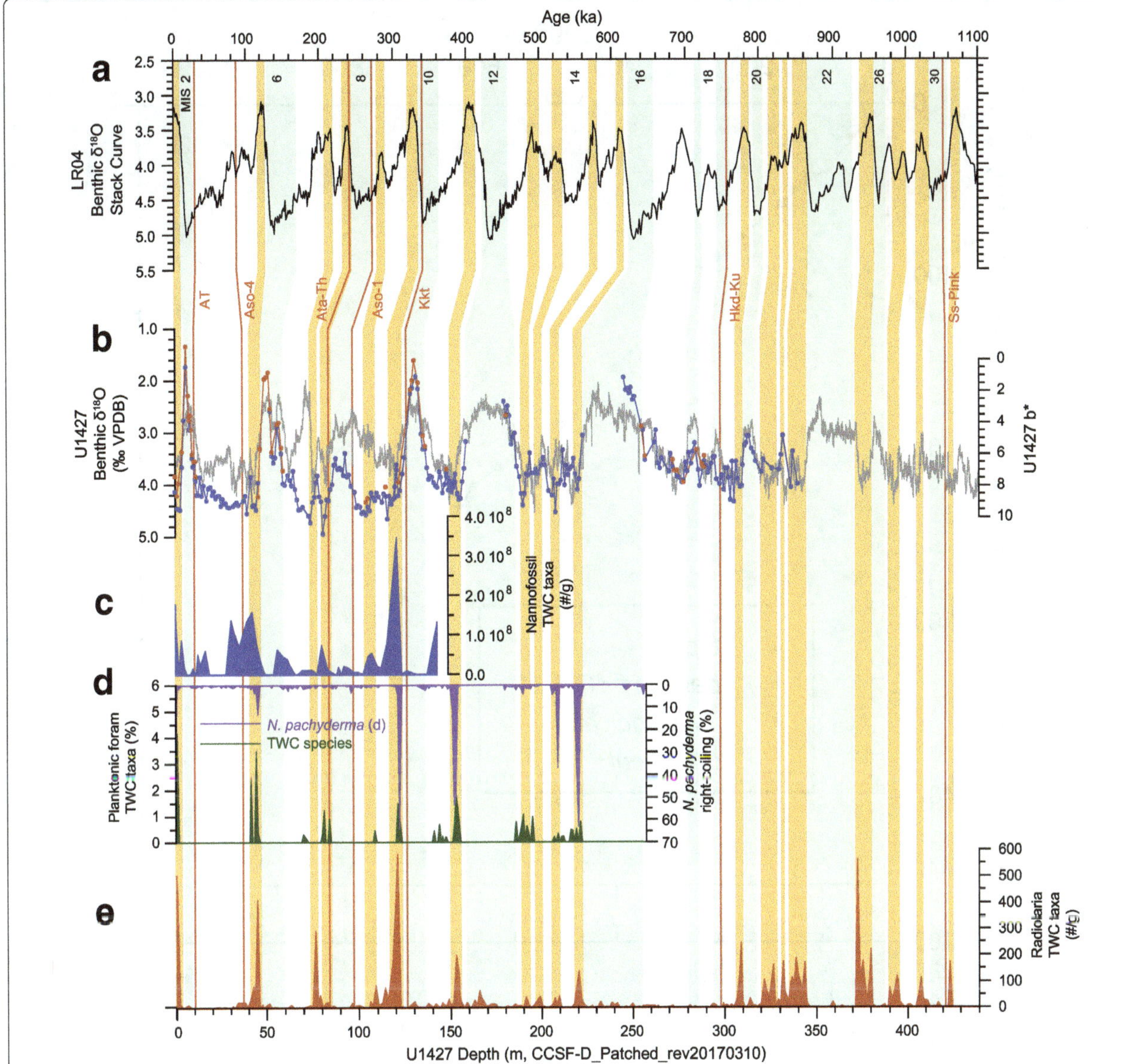

Fig. 7 Summary correlation of Site U1427 records with the LR04 stack. **a** LR04 benthic isotope stack. **b** U1427 benthic foraminiferal $\delta^{18}O$ values and sediment color b* variation. Abundance of TWC taxa of three planktonic groups, **c** calcareous nannofossils, **d** planktonic foraminifera, and **e** radiolaria. Background color indicates correlation of glacial and interglacial maxima, which are defined by benthic $\delta^{18}O$ measurements and TWC indicators, respectively. Red lines indicate age constraints from well-known regional tephra layers

1996). High-resolution chronostratigraphic analyses suggest that the stratigraphic position of the O7 tephra in the Kazusa Group, which is also known as the Ss-Pink tephra (Satoguchi and Nagahashi 2012), is in the early part of MIS 30, and the age of this tephra is ~ 1.050–1.056 Ma (Tsuji et al. 2005).

To further constrain the stratigraphy, we used ingressions of subtropical TWC species of planktonic microfossils to calibrate interglacial maxima. The relative abundance of TWC indicators of these three microfossil groups (calcareous nannofossils, planktonic foraminifera, and radiolarians) generally show good correlations (Fig. 7c–e). The broader peaks of calcareous nannofossil abundance are due to the lower resolution of this part of the study. The higher abundance is only observed in calcareous nannofossils at 12–17, ~ 32, and 57–63 m (Fig. 7c), possibly due to nutrient availability as well as differences in habitat depths between the various microfossil proxies. Calcareous nannoplankton are limited to the photic zone due to photosynthesis, whereas zooplankton dwell in a wide range of depths (e.g., Fairbanks et al. 1980; Itaki 2003; Itaki et al. 2003; Kuroyanagi and Kawahata 2004). Global sea-level records suggest that the amount of sea-level lowering at glacial maxima was comparable to the sill depth of the modern Tsushima Strait (Siddall et al. 2003; Rohling et al. 2014). In other words, the sea level at other periods was sufficiently high to allow

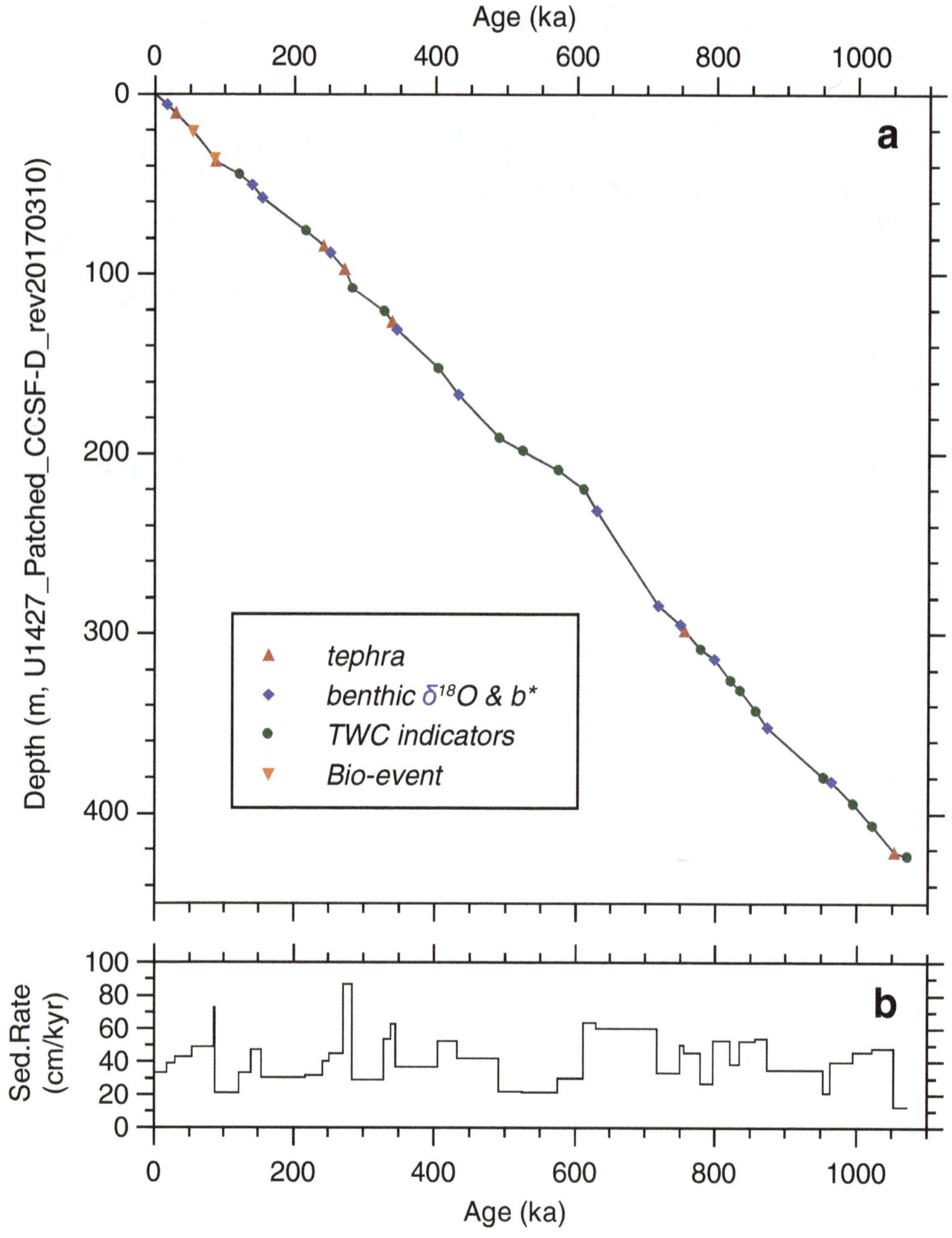

Fig. 8 Age–depth relationship at Site U1427 (**a**) and sedimentation rate (**b**)

surface water to flow in through the strait. Given that the TWC indicators of zooplankton groups show peaks with short durations just after glacial maxima, substantial inflow of the TWC only occurred during interglacial maxima. Therefore, maxima in calcareous nannofossils that are not present in the other microfossil groups could suggest that the sea level was not as high as at the present day and that the depth of the Tsushima Strait was limited to only a few tens of meters, preventing zooplankton from migrating through the strait. Therefore, we interpret the peaks of zooplankton just after the glacial negative $\delta^{18}O$ excursions as interglacial maxima (Fig. 7). All interglacials except for MIS 17 show higher abundances of subtropical species compared with glacial periods. The lack of TWC indicators during MIS 17 might relate to environmental factors outside of the JS; e.g., the northwestern Pacific region and ECS. Radiolarian assemblages in cores from the Kuroshio region yield a relatively low abundance of low- to middle-latitude species during the peak of MIS 17 compared with other interglacial maxima (Matsuzaki et al. 2015), implying that subtropical water masses had less influence on the ECS at this time compared with

Table 4 Age control points for the succession at Site U1427

U1427 Depth (m, Patched_CCSF-D_ rev20170310)	Age (ka)	Source	Datums	Reference
0	0	Core top		
6.01	18	Benthic $\delta^{18}O$	MIS 2	Lisiecki and Raymo (2005)
10.69	30.0	Tephra	AT	Smith et al. (2013)
20.98	54	Bio-event	LO *Lychnocanoma sakaii*	Itaki et al. (2007)
36.09	85	Bio-event	LO *Amphimelissa setosa*	Itaki et al. (2007)
37.54	87.0	Tephra	Aso-4	Aoki (2008)
45.10	123	TWC microfossil	MIS 5e	Lisiecki and Raymo (2005)
50.78	140	Benthic $\delta^{18}O$	MIS 6a	Lisiecki and Raymo (2005)
57.84	155	Benthic $\delta^{18}O$	MIS 6c	Lisiecki and Raymo (2005)
76.67	217	TWC microfossil	MIS 7c	Lisiecki and Raymo (2005)
84.67	242	Tephra (tied to U1429)	Ata-Th	
88.51	252	Benthic $\delta^{18}O$	MIS 8a	Lisiecki and Raymo (2005)
97.68	270.2	Tephra (tied to U1429)	Aso-1	
108.55	285	TWC microfossil	MIS 9a	Lisiecki and Raymo (2005)
121.27	329	TWC microfossil	MIS 9e	Lisiecki and Raymo (2005)
126.74	339.2	Tephra (tied to U1429)	Kkt	
131.00	346	Benthic $\delta^{18}O$	MIS 10a	Lisiecki and Raymo (2005)
153.05	406	TWC microfossil	MIS 11c	Lisiecki and Raymo (2005)
167.2	433	b*	MIS 12a	Lisiecki and Raymo (2005)
191.57	491	TWC microfossil	MIS 13a	Lisiecki and Raymo (2005)
198.94	525	TWC microfossil	MIS 13c	Lisiecki and Raymo (2005)
209.58	575	TWC microfossil	MIS 15a	Lisiecki and Raymo (2005)
220.58	612	TWC microfossil	MIS 15e	Lisiecki and Raymo (2005)
232.0	630	b*	MIS 16a	Lisiecki and Raymo (2005)
284.65	718	Benthic $\delta^{18}O$	MIS 18a	Lisiecki and Raymo (2005)
295.37	750	Benthic $\delta^{18}O$	MIS 18e	Lisiecki and Raymo (2005)
298.37	756	Tephra	Hkd-Ku	Pickering et al. (1999)
309.27	780	TWC microfossil	MIS 19c	Lisiecki and Raymo (2005)
314.11	798	Benthic $\delta^{18}O$	MIS 20b	Lisiecki and Raymo (2005)
326.75	822	TWC microfossil	MIS 21a	Lisiecki and Raymo (2005)
332.13	836	TWC microfossil	MIS 21c	Lisiecki and Raymo (2005)
343.66	858	TWC microfossil	MIS 21 g	Lisiecki and Raymo (2005)
352.29	874	b*	MIS 22	Lisiecki and Raymo (2005)
380.23	954	TWC microfossil	MIS 25e	Lisiecki and Raymo (2005)
382.30	964	b*	MIS 26	Lisiecki and Raymo (2005)
395.08	996	TWC microfossil	MIS 28b	Lisiecki and Raymo (2005)
407.44	1023	TWC microfossil	MIS 29	Lisiecki and Raymo (2005)
421.80	1053	Tephra	Ss-Pink	Tsuji et al. (2005)
424.15	1072	TWC microfossil	MIS 31	Lisiecki and Raymo (2005)

other interglacials. The correlation horizons between TWC microfossil data and LR04 are listed in Table 4.

Four datum levels of radiolarian marker species from Site U1427 were described by Itaki et al. 2018, and three of them are within the period considered in the present study. The last occurrences (LOs) of *Lychnocanomma sakaii* (Morley and Nigrini 1995) and *Amphimelissa setosa* (Cleve) are well-known biohorizons in the

Table 5 Compilation of age estimates for the regional tephra layers of this study

Name of widespread tephra	Abbreviation	Age	Error	Unit	Method*	References
Kikai-Akahoya	K-Ah	ca. 6.5		ka BP	14C	Okuno et al. 2013
		7.7	± 2	ka	TL	Nagatomo et al. 2005
Sakurajima-Satsuma	Sz-S	ca. 12.8		cal ka BP	14C	Okuno 2002
		13.8	± 1.3	ka	TL	Shitaoka et al. 2016
		13.4	± 2	ka	TL	Shitaoka et al. 2016
Aira-Tanzawa	AT	25.12	± 0.27	ka BP	14C	Miyairi et al. 2004
		26	± 4	ka	TL	Nagatomo et al. 2005
		27	± 5	ka	TL	Nagatomo et al. 2005
Aso-4	Aso-4	89	± 7	ka	K-Ar	Matsumoto 1996
Kikai-Tozurahara	K-Tz	98	± 26	ka	FT	Danhara 1995
		96.7	± 7.1	ka	RTL	Miyairi 2007
Ata	Ata	100	± 27	ka	FT	Danhara 1995
		108	± 3	ka	K-Ar	Matsumoto and Ui 1997
Sanbe-Kisuki	SK	100	± 20	ka	FT	Kimura et al. 1999
		110	± 14	ka	TL	Shitaoka et al. 2009
Aso-3	Aso-3	123	± 6	ka	K-Ar	Matsumoto et al. 1991
Ata-Torihama	Ata-Th	240	± 40	ka	FT	Danhara 1995
Aso-1	Aso-1	266	± 14	ka	K-Ar	Matsumoto et al. 1991
Kakuto	Kkt	360	± 80	ka	FT	Danhara 1995
		340	± 10	ka	K-Ar	Machida and Arai 2003
Kobayashi-Kasamori	Kb-Ks	620	± 80	ka	FT	Danhara 1995
Hakkouda-Kokumoto	Hkd-Ku	ca. 650		ka	K-Ar	Muraoka 1991
		650	± 160	ka	FT	Hirosaki City 2001
Shishimuta-Pink	Ss-Pink	990	± 30	ka	K-Ar	Uto and Suto 1985
		1040	± 50	ka	FT	Danhara et al. 1994

*^{14}C radiocarbon dating method, *TL* thermoluminescence dating, *FT* fission track dating, *RTL* red thermoluminescence dating, *K-Ar* Potassium-Argon dating

subarctic Pacific, and their estimated ages in the Japan Sea are 54 and 85 ka, respectively (Itaki et al. 2007). These datums are present at Site U1427 between two widespread tephras, AT and Aso-4. The LO of *Schizodiscus japonicus* Matsuzaki et al. (2014) is the same event as the LO of *Spongodiscus* sp. reported from the North Pacific Ocean, the Bering Sea, and the Sea of Okhotsk. In the Sea of Okhotsk, this event occurs during MIS 9 (287 ka) (Matul et al. 2002) or MIS 8c (272 ka) (Matul et al. 2009). The depth of this event at Site U1427 is between 112.05 and 112.77 m, which is prior to the Aso-1 tephra (272 ka) and the peak of subtropical species at MIS 9a (285 ka). Therefore, the age of ~ 287 ka proposed by Matul et al. (2002) appears to be likely. However, if we use the 287 ka age for this datum, the sedimentation rates rapidly increase by a factor of ~ 10 across this datum, which does not appear to be geologically plausible. This may relate to a large error in the age data of Matul et al. (2002). Therefore, this datum is not included in the age model (Table 4).

The potential sources of error in the age model of Site U1427 with respect to LR04 are uncertainties in the correlation of benthic foraminiferal $\delta^{18}O$ measurements and TWC indicators with LR04, and age estimations of widespread tephras and radiolarian biohorizons. The correlation of $\delta^{18}O$ values and TWC indicators with LR04 would generate errors of a few thousands of years due to the resolution of the records. The errors of age estimation for tephra layers used in this study are less than ± 1 kyr for AT and Aso-4 (Aoki 2008; Smith et al. 2013), and ± 3 kyr for Hkd-Ku and Ss-Pink (Pickering et al. 1999; Tsuji et al. 2005). The ages of biohorizons are calibrated by JS planktonic foraminiferal $\delta^{18}O$ values of MD01-2407 (Itaki et al. 2007), but the errors on these values are unknown. We estimated the errors of these biohorizons as ± 3 kyr, which includes the uncertainty of the age model of MD01-2407 (Kido et al. 2007). Including these sources of error, we estimate the uncertainty of the age model of Site U1427 to be ± 5 kyr for the period 0–400 ka and ± 10 kyr for the period 400–1000 ka.

Age estimation of regional tephra layers using the newly established chronology

Fifteen out of eighteen correlative tephra layers were identified as regional tephra layers. The ages of regional tephra layers have been previously established by several techniques using eruption products near a source volcano or regional deposits (Table 5). The tephrochronology of Pliocene to Pleistocene strata around the Japanese islands was developed using these techniques combined with biostratigraphy, magnetostratigraphy, and oxygen isotope stratigraphy (Tamura and Yamazaki 2010; Satoguchi and Nagahashi 2012). As discussed above, we only used six widespread tephra layers as age constraints, where their relative stratigraphic positions with respect to the oxygen isotope stratigraphy are known. In this section, we discuss the sedimentation ages of tephra layers that are not used as age constraints, by comparing the ages estimated in this study with those reported in previous studies. The ages estimated in this study were calculated by a simple linear interpolation method.

The K-Tz tephra, which is widespread in central and western Japan, was deposited just after MIS 5c (Machida and Arai 2003). The estimated age based on U1429 $\delta^{18}O$ stratigraphy is 91 ka, which corresponds to MIS 5b. This estimate is within error of the results of fission-track (FT) analyses (98 ± 26 ka, Danhara 1995) and the red thermoluminescence method (93.7 ± 7.1 ka, Miyairi 2007) (Table 5).

The stratigraphic position of the Ata tephra in marine sediment was previously reported to be between MIS 5d and 5c, with an estimated age of 103–107 ka (Oba 1991). Our estimate is 98 ka, which is younger than the value suggested by Oba (1991) and the K–Ar age of pyroclastic flow deposits (108 ± 3 ka, Matsumoto and Ui 1997).

Chun et al. (2004) reported the Aso-3 tephra layer in marine sediment cores from the Ulleung Basin, in a bioturbated mud just above a dark finely laminated layer deposited during MIS 6a. Nagahashi et al. (2004) reported the Aso-3 tephra from an interval with an estimated age of 133 ka below MIS 5e in marine clay in cores from the Osaka coast. The stratigraphic position of Aso-3 at Site U1429 agrees with these results, and the estimated age of ~ 137 ka is also consistent with previous studies.

The Ata-Th tephra layer is prior to MIS 7e. In the present study, its age is estimated to be 242 ka. This agrees with its estimated age in lacustrine sediment from Lake Biwa (238 ka, Yoshikawa and Kuwae 2001) and FT analyses (240 ± 40 ka, Danhara 1995).

The Aso-1 tephra lies beneath the maximum $\delta^{18}O$ value in MIS 8 (Fig. 6). We suggest that this $\delta^{18}O$ maximum correlates to MIS 8c rather than 8a, which is the MIS 8 maximum in the LR04 stack, as there is another positive peak at Site U1429 just below 7e (Fig. 6). The estimated age of Aso-1 in this study is 270 ka, which is within error of a K–Ar date (266 ± 14 ka, Matsumoto et al. 1991) and is older than the age estimated from Lake Biwa (249 ka, Yoshikawa and Kuwae 2001). There is a discrepancy of ~ 21 kyr between Site U1429 and Lake Biwa, possibly due to the large error in the Lake Biwa age estimation, as discussed later.

The Kkt tephra is present near the end of MIS 10 at Site U1429 (Fig. 6), and its age is estimated to be 339 ka, consistent with that from Lake Biwa (334 ka, Yoshikawa and Kuwae 2001) and radiometric data (Danhara 1995; Machida and Arai 2003) (Table 5).

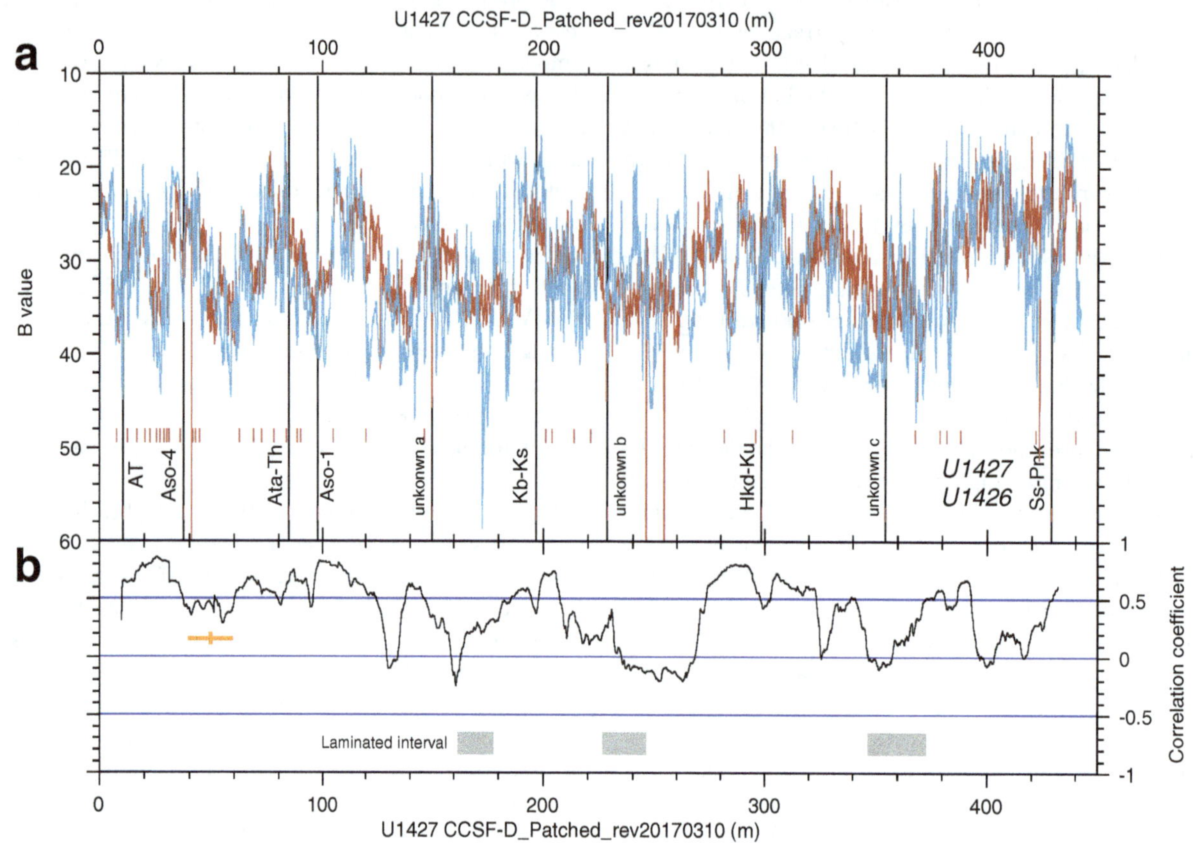

Fig. 9 a Comparison of sediment color profiles at Sites U1427 (red) and U1426 (blue) plotted on the U1427 depth scale. Note that the axis of *B* values is inverted. Control points by correlative tephra layers are shown by vertical black lines. Additional control points from color change correlation are shown by red lines. For additional details, see Additional file 8: Figure S2. **b** Moving correlation coefficient of two color profiles, calculated for a 20 m window with width indicated by orange bar. Intervals of laminated sediment at Site U1427 are shown at the base of the lower panel

The difference in estimated ages between Site U1429 and Lake Biwa is greater for Aso-1 (~ 21 kyr) and smaller for Ata-Th and Kkt (~ 5 kyr). The age estimate for Lake Biwa sediment is based on a change in pollen assemblage, which shows alternations of warm-temperate and cool-boreal species in response to glacial–interglacial cycles (Miyoshi et al. 1999). Ata-Th and Kkt are present at transitions from cool to warm periods, and these can be correlated to terminations. However, the Aso-1 tephra occurs within a period of cool-boreal conditions, equivalent to MIS 8. As the duration of the MIS 8 cold period (~ 30 kyr) is longer than that of terminations (< 8 kyr), the age estimate of Aso-1 may have a greater uncertainty than that of Ata-Th and Kkt.

The ages of the SK and Kb-Ks tephra layers, which are not present at Site U1429, are based on the Site U1427 age model. The SK tephra layer at 41.23 m depth at Site U1427 is ~ 4 m below the Aso-4 tephra (Fig. 5b). The estimated age of the SK tephra is 102 ka in this study, which agrees with radiogenic data (100 ± 20 ka, Kimura et al. 1999; 110 ± 14 ka, Shitaoka et al. 2009) (Table 5). The Kb-Ks tephra layer at 196.86 m is within MIS 13c/13b (Fig. 7), and the estimated age of 516 ka is much younger than that derived from FT analyses (620 ± 80 ka: Danhara 1995). Shirai (2001) interpreted the stratigraphic position of the Kb-Ks tephra at ODP Sites 794 and 797 relative to changes in diatom abundance. The diatom abundance in the deep-sea sediment of the JS is considered to be related to sea-level change (Tada et al. 1992, 1999), which controls nutrient input through the Tsushima Strait. Therefore, variations in diatom abundance follow the $\delta^{18}O$ curve, enabling estimates of sediment age. The age of 520 ± 20 ka (MIS 13c) estimated by Shirai (2001) shows good agreement with the result of this study (516 ka).

In this study, three equivalent tephra layers are present between sites (tentatively called unknown-a, unknown-b, and unknown-c), which were not correlated to regional tephras. The unknown-a tephra is found at three sites, and the unknown-b and unknown-c tephras occur only at Sites U1426 and U1427. The ages of these tephra layers were estimated using the age model of Site

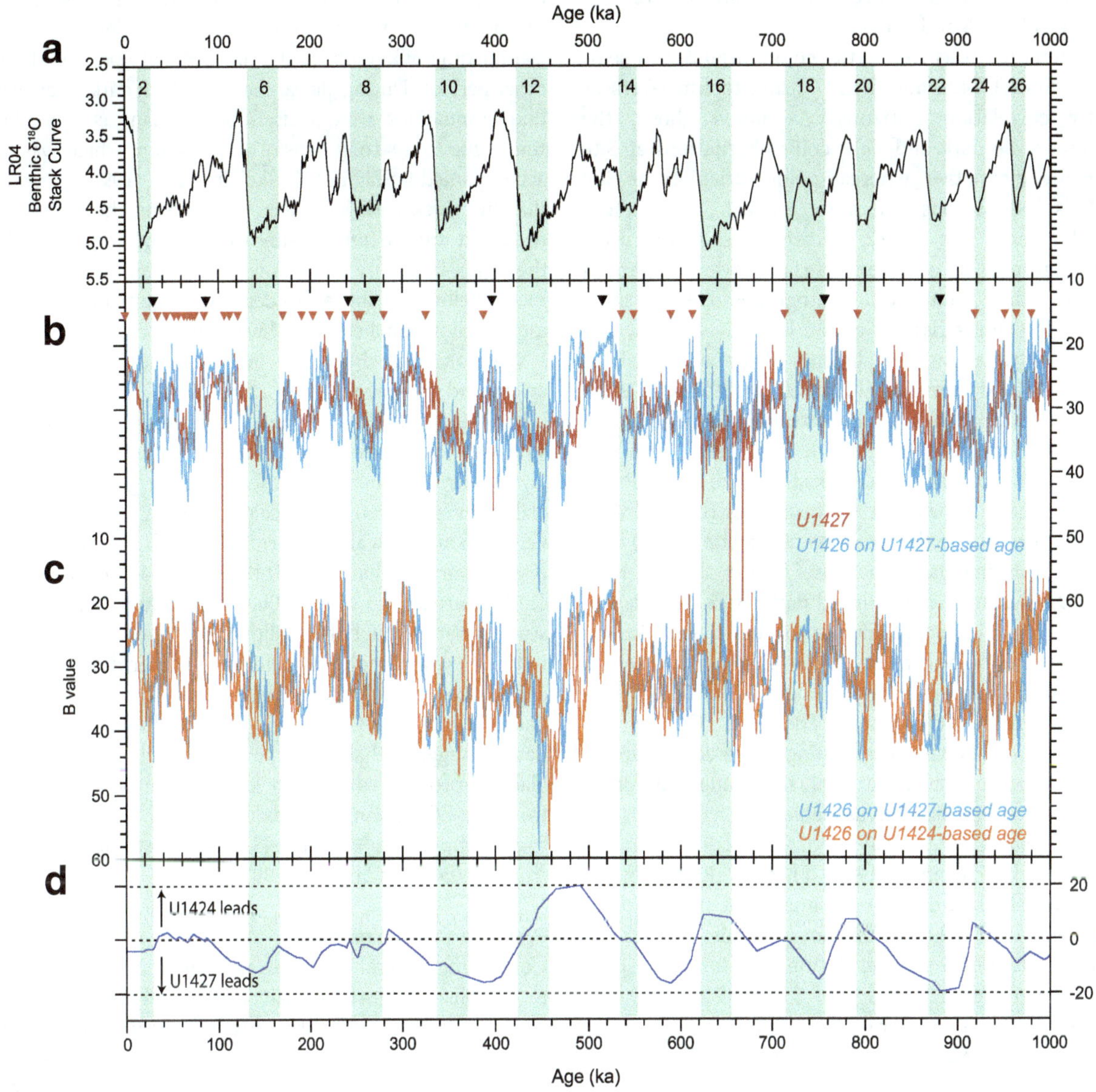

Fig. 10 a LR04 oxygen isotope stack. **b** Color variation at Sites U1426 and U1427 over the past 1.0 Myr. Note that the axis of *B* values is inverted. Control points by correlative tephra layers are shown by black triangles. Additional control points by color correlation are shown by red triangles. **c** Color variations at Site U1426 plotted on a U1427-based age model (blue) and a U1424-based age model (orange). **d** Age difference between U1427-based and U1424-based age estimates for the past 1.0 Myr

U1427. The age of the unknown-a tephra layer is 383 ka, above the distinct peaks of TWC indicators of MIS 11c (Fig. 7). The ages of the unknown-b and unknown-c tephra layers are estimated to be 625 and 880 ka, respectively. Although there is no benthic foraminiferal $\delta^{18}O$ record for these intervals, both tephra layers correspond to b* minima, suggesting that they were erupted during the glacial maxima of MIS 16 and 22. The stratigraphic positions of these three unidentified tephra layers are in MIS 11, 16, and 22, respectively, and they may be useful for correlating sediment successions at other locations.

Correlation of Sites U1426 and U1427

Eight tephra layers are present at Sites U1426 and U1427 (Table 1 and Fig. 2). Although the widespread tephra layers of K-Ah, Aso-4, Aso-3, and Ata-Th are not present at Site U1426, Kido et al. (2007) reported these tephras in core MD01-2427, 10 km away. Since the two sites are adjacent and the color variations of the sediments are similar (Additional file 5: Figure S1), the stratigraphic positions of the tephra layers in core MD01-2407 can be correlated to Site U1426 using sediment color profiles (Additional file 6: Table S5). Therefore, two more widespread tephra layers, Aso-4 and Ata-Th, can also be used to correlate features between Sites U1426 and U1427.

We attempted to correlate the color variations of sediment at Sites U1426 and U1427 using tephra layers as primary correlation horizons. As noted above, the decimeter-scale dark and light color alternations at Site U1426 are distinctive, whereas color variability at Site U1427 is less obvious and more gradual. To highlight color changes at Site U1427, the brightness and contrast in photographs of sediment surfaces were enhanced (Fig. 2). These enhanced photographs clearly show meter-scale color variations at Site U1427.

The color variations of sediments in deeper-water sites have been well documented (Tada et al. 1992; Kido et al. 2007). The fact that dark–light alternation is commonly observed and correlated in the JS indicates that it reflects a basin-wide phenomenon (e.g., Nakajima et al. 1996). Previous studies have shown that the brightness (L*) at deeper sites correlates well to the total organic carbon (TOC) content, except during glacial maxima (Tada et al. 1992; Kido et al. 2007), when finely laminated dark layers were deposited under euxinic conditions due to density stratification in the JS, and TOC was relatively low. Black and Anderson (2017) showed that the L* and B values of Site U1427 correlate well with the TOC content. Therefore, TOC seems to be the major component contributing to the dark sediment color at deeper and shallower sites, suggesting that correlation using color profiles is reasonable.

Figure 9 shows the *B* values of RGB correlation between Sites U1426 and U1427. These cores are correlated using shipboard line-scan images of the sediment surface and RGB data at 0.05 mm intervals, and a 3 cm moving average of 5 mm resolution data for Site U1426 and a 10 cm moving average of 5 mm resolution data for Site U1427. The overall variations in RGB reflect sediment brightness. Therefore, similar to L*, RGB appears to reflect TOC. Ten tephra layers were used as primary tie points in Fig. 9, and additional tie points were selected based on *B* value trends (Additional file 7: Table S6). A detailed comparison of the color profiles at the two sites is shown in Additional file 8: Figure S2. Overall, the color variations appear to correlate well (see the running correlation coefficient calculated using a window with 20 m width in Fig. 9b). The average correlation coefficient over the whole interval is ~ 0.4. Five intervals at ~ 130, 160, 220–265, 345–370, and 395–425 m do not correlate (Fig. 9). Based on our age model, these five intervals correspond to the glacial periods of MIS 10, 12, 16, 22, and 28–30 (Fig. 7). Three glacial periods (MIS 12, 16, and 22) correspond to faintly laminated intervals, where benthic foraminifera are sparse or absent (Tada et al. 2015a), suggesting that intermediate water properties were significantly different during these periods compared with other times. The poor correlation between shallow- and deep-sediment colors during glacial periods is likely due to the fact that the process of dark layer formation during glacial maxima differed from the process that operated during other periods. During glacial maxima, thick dark layers with fine laminations were caused by stagnant euxinic deep water due to a strong density stratification, which was not accompanied by high TOC. Ikehara et al. (1994) suggested that the thick dark layer deposited during the LGM only occurs in water depths greater than 500 m, suggesting that the redox boundary was below ~ 500 m and that environments below and above this depth were quite different during glacial maxima compared with other periods.

We constructed an age model for Site U1427 and attempted to correlate the color profiles between Sites U1427 and U1426. Although the correlation of the two sites is not always significant, ten correlative tephra layers provide robust ties at least for the glacial–interglacial intervals. The depth-scale of Site U1426 was converted to an age scale via correlation with Site U1427. The *B* value variations at Sites U1426 and U1427 over the past 1.0 Myr are shown in Fig. 10b. The *B* values show orbital-scale variations at both shallow and intermediate depths in the JS, where sediment colors are lighter during glacials and darker during interglacials. Given that the sediment color reflects the TOC content of the sediment, the overall pattern suggests that productivity was higher during interglacials, probably due to an increased input of nutrients from the ECS during periods of high sea level in contrast to glacial periods, when the JS was almost entirely cut off from the ECS (Tada et al. 1999).

Evaluation of age models

Combining several kinds of age constraints enabled us to establish the orbital-scale age model of the shallow Site U1427, and this chronology was transferred to the deeper Site U1426 by correlating sediment color profiles (Fig. 9). The sediments at the six deeper sites (U1422–U1426 and U1430) drilled during Expedition 346 are characterized by dark–light alternations. As noted above, the sporadic occurrence of foraminiferal tests made it difficult to establish a stable isotope stratigraphy for the deeper sites. Therefore, age estimation for deeper sites was attempted using a different approach (Tada et al. 2018). To evaluate and understand the limitations of the various approaches, we compared two independent age models that are based on different concepts.

Tada et al. (2018) derived an age model for Site U1424 based on the magnetostratigraphy, tephrochronology, and tuning of physical properties to the LR04 stack. Physical properties, such as gamma ray attenuation (GRA), bulk density, and NGR, in deeper cores are considered to reflect diatom content because of their high-internal porosity (Tada and Iijima 1983) and the effect of dilution of detrital materials. The diatom content in the

JS sediment is thought to be strongly related to eustatic sea-level changes, which constrained nutrient influx into the sea as well as subsequent diatom production (Tada et al. 1992, 1999). Therefore, GRA and NGR data reflect sea-level changes and are useful in identifying marine isotope stages. Correlation of Site U1424 to the LR04 stack was achieved by relating the variability in physical properties to glacial maxima over the past 3 Myr (Tada et al. 2018). Tada et al. (2018) also correlated the six deeper sites using distinct, relatively thick dark layers, magnetostratigraphy, and tephra layers. As a result, 64 horizons can be correlated for the past 1.5 Myr. This inter-site correlation enables us to transfer the age model of Site U1424 to all deeper sites, including Site U1426. Thus, two independent age estimations are available for Site U1426, based on the correlation of dark layers with Site U1424 and the correlation of color variations with Site U1427. Figure 10c–d compares age models and age differences between these approaches. The age difference is calculated by subtracting U1424-based ages from U1427-based ages, and this is plotted against the U1427-based age model. Positive values indicate that the U1424-based age model leads the U1427-based model.

The age estimates for Site U1426 obtained using two different methods show good agreement for the past 300 kyr (Fig. 10), when more tie points of color variations at Sites U1426 and U1427 are available compared with the older intervals (red triangles in Fig. 10b). Large discrepancies (~ 20 kyr) are present at MIS 11, 13–12, 15, and 22. The discrepancies during interglacial stages (MIS 11 and 15) are probably due to the lack of age control points for interglacials at Site U1424. Since the correlation of physical properties between Sites U1424 and LR04 is based on glacial peaks, ages of interglacial maxima were calculated by interpolation using a constant sedimentation rate (Tada et al. 2018). In contrast, the interglacials at Site U1427 are well constrained using the high abundance of TWC taxa (Fig. 7). The poor correlation during interglacials would be improved if the constraints provided by TWC indicators were available for the deeper sites. The age discrepancies during glacial stages (MIS 13–12 and 22) are probably due to lack of color correlation tie points between Sites U1426 and U1427. As discussed above, the sediments of MIS 12, 16, and 22 are faintly laminated at Site U1427 and differ markedly from the equivalent sediments at Site U1426 (Fig. 9). Furthermore, the interval of MIS 13–12 is characterized by large changes in calcium carbonate content at deeper sites (Kido et al. 2007; Khim et al. 2014), which are not present at the shallower Site U1427. Therefore, no tie points are determined for these glacial periods (Fig. 10b). However, there are correlative tephra layers at MIS 22 (unknown-c) and 16 (unknown-a). Therefore, age discrepancies during these glacial periods are attributed not to poor correlation between Sites U1426 and U1427, but to correlation between LR04 and Site U1424 and/or Site U1427. Further investigation is needed to improve the age estimation of these glacial periods. Although there is up to a 20 kyr age difference, the general agreement of the two independent age models implies that the orbital-scale variations in sediment brightness at both shallow and deep sites, which mainly reflect TOC content, were controlled by eustatic sea-level changes.

Conclusions

Two IODP Expedition 346 sites in the Japan Sea and the northern East China Sea are correlated on a glacio-eustatic orbital scale for the first time. Detailed morphological, mineralogical, and elemental-composition analyses of tephras enhance the chronostratigraphic basis of this correlation. Eighteen tephra layers were identified that can be correlated between sites or to other well-known tephras. The results suggest that glacial surface water in the Japan Sea was much fresher than that in the East China Sea. Several microfossil proxy records show strong variability that is directly related to glacial–interglacial sea-level change. The age model for the upper part of Site U1427 suggests that the top 430 m is younger than 1100 ka. Despite the lack of distinct dark–light alternations, the subtle color variations at the shallow Site U1427 generally correlate to color variations at the deeper Site U1426 on an orbital scale, and perhaps on the millennial scale in the 120 kyr interval. The agreement of age estimates based on two different methods supports the age models derived for Sites U1427 and U1424, and the inter-site correlation of JS sites. The inter-site correlation makes it possible to compare proxy records of latitudinally and vertically distributed Expedition 346 Sites in the Japan Sea and to compare between the Japan Sea and the East China Sea.

Additional files

Additional file 1: Table S1. Major element compositions of volcanic glass shards of tephra layers investigated in this study and well-known regional tephra layers.

Additional file 2: Table S2. Petrographic property of tephra layers investigated in this study and well-known regional tephra layers.

Additional file 3: Table S3. Benthic foraminiferal $\delta^{18}O$ results at Site U1429.

Additional file 4: Table S4. Benthic foraminiferal $\delta^{18}O$ results at Site U1427.

Additional file 5: Figure S1. Correlation of sediment color profiles at two adjacent Sites U1426 and MD01 2407.

Additional file 6: Table S5. Correlation tie points at Sites U1426 and MD01-2407.

Additional file 7: Table S6. Correlation tie points at Sites U1426 and U1427.

Additional file 8: Figure S2. Correlation of sediment color profiles at Sites U1426 and U1427. Control poits by correlative tephra layers are shown by black lines. Additional control points are shown by green lines.

Abbreviations

CCD: Carbonate compensation depth; CCSF: Core composite depth below seafloor; ECS: East China Sea; EDS: Energy Dispersive Spectroscopy; FT: Fission track; GRA: Gamma-ray attenuation; IODP: Integrated Ocean Drilling Program; JS: Japan Sea; LGM: Last glacial maximum; LO: Last occurrence; mbsl: Meters below sea level; MIS: Marine isotope stage; MS: Magnetic susceptibility; NGR: Natural gamma radiation; NGRL: Natural gamma radiation logger; NIST: National Institute of Standards and Technology; SEM: Scanning electron microscope; SHIL: Section Half Imaging Logger; SHMSL: Section Half Multisensor Logger; TOC: Total organic carbon; TWC: Tsushima Warm Current; VPDB: Vienna Pee Dee Belemnite; WRMSL: Whole-Round Multisensor Logger; XRF: X-ray fluorescence

Acknowledgements

We are grateful to Yasufumi Iryu and Peter Clift (journal editors) and three anonymous reviewers for their constructive comments on the manuscript. We would like to show our great appreciation to all of the scientists, technicians, and crew of D/V *JOIDES Resolution* during IODP Expedition 346. This research used samples and data provided by the IODP. TS, T Itaki, T Irino, KI, and RT would like to thank J-DESC for financial aid supporting the IODP shipboard scientists. TS thanks Yusuke Okazaki for financial support, Reika Nakamura and Kiyoka Fujino for sample preparation, and Takuya Matsuzaki and technical staff at the Kochi Core Center for their assistance with isotope measurements. This study was performed under the cooperative research program of the Center for Advanced Marine Core Research, Kochi University (14B045, 15A024, 15B021). SJG acknowledges the support of the ANZIC IODP office and the ARC Basin Genesis Hub.

Funding

This work was partly supported by the IODP Exp. 346 After Cruise Research Program, JAMSTEC and by "Collaborative Study of Long-term Changes in the Asian Monsoon" of Program for Advancing Strategic International Networks to Accelerate the Circulation of Talented Researchers, Japan Society for the Promotion of Science. AH gratefully acknowledges institutional funding from Christian-Albrechts-University.

Authors' contributions

TS, T Irino, and RT proposed the topics. YN, YS, and KI conducted tephra analyses and correlation. TS and AH conducted stable isotope measurements. T Itaki, SJG, and MSP conducted microfossil assemblage analyses. All authors contributed to the writing of the manuscript and read, discussed, and approved the final version of the manuscript.

Competing interests

The authors declare that they have no competing interests.

Author details

[1]Institute of Science and Engineering, Kanazawa University, Kakuma-machi, Kanazawa, Ishikawa 920-1192, Japan. [2]Faculty of Symbiotic Systems Science, Fukushima University, 1 Kanayagawa, Fukushima, Fukushima 960-1296, Japan. [3]Lake Biwa Museum, 1091 Oroshimo, Kusatsu, Shiga 525-0001, Japan. [4]Institute of Geosciences, Christian-Albrechts-University, Ludewig-Meyn-Str. 10, 24118 Kiel, Germany. [5]Geological Survey of Japan, National Institute of Advanced Industrial Science and Technology, 1-1-1 Higashi, Tsukuba, Ibaraki 305-8567, Japan. [6]School of Earth Sciences, The University of Melbourne, Melbourne, Victoria 3010, Australia. [7]Department of Geosciences, FB5, University of Bremen, Klagenfurter Strasse, 28359 Bremen, Germany. [8]Faculty of Environmental Earth Science, Hokkaido University, N10W5, Sapporo, Hokkaido 060-0810, Japan. [9]Department of Earth and Planetary Science, The University of Tokyo, 7-3-1 Hongo, Bunkyo-ku, Tokyo 113-0033, Japan.

References

Andruleit H (1996) A filtration technique for quantitative studies of coccoliths. Micropaleontol 42:403–406

Aoki K (2008) Revised age and distribution of ca. 87 ka Aso-4 tephra based on new evidence from the northwest Pacific Ocean. Quat Int 178:100–118

Bintanja R, van de Wal RSW, Oerlemans J (2005) Modelled atmospheric temperatures and global sea levels over the past million years. Nature 437:125–128

Black HD, Anderson WT, (2017) Organic matter contents and paleoproductivity variation within late Pleistocene Japan Sea/East Sea sediments: results from IODP Expedition 346, Abstract [PP23C-1320] presented at 2017 Fall Meeting, AGU, New Orleans, LA, 11–15 Dec.

Bronk Ramsey C, Staff RA, Bryant CL, Brock F, Kitagawa H, van der Plicht J, Schlolaut G, Marshall MH, Brauer A, Lamb HF, Payne RL, Tarasov PE, Haraguchi T, Gotanda K, Yonenobu H, Yokoyama Y, Tada R, Nakagawa T (2012) A complete terrestrial radiocarbon record for 11.2 to 52.8 kyr B.P. Science 338:370–374

Chun J-H, Ikehara K, Sang-Joon H (2004) Evidence in Ulleung basin sediment cores for a termination II (penultimate deglaciation) eruption of the Aso-3 tephra. Quat Res (Daiyonki-Kenkyu) 43:99–112

Danhara T (1995) Towards precise measurement of zircon and glass fission-track geochronology for quaternary tephras. Quat Res (Daiyonki-Kenkyu) 34:221–237 (in Japanese with English abstract)

Danhara T, Kamata H, Hayashida A, Iwano H, Yamashita T (1994) Additional data for age and correlation about OGPK, pink volcanic ash in the Osaka Group, as candidates of standard sample in Japan (part 4), fission-track news letter, pp 18–22 (in Japanese)

Darling KF, Kucera M, Kroon D, Wade CM (2006) A resolution for the coiling direction paradox in *Neogloboquadrina pachyderma*. Paleoceanography 21: PA2011. https://doi.org/10.1029/2005PA001189

Domitsu H, Oda M (2005) Japan Sea planktic foraminifera in surface sediments: geographical distribution and relationships to surface water mass. Paleontol Res 9:255–270

Domitsu H, Oda M (2006) Linkages between surface and deep circulations in the southern Japan Sea during the last 27,000 years: evidence from planktic foraminiferal assemblages and stable isotope records. Mar Micropaleontol 61:155–170

Elderfield H, Ferretti P, Greaves M, Crowhurst S, McCave IN, Hodell D, Piotrowski AM (2012) Evolution of ocean temperature and ice volume through the mid-Pleistocene climate transition. Science 337:704–709

Fairbanks RG, Wiebe PH, Bé AWH (1980) Vertical distribution and isotopic composition of living planktonic foraminifera in the western north Atlantic. Science 207:61–63

Gallagher S, Kitamura A, Iryu Y, Itaki T, Koizumi I, Hoiles P (2015) The Pliocene to recent history of the Kuroshio and Tsushima Currents: a multi-proxy approach. Prog Earth Planet Sci 2:17. https://doi.org/10.1186/s40645-015-0045-6

Harker A (1909) The natural history of igneous rocks. Methuen & co., London, p 458

Hayashida A, Kamata H, Danhara T (1996) Correlation of widespread tephra deposits based on paleomagnetic directions: link between a volcanic field and sedimentary sequences in Japan. Quat Int 34–36:89–98

Hirosaki City (2001) History of Hirosaki city, new edition -overview of history 1- (in Japanese), p 425

Ijiri A, Wang L, Oba T, Kawahata H, Huang CY, Huang CY (2005) Paleoenvironmental changes in the northern area of the East China Sea during the past 42,000 years. Palaeogeogr Palaeoclimatol Palaeoecol 219:239_261

Ikehara K (2015) Marine tephra in the Japan Sea sediments as a tool for paleoceanography and paleoclimatology. Prog Earth Planet Sci 2:36. https://doi.org/10.1186/s40645-015-0068-z

Ikehara K, Kikkawa K, Katayama H, Seto K, (1994) Late Quaternary paleoceanography of the Japan Sea: a tephrochronological and sedimentlogical study. Proc., 29th Int'l. Geol. Congr., Part B:229–235

Irino T, Tada R, Ikehara K, Sagawa T, Karasuda A, Kurokawa S, Seki A, Lu S 2018 Construction of perfectly continuous records of physical properties for dark-light sediment sequences collected from the Japan Sea during Integrated Ocean Drilling Program Expedition 346 and their potential utilities as paleoceanographic studies. Progress in Earth and Planetary Science [in press]. https://doi.org/10.1186/s40645-018-0176-7

Itaki T (2003) Depth-related radiolarian assemblage in the water-column and surface sediments of the Japan Sea. Mar Micropaleontol 47:253–270

Itaki T (2007) Historical changes of deep-sea radiolarians in the Japan Sea during the last 640 kyrs. Fossils 82:43–51 (in Japanese with English abstract)

Itaki T, Ito M, Narita H, Ahagon N, Sakai H (2003) Depth distribution of radiolarians from the Chukchi and Beaufort Seas, western Arctic. Deep-Sea Res, Part I 50:1507–1522

Itaki T, Komatsu N, Motoyama I (2007) Orbital- and millennial-scale changes of radiolarian assemblages during the last 220 kyrs in the Japan Sea. Palaeogeogr Palaeoclimatol Palaeoecol 247:115–130

Itaki T, Sagawa T, Kubota Y, Pleistocene radiolarian biostratigraphy, IODP Expedition 346 Site U1427. In Tada, R., Murray, R.W., Alvarez Zarikian, C.A., and the Expedition 346 Scientists, Proceedings of the Integrated Ocean Drilling Program, 346: College Station, TX (Integrated Ocean Drilling Program) [in press]. https://doi.org/10.2204/iodp.proc.346.202.2018

Khim B-K, Tada R, Itaki T (2014) Different nature of glacial CaCO3 constituents between MIS 2 and MIS 12 in the East Sea/Japan Sea and its paleoceanographic implication. Quat Int 333:130–138

Kido Y, Minami I, Tada R, Fujine K, Irino T, Ikehara K, Chun JH (2007) Orbital-scale stratigraphy and high-resolution analysis of biogenic components and deep-water oxygenation conditions in the Japan Sea during the last 640 kyr. Palaeogeogr Palaeoclimatol Palaeoecol 247:32–49

Kimura J-I, Okada S, Nakayama K, Umeda K, Kusano T, Asahara Y, Tateno M, Danhara T (1999) Fission track ages of tephras from Daisen and Sambe volcanoes and their volcanological implications. Quat Res (Daiyonki-Kenkyu) 38:145–155 (in Japanese with English abstract)

Kubota Y, Kimoto K, Tada R, Oda H, Yokoyama Y, Matsuzaki H (2010) Variations of East Asian summer monsoon since the last deglaciation based on Mg/Ca and oxygen isotope of planktic foraminifera in the northern East China Sea. Paleoceanography 25:PA4205. https://doi.org/10.1029/2009PA001891

Kuroyanagi A, Kawahata H (2004) Vertical distribution of living planktonic foraminifera in the seas around Japan. Mar Micropaleontol 53:173–196

Lisiecki LE, Raymo ME (2005) A Pliocene-Pleistocene stack of 57 globally distributed benthic $\delta^{18}O$ records. Paleoceanography 20:1–17

Locarnini RA, Mishonov AV, Antonov JI, Boyer TP, Garcia HE, Baranova OK, Zweng MM, Paver CR, Reagan JR, Johnson DR, Hamilton M, Seidov D (2013) World ocean atlas 2013, volume 1: temperature. U.S. Government Printing Office, Washington, D.C., p 40

Machida H (1999) The stratigraphy, chronology and distribution of distal marker-tephras in and around Japan. Glob Planet Chang 21:71–94

Machida H, Arai F (2003) Atlas of tephra in and around Japan: revised edition Tokyo University Press, Tokyo (in Japanese), p 336

Matsui H, Tada R, Oba T (1998) Low-salinity isolation event in the Japan Sea in response to eustatic sea-level drop during LGM: reconstruction based on salinity-balance model. Quat Res (Daiyonki Kenkyu) 37:221–233 (in Japanese with English abstract)

Matsumoto A (1996) K-Ar age determinations of young volcanic rocks-- correction for initial 40Ar/36Ar ratios and its application. Chishitsu News 501: 12–17 (in Japanese)

Matsumoto A, Ui T (1997) K-Ar age of Ata pyroclastic flow deposit, southern Kyushu, Japan. Bull Volc Soc Japan 42:223–225 (in Japanese)

Matsumoto A, Uto K, Ono K, Watanabe K (1991) K-Ar age determinations for Aso volcanic rocks--concordance with volcanostratigraphy and application to pyroclastic flows. Program Abstr Volc Soc Japan 2:73 (in Japanese)

Matsuzaki KM, Suzuki N, Nishi H, Hayashi H, Gyawali BR, Takashima R, Ikehara M (2015) Early to Middle pleistocene paleoceanographic history of southern Japan based on radiolarian data from IODP Exp. 314/315 Sites C0001 and C0002. Mar Micropaleontol 118:17–33

Matsuzaki KM, Suzuki N, Nishi H, Takashima R, Kawate Y, Sakaii T (2014) Middle to Late pleistocene radiolarian biostratigraphy in the water-mixed region of the Kuroshio and Oyashio currents, northeastern margin of Japan (JAMSTEC Hole 902-C9001C). J Micropalaeontol 33:205–222

Matul A, Abelmann A, Tiedemann R, Kaiser A, Nürnberg D (2002) Late Quaternary polycystine radiolarian datum events in the Sea of Okhotsk. Geo-Mar Lett 22:25–32

Matul AG, Abelmann A, Nürnberg D, Tiedemann R (2009) Stratigraphy and major paleoenvironmental changes in the Sea of Okhotsk during the last million years inferred from radiolarian data. Oceanology 49:93–100

Mix AC, Bard E, Schneider R (2001) Environmental processes of the ice age: land, oceans, glaciers (EPILOG). Quat Sci Rev 20:627–657

Miyairi Y, (2007) New measure for determination Quaternary volcanic eruption age using accelerator mass spectrometric radiocarbon dating and red thermoluminescence dating Ph.D thesis, 168, The University of Tokyo

Miyairi Y, Yoshida K, Miyazaki Y, Matsuzaki H, Kaneoka I (2004) Improved ^{14}C dating of a tephra layer (AT tephra, Japan) using AMS on selected organic fractions. Nucl Instrum Methods Phys Res B 223–224:555–559

Miyoshi N, Fujiki T, Morita Y (1999) Palynology of a 250-m core from Lake Biwa: a 430,000-year record of glacial–interglacial vegetation change in Japan. Rev Palaeobot Palynol 104:267–283

Moriwaki H, Suzuki T, Murata M, Ikehara M, Machida H, Lowe DJ (2011) Sakurajima-Satsuma (Sz-S) and Noike-Yumugi (N-Ym) tephras: new tephrochronological marker beds for the last deglaciation, southern Kyushu, Japan. Quat Int 246:203–212

Motoyama I, Yamada Y, Hoshiba M, Itaki T (2016) Radiolarian assemblages in surface sediments of the Japan Sea. Paleontol Res 20:176–206

Muraoka H, (1991) Heat source systems in the Hakkoda geothermal field, Northeast Japan: Report Geological Survey of Japan 113–134 (in Japanese with English abstract)

Muza JP (1992) Calcareous nannofossil biostratigraphy from the Japan Sea, Sites 798 and 799: evidence for an oscillating Pleistocene oceanographic frontal boundary. Proc ODP Sci Results 127/128(Pt. 1):155–169

Nagahashi Y, Nakai S, Kikkawa K, Okudaira T, Yoshikawa S, Yoshida T (2015) Petrological properties of tephra beds based on the major and trace element composition of volcanic glass shards : a case study of the Osaka Group and the Takashima-oki drilling core sediments in Lake Biwa, Kinki district, Japan. Chikyu-Kagaku 69:205–222 (in Japanese with English abstract)

Nagahashi Y, Sato T, Takeshita Y, Tawara T, Kumon F (2007) Stratigraphy and chronology of widespread tephra beds intercalated in the TKN-2004 core sediment obtained from the Takano formation, Central Japan. Quat Res (Daiyonki-Kenkyu) 46:305–325 (in Japanese with English abstract)

Nagahashi Y, Yoshida T, Nakai S, Okudaira T (2003) Evaluation and correction of EDS results of the glass shards from some representative tephra by comparison with XRF analysis. Quat Res (Daiyonki-Kenkyu) 42:265–277 (in Japanese with English abstract)

Nagahashi Y, Yoshikawa S, Miyakawa C, Uchiyama T, Inouchi Y (2004) Stratigraphy and chronology of widespread tephra layers during the past 430 ky in the Kinki District and the Yatsugatake mountains: major element composition of the glass shards using EDS analysis. Quat Res (Daiyonki-Kenkyu) 43:15–35 (in Japanese with English abstract)

Nagatomo T, Shitaoka Y, Nishimura S (2005) Comparison of results between TL and OSL dating of the same tephras. Bull Nara Univ Educ 52:1–9 (in Japanese with English abstract)

Nakagawa M, Fujioka K, Furuta T, Koizumi S, (1994) Volcanic ash beds in the cores of the Leg.127 and 128 under Japan sea--volcanic ash correlations and estimation of active history of Quaternary volcanoes from major element compositions of volcanic glass shards. Chikyu Monthly 16 (in Japanese)

Nakajima T, Kikkawa K, Ikehara K, Katayama H, Kikawa E, Joshima M, Seto K (1996) Marine sediments and late Quaternary stratigraphy in the southeastern part of the Japan sea--concerning the timing of dark layer deposition. J Geol Soc Jpn 102:125–138 (in Japanese with English abstract)

Morley JJ. and Nigrini C. (1995) Miocene to Pleistocene Radiolarian Biostratigraphy of North Pacific Sites 881, 884, 885, 886, and 887. In Rea, D.K., Basov, I.A., Scholl, D.W. and Allan, J.F. (eds.) Proceedings of the Ocean Drilling Program Proc. ODP, Sci. Results, College Station TX (Ocean Drilling Program), 145:55–91 https://doi.org/10.2973/odp.proc.sr.145.107.1995

Oba T (1991) Aso-4 and Ata tephra from the viewpoint of oxygen isotope stratigraphy. Chikyu Mon 13:224–227 (in Japanese)

Oba T, Kato M, Kitazato H, Koizumi I, Omura A, Sakai T, Takayama T (1991) Paleoenvironmental changes in the Japan Sea during the last 85 000 years. Paleoceanography 6:499–518

Okuno M (2002) Chronology of tephra layers in southern Kyushu, SW Japan, for the last 30,000 years. Quat Res (Daiyonki-Kenkyu) 41:225–236 (in Japanese with English abstract)

Okuno M, Nakamura T, Geshi N, Kimura K, Saito-Kokubu Y, Kobayashi T (2013) AMS radiocarbon dating of wood trunks in the pumiceous deposits of the Kikai-Akahoya eruption in Yakushima Island, SW Japan. Nucl Instru Meth Phys Res B 294:602–605

Pickering KT, Souter C, Oba T, Taira A, Schaaf M, Platzman E (1999) Glacio-eustatic control on deep-marine clastic forearc sedimentation, Pliocene-mid-Pleistocene (c. 1180–600 ka) Kazusa Group, SE Japan. J Geol Soc 156:125–136

Pouclet A, Scott S (1992) Volcanic ash layers in the Japan Sea: tephrochronology of Site 798 and 799. Proc ODP Sci Results 127/128(Pt. 2):791–803

Railsback LB, Gibbard PL, Head MJ, Voarintsoa NRG, Toucanne S (2015) An optimized scheme of lettered marine isotope substages for the last 1.0 million years, and the climatostratigraphic nature of isotope stages and substages. Quat Sci Rev 111:94–106

Rohling EJ, Foster GL, Grant KM, Marino G, Roberts AP, Tamisiea ME, Williams F (2014) Sea-level and deep-sea-temperature variability over the past 5.3 million years. Nature 508:477–482

Satoguchi Y, Nagahashi Y (2012) Tephrostratigraphy of the Pliocene to Middle Pleistocene series in Honshu and Kyushu Islands, Japan. Island Arc 21:149–169

Shirai M (2001) Widespread tephra of the middle Pleistocene found in the eastern Japan Sea. Chikyu Mon 23:600–604 (in Japanese)

Shirai M, Tada R, Fujioka K (1997) Identification and chronostratigraphy of middle to upper Quaternary marker tephras occurring in the Anden coast based on comparison with ODP cores in the sea of Japan. Quat Res (Daiyonki-Kenkyu) 36:183–196 (in Japanese with English abstract)

Shitaoka Y, Fukuoka T, Hasegawa A, Kusano T, Nagatomo T (2009) Thermoluminescence dating of the pyroclastic deposits of the Sanbe Volcano. Bull Shimane Nat Mus Mt Sanbe (Sahimel) 7:15–24

Shitaoka Y, Moriwaki H, Akai F, Nakamura N, Miyoshi M, Yamamoto J (2016) Eruption age of Sakurajima-Satsuma tephra using thermoluminescence dating. Bull Geol Environ Sci 18:29–35

Siddall M, Rohling EJ, Almogi-Labin A, Hemleben C, Meischner D, Schmelzer I, Smeed DA (2003) Sea-level fluctuations during the last glacial cycle. Nature 423:853–858

Smith VC, Staff RA, Blockley SPE, Bronk Ramsey C, Nakagawa T, Mark DF, Takemura K, Danhara T (2013) Identification and correlation of visible tephras in the Lake Suigetsu SG06 sedimentary archive, Japan: chronostratigraphic markers for synchronising of east Asian/west Pacific palaeoclimatic records across the last 150 ka. Quat Sci Rev 67:121–137

Sparks RSJ, Walker GPL (1977) The significance of vitric-enriched air-fall ashes associated with crystal enriched ignimbrite. J Volcanol Geotherm Res 2:329–341

Tada R, Iijima A (1983) Petrology and diagenetic changes of Neogene siliceous rocks in northern Japan. J Sediment Res 53:911–930

Tada R, Irino T, Koizumi I (1999) Land-ocean linkages over orbital and millennial timescales recorded in late Quaternary sediments of the Japan Sea. Paleoceanography 14:236–247

Tada R, Irino T, Ikehara K, Karasuda A, Sugisaki S, Xuan C, Sagawa T, Itaki T, Kubota Y, Lu S, Seki A, Murray RW, Alvarez-Zarikian C, Anderson WT Jr, Bassetti M-A, Brace BJ, Clemens SC, da Costa Gurgel MH, Dickens GR, Dunlea AG, Gallagher SJ, Giosan L, Henderson ACG, Holbourn AE, Kinsley CW, Lee GS, Lee KE, Lofi J, Lopes CICD, Pellitero MS, Peterson LC, Singh RK, Toucanne S, Wan S, Zheng H, Ziegler M, High-resolution and -precision correlation of dark and light layers in the Quaternary hemipelagic sediments of the Japan Sea recovered during IODP Expedition 346. Progress in Earth and Planetary Science [in press]. https://doi.org/10.1186/s40645-018-0167-8

Tada R, Koizumi I, Cramp A, Rahman A (1992) Correlation of dark and light layers, and the origin of their cyclicity in the Quaternary sediments from the Japan Sea. Proc ODP Sci Results 127/127(Pt. 1):577–601

Tada R, Murray R, Alvarez Zarikian C, Anderson W, Bassetti M-A, Brace B, Clemens S, da Costa Gurgel M, Dickens G, Dunlea A, Gallagher S, Giosan L, Henderson A, Holbourn A, Ikehara K, Irino T, Itaki T, Karasuda A, Kinsley C, Kubota Y, Lee G, Lee K, Lofi J, Lopes C, Peterson L, Saavedra-Pellitero M, Sagawa T, Singh R, Sugisaki S, Toucanne S, Wan S, Xuan C, Zheng H, Ziegler M, (2015a) Site U1427 *In* Tada, R., Murray, R.W., Alvarez Zarikian, C.A., and the Expedition 346 Scientists, Proc. IODP, 346, College Station, TX (Integrated Ocean Drilling Program), https://doi.org/10.2204/iodp.proc.346.108.2015

Tada R, Murray R, Alvarez Zarikian C, Anderson W, Bassetti M-A, Brace B, Clemens S, da Costa Gurgel M, Dickens G, Dunlea A, Gallagher S, Giosan L, Henderson A, Holbourn A, Ikehara K, Irino T, Itaki T, Karasuda A, Kinsley C, Kubota Y, Lee G, Lee K, Lofi J, Lopes C, Peterson L, Saavedra-Pellitero M, Sagawa T, Singh R, Sugisaki S, Toucanne S, Wan S, Xuan C, Zheng H, Ziegler M, (2015b) Site U1426 *In* Tada, R., Murray, R.W., Alvarez Zarikian, C.A., and the Expedition 346 Scientists, Proc. IODP, 346, College Station, TX (Integrated Ocean Drilling Program), https://doi.org/10.2204/iodp.proc.346.107.2015

Tada R, Murray R, Alvarez Zarikian C, Anderson W, Bassetti M-A, Brace B, Clemens S, da Costa Gurgel M, Dickens G, Dunlea A, Gallagher S, Giosan L, Henderson A, Holbourn A, Ikehara K, Irino T, Itaki T, Karasuda A, Kinsley C, Kubota Y, Lee G, Lee K, Lofi J, Lopes C, Peterson L, Saavedra-Pellitero M, Sagawa T, Singh R, Sugisaki S, Toucanne S, Wan S, Xuan C, Zheng H, Ziegler M, (2015c) Sites U1428 and U1429 *In* Tada, R., Murray, R.W., Alvarez Zarikian, C.A., and the Expedition 346 Scientists, Proc. IODP, 346, College Station, TX (Integrated Ocean Drilling Program), https://doi.org/10.2204/iodp.proc.346.109.2015

Tada R, Murray R, Alvarez Zarikian C, Anderson W, Bassetti M-A, Brace B, Clemens S, da Costa Gurgel M, Dickens G, Dunlea A, Gallagher S, Giosan L, Henderson A, Holbourn A, Ikehara K, Irino T, Itaki T, Karasuda A, Kinsley C, Kubota Y, Lee G, Lee K, Lofi J, Lopes C, Peterson L, Saavedra-Pellitero M, Sagawa T, Singh R, Sugisaki S, Toucanne S, Wan S, Xuan C, Zheng H, Ziegler M, (2015d) Methods, *In* Tada, R., Murray, R.W., Alvarez Zarikian, C.A., and the Expedition 346 Scientists, Proc. IODP, 346, College Station, TX (Integrated Ocean Drilling Program), https://doi.org/10.2204/iodp.proc.346.102.2015

Tamura I, Yamazaki H (2010) Significance of the remarkable unconformity in the Plio–Pleistocene of the Japanese Islands. Quat Int 219:45–54

Torii M, Yoshikawa S, Itihara M (1974) Paleomagnetism in the water-laid volcanic ash layers in the Osaka Group, Sennan and Senpoku hills, southwest Japan. Rock Magn Paleogeophysics 2:34–37

Tsuji T, Miyata Y, Okada M, Mita I, Nakagawa H, Sato Y, Nakamizu M (2005) High-resolution chronology of the lower Pleistocene Otadai and Umegase Formations of the Kazusa Group, Boso Peninsula, central Japan-Chronostratigraphy of the JNOC TR-3 cores based on oxygen isotope, magnetostratigraphy and calcareous nannofossil. J Geol Soc Jpn 111:1–20 (in Japanese with English abstract)

Uto K, Suto S, (1985) K-Ar age determination of volcanic rocks from the Hohi geothermal area, Kyushu, Japan. Geological Survey of Japan, pp. 67–83 (in Japanese with English abstract)

Yokoyama Y, Kido Y, Tada R, Minami I, Finkel RC, Matsuzaki H (2007) Japan Sea oxygen isotope stratigraphy and global sea-level changes for the last 50,000 years recorded in sediment cores from the Oki Ridge. Palaeogeogr Palaeoclimatol Palaeoecol 247:5–17

Yokoyama Y, Lambeck K, De Deckker P, Johnston P, Fifield LK (2000) Timing of the Last Glacial Maximum from observed sea-level minima. Nature 406:713–716

Yoshikawa S, Inouchi Y (1993) Middle Pleistocene to Holocene explosive volcanism revealed by ashes of the Takashima-oki core samples from Lake Biwa, central Japan. Chikyu-Kagaku 47:97–109 (in Japanese with English abstract)

Yoshikawa S, Kuwae M (2001) High precision tephrochronology for the past 400 ky using Lake Biwa sediment. Chikyu Mon 23:594–599 (in Japanese)

Young J, Geisen M, Cros L, Kleijne A, Sprengel C, Probert I, Ostergaard J (2003) A guide to extant coccolithophore taxonomy. J Nannoplankton Res Spec Issue 1:1–125

6

$\delta^{18}O_{sw}$ estimate for *Globigerinoides ruber* from core-top sediments in the East China Sea

Keiji Horikawa[1*], Tomohiro Kodaira[2], Jing Zhang[1] and Masafumi Murayama[3]

Abstract

The paired analyses of the Mg/Ca ratio and oxygen isotopic composition ($\delta^{18}O_c$) of surface-dwelling planktonic foraminifera have become a widely used method for reconstructing the oxygen isotopic composition of ambient seawater ($\delta^{18}O_{sw}$) as a robust proxy for surface salinity. *Globigerinoides ruber* (*G. ruber*) is a mixed-layer dweller, and its fossil shell is an ideal archive for recording past sea surface water conditions, such as those caused by variability in the East Asian summer monsoon (EASM). Here, we investigate the validity of shell-derived $\delta^{18}O_{sw}$ estimates for *G. ruber* using core-top sediments from the East China Sea (ECS). First, we determined a local $\delta^{18}O_{sw}$–salinity equation for the eastern part of the ECS in July [$\delta^{18}O_{sw} = -7.74 + 0.23 \times salinity$]. Then, we calculated $\delta^{18}O_{sw}$ from core-top $\delta^{18}O_c$ and Mg/Ca values in *G. ruber* using the $\delta^{18}O_c$–temperature equation of Bemis et al. (Paleoceanography 13(2):150–160, 1998) and the Mg/Ca–temperature equation of Hastings et al. (EOS 82:PP12B-10, 2001). The core-top $\delta^{18}O_{sw}$ and salinity were estimated to be in the ranges of −0.2 to +0.39 ‰ and 33.7 to 34.5, respectively, which fall close to the local $\delta^{18}O_{sw}$–salinity regression line. The core-top data showed that the Mg/Ca–temperature calibration by Hastings et al. (EOS 82:PP12B-10, 2001) and the $\delta^{18}O_c$–temperature equation by Bemis et al. (Paleoceanography 13(2):150–160, 1998) are appropriate for calculating $\delta^{18}O_{sw}$ in the ECS. Furthermore, we measured core-top Ba/Ca ratios of *G. ruber* ($Ba/Ca_{G.\ ruber}$), which ranged from 0.66 to 2.82 μmol mol^{-1}. There was not a significant relationship between the salinity and $Ba/Ca_{G.\ ruber}$ ratios due to the highly variable $Ba/Ca_{G.\ ruber}$ data. Given the seawater Ba/Ca data and the published partition coefficient for Ba ($D_{Ba} = 0.15$–0.22), pristine $Ba/Ca_{G.\ ruber}$ ratios at northern Okinawa Trough sites should be less than 0.84 μmol mol^{-1}. Anomalously high $Ba/Ca_{G.\ ruber}$ ratios (>0.84 μmol mol^{-1}) might be attributable to contamination by sedimentary barite adherent on fossil shells. Therefore, further evaluation of the $Ba/Ca_{G.\ ruber}$ ratio as a paleo-salinity proxy requires diethylene triamine pentaacetic acid (DTPA)-cleaned Ba/Ca data that can minimize the influence of barite contamination.

Keywords: Oxygen isotope composition of seawater; Reconstruction of sea surface salinity; Mg/Ca–temperature; *Globigerinoides ruber*; East China Sea

Background

An East Asian summer monsoon (EASM) precipitation zone is formed by the convergence of a westerly flow to the north of the Tibetan Plateau and a southerly monsoon flow over eastern China, which gradually moves northward from South China and reaches northern China in late July (Qian and Lee 2000). This EASM precipitation has caused anomalous climatic events (both wet and dry) on an inter-decadal scale, influencing the agricultural production and economy in Asia (Wang 2006). Paleoclimate records show that anomalous dry periods, when summer monsoons weakened, caused crop failure and led to the collapse of some Chinese dynasties (Yancheva et al. 2007). Therefore, EASM variability, its effect on remote areas, and the mechanisms behind this variability have been investigated intensively as important climate change issues (Wang et al. 2005; Chen et al. 2008; Wang et al. 2008; Chang et al. 2009; Liu et al. 2014).

The EASM brings a large amount of precipitation to South China and the drainage area of the Changjiang River (Yangtze River) and induces a large influx of Changjiang River runoff to the East China Sea (ECS), forming Changjiang diluted water (CDW) in the

* Correspondence: horikawa@sci.u-toyama.ac.jp
[1]Graduate School of Science and Engineering for Research, University of Toyama, Gofuku 3190, Toyama 930-8555, Japan
Full list of author information is available at the end of the article

Changjiang River Estuary (Zhang et al. 1990; Chen et al. 1994) (Fig. 1). The CDW extends farther offshore, driven by the southerly and southwesterly monsoons from June to August, and advection of the CDW causes a significant decrease in the sea surface salinity (SSS) of the northern Okinawa Trough (Chang and Isobe 2003; Lee and Chao 2003). Consequently, the northern Okinawa Trough is considered an oceanic area where EASM-related variability in SSS can be seen (Chang and Isobe 2003).

In such areas, the oxygen isotopic compositions of seawater ($\delta^{18}O_{sw}$) have been estimated from paired analyses

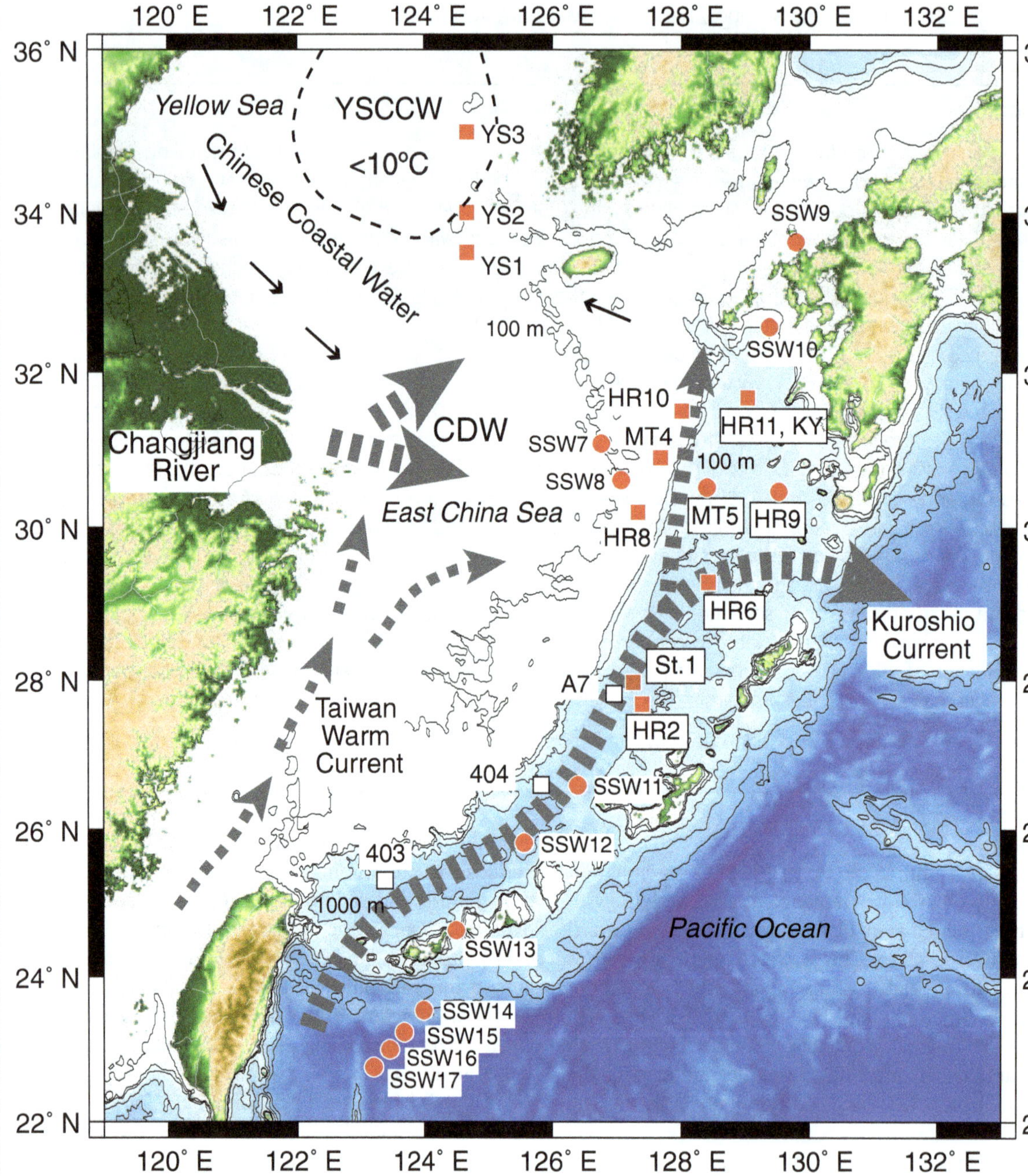

Fig. 1 Map of sample locations. Locations for surface seawater samples (*red circles*) and surface–subsurface seawater samples (*red squares*) are shown together with the present current system (Ichikawa and Beardsley 2002). Surface sediments were taken from HR11, MT5, HR9, HR6, HR2, and St.1. Locations for sediment core at KY, A7, 403, and 404 (Sun et al. 2005; Lin et al. 2006; Chen et al. 2010; Kubota et al. 2010) are also shown. CDW and YSCCW represent the Changjiang diluted water and Yellow Sea Central Cold Water, respectively. The bathymetric contours represent 100, 200, 300, 1000, 2000, and 3000 m

of Mg/Ca temperatures and oxygen isotopes ($\delta^{18}O_c$) in the surface-dwelling foraminifera *Globigerinoides ruber* (*G. ruber*) to reconstruct EASM-related variability in SSS (Sun et al. 2005; Kubota et al. 2010). However, $\delta^{18}O_{sw}$ estimates from paired analyses of Mg/Ca temperatures and $\delta^{18}O_c$ in planktonic foraminifera entail high levels of uncertainty (>0.2 ‰) due to the propagation error of $\delta^{18}O_c$ and Mg/Ca measurements and Mg/Ca–temperature calibration (Schmidt 1999; Rohling 2007). Furthermore, the lack of a local $\delta^{18}O_{sw}$–salinity equation causes additional uncertainties in salinity estimates calculated from $\delta^{18}O_{sw}$ values. Additionally, if the habitat of *G. ruber* shifts seasonally, independent of EASM variability, shell-derived $\delta^{18}O_{sw}$ estimates also change, and interpreting EASM variability will be complicated. Therefore, a reliable reconstruction of SSS with respect to EASM variability requires adequate Mg/Ca–temperature, $\delta^{18}O_c$–temperature, and local $\delta^{18}O_{sw}$–salinity calibration equations and an understanding of the habitat season and depth of *G. ruber* in the ECS.

In this study, we first derive a local $\delta^{18}O_{sw}$–salinity equation for the northern Okinawa Trough to provide a basis for salinity estimates calculated from $\delta^{18}O_{sw}$ for *G. ruber*. Then, using ECS core-top sediments, we calculate the core-top $\delta^{18}O_{sw}$ from Mg/Ca ratios and $\delta^{18}O_c$ for *G. ruber* and compare them with the local $\delta^{18}O_{sw}$–salinity line. We find that the Mg/Ca–temperature calibration by Hastings et al. (2001) and the $\delta^{18}O_c$–temperature equation by Bemis et al. (1998) are appropriate for calculating $\delta^{18}O_{sw}$ in the ECS. Finally, we present the Ba/Ca ratios of *G. ruber* from ECS core-top sediments for the first time and discuss the possibility of using shell Ba/Ca ratios as an independent paleo-salinity proxy in the ECS.

Oceanographic setting

The ECS is a marginal sea in the northwestern Pacific bounded by China, Taiwan, Ryukyu Island, Kyushu, and the Korean Peninsula (Fig. 1). The Yellow Sea is located in the northern part of the ECS, with most (>70 %) of both the Yellow Sea and ECS located above the continental shelf (<200 m water depth). The deep Okinawa Trough (~2000 m water depth) is a back-arc basin that occupies the remaining southeastern part of the ECS (Ichikawa and Beardsley 2002).

The Yellow Sea and the ECS receive the saline and warm Kuroshio water that enters the ECS along the east of Taiwan and flows northeastward along the shelf slope in the ECS (Fig. 1). The annual cycle of Kuroshio volume transport has an estimated maximum of 24 Sv (1 Sv = 1×10^6 m^3 s^{-1}) in the summer and a minimum of 20 Sv in the autumn (Lee et al. 2001). In addition, the Taiwan warm current (TWWC), which originates from the Kuroshio, flows into the ECS from the South China Sea (SCS) through the Taiwan Strait off western Taiwan, with an annual mean northward transport of 0.78 Sv (Jan et al. 2006; Zhang et al. 2014) (Fig. 1). In the autumn and winter, Changjiang River runoff decreases (1.0×10^4 m^3 s^{-1} in January) (Yanagi 1994), and the CDW flows southwestward along the Chinese coast as a narrow band; the saline water (>34) of Kuroshio origin enters the Yellow Sea (Chang and Isobe 2003). However, during the summer, when the Changjiang River supplies a large amount of freshwater to the ECS (4.8×10^4 m^3 s^{-1} in July) (Yanagi 1994), the CDW starts to extend offshore and spread east- and northeastward. Then, the eastward-flowing CDW mixes with the TWWC and the Kuroshio water in the southern and eastern parts of the ECS, respectively (Chang and Isobe 2003) (Fig. 2a).

Since the annually averaged precipitation and evaporation are near equilibrium in the ECS and Yellow Sea and the Changjiang River runoff accounts for 90 % of the total river discharge (with an annual mean of ~3.0×10^4 m^3 s^{-1}), the Changjiang River discharge is considered the dominant freshwater source in the ECS (Beardsley et al. 1985; Chen et al. 1994; Yanagi 1994). This Changjiang River water extends offshore, and therefore, the water mass characteristics of the offshore ECS can be roughly described by the binary mixing of the cooler, less saline CDW and the warmer, more saline Kuroshio water (Zhang et al. 1990; Ichikawa and Beardsley 2002; Zhang et al. 2007). In contrast, there are other local water masses in the coastal ECS and the Yellow Sea, such as the TWWC and the Yellow Sea Central Cold Water (YSCCW) (Fig. 1). Therefore, the water mass characteristics and $\delta^{18}O_{sw}$–salinity relationships of the coastal ECS are more complicated than those of the offshore ECS, as discussed in a later section.

The northern Okinawa Trough, which is the main focus of this study, shows distinct seasonal changes in SSS (e.g., site HR11, SSS = 33.2–34.7) due to the advection of the CDW, in comparison with the central Okinawa Trough (e.g., site HR2, SSS = 34.4–34.8) (Figs. 1 and 2a). Sea surface temperature (SST) distribution also varies seasonally owing to monsoonal winds (Lee and Chao 2003). The climatological mean annual SST is 22.2 °C near site HR11 in the northern Okinawa Trough, with a maximum in August (28.1 °C) and a minimum in February (17.5 °C). In the central Okinawa Trough, the mean annual SST is 25.0 °C near site HR2, with a maximum in August (28.9 °C) and a minimum in February (21.5 °C) (Japan Oceanographic Data Center (JODC), available at http://www.jodc.go.jp/).

Methods

Seawater and sediment samples

Seawater samples were collected using a CTD (conductivity, temperature, and depth) Carousel multisampling

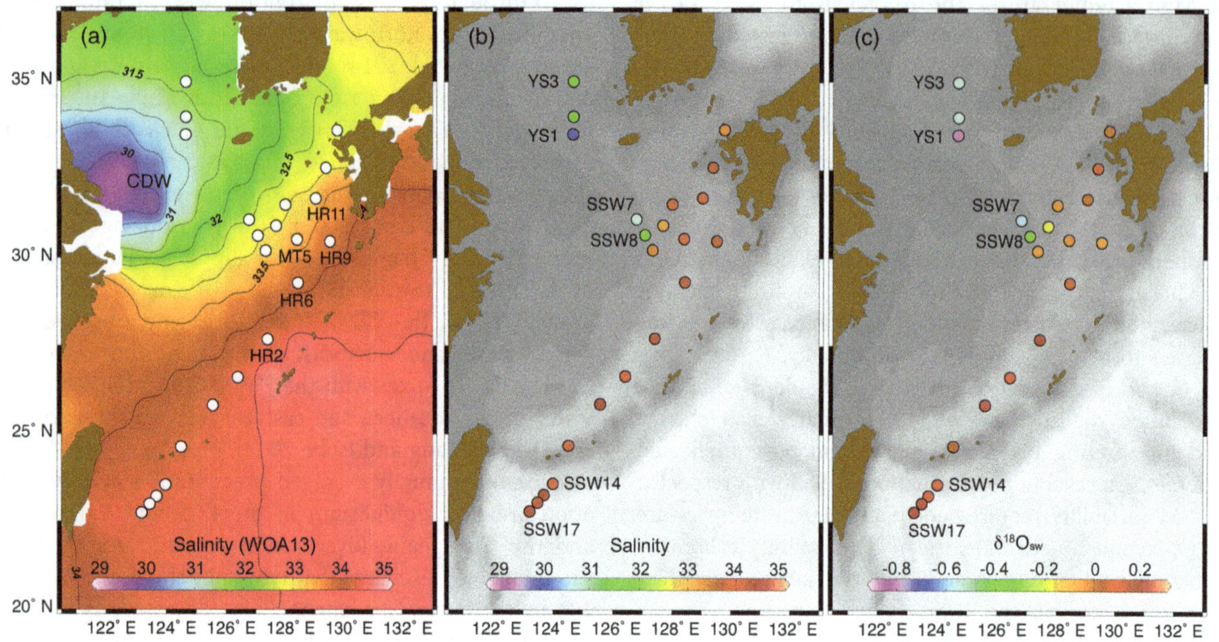

Fig. 2 Surface distribution of salinity and $\delta^{18}O_{sw}$. **a** SSS in July–September averaged statistically for 1955–2012. The data is from the World Ocean Atlas 2013 (Zweng et al. 2013). *Circles* represent locations for surface seawater samples (see Fig. 1 and Table 1). **b** SSS and **c** $\delta^{18}O_{sw}$ of seawater in early July 2013. The lowest SSS and $\delta^{18}O_{sw}$ values were found at YS1 within the CDW

system (Sea-Bird, model SBE 9plus) during the KH-13-4 cruise in early July 2013 (R/V Hakuho Maru) (Table 1). The CTD system has 24 attached Niskin bottles, each with a volume of 12 L. Subsurface seawater samples were taken with the Niskin bottles, and surface water samples were taken using a bucket (~0 m depth) or pumped from below the ship (~5 m depth). All seawater samples were transferred into glass vials for salinity and $\delta^{18}O_{sw}$ measurements, and the samples for the $\delta^{18}O_{sw}$ measurements were stored at 4 °C until laboratory analysis. Salinity was measured with an onboard Autosal laboratory salinometer (Model 8400B, Guildline Instruments Ltd., Canada) and is reported using the practical salinity scale. Sampling vials for salinity were prepared according to Joint Global Ocean Flux Study (JGOFS) protocols. The Autosal was standardized using International Association for the Physical Sciences of the Oceans (IAPSO) standard seawater.

Three sampling stations (YS1–3) are located in the southwest area off the Korean Peninsula, and 19 other stations (MT, HR, and SSW sites) are located in the continental shelf slope area of the ECS and in the Pacific Ocean (Fig. 1). We report salinity and $\delta^{18}O_{sw}$ data from these stations, and the vertical profiles of both datasets were obtained for nine stations (Fig. 1 and Table 1). Surface sediment samples were taken by a multiple corer at six sites along the central to northern parts of the Okinawa Trough from water depths of 709 to 1675 m during the KH-13-4 and KT-12-25 cruises (Fig. 1 and Table 2). Sediment samples were subsampled onboard with a 1 cm resolution and stored in a refrigerator until analysis in the laboratory.

The lithology of the sediment cores (~30 cm core length) was characterized as foraminifer-bearing silty clay or clayey silt with no visible turbidities or erosional surfaces. In general, the central to northern parts of the Okinawa Trough preserve thick late Holocene sediments due to their high sedimentation rates (10–50 cm kyr^{-1}), except in shelf slope regions (Ujiié et al. 2003; Sun et al. 2005; Kubota et al. 2010). All core-top $\delta^{18}O_c$ values of *G. ruber* were isotopically lighter than −2.0 ‰, which is consistent with the late Holocene values in the central to northern Okinawa Trough (Table 2) (Sun et al. 2005; Kubota et al. 2010). Therefore, we conclude that the core-top sediments used in this study represent at least late Holocene sediments. Site HR2 was located at a depth below the present calcite saturation horizon (CSH) in the ECS (~1600 m) (Fig. 6a). The Mg/Ca ratios from this site should be treated with caution regarding the partial dissolution of *G. ruber* shells, which will be discussed in a later section.

Analysis of the oxygen isotopic composition of seawater ($\delta^{18}O_{sw}$)

We measured $\delta^{18}O_{sw}$ with a stable isotope mass spectrometer (PRISM, Micromass UK, Ltd.) at the University of Toyama. Oxygen isotopic analysis was carried out using the automated H_2O–CO_2 equilibrium method

Table 1 Salinity and $\delta^{18}O_{sw}$ data at each station in the ECS

Depth (m)	Temperature (°C)	Salinity	$\delta^{18}O_{sw}$ (‰, VSMOW)	Std.dev (‰)	Site area
YS1 (30 June 2013, 124° 39.66′ E, 33° 29.80′ N)					Yellow Sea–ECS shelf site
0	23.40	29.89	−0.81	0.03	
5	22.95	30.60	−0.67	0.03	
10	22.83	31.51	−0.47	0.02	
20	17.10	32.82	−0.18	0.02	
40	10.17	33.05	−0.12	0.02	
YS2 (30 June 2013, 124° 40.07′ E, 34° 00.40′ N)					Yellow Sea–ECS shelf site
0	22.90	31.37	−0.54	0.03	
5	22.80	31.50	−0.52	0.02	
10	22.76	31.86	−0.43	0.03	
20	17.14	32.35	−0.35	0.04	
30	13.54	32.57	−0.29	0.03	
YS3 (1 July 2013, 124° 40.04′ E, 34° 59.64′ N)					Yellow Sea–ECS shelf site
0	23.90	31.88	−0.53	0.03	
5	23.12	31.86	−0.52	0.03	
10	17.29	31.87	−0.48	0.02	
20	13.73	32.09	−0.44	0.02	
30	7.71	32.21	−0.42	0.02	
HR8 (2 July 2013, 127° 19.84′ E, 30° 12.68′ N)					Yellow Sea–ECS shelf site
0	26.70	33.40	−0.06	0.02	
20	25.50	33.90	0.07	0.02	
40	24.09	34.34	0.15	0.02	
60	21.33	34.58	0.27	0.02	
70	19.44	34.61	0.26	0.02	
MT4 (2 July 2013, 127° 41.25′ E, 30° 54.42′ N)					Yellow Sea–ECS shelf site
0	26.10	33.10	−0.18	0.02	
6	25.81	33.16	−0.19	0.02	
10	25.79	33.37	−0.14	0.04	
30	24.44	33.81	0.00	0.03	
60	21.56	34.19	0.10	0.02	
100	19.08	34.64	0.20	0.02	
HR10 (2 July 2013, 128° 00.21′ E, 31° 30.22′ N)					Yellow Sea–ECS shelf site
0	26.10	33.75	−0.04	0.02	
5	25.87	33.68	−0.08	0.02	
10	25.80	33.82	−0.02	0.02	
20	25.49	34.07	0.05	0.02	
40	24.96	34.20	0.09	0.02	
59	22.91	34.39	0.17	0.04	
80	21.01	34.59	0.22	0.02	
99	19.86	34.65	0.24	0.03	
HR11 (2 July 2013, 129° 01.85′ E, 31° 40.60′ N)					Okinawa Trough site
0	26.90	33.88	0.01	0.02	
49	23.76	34.30	0.15	0.02	

Table 1 Salinity and $\delta^{18}O_{sw}$ data at each station in the ECS *(Continued)*

75	21.96	34.56	0.23	0.02	
99	20.27	34.69	0.24	0.04	
HR6 (8 July 2013, 128° 25.43′ E, 29° 17.42′ N)					Okinawa Trough site
0	29.70	34.22	0.12	0.05	
20	29.57	34.23	0.11	0.02	
50	28.63	34.35	0.23	0.02	
75	27.21	34.52	0.27	0.04	
99	23.69	34.66	0.33	0.04	
HR2 (9 July 2013, 127° 24.03′ E, 27° 40.97′ N)					Okinawa Trough site
0	29.40	34.25	0.18	0.02	
5	29.63	34.24	0.13	0.02	
10	29.36	34.19	0.14	0.03	
20	29.01	34.17	0.14	0.02	
50	26.32	34.40	0.24	0.02	
75	23.86	34.61	0.32	0.02	
100	22.47	34.74	0.35	0.02	
HR9 (7 July 2013, 129° 31.29′ E, 30° 28.26′ N)					Okinawa Trough site
0	28.50	33.73	−0.07	0.03	
MT5 (8 July 2013, 128° 23.41′ E, 30° 31.22′ N)					Okinawa Trough site
0	28.50	33.91	−0.03	0.02	
SSW07 (2 July 2013, 126° 45.99′ E, 31° 04.86′ N)					Yellow Sea–ECS shelf site
5	23.20	30.80	−0.59	0.04	
SSW08 (2 July 2013, 127° 04.33′ E, 30° 37.06′ N)					Yellow Sea–ECS shelf site
5	25.30	31.80	−0.36	0.02	
SSW09 (7 July 2013, 129° 47.01′ E, 33° 38.63′ N)					Off-Kyushu site
5	24.10	33.33	0.00	0.02	
SSW10 (7 July 2013, 129° 22.50′ E, 32° 34.00′ N)					Off-Kyushu site
5	23.60	33.71	0.12	0.05	
SSW11 (9 July 2013, 126° 24.20′ E, 26° 36.00′ N)					Okinawa Trough site
5	28.80	34.00	0.12	0.02	
SSW12 (9 July 2013, 125° 33.84′ E, 25° 49.26′ N)					Okinawa Trough site
5	29.20	34.38	0.21	0.03	
SSW13 (9 July 2013, 124° 30.06′ E, 24° 38.41′ N)					Okinawa Trough site
5	28.80	33.81	0.03	0.02	
SSW14 (10 July 2013, 123° 59.02′ E, 23° 32.27′ N)					Kuroshio site
5	29.30	34.06	0.06	0.02	
SSW15 (10 July 2013, 123° 40.09′ E, 23° 14.07′ N)					Kuroshio site
5	29.30	34.33	0.12	0.09	
SSW16 (10 July 2013, 123° 26.97′ E, 22° 59.89′ N)					Kuroshio site
5	29.40	34.23	0.15	0.02	
SSW17 (10 July 2013, 123° 11.60′ E, 22° 45.00′ N)					Kuroshio site
5	29.40	34.34	0.20	0.02	

Table 2 Details of the core-top sediments and metal/Ca and $\delta^{18}O$ values from *Globigerinoides ruber* (sensu stricto)

Core site	Latitude	Longitude	Water depth (m)	$\delta^{18}O$ (VPDB ‰)		Mg/Ca (mmol mol^{-1})	±SD	Ba/Ca (μmol mol^{-1})	±SD	Mn/Ca (μmol mol^{-1})	±SD
KH13-4-HR2MC	27° 40.79′ N	127° 23.95′ E	1675	−2.93	Ave	4.09	0.18	0.72	0.74	7.88	5.5
					Run #1	4.05		0.66		0.74	
					Run #2	3.94		0.78		6.50	
					Run #3	4.03		1.53		11.58	
					Run #4	4.35		2.25		12.71	
KT12-25-St.1	27° 58′ N	127° 16′ E	1153	−2.52	Ave.	4.20	0.03	0.77	0.01	23.25	21.3
					Run #1	4.22		0.78		8.18	
					Run #2	4.18		0.77		38.31	
KH13-4-HR6MC	29° 17.35′ N	128° 25.18′ E	1065	−2.47	Ave.	3.97	0.25	0.79	0.20	4.37	4.4
					Run #1	3.68		1.04		1.92	
					Run #2	4.13		0.66		9.49	
					Run #3	4.09		0.92		1.72	
KH13-4-HR9MC	30° 28.19′ N	129° 31.06′ E	709	−2.78	Ave.	4.19	0.25	0.81	1.09	53.01	13.7
					Run #1	3.91		0.81		48.40	
					Run #2	4.36		1.11		68.44	
					Run #3	4.30		2.82		42.18	
KH13-4-MT5MC	30° 31.23′ N	128° 23.38′ E	823	−2.27	Ave.	3.61	0.22	0.88	0.11	44.50	7.7
					Run #1	3.41		0.72		51.73	
					Run #2	3.44		0.95		46.37	
					Run #3	3.68		0.97		46.21	
					Run #4	3.89		0.87		33.67	
KH13-4-HR11MC	31° 40.57′ N	129° 01.99′ E	725	−2.42	Ave.	3.30	0.02	0.85	0.06	44.86	37.0
					Run #1	3.31		0.89		18.67	
					Run #2	3.29		0.80		71.05	
A7	27° 49.2′ N	126° 58.7′ E	1264	−2.35		4.06					
MD012403	25° 17′ N	123° 10′ E	1420	−2.51		4.40					
MD012404	26° 38.84′ N	125° 48.75′ E	1397	−2.60		4.24					
KY07-04-01	31° 33.35′ N	128° 56.64′ E	725	−2.50		3.50					

Table 2 Details of the core-top sediments and metal/Ca and $\delta^{18}O$ values from *Globigerinoides ruber* (sensu stricto)

No dissolution-corrected Mg/Ca-derived SST			Dissolution-corrected Mg/Ca-derived SST							
SST[a] (°C)	SST[b] (°C)	SST[c] (°C)	SST[a] (°C)	$\Delta[CO_3^{2-}]$ (μmol kg^{-1})	Corrected Mg/Ca (mmol mol^{-1})	Calcification SST	Shell-derived $\delta^{18}O_{sw}$ (VSMOW ‰)	Calcification salinity	Calcification season	Calcification depth (m)
26.7	24.4	29.4	28.7	−2.3	4.90	28.5	0.22	34.43	July–Sept	0–10
27.0	24.6	29.7	28.0	5.9	4.60	26.6	0.27	34.50	May–Oct	0–30
26.4	24.1	29.0	27.3	7.3	4.30	26.3	0.19	34.41	May–Oct	0–30
27.0	24.6	29.6	27.1	12.9	4.24	27.3	0.01	34.01	June–Sept	0–30
25.3	23.2	27.9	25.7	11.1	3.75	25.1	0.17	34.17	May–Oct	0–30
24.3	22.3	26.9	24.5	12.7	3.37	25.3	−0.20	33.76	June–Sept	0–30
26.6	24.3	29.3	27.9	4.2	4.55	25.9	0.37	34.54	May–Oct	0–30
27.5	25.1	30.2	29.0	1.7	5.01	26.5	0.39	34.43	May and Oct	0–30
27.1	24.7	29.8	28.6	2.1	4.84	26.9	0.21	34.38	May–Oct	0–30
24.9	22.9	27.6	25.2	12.7	3.57	25.6	−0.14	33.68	June–Sept	0–20

Decreases in Mg/Ca ratios due to calcite dissolution were calculated using critical thresholds for the dissolution (14 μmol kg^{-1} $\Delta[CO_3^{2-}]$) and sensitivity of Mg/Ca ratio to $\Delta[CO_3^{2-}]$ (0.05 mmol mol^{-1} per μmol^{-1} kg^{-1}) (Johnstone et al. 2011). Core-top Mg/Ca and $\delta^{18}O$ data from A7, MD012403, MD012404, KY07-04-01 were from Sun et al. (2005), Lin et al. (2006), Chen et al. (2010), and Kubota et al. (2010), respectively

[a]SST = ln(Mg/Ca/0.38)/0.089, Hastings et al. (2001)

[b]SST = ln(Mg/Ca/0.34)/0.102, Anand et al. (2003)

[c]SST = ln(Mg/Ca/0.3)/0.089, Lea et al. (2000)

(Epstein and Mayeda 1953). The stable isotope ratios are given as conventional δ values (‰), and the analytical precision was ±0.02 ‰ (1σ) for $\delta^{18}O$. Oxygen isotopic composition is expressed relative to Vienna Standard Mean Ocean Water (VSMOW).

$\delta^{18}O$ and Mg/Ca analyses of *G. ruber*

Core-top sediments (0–1 cm) were wet washed through a 63 μm sieve and then dried in an oven at 50 °C. To minimize ontogenic and growth rate effects on shell geochemistry, specimens of *G. ruber* (sensu stricto form; 20–30 shells) were handpicked from the 250–355 μm size fraction. Planktonic foraminifera were gently crushed and rinsed three times with ultrapure water and methanol (super special grade, Wako Pure Chemical Industries, Ltd.) to remove adherent clay particles. The shells were then split into two fractions for $\delta^{18}O_c$ and metal/Ca measurements. $\delta^{18}O_c$ values were obtained using a Finnigan MAT 253 mass spectrometer at the Center for Advanced Marine Core Research (CMCR), Kochi University and calibrated in accordance with standard NBS 19. The precision of these measurements was better than ±0.08 ‰ (1σ) for $\delta^{18}O_c$.

The samples for metal/Ca analysis were cleaned according to the procedure developed for trace element analysis (Boyle and Keigwin 1985; Rosenthal et al. 1997). In brief, samples underwent a multistep process consisting of initial rinses in ultrapure water and methanol, followed by treatments with hot reducing and oxidizing solutions, transferred into new acid-leached microcentrifuge tubes (1.5 mL), and finally leached with a dilute ultrapure nitric acid solution (0.001 M HNO_3, TAMAPURE-AA-100 from Tama Chemicals, Ltd.). The sample solution was then dissolved with a Sc-spiked

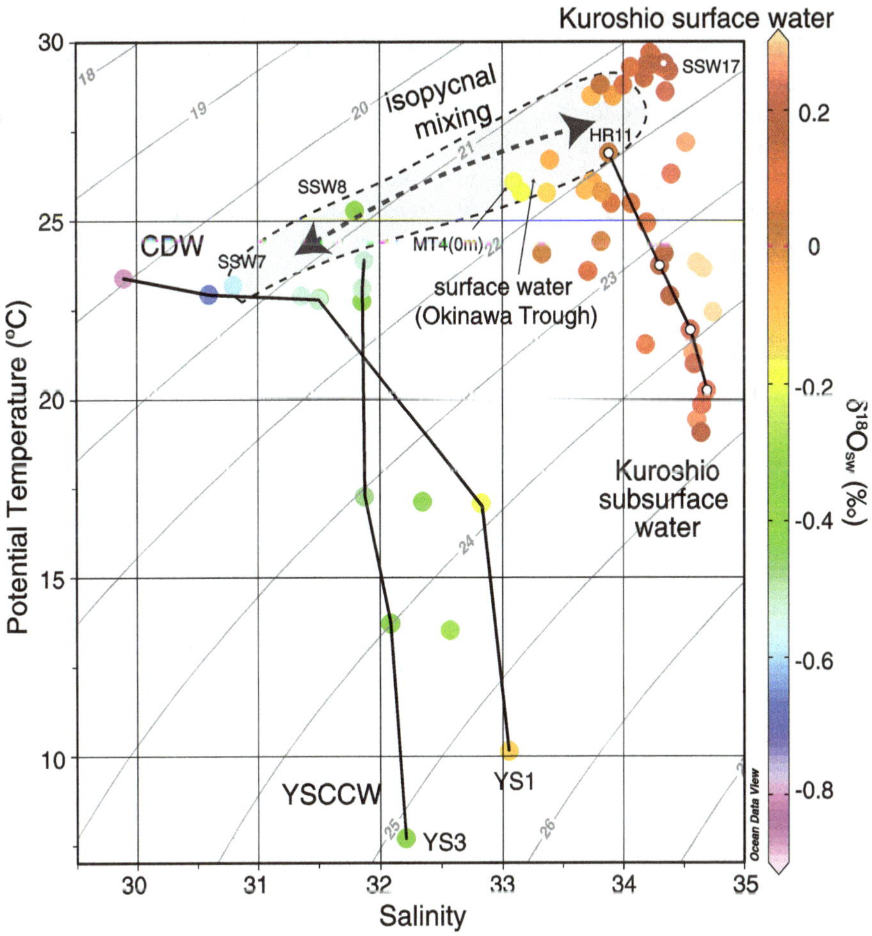

Fig. 3 Potential temperature and salinity (*T–S*) plot in the ECS. The *T–S* diagram shows four distinct water masses in early July: the Kuroshio surface water, Kuroshio subsurface water, YSCCW, and CDW. The *color in circles* indicates the $\delta^{18}O_{sw}$ value of the seawater. The Kuroshio surface and subsurface waters are characterized by higher $\delta^{18}O_{sw}$ values, whereas the CDW is characterized by relatively low $\delta^{18}O_{sw}$ values. Surface water properties in the northern part of the Okinawa Trough can be explained by isopycnal mixing with the CDW and Kuroshio surface water. The potential density curves (σ_θ = 19–27) are also shown. The *T–S* diagram was made using Ocean Data View (Schlitzer R, Ocean Data View, http://odv.awi.de, 2014)

dilute ultrapure nitric acid solution (2 % HNO_3) to obtain a Ca concentration of 10 ± 1 μg g^{-1}. All clean work was conducted in laminar flow benches or a clean room under trace metal clean conditions.

The metal/Ca ratio was determined with a Thermo Scientific ELEMENT 2 sector field inductively coupled plasma mass spectrometer (ICP–MS) at the University of Toyama, operated in a low-resolution mode ($m/\Delta m$ = 300) (Marchitto 2006). We analyzed ^{24}Mg, ^{26}Mg, ^{43}Ca, ^{44}Ca, ^{88}Sr, ^{137}Ba, and ^{138}Ba during the same run to determine Mg/Ca for temperature reconstructions and measured ^{56}Fe (in a mid-resolution mode) and ^{55}Mn to monitor contamination by clay minerals and diagenetic coatings (Table 2). Element counts were converted into molar ratios by the intensity ratio method based on a series of matrix-matched standard solutions. The accuracy and precision of Mg/Ca ratios were confirmed by analyses of the $CaCO_3$ reference materials BAM RS3 and ECRM 752-1. The repeated analyses gave Mg/Ca ratios of 0.786 ± 0.008 (n = 100, 1σ, RSD = 1.0 %) and 3.92 $\pm$ 0.06 (n = 24, 1σ, RSD = 1.5 %), respectively, which were within the reported value for BAM RS3 (0.791 ± 0.03) and slightly higher than the reported value for ECRM 752-1 (3.824 ± 0.095) (Greaves et al. 2008). Samples often show high Mn/Ca ratios (>100 μmol mol^{-1}) due to the presence of diagenetic coatings that were not removed during the cleaning process. Such samples were rejected. To further minimize the impact of sample heterogeneity and the analytical error of metal/Ca ratios, we report replicate measurements of metal/Ca ratios on re-picked *G. ruber* shells (Table 2).

Results and discussion

$\delta^{18}O_{sw}$–salinity relationship in the eastern ECS

In early July, surface water samples showed the highest SSS (34.3) at site SSW17 in the Kuroshio water area and the lowest SSS (29.9) at YS1 in the Yellow Sea (Fig. 2b and Table 1). The $\delta^{18}O_{sw}$ ranged from −0.81 to 0.21 ‰, and relatively low $\delta^{18}O_{sw}$ values were observed in the Yellow Sea, the ECS shelf, and the northern part of the Okinawa Trough (Fig. 2c). Spatial distributions of salinity and $\delta^{18}O_{sw}$ values were almost similar to the summer (July–September) salinity distribution averaged statistically over 1955–2012 (Zweng et al. 2013) (Fig. 2a), suggesting that the CDW spread east- and northeastward in the southern Yellow Sea and the northern Okinawa Trough during the sampling period.

A temperature and salinity plot (T–S diagram) of the surface and subsurface waters (0–100 m) in the Yellow Sea and the ECS shows four typical water masses: (1) the Kuroshio surface water in the subtropical Pacific Ocean (e.g., SSW14–17), (2) the Kuroshio subsurface

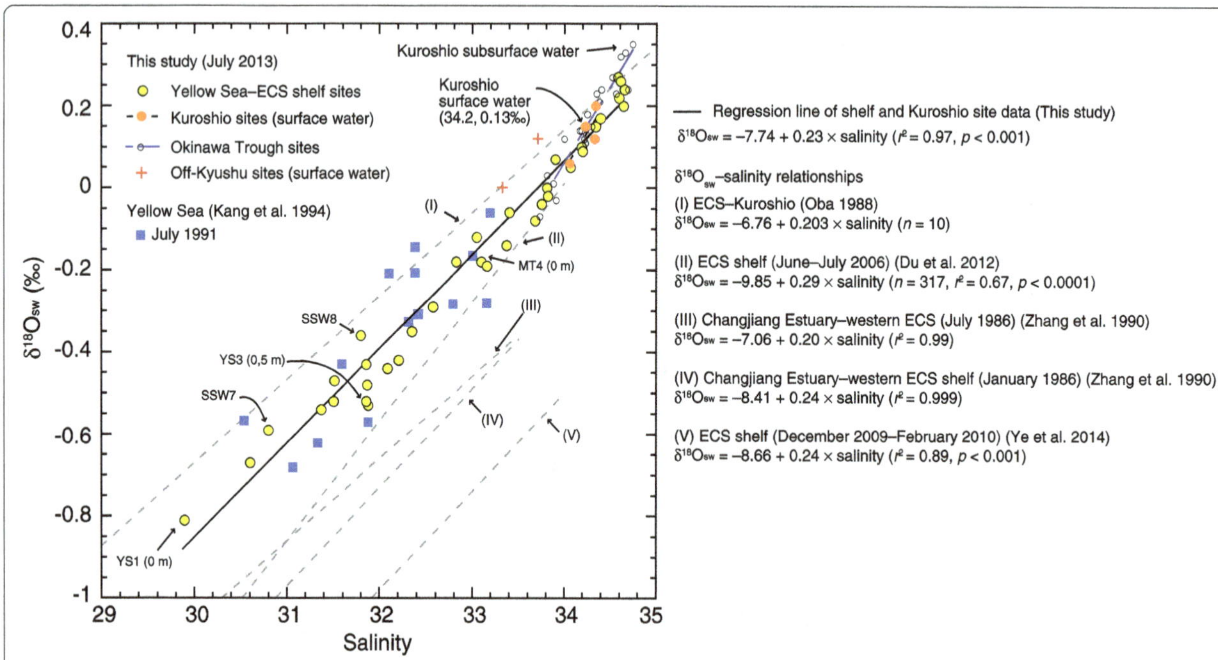

Fig. 4 $\delta^{18}O_{sw}$–salinity plot of shallow seawaters in the ECS. $\delta^{18}O_{sw}$–salinity plot of all shallow seawater (0–100 m) data in this study (early July 2013). The data are classified into four site groups: Yellow Sea–ECS shelf, Kuroshio, Okinawa Trough, and off-Kyushu sites (Table 1). The *black* regression line is derived from the data in the Yellow Sea–ECS shelf and Kuroshio sites. Since the surface waters in the northern Okinawa Trough are a mixture of the CDW and Kuroshio surface water, $\delta^{18}O_{sw}$ and salinity data from northern Okinawa Trough sites fall on this regression line. Regression lines (*dotted lines*) and equations from the literature are also shown. Regression lines III, IV, and V were developed in mainly coastal and shelf waters near the Changjiang Estuary. Although these regression lines represent freshwater end-members that are similar to our regression line, the end-member of the seawater shows lower $\delta^{18}O_{sw}$ values than that of the Kuroshio surface water (34.2 and 0.13 ‰; SSW14–17)

water in the Okinawa Trough (identified by a salinity of ~34.4 and a $\delta^{18}O_{sw}$ of 0.2–0.5 ‰) (Kang et al. 1994; Kim et al. 2005), (3) the YSCCW (<10 °C) in the bottom water (30 m) at YS3, and (4) the less saline CDW in the surface water at YS1 (Fig. 3). The *T–S* diagram also shows that the majority of the shallow water data in the Okinawa Trough fall on the isopycnal mixing line connecting the CDW and Kuroshio surface water (Fig. 3). Indeed, the $\delta^{18}O_{sw}$ values and salinity of the shallow seawater (0–100 m) are strongly correlated, indicating that the data can be explained by a binary mixing of freshwater and saline seawater (Fig. 4).

To derive a local $\delta^{18}O_{sw}$–salinity relationship around the northern Okinawa Trough, sampling sites were classified by geographic and oceanographic characteristics (Table 1). Sites SSW14–17, which are outside the ECS and under the influence of the Kuroshio surface water, were defined as Kuroshio sites. Data from the Yellow Sea and the ECS were divided into three groups: Yellow Sea–ECS shelf sites at depths shallower than 150 m, Okinawa Trough sites deeper than 150 m, and off-Kyushu sites influenced by local freshwater inputs from Kyushu (Fig. 4 and Table 1). Given that the eastward-flowing CDW mixes with the warmer, more saline Kuroshio water near the shelf edge in the summer (Chang and Isobe 2003), salinity and $\delta^{18}O_{sw}$ values in the Kuroshio surface water should reflect the end-member for saline water (Kang et al. 1994). Therefore,

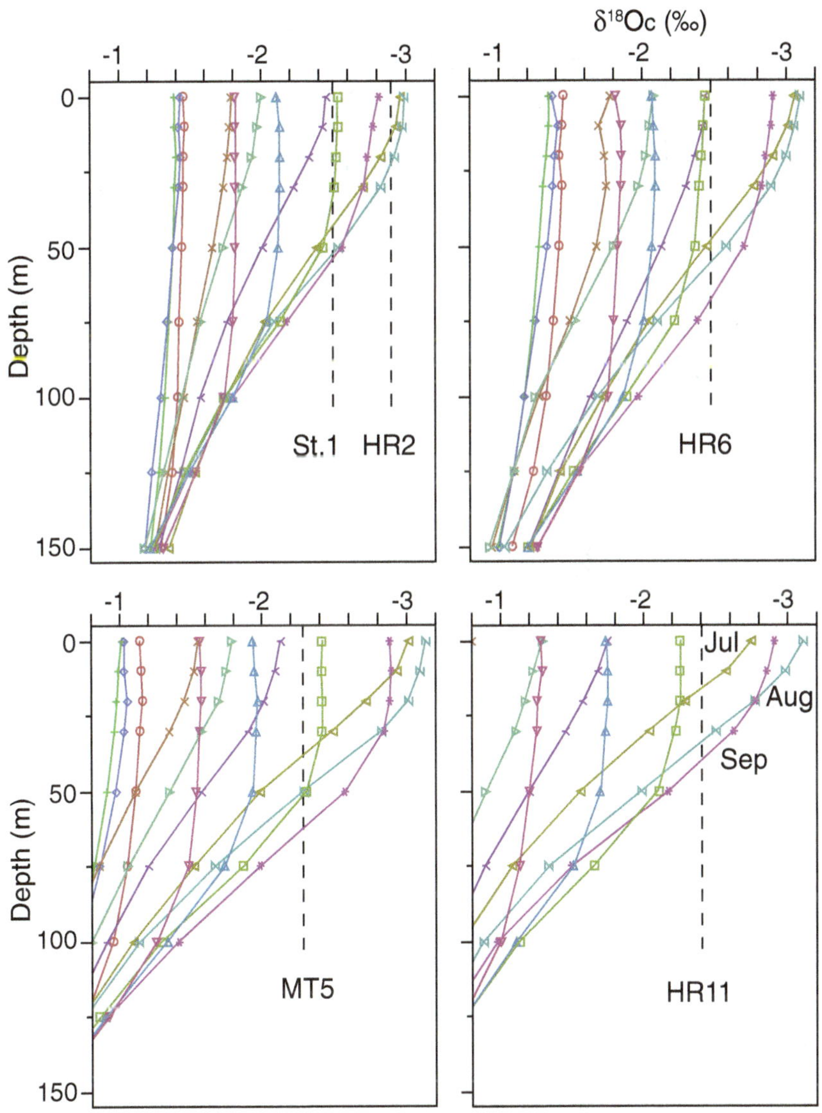

Fig. 5 Comparison of monthly variation in the predicted $\delta^{18}O_c$ values with the core-top $\delta^{18}O_c$ values. The *solid lines* show the monthly variation of the predicted $\delta^{18}O_c$ values within the water column. The *dotted lines* show the core-top $\delta^{18}O_c$ values of *G. ruber*. The predicted $\delta^{18}O_c$ values are calculated from temperature and salinity data (http://www.jodc.go.jp/) using a calcite equilibrium equation (Bemis et al. 1998). See details in the text

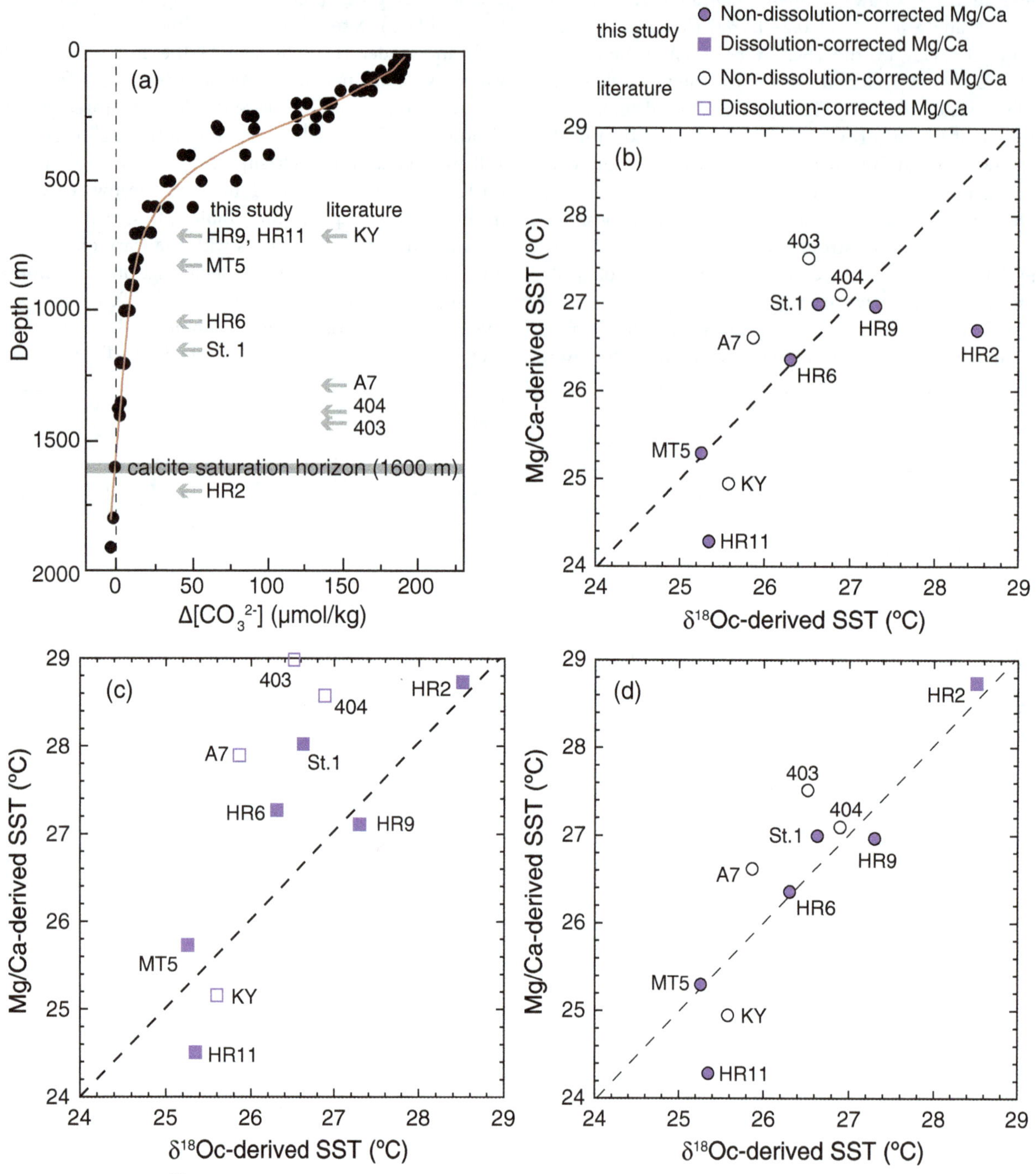

Fig. 6 Comparison of $\delta^{18}O_c$-derived calcification temperature with Mg/Ca-derived temperature. **a** Water column profile of $\Delta[CO_3^{2-}]$ in the ECS. Calculation of the calcite saturation state $\Delta[CO_3^{2-}]$ was accomplished via subtracting the carbonate ion concentration $[CO_3^{2-}]$ at saturation from the in situ $[CO_3^{2-}]$. To compute the in situ $[CO_3^{2-}]$ with the program CO2sys.xls (Lewis and Wallace 1998), we used total alkalinity, TCO_2, pH, and hydrographic water column data in the ECS (NODC Accession 0109919, http://data.nodc.noaa.gov/nodc/archive/metadata/approved/). *Arrows* indicate the water depths of cores analyzed in this study and by Sun et al. (2005), Lin et al. (2006), Chen et al. (2010), and Kubota et al. (2010). **b** $\delta^{18}O_c$-derived calcification temperature and non-dissolution-corrected Mg/Ca-derived temperature plot. **c** $\delta^{18}O_c$-derived calcification temperature and dissolution-corrected Mg/Ca-derived temperature plot. **d** $\delta^{18}O_c$-derived calcification temperature and Mg/Ca-derived temperature plot. The dissolution-corrected Mg/Ca ratio was used for the deepest site, HR2, where $\Delta[CO_3^{2-}] < 0$ μmol kg^{-1}. The other data were non-dissolution-corrected Mg/Ca ratios

taking into consideration this mixing of two water masses in the eastern part of the ECS, we derived the following $\delta^{18}O_{sw}$–salinity equation from the summer Yellow Sea–ECS shelf and Kuroshio site data:

$$\delta^{18}O_{sw} = -7.74\,(\pm 0.4) + 0.23\,(\pm 0.01) \times \textit{salinity}\,(r^2 = 0.97, p < 0.001, n = 40) \quad (1)$$

The quoted errors of the slope and intercept are 95 % confidence intervals. This equation includes the surface (0–30 m water depth) and subsurface (40–100 m water depth) data. The surface data (0–30 m) alone result in the same equation within the 95 % confidence intervals of Eq. (1).

Equation (1) indicates a projected freshwater end-member of −7.74 ‰ (±0.4 ‰). This value is consistent with the $\delta^{18}O$ values (−8.4 to −7.1 ‰) for Changjiang River water in January and July (Zhang et al. 1990). However, the regression lines of the $\delta^{18}O_{sw}$–salinity relationships developed for the Changjiang River mouth and the western part of the ECS indicate lower $\delta^{18}O_{sw}$ values in the salinity range of 30 to 34 [$\delta^{18}O_{sw} = -8.41 + 0.24 \times$ salinity in January (line IV of Fig. 4) and $\delta^{18}O_{sw} = -7.06 +$ $0.20 \times$ salinity in July (line III of Fig. 4)]. Similar regression lines were developed for the Yellow Sea and the ECS [$\delta^{18}O_{sw} = -8.66 + 0.24 \times$ salinity in winter (line V of Fig. 4) and $\delta^{18}O_{sw} = -10.7 + 0.27 \times$ salinity in summer] (Ye et al. 2014). These regression lines do not pass through the end-member values of the Kuroshio surface water (salinity = 34.2–34.4 and $\delta^{18}O_{sw}$ = 0.1–0.2 ‰), indicating that the Kuroshio surface water was not the source of saline water for these areas (Fig. 4). Since these data were mainly from coastal and shelf waters near the Changjiang Estuary, the less saline, $\delta^{18}O$-depleted TWWC may be the dominant source of saline water rather than the Kuroshio water. Therefore, $\delta^{18}O_{sw}$–salinity relationships in the coastal ECS are highly variable and differ from our equation.

Like our regression line, regression lines that pass through the saline Kuroshio surface water were derived for the ECS shelf covering a broad sampling area for June–July (line II of Fig. 4) (Du et al. 2012), for the southern Yellow Sea in July (Kang et al. 1994), and for the area from the ECS to off the southern Japan Islands (Oba 1990) (line I of Fig. 4). Our regression line occupied an intermediate area between these regression lines. Furthermore, our equation is derived from the data obtained in the eastern part of the ECS where the eastward-flowing CDW mixes with the Kuroshio water in the summer. Therefore, our $\delta^{18}O_{sw}$–salinity equation should be representative of the northern Okinawa Trough during the EASM season.

Even if the freshwater end-member changed from −7.74 to −9.0 ‰ (given the averaged *y*-intercept of lines II, IV, and V as the potential change in the riverine $\delta^{18}O$), this change in amplitude does not yield a significant difference between the two calibration lines within the salinity range of 33 to 34.2. In addition, since end-member values of the CDW and the Kuroshio surface water may not have changed significantly during the Holocene in comparison with the glacial–interglacial cycles (Wang et al. 2008), our local $\delta^{18}O_{sw}$–salinity equation may provide an approximate relationship between the salinity and $\delta^{18}O_{sw}$ in the eastern part of the ECS, at least during the Holocene.

Core-top $\delta^{18}O_c$ and Mg/Ca ratios for *G. ruber*

The core-top $\delta^{18}O_c$ values for *G. ruber* ranged from −2.9 to −2.3 ‰ in the central to northern parts of the Okinawa Trough (Fig. 1 and Table 2). The lightest value was observed at HR2, the southernmost site, whereas the heaviest value was found at MT5 near the CDW advection area. Core-top Mg/Ca ratios of *G. ruber* ranged from 3.29 to 4.35 mmol mol^{-1} (Table 2). The Mg/Ca value at HR2 (4.09 mmol mol^{-1}, 1675 m water depth), which is the deepest site, was lower than that at the nearby shallower site St.1 (4.20 mmol mol^{-1}, 1153 m), even though HR2 had the lowest $\delta^{18}O_c$ value of *G. ruber* (−2.9 ‰) (Table 2). Since site HR2 is below the CSH (~1600 m) (Fig. 6a), the Mg/Ca ratio at HR2 seems to be lowered due to calcite dissolution (discussed below). Core-top Mg/Ca ratios, including the data from the literature (Sun et al. 2005; Lin et al. 2006; Chen et al. 2010; Kubota et al. 2010), tend to be higher in the southern sites (e.g., HR2 and St.1) and lower in the northern sites (HR11 and MT5) (Table 2), which is consistent with the distribution of the summer SST. In contrast, core-top $\delta^{18}O_c$ values correlate poorly with the Mg/Ca ratios ($r^2 = 0.19$), indicating that $\delta^{18}O_c$ values are strongly influenced by $\delta^{18}O_{sw}$.

Calcification season and calcification temperature of *G. ruber*

The foraminifera shell Mg/Ca response to temperature is biologically modulated and dependent on pH and/or [CO_3^{2-}] (Lea et al. 1999; Russell et al. 2004). Therefore, many species- and basin-specific Mg/Ca–temperature equations have been developed, some of which take into consideration the potential effect of dissolution (Lea et al. 2000; Dekens et al. 2002; Anand et al. 2003; McConnell and Thunell 2005). There is, however, no equation for *G. ruber* in the ECS, and previous studies conducted in the ECS have converted Mg/Ca values of *G. ruber* to SST using a Mg/Ca–temperature calibration equation [T (°C) = ln(Mg/Ca/0.38)/0.089] developed for *G. ruber* based on the SCS core-top sediments by Hastings et al. (2001) (Sun et al. 2005; Lin et al. 2006; Chen et al. 2010; Kubota et al. 2010). Although this calibration equation yields temperatures corresponding to the warm summer months, it has not been fully investigated whether there is consistency between $\delta^{18}O_c$- and Mg/Ca-derived temperatures, even though reconciled estimates of temperatures are crucial for calculating $\delta^{18}O_{sw}$.

First, to calculate calcification temperatures ($\delta^{18}O_c$-derived SSTs), we identified the calcification season and depth based on a comparison of core-top $\delta^{18}O_c$ values with predicted $\delta^{18}O_c$ values. Predicted $\delta^{18}O_c$ values were calculated using the following $\delta^{18}O_c$–temperature equation assuming oxygen isotopic equilibrium in foraminiferal calcite:

$$T(^{\circ}C) = 14.9\text{–}4.8\left(\delta^{18}O_c\text{–}\left(\delta^{18}O_{sw}\text{–}0.27‰\right)\right) \quad (2)$$

This $\delta^{18}O_c$–temperature equation was determined from culturing experiments of a symbiont-bearing species, *Orbulina universa*, grown under high-light conditions (Bemis et al. 1998). The high-light conditions (>380 μEinstein m^{-2} s^{-1}) in the culturing experiments take into consideration symbiont effects on shell $\delta^{18}O_c$ (i.e., high photosynthetic activity forms ^{18}O-depleted shells). The applicability of this equation to the surface-dwelling, symbiont-bearing species *G. ruber* has been confirmed (Bemis et al. 1998; Benway et al. 2006). Temperature and salinity data averaged statistically for 1930–2003 in a 1° × 1° (latitude × longitude) grid provided by JODC were used for the calculation. Salinity was converted to $\delta^{18}O_{sw}$ values using Eq. (1). The term −0.27 ‰ corrects for the $\delta^{18}O$ difference between VSMOW and Pee Dee Belemnite (PDB).

Figure 5 shows a comparison of the predicted vertical profiles of $\delta^{18}O_c$ values for each month with core-top $\delta^{18}O_c$ values of *G. ruber* at each site. Among the six core-top datasets, HR2 and HR9 showed relatively lighter $\delta^{18}O_c$ values that are approximately equal to the $\delta^{18}O_c$ values in the surface water for July–September. In contrast, the other sites (St.1, HR6, MT5, and HR11) were marked by isotopically heavier values than the peak summer $\delta^{18}O_c$ in the surface water, suggesting a large contribution of *G. ruber* shells formed in deeper waters (>50 m) or in surface water in May, June, and/or October (Fig. 5). Since *G. ruber* (sensu stricto) lives predominantly in the top 20 m of the subtropical North Pacific (Kuroyanagi and Kawahata 2004), calcification of shells in deeper water (>50 m) should be negligible. Furthermore, sediment trap data from the ECS, which unfortunately do not include data for September, show that the fluxes of *G. ruber* from May to October account for 80 % of the total fluxes each year and the fluxes in May alone account for 20 % of the yearly fluxes (Xu et al. 2005). Based on these data, we regard the dominant habitat depth of *G. ruber* in the ECS as the top 0 to 30 m (mixed layer) and contend that the *G. ruber* shell geochemistry in the ECS records the weighted mean surface ocean conditions from May to October.

To calculate the calcification temperatures and salinity of the seawater in which *G. ruber* shells were formed, we computed seasonal average (May–October or July–September) $\delta^{18}O_c$ values and compared these predicted values to core-top $\delta^{18}O_c$ values (Table 2). The calcification seasons and depths at HR2 and HR11 were identified as July–September and 0–10 m and June–September and 0–30 m, respectively. The $\delta^{18}O_c$-derived SSTs at HR2 and HR11 were calculated as 28.5 and 25.3 °C, respectively (Fig. 6 and Table 2). The calculated calcification SSTs at all sites ranged from 25.1 to 28.5 °C in the central to northern Okinawa Trough (Table 2). It is noteworthy that the calcification season of *G. ruber* was not necessarily uniform across the ECS; data from sites HR2 and HR9 clearly show a warmer seasonal bias (close to the peak summer SST), in contrast to other site data (Figs. 1 and 5). The core-top $\delta^{18}O_c$ data shows that Eq. (2) provides better estimates for calcification depth, season, and temperature that agree with sediment trap and plankton tow data (Kuroyanagi and Kawahata 2004; Xu et al. 2005).

Although we applied the $\delta^{18}O_c$–temperature equation by Bemis et al. (1998) to the core-top data, there are other $\delta^{18}O_c$–temperature relationships that yield various estimates for calcification temperature. For instance, $\delta^{18}O_c$–temperature equations by Kim and O'Neil (1997) and Mulitza et al. (2003) yield an offset of approximately 1 °C to the calcification temperatures calculated by Eq. (2). If these equations are applied to our core-top data with the assumption of $\delta^{18}O$ equilibrium, calcification temperatures at HR2 and HR9 will exceed the peak summer SST. Therefore, we should consider $\delta^{18}O$ disequilibrium in foraminiferal calcite (i.e., vital effect) for at least these sites. However, since the vital effect for *G. ruber* has not been determined precisely (0 to −1.0 ‰) (Niebler et al. 1999), applying these equations will introduce significant uncertainties in calcification temperature estimates.

Assessment of *G. ruber* Mg/Ca-derived temperatures

G. ruber is one of the species of planktonic foraminifera most susceptible to dissolution (Thunell and Honjo 1981). Partial dissolution of *G. ruber* shells can decrease Mg/Ca ratios owing to the higher solubility of Mg-rich calcite (Dekens et al. 2002). The ECS is the most undersaturated marginal sea with respect to calcite, and the CSH is shallow (~1600 m) (Fig. 6a). In our dataset, site HR2 is below the CSH, and the published core-top data from site A7, MD012404 (hereafter 404), and MD012403 (hereafter 403) were also derived from bottom water conditions with low calcite saturation states of <5 μmol kg^{-1} ($\Delta[CO_3^{2-}] = [CO_3^{2-}]_{in\ situ} - [CO_3^{2-}]_{saturation}$) (Fig. 6a; see caption of Fig. 6a for more details). According to data from core-top *G. ruber* tests of a depth transect in the Ontong Java Plateau, Mg/Ca ratios of *G. ruber* started to decrease with decreasing $\Delta[CO_3^{2-}]$ (−0.05 mmol mol^{-1} per μmol^{-1} kg^{-1}) from an initial $\Delta[CO_3^{2-}]$ of 14 μmol kg^{-1} (Johnstone et al. 2011). This finding is a cause for concern that ECS Mg/Ca data obtained under bottom water conditions with low $\Delta[CO_3^{2-}]$ may be influenced by dissolution (Fig. 6a).

Therefore, we assessed the impact of calcite dissolution on Mg/Ca ratios in the ECS by applying critical thresholds for the dissolution (14 μmol kg^{-1} $\Delta[CO_3^{2-}]$) and the sensitivity of the Mg/Ca ratio to $\Delta[CO_3^{2-}]$ (0.05 mmol mol^{-1} per μmol^{-1} kg^{-1}) (Johnstone et al. 2011). We calculated $\Delta[CO_3^{2-}]$ for the bottom waters at each core site and estimated the dissolution-corrected Mg/Ca ratios for all core sites (Table 2). Figure 6 shows a comparison of the $\delta^{18}O_c$-derived SSTs with non-dissolution-corrected and dissolution-corrected Mg/Ca-derived SSTs (Fig. 6b, c). Both Mg/Ca ratios were converted to SSTs using the SCS calibration equation of Hastings et al. (2001).

The shallowest site, HR9, where the $\Delta[CO_3^{2-}]$ value is 12.9 μmol kg^{-1}, showed good agreement with the non-dissolution-corrected Mg/Ca- and $\delta^{18}O_c$-derived SSTs (Fig. 6b). It is noteworthy that core-top data at depths ranging from 1000 to ~1400 m (HR6, St.1, A7, 404, and 403) also showed agreement with the non-dissolution-corrected Mg/Ca- and $\delta^{18}O_c$-derived SSTs, even though the $\Delta[CO_3^{2-}]$ values were lower than the critical threshold for dissolution (10–20 μmol kg^{-1}) (Regenberg et al. 2009; Johnstone et al. 2011) (Fig. 6b). The deepest site, HR2 (1675 m water depth), showed that the non-dissolution-corrected Mg/Ca derived SST underestimates the $\delta^{18}O_c$-derived SST by 1.8 °C (Fig. 6b), and the dissolution-corrected Mg/Ca-derived SST agrees well with the $\delta^{18}O_c$-derived SST (Fig. 6c). However, this dissolution correction induced a larger offset to the $\delta^{18}O_c$-derived SSTs for other sites (1000 to ~1400 m water depths) (Fig. 6c).

The Mg/Ca–temperature calibration equation of Hastings et al. (2001) was developed for *G. ruber* from SCS core-top samples recovered from depths above 2000 m. As in the ECS, the deep waters in the SCS are characterized by low $\Delta[CO_3^{2-}]$ with the CSH at a water depth of 2500 m (Regenberg et al. 2009). Presumably, samples recovered from depths of 1500 to 2000 m (<15 μmol kg^{-1} $\Delta[CO_3^{2-}]$) for the calibration equation might have been influenced by partial dissolution under low $\Delta[CO_3^{2-}]$. Therefore, we consider that the SCS calibration equation potentially involves a correction for calcite dissolution under low $\Delta[CO_3^{2-}]$. Consequently, the correction for calcite dissolution is required only for samples obtained below the CSH in the ECS.

Thus, in this study, we integrated data of dissolution-corrected (for HR2 from below the CSH) and non-dissolution-corrected (for other cores above the CSH) Mg/Ca values to obtain Mg/Ca-derived SSTs (Fig. 6d). Although HR11 and 403 are still offset from the $\delta^{18}O_c$-derived SST by 1 °C, the other core-top data showed good agreement with both SST estimates ($r^2 = 0.83$, $p < 0.001$, except HR11 and 403) (Fig. 6d). This SCS calibration equation reconciles Mg/Ca-derived and $\delta^{18}O_c$-derived SSTs in the ECS, both of which reflect SSTs in the warmer season months. However, other calibration equations (Lea et al. 2000; Anand et al. 2003; McConnell and Thunell 2005) do not yield temperatures corresponding to the warmer season months and do not agree well with $\delta^{18}O_c$-derived SSTs (Table 2).

$\delta^{18}O_{sw}$ estimate for *G. ruber* in core-top sediments

Core-top $\delta^{18}O_{sw}$ values were calculated from Mg/Ca-derived SSTs and core-top $\delta^{18}O_c$ values using Eq. (2). The core-top data of the shell-derived $\delta^{18}O_{sw}$ and the salinity, including the published data from KY, A7, 404, and 403 (Sun et al. 2005; Lin et al. 2006; Chen et al. 2010; Kubota et al. 2010), are plotted in Fig. 7a. Since the estimation of the shell-derived $\delta^{18}O_{sw}$ is affected by the uncertainty in the $\delta^{18}O_c$ (±0.11 ‰) and the Mg/Ca-derived temperature (±1 °C, corresponding to approximately ±0.26 ‰) (Schmidt 1999; Rohling 2007), the reconstructed $\delta^{18}O_{sw}$ introduces a large propagation error of ±0.28 ‰ (1σ). Although the regression line of the core-top $\delta^{18}O_{sw}$ data has a steeper gradient and a lower *y*-intercept, given 1σ error bars of $\delta^{18}O_{sw}$ estimates the data points fall primarily on the local $\delta^{18}O_{sw}$–salinity regression line of the eastern part of the ECS (Fig. 7a). Indeed, sites KY and HR11 in the northern Okinawa Trough were marked by the lowest shell-derived $\delta^{18}O_{sw}$ values, whereas St.1 and A7 in the central Okinawa Trough showed the highest shell-derived $\delta^{18}O_{sw}$ values. Sites HR9 and MT5 had intermediate $\delta^{18}O_{sw}$ values between the above two datasets (Fig. 7a). The core-top $\delta^{18}O_{sw}$ values were roughly distributed as expected from the peak summer SSS at each site.

Although site MT5 is in an area with a lower summer SSS than that of site HR9, the shell-derived $\delta^{18}O_{sw}$ value at MT5 (0.17 ‰) was higher than that at HR9 (0.01 ‰) (Fig. 7a). One potential explanation is that *G. ruber* shells at HR9 were formed in the surface waters during the peak summer season when SSS decreases distinctly, whereas *G. ruber* at MT5 was not abundant in the peak summer season, as inferred from the $\delta^{18}O_c$ data (Fig. 5). These data tell us that the difference of 0.16 ‰ in the shell-derived $\delta^{18}O_{sw}$ may be caused by changes in the seasonal habitat of *G. ruber*.

This core-top investigation shows that realistic $\delta^{18}O_{sw}$ values can be reconstructed by the Mg/Ca–temperature calibration of Hastings et al. (2001) and the $\delta^{18}O_c$–

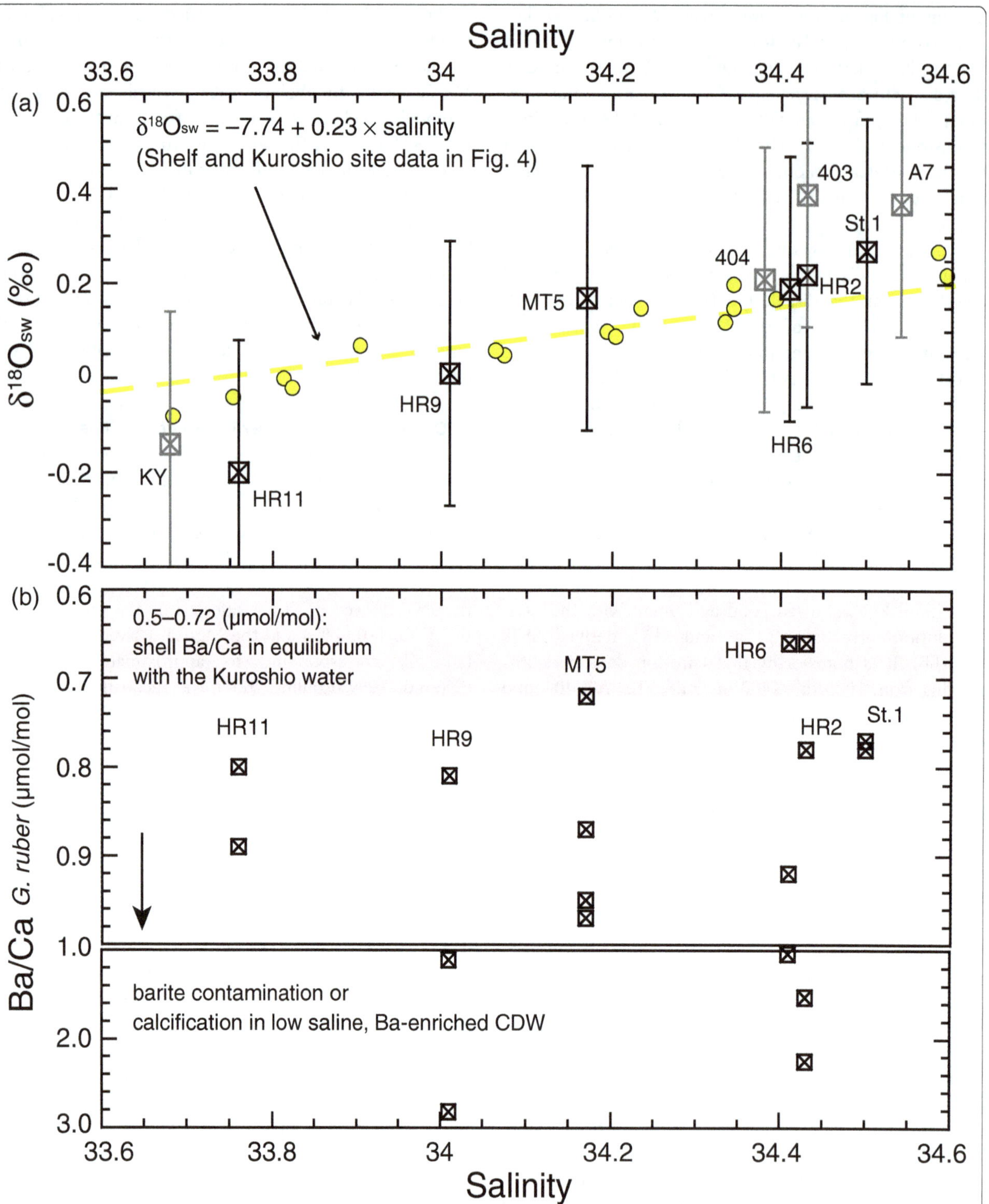

Fig. 7 Salinity, shell-derived $\delta^{18}O_{sw}$, and Ba/Ca ratios of core-top *G. ruber* tests. **a** Shell-derived $\delta^{18}O_{sw}$ and calcification salinity plot of core-top samples in this study and the literature. Shell-derived $\delta^{18}O_{sw}$ values are calculated using the Mg/Ca-derived temperature and core-top $\delta^{18}O_c$ of *G. ruber*. The data from KY, A7, 403, and 404 (*gray squares*) are from Kubota et al. (2010), Sun et al. (2005), Lin et al. (2006), and Chen et al. (2010), respectively. *Error bars* show the propagation error of ±0.28 ‰ (1σ). Seawater data (*green circles*) and their regression line (green broken line, Fig. 4) are also shown with the core-top data. Core-top data fall close to the local $\delta^{18}O_{sw}$–salinity regression line. **b** Ba/Ca$_{G.\ ruber}$ ratios and calcification salinity plot of all core-top samples in this study. The replicated analyses of Ba/Ca$_{G.\ ruber}$ ratios at each core site show high variability, resulting in the range from 0.66 to 2.82 μmol mol^{-1}. The calculated shell Ba/Ca ratio in equilibrium with the Kuroshio water is 0.5–0.72 μmol mol^{-1}. Lower values at HR2 and HR6 are within this range. Higher shell Ba/Ca ratios than this Kuroshio water range might be attributable to barite contamination or calcification in low saline, Ba-enriched CDW

temperature equation of Bemis et al. (1998), although $\delta^{18}O_{sw}$ estimates introduce a large error. Furthermore, this finding indicates that robust reconstruction of EASM-related variability in SSS might be possible by conducting multiple replicates of paired $\delta^{18}O_c$ and Mg/Ca analyses to decrease the large uncertainty. In addition, the utility of potential paleo-salinity proxies, such as the foraminiferal Ba/Ca ratio and the organic compound-specific δD, should be investigated further (Rohling 2007).

Possibility of using Ba/Ca ratios of *G. ruber* as a proxy for surface salinity

In general, the desorption of Ba^{2+} from suspended sediments in rivers results in a very high riverine [Ba^{2+}] relative to seawater. If there is a single river source, conservative mixing of river water and ambient seawater produces a linear inverse correlation between salinity and [Ba^{2+}] (Bahr et al. 2013). Ba^{2+} incorporation into living planktonic foraminifera shells is linearly dependent on the [Ba^{2+}] of the water, with a constant partition coefficient for Ba ($D_{Ba} = 0.15$) (Lea and Spero 1994; Hönisch et al. 2011). *G. ruber* populations can thrive in low-salinity surface water influenced by river discharge (Schmuker and Schiebel 2002; Arnold and Parker 2003). This ecological preference of *G. ruber* and the incorporation of Ba into foraminifera shells makes Ba/Ca ratios in fossil *G. ruber* a proxy for the influence of river discharge on salinity (Weldeab et al. 2007; Schmidt and Lynch-Stieglitz 2011).

In the case of the ECS, the Changjiang River accounts for approximately 90 % of the total river discharge into the ECS, with large quantities of Ba^{2+} (~360 nmol L^{-1}) (Beardsley et al. 1985; Qu et al. 1993; Yanagi 1994). Based on our preliminary [Ba^{2+}] data in the ECS measured by an isotope dilution method (Klinkhammer and Chan 1990), the surface water at SSW8 (30.80 salinity) showed a high [Ba^{2+}] of 57.0 nmol L^{-1} (Ba/Ca ratio of 6.31 μmol mol^{-1}), and the Kuroshio surface water (SSW14–17, average 34.2) showed a low [Ba^{2+}] of 32.9 nmol L^{-1} (Ba/Ca of 3.27 μmol mol^{-1}). The [Ba^{2+}] data for the Kuroshio water were consistent with previously reported data from Kuroshio waters (29.9–33.5 nmol L^{-1}) (Sugiyama et al. 1984). Other surface samples from the northern Okinawa Trough fall within these [Ba^{2+}] and Ba/Ca ranges, showing a linear relationship between Ba/$Ca_{seawater}$ and salinity (Horikawa et al., personal communication).

Our Ba/Ca ratios in *G. ruber* (hereafter Ba/$Ca_{G.\ ruber}$) from core-top sediments ranged from 0.66 to 2.82 μmol mol^{-1} (Fig. 7b). The lowest Ba/Ca value of 0.66 μmol mol^{-1} was observed at sites HR2 and HR6 in the central Okinawa Trough. Given the Ba/$Ca_{seawater}$ ratio of 3.27 μmol mol^{-1} in the Kuroshio surface water and the likely D_{Ba} of 0.15–0.22 (Hall and Chan 2004; Hönisch et al. 2011), the calculated shell Ba/Ca ratio in equilibrium with the Kuroshio water should be 0.5 to 0.72 μmol mol^{-1}. Indeed, HR2 and HR6 values were within the expected range for Kuroshio surface water. In contrast, HR11, where the lowest SSS among the studied sites was observed, had relatively higher Ba/$Ca_{G.\ ruber}$ ratios (0.80–0.89 μmol mol^{-1}) (Fig. 7b). Given the Ba/$Ca_{seawater}$ ratio of 3.8 μmol mol^{-1} (at HR10, 0 m) and the D_{Ba} of 0.15 to 0.22, the shell Ba/Ca ratio in equilibrium should be 0.57 to 0.84 μmol mol^{-1}. The core-top Ba/$Ca_{G.\ ruber}$ ratios observed at HR11 were consistent with the upper range of expected values, probably reflecting the influence of the Ba-enriched CDW (i.e., low-salinity waters).

However, all core-top Ba/$Ca_{G.\ ruber}$ data at six sites do not show a significant relationship with the salinity, but replicate data at each site showed highly variable Ba/$Ca_{G.\ ruber}$ ratios, some of which exceeded the values at site HR11 (Fig. 7b). Although we are unable to adequately explain such highly variable Ba/$Ca_{G.\ ruber}$ ratios at this time, one likely reason is sample heterogeneity. In this study, to obtain averaged Ba/$Ca_{G.\ ruber}$ ratios for each site, we re-picked 20–30 shells of *G. ruber* more than twice after picking the samples labeled run #1 (Table 2). Although the samples from run #1 gave reasonable Ba/$Ca_{G.\ ruber}$ ratios (0.66–0.89 μmol mol^{-1}) at each site except HR6, the samples from runs #2 to #4 yielded anomalously high Ba/$Ca_{G.\ ruber}$ ratios compared with the samples from run #1. Given that there were few well-preserved, clean *G. ruber* tests for these sites and that replicate measurements of Mn/Ca and Mg/Ca ratios were within an acceptable range, we argue that samples from runs #2 to #4 may have involved impure *G. ruber* tests in which barite crystallized on fossil shell surfaces.

Finally, based on the available dataset of Ba/$Ca_{seawater}$ in the ECS, we propose that pristine Ba/$Ca_{G.\ ruber}$ ratios at northern Okinawa Trough sites would be less than 0.84 μmol mol^{-1}. A possible reason for higher core-top Ba/$Ca_{G.\ ruber}$ ratios (>0.84 μmol mol^{-1}) is contamination by sedimentary barite adherent on fossil shells. Our preliminary data suggest that Ba/$Ca_{G.\ ruber}$ data as a paleo-salinity proxy should be derived from well-preserved, clean *G. ruber* tests or diethylene triamine pentaacetic acid (DTPA)-cleaned *G. ruber* tests that can minimize the influence of barite contamination (Lea and Boyle 1991; Hall and Chan 2004). Since the Ba/$Ca_{G.\ ruber}$ records from the northern Okinawa Trough sediments may potentially identify high river discharge events related to stronger EASMs, assessment of foraminiferal Ba/Ca should be continued in the ECS.

Conclusions

The main findings of this study are as follows:

(1) Shallow seawater samples (0–100 m) taken from the ECS in early July showed a strong correlation between $\delta^{18}O_{sw}$ and salinity. The T–S diagram in the surface waters in the ECS shows that the eastward-flowing CDW mixes with the saline Kuroshio waters. Based on this finding, we derived the following $\delta^{18}O_{sw}$–salinity equation from the Yellow Sea–ECS shelf and Kuroshio site data:

$$\delta^{18}O_{sw} = -7.74(\pm 0.4) + 0.23(\pm 0.01) \times salinity\left(r^2 = 0.97, p < 0.001, n = 40\right).$$

This local $\delta^{18}O_{sw}$–salinity equation might be representative of the northern Okinawa Trough during the EASM season.

(2) We found that the dominant habitat depth of *G. ruber* is within the top 0 to 30 m in the ECS. The calcification season of *G. ruber* is mainly during the warm summer months (May–October) but may not necessarily be uniform across the ECS. We confirmed that the Mg/Ca–temperature calibration by Hastings et al. (2001) yields temperatures corresponding to the warmer season months, as expected from sediment trap data. The Mg/Ca-derived SSTs agreed with calcification temperatures calculated by Bemis et al. (1998). Site HR2, where the bottom water is undersaturated with calcite, required a correction for calcite dissolution.

(3) We found that core-top data for shell-derived $\delta^{18}O_{sw}$ and salinity fall primarily on our local $\delta^{18}O_{sw}$–salinity regression line giving 1σ error of $\delta^{18}O_{sw}$ estimates. The Mg/Ca–temperature calibration by Hastings et al. (2001) and the $\delta^{18}O_c$–temperature equation by Bemis et al. (1998) should be appropriate for calculating $\delta^{18}O_{sw}$ in the ECS.

(4) $Ba/Ca_{G.\ ruber}$ from core-top sediments ranged from 0.66 to 2.82 μmol mol^{-1}. There was not a significant relationship between salinity and $Ba/Ca_{G.\ ruber}$ due to the highly variable $Ba/Ca_{G.\ ruber}$ data. Given the seawater Ba/Ca data and the published partition coefficient for Ba (D_{Ba} = 0.15–0.22), pristine $Ba/Ca_{G.\ ruber}$ ratios at northern Okinawa Trough sites should be less than 0.84 μmol mol^{-1}. One possible reason for higher $Ba/Ca_{G.\ ruber}$ ratios (>0.84 μmol mol^{-1}) is contamination by sedimentary barite adherent on fossil shells. Further evaluation of the $Ba/Ca_{G.\ ruber}$ ratio as a paleo-salinity proxy requires DTPA-cleaned Ba/Ca data that can minimize the influence of barite contamination.

Abbreviations

EASM: East Asian summer monsoon; ECS: East China Sea; CDW: Changjiang diluted water; SSS: sea surface salinity; TWWC: Taiwan warm current; SCS: South China Sea; YSCCW: Yellow Sea Central Cold Water; SST: sea surface temperature; JODC: Japan Oceanographic Data Center; CTD: conductivity, temperature, and depth; JGOFS: Joint Global Ocean Flux Study; IAPSO: International Association for the Physical Sciences of the Oceans; CSH: calcite saturation horizon; VSMOW: Vienna Standard Mean Ocean Water; CMCR: Center for Advanced Marine Core Research; PDB: Pee Dee Belemnite; DTPA: diethylene triamine pentaacetic acid.

Competing interests

The authors declare that they have no competing interests.

Authors' contributions

KH conceived and designed the study. TK carried out the measurements of the seawater $\delta^{18}O$, foraminiferal $\delta^{18}O_c$, and metal/Ca ratios under the supervision of Professors MM and JZ. All authors analyzed and discussed the data. All authors contributed toward writing the manuscript and read and approved the final manuscript.

Authors' information

KH is an associate professor at the University of Toyama. He is a paleoceanographer and a geochemist who has worked on SST reconstruction around Japan using the Mg/Ca–temperature proxy and the analysis of deep-sea water mass circulation using neodymium isotopes. JZ is also a geochemist and a professor at the University of Toyama with considerable experience in geochemical analysis. TK is a Ph.D. student supervised by KH and JZ. MM is a professor at the Center for Advanced Marine Core Research (CMCR), Kochi University.

Acknowledgements

We acknowledge the tremendous support of the KH-13-4 and KT-12-25 shipboard scientists and staff. This study was performed under the cooperative research program of the Center for Advanced Marine Core Research (CMCR), Kochi University (#13B050). We thank the chief editors Ryuji Tada, Hodaka Kawahata, Yasufumi Iryu, and two anonymous reviewers for their significant help in improving this manuscript.

Author details

[1]Graduate School of Science and Engineering for Research, University of Toyama, Gofuku 3190, Toyama 930-8555, Japan. [2]Graduate School of Science and Engineering for Education, University of Toyama, Gofuku 3190, Toyama 930-8555, Japan. [3]Center for Advanced Marine Core Research, Kochi University, B200 Monobe, Nankoku 783-8502, Japan.

References

Anand P, Elderfield H, Conte MH (2003) Calibration of Mg/Ca thermometry in planktonic foraminifera from a sediment trap time series. Paleoceanography 18(2):1050. doi:10.1029/2002PA000846

Arnold A, Parker W (2003) Biogeography of planktonic Foraminifera. In: Gupta BKS (ed) Modern Foraminifera. Springer, Netherlands, pp 103–122. doi:10.1007/0-306-48104-9_7

Bahr A, Schönfeld J, Hoffmann J, Voigt S, Aurahs R, Kucera M, Flögel S, Jentzen A, Gerdes A (2013) Comparison of Ba/Ca and $\delta^{18}O_{WATER}$ as freshwater proxies: A multi-species core-top study on planktonic foraminifera from the vicinity of the Orinoco River mouth. Earth Planet Sci Lett 383:45–57. doi:10.1016/j.epsl.2013.09.036

Beardsley RC, Limeburner R, Yu H, Cannon GA (1985) Discharge of the Changjiang (Yangtze River) into the East China Sea. Cont Shelf Res 4(1–2):57–76. doi:10.1016/0278-4343(85)90022-6

Bemis BE, Spero HJ, Bijima J, Lea DW (1998) Reevaluation of the oxygen isotopic composition of planktonic foraminifera: Experimental results and revised paleotemperature equations. Paleoceanography 13(2):150–160

Benway HM, Mix AC, Haley BA, Klinkhammer GP (2006) Eastern Pacific Warm Pool paleosalinity and climate variability: 0–30 kyr. Paleoceanography 21(3):PA3008. doi:10.1029/2005PA001208

Boyle EA, Keigwin LD (1985) Comparison of Atlantic and Pacific paleochemical records for the last 215,000 years: changes in deep ocean circulation and chemical inventories. Earth Planet Sci Lett 76(1–2):135–150

Chang P-H, Isobe A (2003) A numerical study on the Changjiang diluted water in the Yellow and East China Seas. J Geophys Res 108(C9):3299. doi:10.1029/2002JC001749

Chang Y-P, Chen M-T, Yokoyama Y, Matsuzaki H, Thompson WG, Kao S-J, Kawahata H (2009) Monsoon hydrography and productivity changes in the East China Sea during the past 100,000 years: Okinawa Trough evidence (MD012404). Paleoceanography 24(3):PA3208. doi:10.1029/2007PA001577

Chen C, Beardsley RC, Limeburner R, Kim K (1994) Comparison of winter and summer hydrographic observations in the Yellow and East China Seas and adjacent Kuroshio during 1986. Cont Shelf Res 14(7–8):909–929. doi:10.1016/0278-4343(94)90079-5

Chen F, Yu Z, Yang M, Ito E, Wang S, Madsen DB, Huang X, Zhao Y, Sato T, Birks HJB, Boomer I, Chen J, An C, Wünnemann B (2008) Holocene moisture evolution in arid central Asia and its out-of-phase relationship with Asian monsoon history. Quat Sci Rev 27(3–4):351–364. doi:10.1016/j.quascirev.2007.10.017

Chen MT, Lin XP, Chang YP, Chen YC, Lo L, Shen CC, Yokoyama Y, Oppo DW, Thompson WG, Zhang R (2010) Dynamic millennial-scale climate changes in the northwestern Pacific over the past 40,000 years. Geophys Res Lett 37(23):L23603. doi:10.1029/2010GL045202

Dekens PS, Lea DW, Pak DK, Spero HJ (2002) Core top calibration of Mg/Ca in tropical foraminifera: Refining paleotemperature estimation. Geochem Geophys Geosyst 3(4):1–29. doi:10.1029/2001GC000200

Du J, Chen M, Cao J, Qiu Y, Tong J, Ma Q, Yang J (2012) Oxygen isotope in seawater and its hydrological implication in the southern Yellow Sea and the East China Sea. Oceanol Limnol Sin 43(6):1057–1066 (in Chinese with English abstract)

Epstein S, Mayeda T (1953) Variation of O18 content of waters from natural sources. Geochim Cosmochim Acta 4(5):213–224. doi:10.1016/0016-7037(53)90051-9

Greaves M, Caillon N, Rebaubier H, Bartoli G, Bohaty S, Cacho I, Clarke L, Cooper M, Daunt C, Delaney M, deMenocal P, Dutton A, Eggins S, Elderfield H, Garbe-Schoenberg D, Goddard E, Green D, Groeneveld J, Hastings D, Hathorne E, Kimoto K, Klinkhammer G, Labeyrie L, Lea DW, Marchitto T, Martinez-Boti MA, Mortyn PG, Ni Y, Nuernberg D, Paradis G et al (2008) Interlaboratory comparison study of calibration standards for foraminiferal Mg/Ca thermometry. Geochem Geophys Geosyst 9(8):Q08010. doi:10.1029/2008gc001974

Hall JM, Chan LH (2004) Ba/Ca in Neogloboquadrina pachyderma as an indicator of deglacial meltwater discharge into the western Arctic Ocean. Paleoceanography 19(1):PA1017. doi:10.1029/2003PA000910

Hastings DW, Whitko A, Kienast M, Steinke S (2001) A comparison of three independent paleotemperature estimates from a high resolution record of deglacial SST records in the tropical South China Sea. Eos 82:PP12B–10

Hönisch B, Allen KA, Russell AD, Eggins SM, Bijma J, Spero HJ, Lea DW, Yu J (2011) Planktic foraminifers as recorders of seawater Ba/Ca. Mar Micropaleontol 79(1–2):52–57. doi:10.1016/j.marmicro.2011.01.003

Ichikawa H, Beardsley R (2002) The Current System in the Yellow and East China Seas. J Oceanogr 58(1):77–92. doi:10.1023/A:1015876701363

Jan S, Sheu DD, Kuo H-M (2006) Water mass and throughflow transport variability in the Taiwan Strait. J Geophys Res 111(C12):C12012. doi:10.1029/2006JC003656

Johnstone HJH, Yu J, Elderfield H, Schulz M (2011) Improving temperature estimates derived from Mg/Ca of planktonic foraminifera using X-ray computed tomography–based dissolution index, XDX. Paleoceanography 26(1):PA1215. doi:10.1029/2009PA001902

Kang D-J, Chung CS, Kim SH, Hong GH, Kim K-R (1994) Oxygen isotope characteristics of seawaters in the Yellow Sea. La mer 32:279–284

Kim S-T, O'Neil JR (1997) Equilibrium and nonequilibrium oxygen isotope effects in synthetic carbonates. Geochim Cosmochim Acta 61(16):3461–3475. doi:10.1016/S0016-7037(97)00169-5

Kim K-R, Cho Y-K, Kang D-J, Ki J-H (2005) The origin of the Tsushima Current based on oxygen isotope measurement. Geophys Res Lett 32(3):L03602. doi:10.1029/2004GL021211

Klinkhammer GP, Chan LH (1990) Determination of barium in marine waters by isotope dilution inductively coupled plasma mass spectrometry. Anal Chim Acta 232(0):323–329. doi:10.1016/S0003-2670(00)81249-0

Kubota Y, Kimoto K, Tada R, Oda H, Yokoyama Y, Matsuzaki H (2010) Variations of East Asian summer monsoon since the last deglaciation based on Mg/Ca and oxygen isotope of planktic foraminifera in the northern East China Sea. Paleoceanography 25(4):PA4205. doi:10.1029/2009pa001891

Kuroyanagi A, Kawahata H (2004) Vertical distribution of living planktonic foraminifera in the seas around Japan. Mar Micropaleontol 53(1–2):173–196. doi:10.1016/j.marmicro.2004.06.001

Lea DW, Boyle EA (1991) Barium in planktonic foraminifera. Geochim Cosmochim Acta 55(11):3321–3331

Lea DW, Spero HJ (1994) Assessing the reliability of paleochemical tracers: Barium uptake in the shells of planktonic foraminifera. Paleoceanography 9(3):445–452. doi:10.1029/94PA00151

Lea DW, Mashiotta TA, Spero HJ (1999) Controls on magnesium and strontium uptake in planktonic foraminifera determined by live culturing. Geochim Cosmochim Acta 63(16):2369–2379

Lea DW, Pak DK, Spero HJ (2000) Climate impact of Late Quaternary equatorial Pacific sea surface temperature variations. Science 289(5485):1719–1724. doi:10.1126/science.289.5485.1719

Lee H-J, Chao S-Y (2003) A climatological description of circulation in and around the East China Sea. Deep-Sea Res PT II 50(6–7):1065–1084. doi:10.1016/S0967-0645(03)00010-9

Lee TN, Johns WE, Liu C-T, Zhang D, Zantopp R, Yang Y (2001) Mean transport and seasonal cycle of the Kuroshio east of Taiwan with comparison to the Florida Current. J Geophys Res 106(C10):22143–22158. doi:10.1029/2000JC000535

Lewis E, Wallace DWR (1998) CO2SYS–Program Developed for CO_2 System Calculations ORNL/CDIAC-105 (Carbon Dioxide Information Analysis Center, Oak Ridge National Laboratory, U.S. Department of Energy, Oak Ridge, TN)

Lin Y-S, Wei K-Y, Lin I-T, Yu P-S, Chiang H-W, Chen C-Y, Shen C-C, Mii H-S, Chen Y-G (2006) The Holocene Pulleniatina Minimum Event revisited: Geochemical and faunal evidence from the Okinawa Trough and upper reaches of the Kuroshio current. Mar Micropaleontol 59(3–4):153–170. doi:10.1016/j.marmicro.2006.02.003

Liu Z, Wen X, Brady EC, Otto-Bliesner B, Yu G, Lu H, Cheng H, Wang Y, Zheng W, Ding Y, Edwards RL, Cheng J, Liu W, Yang H (2014) Chinese cave records and the East Asia Summer Monsoon. Quat Sci Rev 83:115–128. doi:10.1016/j.quascirev.2013.10.021

Marchitto TM (2006) Precise multielemental ratios in small foraminiferal samples determined by sector field ICP-MS. Geochem Geophys Geosyst 7(5):Q05P13. doi:10.1029/2005GC001018

McConnell MC, Thunell RC (2005) Calibration of the planktonic foraminiferal Mg/Ca paleothermometer: Sediment trap results from the Guaymas Basin, Gulf of California. Paleoceanography 20(2):PA2016. doi:10.1029/2004PA001077

Mulitza S, Boltovskoy D, Donner B, Meggers H, Paul A, Wefer G (2003) Temperature:$\delta^{18}O$ relationships of planktonic foraminifera collected from surface waters. Palaeogeogr Palaeocl 202(1–2):143–152. doi:10.1016/S0031-0182(03)00633-3

Niebler HS, Hubberten HW, Gersonde R (1999) Oxygen isotope values of planktic foraminifera: A tool for the reconstruction of surface water stratification. In: Wefer G (ed) Fischer G. Use of Proxies in Paleoceanography, Springer Berlin Heidelberg, pp 165–189. doi:10.1007/978-3-642-58646-0_6

Oba T (1990) Paleoceanographic information obtained by isotopic measurement of individual foraminiferal specimens. In: Wang P, Lao Q, He Q (eds) Proceedings of the First International Conference on Asian Marine Geology. China Ocean Press, Beijing, pp 169–180

Qian W, Lee D-K (2000) Seasonal march of Asian summer monsoon. Int J Climatol 20:1371–1386

Qu CH, Chen CZ, Yang JR, Wang LZ, Lu YL (1993) Geochemistry of dissolved and particulate elements in the major rivers of China (The Huanghe, Changjiang, and Zhunjiang rivers). Estuaries 16(3):475–487. doi:10.2307/1352595

Regenberg M, Steph S, Nürnberg D, Tiedemann R, Garbe-Schönberg D (2009) Calibrating Mg/Ca ratios of multiple planktonic foraminiferal species with $\delta^{18}O$-calcification temperatures: Paleothermometry for the upper water column. Earth Planet Sci Lett 278(3–4):324–336. doi:10.1016/j.epsl.2008.12.019

Rohling EJ (2007) Progress in paleosalinity: Overview and presentation of a new approach. Paleoceanography 22(3):PA3215. doi:10.1029/2007PA001437

Rosenthal Y, Boyle EA, Labeyrie L (1997) Last glacial maximum paleochemistry and deepwater circulation in the Southern Ocean: Evidence from foraminiferal cadmium. Paleoceanography 12(6):787–796

Russell AD, Hönisch B, Spero HJ, Lea DW (2004) Effects of seawater carbonate ion concentration and temperature on shell U, Mg, and Sr in cultured planktonic foraminifera. Geochim Cosmochim Acta 68(21):4347–4361. doi:10.1016/j.gca.2004.03.013

Schmidt GA (1999) Error analysis of paleosalinity calculations. Paleoceanography 14(3):422–429. doi:10.1029/1999PA900008

Schmidt MW, Lynch-Stieglitz J (2011) Florida Straits deglacial temperature and salinity change: Implications for tropical hydrologic cycle variability during the Younger Dryas. Paleoceanography 26(4):PA4205. doi:10.1029/2011PA002157

Schmuker B, Schiebel R (2002) Planktic foraminifers and hydrography of the eastern and northern Caribbean Sea. Mar Micropaleontol 46(3–4):387–403. doi:10.1016/S0377-8398(02)00082-8

Sugiyama M, Matsui M, Nakayama E (1984) Direct determination of barium in sea water by inductively coupled plasma emission spectrometry. J Oceanogr 40:295–302

Sun Y, Oppo DW, Xiang R, Liu W, Gao S (2005) Last deglaciation in the Okinawa Trough: Subtropical northwest Pacific link to Northern Hemisphere and tropical climate. Paleoceanography 20(4):PA4005. doi:10.1029/2004PA001061

Thunell RC, Honjo S (1981) Calcite dissolution and the modification of planktonic foraminiferal assemblages. Mar Micropaleontol 6(2):169–182. doi:10.1016/0377-8398(81)90004-9

Ujiié Y, Ujiié H, Taira A, Nakamura T, Oguri K (2003) Spatial and temporal variability of surface water in the Kuroshio source region, Pacific Ocean, over the past 21,000 years: evidence from planktonic foraminifera. Mar Micropaleontol 49(4):335–364

Wang B (2006) The Asian Monsoon. Springer Praxis, Chichester

Wang Y, Cheng H, Edwards RL, He Y, Kong X, An Z, Wu J, Kelly MJ, Dykoski CA, Li X (2005) The Holocene Asian Monsoon: Links to solar changes and North Atlantic climate. Science 308(5723):854–857. doi:10.1126/science.1106296

Wang Y, Cheng H, Edwards RL, Kong X, Shao X, Chen S, Wu J, Jiang X, Wang X, An Z (2008) Millennial- and orbital-scale changes in the East Asian monsoon over the past 224,000 years. Nature 451(7182):1090–1093

Weldeab S, Lea DW, Schneider RR, Andersen N (2007) 155,000 years of west African monsoon and ocean thermal evolution. Science 316(5829):1303–1307. doi:10.1126/science.1140461

Xu X, Yamasaki M, Oda M, Honda MC (2005) Comparison of seasonal flux variations of planktonic foraminifera in sediment traps on both sides of the Ryukyu Islands, Japan. Mar Micropaleontol 58(1):45–55. doi:10.1016/j.marmicro.2005.09.002

Yanagi T (1994) Material transport in the Yellow/East China Seas. Bull Coast Oceanogr 31:239–256

Yancheva G, Nowaczyk NR, Mingram J, Dulski P, Schettler G, Negendank JFW, Liu J, Sigman DM, Peterson LC, Haug GH (2007) Influence of the intertropical convergence zone on the East Asian monsoon. Nature 445(7123):74–77

Ye F, Deng W, Xie L, Wei G, Jia G (2014) Surface water $\delta^{18}O$ in the marginal China seas and its hydrological implications. Est Coast Shelf Sci 147:25–31. doi:10.1016/j.ecss.2014.05.033

Zhang J, Letolle R, Martin JM, Jusserand C, Mouchel JM (1990) Stable oxygen isotope distribution in the Huanghe (Yellow River) and the Changjiang (Yangtze River) estuarine systems. Cont Shelf Res 10(4):369–384. doi:10.1016/0278-4343(90)90057-S

Zhang L, Liu Z, Zhang J, Hong GH, Park Y, Zhang HF (2007) Reevaluation of mixing among multiple water masses in the shelf: An example from the East China Sea. Cont Shelf Res 27(15):1969–1979. doi:10.1016/j.csr.2007.04.002

Zhang W-Z, Chai F, Hong H-S, Xue H (2014) Volume transport through the Taiwan Strait and the effect of synoptic events. Cont Shelf Res. doi:10.1016/j.csr.2014.07.010

Zweng MM, Reagan JR, Antonov JI, Locarnini RA, Mishonov AV, Boyer TP, Garcia HE, Baranova OK, Johnson DR, Seidov D, Biddle MM (2013) World Ocean Atlas 2013, Volume 2: Salinity. NOAA Atlas NESDIS 74. Silver Spring, MD

7

Arsenic cycling in the Earth's crust and hydrosphere: interaction between naturally occurring arsenic and human activities

Harue Masuda

Abstract

Field-based research on naturally occurring arsenic contamination of surface waters and groundwaters and the mechanisms of contamination are reviewed. The distribution of arsenic is strongly related to areas of active plate tectonics, magmatism and associated hydrothermal activity, and high rates of erosion. Sources of arsenic contamination are mainly hydrothermal water, sulfide and arsenide minerals, volcanic ash, and iron oxyhydroxide/oxide as weathering products. The promotion of the reduction and oxidation of arsenic source minerals by in situ microbial activity is an important secondary mechanism that often determines arsenic levels in groundwater. Anthropogenic activities, such as geothermal and mining operations, as well as excess pumping of shallow groundwaters, disperse arsenic in the environment, thereby expanding areas of arsenic contamination.

Keywords: Arsenic contamination, Geologic cycle, Plate tectonics, Reduction, Oxidation

Introduction

Human beings are facing a crisis of an increasing demand for freshwater in the face of decreasing freshwater resources. One of the most serious problems threatening water resources is contamination with arsenic, which is highly toxic. Since the late 1980s, when the arsenicosis was reported in West Bengal, India (Mandal and Suzuki 2002), arsenic-contaminated groundwater and associated health disasters have occurred throughout the world, but especially in Asian countries. Arsenic is the first toxic element known to spread globally in the hydrosphere.

The World Health Organization (WHO) guidelines have recommended since 1993 that the arsenic level in aquatic solutions, including drinking water, should be < 10 μg/L. This threshold was selected not so much because it is an epidemiologically safe level but because of the limits of analytical and treatment techniques (WHO 2006). For example, the quantification limit of arsenic by the commonly used atomic absorption spectrophotometry method is 0.001 to 0.005 mg/L, much higher than the limit of 0.01 μg/L for inductively coupled plasma mass spectroscopy. In addition, the guideline followed differs by country: Japan (since 1993), the USA, Canada (both since 2006; EPA 2018; Health Canada 2017), and Australia (since 2011; NHMRC and NRMMC 2011) follow the WHO guideline of 10 μg/L, but many developing countries still apply the older standard of 50 μg/L.

Arsenic is a minor but ubiquitous element of the Earth's crust, and some chemical species of this amphoteric element are volatile and highly soluble in aquatic solution. As a result, arsenic is easily released from the crust into natural waters in conjunction with changes to the chemistry of the aquatic environment. The estimated arsenic reservoirs and fluxes at the Earth's surface, including in the shallow crust (Fig. 1), show that anthropogenic activities play an important role in dispersing arsenic contamination to the hydrosphere, pedosphere, and atmosphere. Arsenic has been used for many purposes, but especially for pesticides and wood preservation. For example, sodium arsenite solution applied as a pesticide in cattle dip from 1950s to 1980s led to groundwater in Australia being polluted with up to 3000 mg/L arsenic (McLaren et al. 1998), and in the USA, applications of arsenic-containing pesticides to cotton fields until 1990s increased the arsenic concentration in the soil to 23 mg/kg, compared with an average concentration of 5 mg/kg in surrounding soils (Cox et al. 1996). Since these studies were published, the production

Correspondence: harue@sci.osaka-cu.ac.jp
Graduate School of Science, Osaka City University, 3-3-138, Sugimoto, Sumiyoshi-ku, Osaka 558-8585, Japan

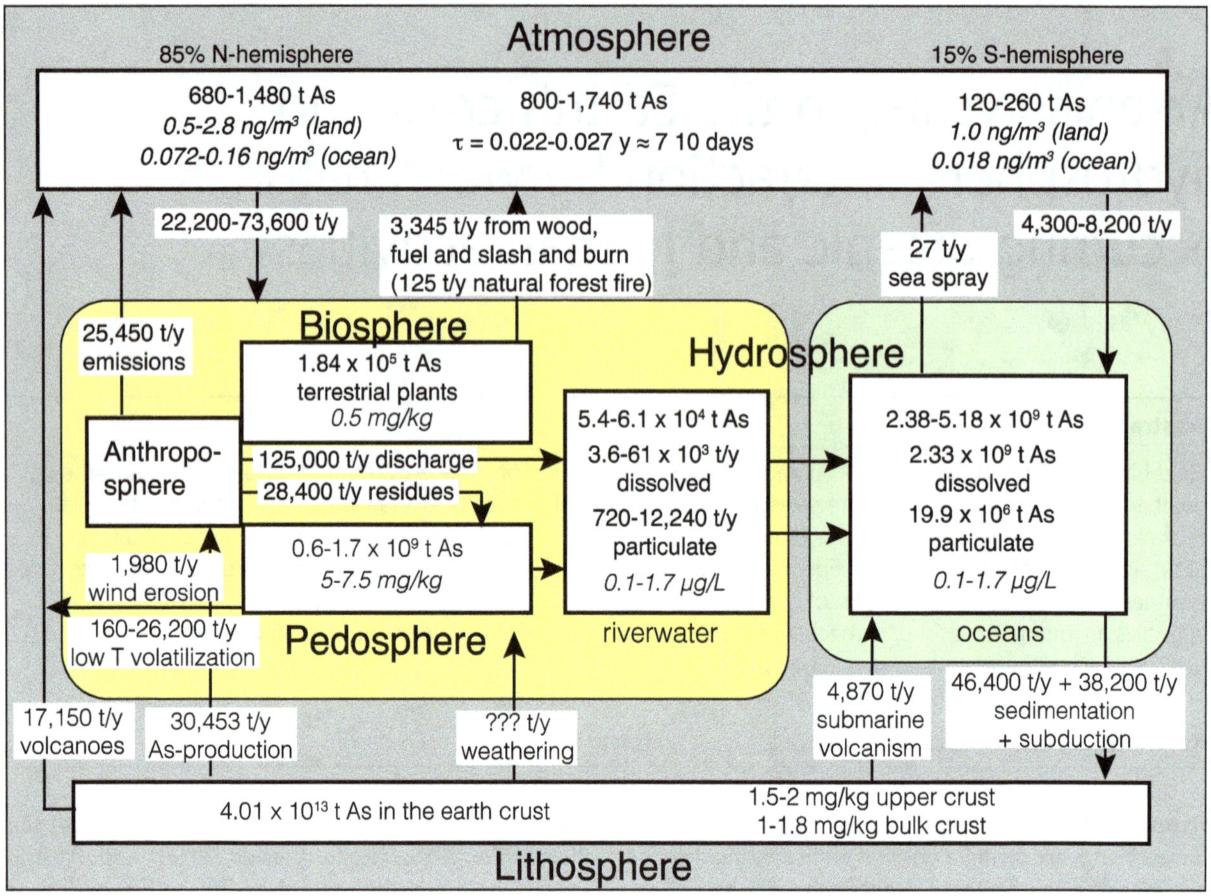

Fig. 1 Global arsenic cycle. Fluxes and reservoir sizes in different parts of the geosphere (Matschullat 2000)

of toxic arsenic compounds such as pesticides has decreased, although alloys containing arsenic continue to be used for semiconductors (JOGMEC 2012). Although anthropogenic contamination due to the application of arsenic compounds to agricultural fields is probably decreasing at least in more developed countries, activities to develop natural resources such as mining and geothermal energy extraction are a continuing cause of environmental arsenic contamination.

This review focuses on naturally occurring arsenic contamination from the points of views of geologic arsenic cycling and anthropogenic impacts on the spread of arsenic contamination.

Global distribution of arsenic contamination

There are three major natural sources of arsenic contamination: hydrothermal activity, ore deposits, and Cenozoic sediments. Other sources include coal and coal combustion. Arsenic-contaminated areas in Latin America (14 countries in Central and South America) are concentrated in the circum-Pacific region, where they are associated with the "Ring of Fire" (Fig. 2). The arsenic sources in this area are related to magmatic activity (volcanoes including volcanic ash and sulfide deposits). In addition, more than 100 aquifers along river basins in this region are contaminated with arsenic (Bundschuh et al. 2012).

High arsenic concentrations in geothermal waters reflect the fact that arsenic is a volatile component of magma and highly soluble in water. Hot-spring waters sometimes contain high concentrations of arsenic; for example, many hot springs in Yellowstone National Park, USA, are contaminated with up to 10 mg/L arsenic (Stauffer and Thompson 1984; Webster and Nordtrom 2003). Arsenic from hot springs can be transported in surface waters and enter riverbed sediments; for example, along the Owens River, arsenite released from hot springs and oxidized to arsenate in the river water by microbial activity precipitates and settles to the bottom, where it becomes a component of the riverbed sediments (0–60 mg/kg; Nimick et al. 1998). Geothermal well fluids occasionally contain higher arsenic concentrations than hot-spring waters, from >1 up to 50 mg/L (Webster and Nordtrom 2003 and references therein). Prior to 2000, the outflow of arsenic from Waikato Power Station, New Zealand, was 145 t/year, but since then, it has been reduced to ~70 t/year (Webster-Brown and Lane 2005 and references therein). The large quantity

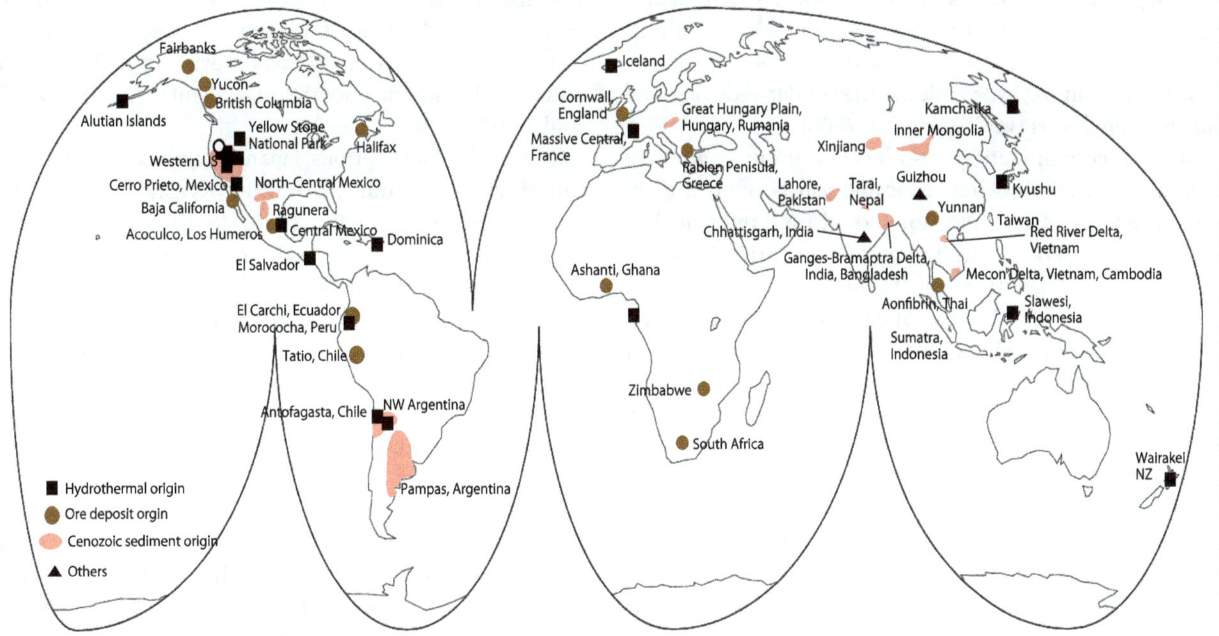

Fig. 2 World arsenic contamination map. This figure was produced based on the following reports and those in the text: Smedley and Kinniburgh (2002); Acharyya et al. (2005); An et al. (1997); Blanes et al. (2011); Camm et al. (2004); Gutiérrez-Caminero et al. (2015); Razo et al. (2004)

of arsenic in the power station drainage water has led to arsenic concentrations of 61–1790 mg/kg in the Waikato River (Webster-Brown et al. 2000).

Ore deposits are the second main source of arsenic. For example, in New Zealand, arsenic released from arsenic-bearing marcasite found in hydrothermal ore deposits formed 5 Ma causes groundwater concentrations of > 100 μg/L (Craw et al. 2000). Although the distribution of naturally occurring arsenic contamination around ore bodies is generally localized, anthropogenic materials associated with the mining of ore bodies, such as tailings and sludge, expand the area of contamination into the hydrosphere and pedosphere of the surrounding area. The excavated ore bodies and the ensuing chemical weathering produce sediments highly contaminated with arsenic that then become sources of potential arsenic contamination.

The third main source of arsenic contamination is aquifers in Cenozoic sediments, which occur widely around the world (Fig. 2). The total arsenic concentration in contaminated groundwater is not extremely high compared with the average concentration in soils and sediments. For example, in a Holocene aquifer in Bangladesh, the total arsenic concentration in bulk aquifer sediments ranges from 5 to 16 mg/kg whereas > 1 mg/L arsenic has been detected in the groundwater (Seddique et al. 2011). Thus, contamination of the groundwater depends mainly on the chemical form of the arsenic host phase and the occurrence of changes to the aquatic environment that lead to the release of arsenic from the aquifer sediments rather than on the arsenic concentration in the sediments. The mechanisms by which arsenic is released from Cenozoic, especially Holocene, aquifer sediments are described in the next section.

Release mechanisms of arsenic from aquifer sediments

Naturally occurring arsenic contamination of surface waters and groundwater is associated with changes in ambient physicochemical conditions. The most important factors controlling the release of arsenic from sediments to the aquatic environment are the redox conditions and the pH of the water.

Arsenic contamination of groundwater is most serious, from the point of view of spatial scale and the number of affected individuals, in the Holocene aquifers along large rivers that originate in the Himalayas and on the Tibetan Plateau: for example, along the Ganges–Brahmaputra river system in Nepal, India, and Bangladesh; along the Red River in Vietnam; along the Mekong River in Cambodia and Vietnam (e.g., Charlet and Polya 2006; Fendorf et al. 2010); and along the Yellow River in Inner Mongolia, China (e.g., Smedley et al. 2003; Guo et al. 2016). Arsenic-contaminated aquifers in which ferrous and ammonium ions are dissolved in the groundwater are characterized by reducing conditions and microbial respiration (e.g., Radloff et al. 2008; Quichsall et al. 2008). Reduction of groundwater is the main mechanism causing the dissolution of arsenic in Cenozoic aquifers; the major host phase is iron oxyhydroxides/oxides, which strongly adsorb arsenic (e.g., Nickson et al. 2000; van Geen et al. 2004).

Iron oxyhydroxides/oxides are stable under oxic (positive Eh) and weakly acid to alkaline (pH > 4) conditions, whereas arsenic is released by desorption from or decomposition of iron oxyhydroxides/oxides under reducing and alkaline conditions (van Geen et al. 2006).

Arsenic contamination also occurs in oxic aquifers (Amini et al. 2008), where volcanic ash is often the arsenic source. For example, aquifers of the Ethiopian Rift Valley, Africa (0.6–190 μg/L; Rango et al. 2013), have become contaminated through the dissolution of volcanic ash under oxic conditions. Such arsenic-contaminated groundwaters formed via dissolution of volcanic ash in oxic environments are widespread in South America, especially in Argentina and Bolivia. Groundwaters in the Pampas of Argentina contain a few hundred micrograms arsenic per liter, and the mean Eh, dissolved O_2, and pH of the contaminated groundwaters are + 73 ± 65 mV, 6.6 ± 2.2 mg/L, and 6.9–8.9, respectively. The level of contamination is in general lower in oxic than in anoxic aquifers, although the arsenic concentration in the aquifers of the Pampas can be as high as 15,000 μg/L (Bundschuh et al. 2012 and references therein; Nicolli et al. 2012). In a semi-arid climate, intense chemical weathering of volcanic ash during the rainy season and carbonate precipitation from the water during the dry season produce alkaline conditions. Arsenic concentrations in South America are secondarily controlled by the adsorption–desorption affinity of iron oxyhydroxides produced by chemical weathering (Bundschuh et al. 2012). An alkaline arsenic-contaminated water was observed in the pit lake of an abandoned gold mine (Savage et al. 2009) and in groundwaters in Punjab, Pakistan, where the arsenic concentration can reach 1900 μg/L, plausibly as a result of chlorite weathering at a pH range from 6.9 to 8.0 under oxic conditions without dissolved iron (Farooqi et al. 2007a, 2007b; Masuda et al. 2010).

Distribution of arsenic in Japan

A geochemical map of arsenic in the Japanese islands (Imai et al. 2004) was produced using data from elemental analyses of riverbed sediments, which generally reflect the average concentrations in the surrounding geology (Fig. 3). It is notable that arsenic is distributed along the active volcanic front in Northeast Japan and Kyushu, but it is associated with Cretaceous to Neogene igneous rock bodies on western Honshu Island.

Hydrothermal activity has been reported to cause arsenic contamination in many hot spring areas of Japan since the 1950s; for example, in Beppu (Kyushu) (Kawakami et al. 1956), Tamagawa (Akita, Northeast Japan) (Minami et al. 1958), Hakone, Yugawara (Kanagawa) (Awaya et al. 2002), and Toyohira (Hokkaido) (Jin et al. 2012). In Kusatsu-Shirane (Gunma), hot-spring water with an arsenic concentration of up to 7 mg/L flowed into the Yukawa River from the middle 1980s to the 1990s (Kikawada et al. 2006, 2009). A maximum arsenic concentration of 3.4 mg/L was reported in hydrothermal water at Kakkonda Geothermal Plant (Akita) (Okada et al. 2002).

One of the most serious Japanese arsenic disasters occurred at Toroku mine, Miyazaki (Kyushu), where arsenite was mined from the 1820s to 1962; in 1990, 146 people were found to be suffering from arsenicosis due to inhalation of refining soot and drinking contaminated water (Shimada 2009). Mining and smelting activities often cause arsenic contamination to be spread to air, sediments, and streams, whereas in the absence of such activities, it is usually confined to a small area (as noted in the "Global distribution of arsenic contamination" section). In a study in the Hokusetsu Mountains, Osaka Prefecture, on the eastern edge of the Cretaceous–Neogene magmatic bodies of western Honshu, Even et al. (2017) observed that hornfels with a high arsenic content near the granitic complex led to arsenic contamination of the surrounding ground and surface waters. Although many sulfide mines are present in this area, they are not as important as sources of arsenic contamination as the arsenide disseminated in hornfels and shales. Similarly, leaching and oxidation of arsenic in disseminated hydrothermal gold ore deposits hosted in Miocene sedimentary rock led to arsenic contamination of shallow groundwater in Sulawesi, Indonesia (Iskandar et al. 2012). The mechanism of arsenic contamination in both cases was probably oxidation and decomposition reactions of oxic water infiltrating through a fault system with diffusively distributed sulfides and arsenides.

In addition, Shimada (1996) reported that 2635 of 11,673 wells (22.6%) in the Cenozoic aquifer beneath Tsukushi Plain, Fukuoka Prefecture (northeast Japan), contained > 10 μg/L of arsenic. Fortunately, however, no cases of arsenicosis have been reported from this area.

Geologic cycling of arsenic

Arc magmatism and subsequent hydrothermal activity, together with high rates of erosion and transport to sedimentary basins downstream, contribute to contamination via in situ decomposition of arsenic-containing minerals. Tectonics plays a major role in determining erosion rates, which is partly ascribable to steeper slopes produced by active crustal movement (Hecht and Oguchi 2017). Thus, plate tectonics play an important role in the geologic cycling of arsenic. In this section, the geologic cycling of arsenic is documented (Fig. 4).

Arsenic is both a siderophile and chalcophile element that can behave as a metal and combine with sulfur; however, in molten rock, it shows volatile behavior, similar to the light rare earth elements (Sims et al. 1990). As of 2014, arsenic has been shown to occur in 568

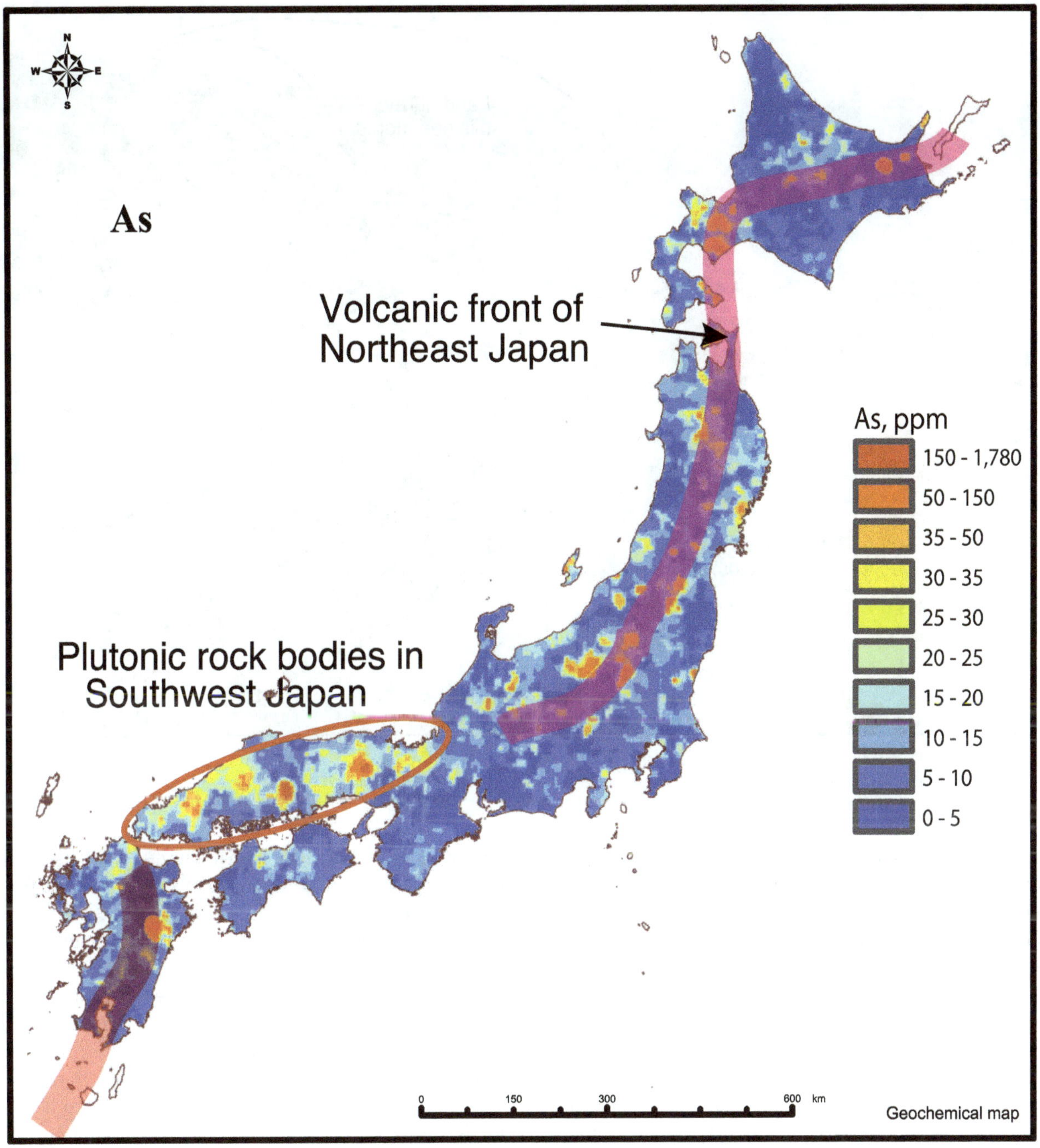

Fig. 3 Geochemical map showing arsenic distribution of riverbed sediments of Japan (Imai et al. 2004). The areas indicating volcanic fronts and plutonic rock ranges are added for this report

different mineral species (Nordstrom et al. 2014). The primary arsenic-containing mineral species are sulfides and arsenides, which form abundantly in association with magmatism and related hydrothermal activities. These include realgar (As_4S_4), orpiment (As_2S_3), arsenopyrite (FeAsS), and enargite (CuAsS). Other sulfide minerals are also primary sources of arsenic, especially pyrite, which can contain > 2% arsenic (e.g., pyrite in the Mother Lode Gold District of California; Savage et al. 2000). Pyrite is commonly distributed not only in igneous rock bodies but also in sedimentary formations; for example, under strongly reducing conditions, arsenic is incorporated into authigenic pyrite during sediment diagenesis (e.g., Masuda et al. 2005; Saunders et al. 2008).

The partition coefficient of arsenic between sulfide and silicate melts varies as a function of oxygen fugacity; at a

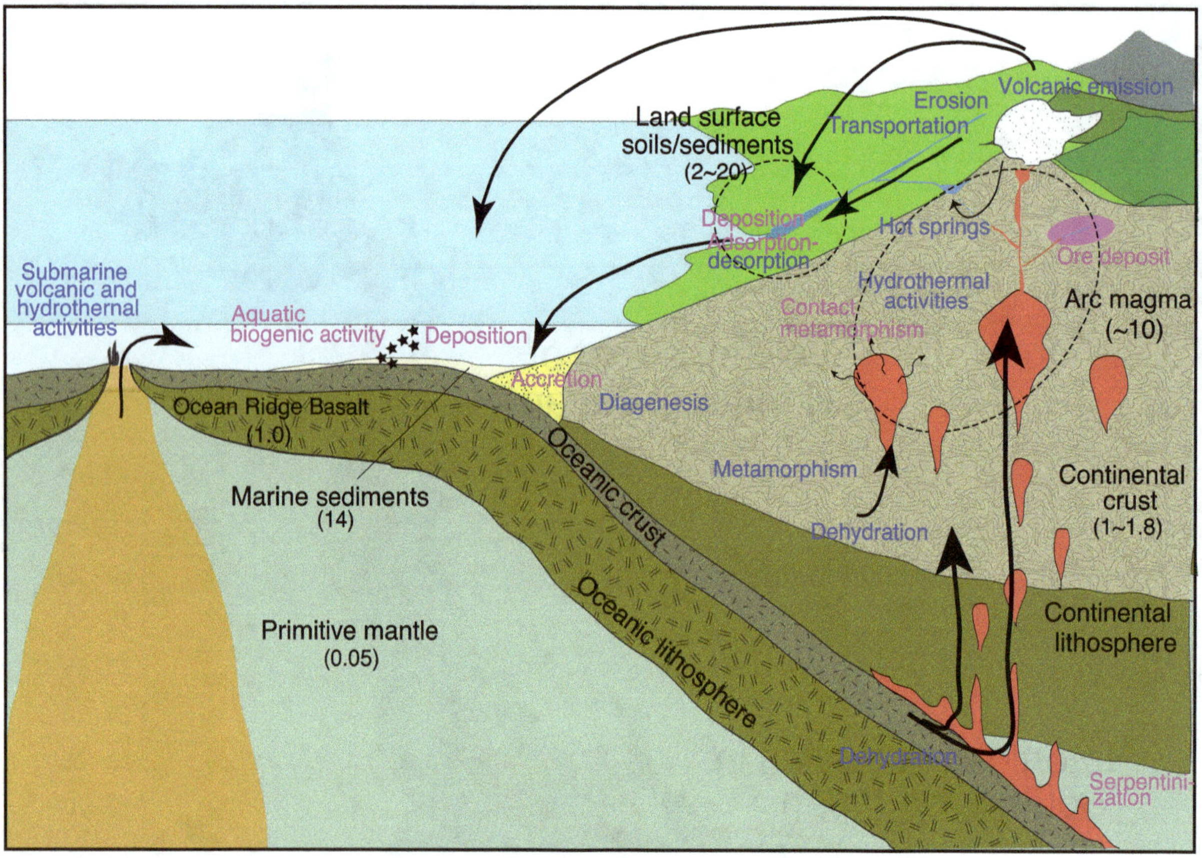

Fig. 4 Geologic cycle concerning redistribution of arsenic. Digit in parenthesis gives the average concentrations of each geologic body (references are in the text). Purple and blue colored letters indicate the releasing and fixing mechanism of arsenic in the cycle

given temperature and FeO content, the proportion of arsenic in silicate and monosulfide melts increases with oxygen fugacity (Li and Audétat 2012, 2015). Although silicate and carbonate mineral phases do not host arsenic as a major element (Majzlan et al. 2014), under conditions of low temperature, high oxygen fugacity, and low sulfur fugacity, silicates and carbonates can incorporate arsenic; for example, filatovite ($K[Al, Zn]_2(As, Si)_2O_8$), a variety of feldspar, has been found to be associated with volcanism in Kamchatka (Vergasova et al. 2004). Charnock et al. (2007) also found that arsenic was incorporated into garnet in the overgrowth of a hydrothermal vein, and Pascua et al. (2005) found arsenic-enriched smectite in hydrothermal precipitates at a geothermal plant in Akita, Japan. In addition, Ryan et al. (2011) reported that groundwater in the fracture zone of ultramafic rocks was contaminated with arsenic derived from serpentine and magnesite via chemical weathering.

McDonough and Sun (1995) estimated the arsenic concentration of Earth's primitive mantle to be 0.05 mg/kg. Hattori et al. (2002) analyzed mantle xenoliths and found that arsenic was highly concentrated in microscale sulfide minerals, and they reported that its concentration was higher in lherzolite than in harzburgite from the same area. For example, concentrations in sulfide minerals from lherzolite and harzburgite in Lesotho were > 200 mg/kg and < 14 mg/kg, respectively. In addition, comparative analyses of mantle xenoliths of Ichinomegata, Japan, showed that arsenic was depleted in wedge mantle, where microscale sulfide minerals had arsenic concentrations of < 50 mg/kg (Hattori et al. 2002).

Norr et al. (1996) reported arsenic concentrations of ~ 10 mg/kg in arc magmas enriched by hydrothermal fluids derived from metamorphic dehydration of the subducting slab, and they confirmed the release of arsenic from the subducting slab by both laboratory experiments and analyses of field-collected samples. You et al. (1996) conducted laboratory hydrothermal experiments in which they reacted sediment from the Nankai Trough, where the Philippine Sea Plate is subducting beneath the Eurasian Plate, with a synthetic $NaCl–CaCl_2$ solution and showed enrichment of the hydrothermal solution with arsenic at the relatively low temperature of ~ 300 °C. However, in relatively cool subduction zones such as that in which the Catalina Schist formed, less fore-arc devolatilization of arsenic occurs, and arsenic is more likely to be deeply subducted, resulting in arsenic-enriched arc lavas (Bebout et al. 1999). Arsenic-bearing serpentine, possibly formed by alteration of fore-arc mantle, has been found in the Himalayas, where the arsenic was derived from fluid from the subducting slab (Hattori et al. 2005). Norr et al. (1996) documented a decrease in the arsenic concentration of arc lavas on the order of 85%, in some cases as much as 95%, from fore-arc to back-arc along convergent margins. Arsenic is also released from deep continental and island-arc crust during the low-temperature and high-pressure metamorphism that forms pelitic schists; in rocks from the Sambagawa metamorphic belt, Japan, the arsenic concentration decreased from ~ 10 mg/kg to 0 as the temperature of metamorphism increased from 300 to 450 °C (Fujiwara et al. 2011).

Submarine hydrothermal fluids from the East Pacific Rise and the Mid-Atlantic Ridge contain up to 80.5 µg/L and 24 µg/L of arsenic, respectively, whereas those from back-arc basin and island-arc locations are much higher (1386 µg/L and 5850 µg/L, respectively; Breuer and Pichler 2013). Most arsenic in hydrothermal solutions hosted by basalt in Iceland is complexed with sulfur (i.e., $As(OH)S_2^{2-}$, AsS_3H^{2-}, AsS_3^{3-}, $As(SH)_4^-$; Sigufusson et al. 2011). Arsenic and sulfur form covalent bonds, and these chemical species precipitate from the hydrothermal solution as arsenide and sulfide minerals. Thus, many sulfide minerals, such as pyrite, contain arsenic as an impurity. As the temperature of the hydrothermal solution decreases and its oxygen fugacity increases, arsenic–sulfur complexes are decomposed and oxidized to arsenites, which are commonly found in hot-spring water where it flows out at the surface.

The arsenic concentration of Earth's crust is 1 to 1.8 mg/kg on average (e.g., Taylor and McLennan 1985; Wedepohl 1995), and all types of rocks and sediments contain arsenic at concentrations on the order of milligrams per kilogram (Table 1). Surface sediments and soils in various parts of the world contain higher and more variable concentrations, ranging from 2 to 20 mg/kg on average, than the average crustal concentration (Mandal and Suzuki 2002 and references therein). The average arsenic concentration in 3024 marine sediment samples from coastal Japan was determined to be 14 mg/kg, with a maximum concentration of ~ 580 mg/kg; however, 65% of the sediments had < 10 mg/kg, and 87.5% had < 20 mg/kg (AIST: https://gbank.gsj.jp/geochemmap/ocean/data/ganyuryo/ocean-noudo.csv). Sediments with high concentrations of arsenic in inland areas of Japan are associated with coal and sulfide mines (Ohta et al. 2010).

Solubilities of sulfide and arsenide minerals in water are generally low, and in anoxic environments, these minerals are insoluble. For example, under standard

Table 1 Arsenic concentrations of rocks and sediments

Rock type	Average As concentration (mg/kg)	Range of concentration (mg/kg)
Igneous rock		
Basalt	2.3	0.18–113
Andesite	2.7	0.5–5.8
Granite	1.3	0.2–15
Metamorphic rock		
Slate	18	0.5–143
Sedimentary rock, sediments		
Shale/submarine shale	3–15	< 490
Marine sediments[a]	14	0~580
Sandstone	4.1	0.6–120
Limestone/dolomite	2.6	0.1–20.1
Evaporite	3.5	0.1–10
Ferrigenous sedimentary rocks		1–2900
River and lake sediments	5	< 1–13,000
Soli		0.4–100
Peaty soil	13	2–36
Coal		0.3–35,00

Edited data from Boyle and Jonasson (1973) and Smedley and Kinniburgh (2002)
[a]Coastal area of Japan Islands,
AIST https://gbank.gsj.jp/geochemmap/ocean/data/ganyuryo/ocean-noudo.csv

temperature and pressure conditions, the solubility coefficient (as log K) of realgar (AsS) is − 83.13, and that of orpiment (As_2S_2) is − 180.43 (Boyle and Jonasson 1973). However, in oxic waters, those minerals are easily oxidized to dissolve. Typically, during chemical weathering of arsenic ore deposits and subsequent soil formation, arsenic host phases change as follows: arsenopyrite and löllinite in the ore body become scorodite, pharmacosiderite, and arsenosiderite in the weathered rock, and then arsenic-bearing goethite in soil (Majzlan et al. 2014). In the alteration products, arsenate ion substitutes for sulfate ion under neutral to alkaline conditions (Fernández-Martínez et al. 2008). Because in general arsenite and arsenate minerals (e.g., scorodite) are highly soluble in aqueous solution, except under strongly acidic conditions, these minerals are often sources of arsenic contamination. In fact, weathering products can be more important sources of arsenic contamination than primary source rocks and minerals as a result of secondary enrichment and their higher solubilities. For example, bulk arsenic concentrations in soils, sediments, and weathered crust from river basins in the Mother Lode District, California, are higher (> 2000 mg/kg at the maximum) than concentrations in nearby host rocks (< 600 mg/kg) (Savage et al. 2000).

In arsenic-contaminated Cenozoic aquifers, arsenic is hosted mainly in iron oxyhydroxides/oxides, especially goethite, which is a weakly crystallized product from ferrihydrite (amorphous iron hydroxides) via chemical weathering of iron-bearing minerals. However, few studies have examined the site of iron oxyhydroxides/oxide precipitation and the incorporation of arsenic into the aquifer sediments. Dissolved arsenic in the Red River of Vietnam is probably derived from sulfide mineral smelting operations at upstream locations in China (Cui et al. 2014). However, waste arsenic from smelting operations cannot be the source of the arsenic in the Quaternary sediment aquifers of the delta. In the Red River delta, the major mechanism of arsenic contamination of groundwater is probably reduction of arsenic-bearing iron oxyhydroxide minerals (e.g., Berg et al. 2007). Instead, arsenic-bearing iron hydroxides/oxides in the delta sediments are probably chemical weathering products transported from an upstream source area. In situ decomposition of arsenic-bearing pyrite occurring in association with seasonal changes of the groundwater level was presumed to be the cause of arsenic contamination in Munshuganj, Bangladesh (Pollizzotto et al. 2005), whereas in situ chemical weathering of chlorite was inferred to be the main mechanism in Sonragaon, Bangladesh (Masuda et al. 2012, 2013). The results of these studies imply that the source of arsenic contamination in aquifers of the Ganges–Brahmaputra delta is in situ dissolution of arsenic from primary minerals rather than dissolution of or desorption from arsenic-bearing iron oxyhydroxides. Moreover, excess use of groundwater would accelerate such dissolution by promoting inductive recharge of surface water and rapid changes in chemical conditions.

Biogeochemical reactions related to arsenic contamination

Biogeochemical reactions are another factor controlling arsenic behavior, especially in the hydrosphere and pedosphere. Primary production affects arsenic behavior, especially in surface waters, because of the similarity of the chemical characteristics of As(V) and phosphorous. Clean open-ocean water contains 1.5 μg/L arsenic on average (Andreae 1978), whereas in terrestrial waters, the background level of arsenic is < 0.1 μg/L (Matschullat 2000), but the concentration varies widely, depending on the geology and on the presence of anthropogenic sources along the channels. Biological alteration of arsenic compounds in surface waters has been well documented. Andreae (1978) showed that As(III), DMAA(V) (dimethylarsinic acid), and MMAA (V) (monomethylarsonic acid) are produced in association with increasing phytoplankton activity in Donner and Squaw lakes in California, and since then, numerous studies have documented that increasing eutrophication of estuaries is globally associated with the reduction of arsenate and organo-arsenical production. For example, arsenite and methyl-arsenic species became dominant in Lake Biwa water as biological activity increased (Hasegawa et al. 2010). Seawater contains as much as 1.8 μg/L arsenic, and the arsenic is sometimes concentrated by marine biological activity. For example, arseno-sugar compounds concentrate in phytoplankton and algae, and arseno-betaine concentrates in marine animals (Edmonds et al. 1997). Derivatives such as monomethylarsonic acid can also form by in situ microbial methylation (Reimer and Thompson 1988). In coastal areas, organo-arsenic compounds, mostly derived from marine biological activity, account for < 1% of total arsenic in surface sediments (Takeuchi et al. 2005). Takeuchi et al. (2005) noted that organo-arsenic compounds in marine sediments are decomposed by microbial activity within several decades; therefore, the formation rate of new organo-arsenic compounds is smaller than their degradation rate.

Microbial metabolic activity also plays a key role in the cycling of arsenic in the geosphere, including in groundwater aquifers, because microorganisms transform the arsenic into more mobile forms (Oremland and Stolz 2003; Yamamura and Amachi 2014). Microbially mediated metabolic activity involving arsenic can be categorized into four processes: (1) arsenite oxidation, (2) arsenate respiration, (3) arsenate reduction, and (4) arsenite methylation (Mukhopadhyay et al. 2002; Cai et al. 2013). Arsenic-metabolizing microbes were first identified in geologic bodies in which arsenic had become concentrated; for example, arsenite-oxidizing and arsenate-reducing bacteria were found in riverbed sediments containing mine tailings in France (Quéméneur et al. 2010). Arsenic-metabolizing bacteria are now known to be widely distributed in both contaminated and uncontaminated sediments (Cai et al. 2013).

Incubation experiments conducted with sediments from arsenic-contaminated groundwater aquifers of the Ganges–Brahmaputra delta demonstrated a relation between microbial activity and arsenic behavior; for example, arsenic was fixed in authigenic framboidal pyrite (Akai et al. 2004), and anaerobic metal-reducing bacteria worked to release arsenic from the sediments via reduction of Fe (III) (Islam et al. 2004). Zhu et al. (2008) showed that arsenic-bearing pyrite in black shale decomposed seven times faster when incubated with microbes than when incubated without microbes. These experimental results demonstrated that microbial activity can promote arsenic contamination of groundwater aquifers by changing the arsenic phase from insoluble to soluble.

Arsenic-oxidizing (Ghosh et al. 2014) and arsenic-reducing bacteria (Paul et al. 2015) identified by genetic analyses of microorganisms in arsenic-contaminated groundwaters from West Bengal, India, also suggest that microbial activity plays an essential role in releasing arsenic into groundwater aquifers in that area.

Conclusions

Although natural arsenic contamination processes of surface and groundwaters are not simple but involve various interactions among the geosphere, hydrosphere, and biosphere, contamination mechanisms have been well documented in recent decades. Arsenic contamination is widespread globally, but contaminated areas tend to concentrate in and around active tectonic zones. Large-scale contamination does not simply reflect the arsenic concentration in host rocks and sediments; the dispersal of arsenic phases is important for promoting water–mineral interactions because of the large reaction interface between minerals/materials and water. In addition, secondary arsenic minerals formed by chemical weathering likely expand the area and increase the level of contamination in association with biogeochemically promoted changes of redox and pH conditions in the hydrosphere.

The arsenic biogeochemical cycle affects the food chain. Inorganic arsenic compounds not only contaminate the hydrosphere, they also accumulate in biological organisms. Because inorganic arsenic compounds are more toxic than most organic arsenic compounds, their presence in agricultural products is concerning. For example, in Japan, rice is a major source of inorganic arsenic (Oguri et al. 2014). Research aimed at enhancing the understanding of the metabolic cycling of arsenic by humans would facilitate prevention and mitigation of arsenic poisoning.

Not all arsenic contamination can be mitigated, because some contamination occurs as a result of natural processes. However, by acquiring more knowledge of the natural arsenic cycle, we can find ways to minimize the area and number of people affected by arsenic contamination. Many studies have documented the chemical properties and reaction paths of arsenic species and identified arsenic-metabolizing bacteria with the aim of developing chemical treatment and bioremediation methods for arsenic-contaminated groundwaters (e.g., Takeuchi et al. 2007; Yamamura and Amachi 2014). International cooperation with regard to both scientific and practical aspects is key to preventing and mitigating arsenic disasters, especially in developing countries, where they can be particularly severe.

Acknowledgements

I thank the session conveners of the 2017 JpGU annual meeting, especially Prof. Y. Himiyama, for giving me the opportunity to present this report, and two anonymous reviewers, for the valuable comments to improve the manuscript.

Funding

This work was supported by KAKENHI (JSPS scientific fund) #17H01863.

Author's contributions

The author read and approved the final manuscript.

Competing interests

The author declares that there are no competing interests.

References

Acharyya SK, Shah BA, Ashyiya ID, Pandy Y (2005) Arsenic contamination in groundwater from parts of Ambagarh-Chowki block, Chhattisgarh, India: source and release mechanism. Environ Geol 49:148–158

Akai J, Izumi K, Fukuhara H, Masuda H, Nakano H, Yoshimura T, Ohfuji H, HMd A, Akai K (2004) Mineralogical and geomicrobiological investigations on groundwater arsenic enrichment in Bangladesh. Appl Geochem 19:215–230

Amini M, Abbaspour KC, Berg M, Winkel L, Hug SJ, Hoehn E, Yang H, Johnson CA (2008) Statistical modeling of global geogenic arsenic contamination in groundwater. Environ Sci Technol 42:3669–3675

An D, He YG, Hu QX (1997) Poisoning by coal smoke containing arsenic and fluoride. Fluoride 30:29–32

Andreae MO (1978) Distribution and speciation of arsenic in natural waters and some marine algae. Deep Sea Res 25:391–402

Awaya T, Oyama M, Ishizaka N, Ita K (2002) The amount of arsenic loads of river waters and hot springs in the Hakone-Yugawara area. Ann Rep Kanagawa Onsen Chigaku Kenkyujo 33:49–70 Japanese

Bebout GE, Ryan JG, Leeman WP, Bebaout AE (1999) Fractionation of trace elements by subduction-zone metamorphism – effect of convergent-margin thermal evolution. Earth Planet Sci Lett 171:63–81

Berg M, Stengel C, Trang PTK, Viet PH, Sampson ML, Leng M, Samreth S, Fredricks D (2007) Magnitude of arsenic pollution in the Mekong and Red River deltas – Cambodia and Vietnam. Sci Total Environ 372:413–425

Blanes FS, Buchhamer EE, Giménez MC (2011) Natural contamination with arsenic and other trace elements in groundwater of the Central-West region of Chaco, Argentina. J Environ Sci Health, Part A 46:1197–1206

Boyle RW, Jonasson IR (1973) The geochemistry of arsenic and its use as an indicator element in geochemical prospecting. J Geochem Explor 2:251–296

Breuer C, Pichler T (2013) Arsenic in marine hydrothermal fluids. Chem Geol 348:2–14

Bundschuh J, Litter MI, Parvez F, Roman-Ross G, Nicolli HB, Jean JS, Liu CW, Lopez D, Armienta MA, Guilherme LRG (2012) One century of arsenic exposure in Latin America: a review of history and occurrence from 14 countries. Sci Total Environ 429:2–35

Cai L, Yu K, Yang Y, Chen B-W, Li X-D, Zhang T (2013) Metanogenic exploration reveals high levels of microbial arsenic metabolism genes in activated sludge and coastal sediments. Environ Biotechnol 97:9579–9588

Camm GS, Glass HJ, Bryce DW, Butcher AR (2004) Characterisation of a mining-related arsenic-contaminated site, Cornwall, UK. J Geochem Explor 82:1–15

Charlet L, Polya DA (2006) Arsenic in shallow, reducing groundwaters in southern Asia: an environmental health disaster. Elements 2:91–96

Charnock JM, Polya DA, Gault AG, Wogelius RA (2007) Direct EXAFS evidence for incorporation of As^{5+} in the tetrahedral site of natural andraditic garnet. Am Mineral 92:1856–1861

Cox MS, Bell PF, Kover JL (1996) Arsenic supply characteristics of four cotton-producing soils. Plant Soil 180:11–17

Craw D, Chappell D, Reay A (2000) Environmental mercury and arsenic sources in fossil hydrothermal systems, Northland, New Zealand. Environ Geol 38:875–887

Cui B, Zhu M, Yang J, Yanxue J (2014) Identification of the sources of metals and arsenic in river sediments by multivariate analysis and geochemical approaches. J Soils Sed 14:1456–1468

Edmonds JS, Francesconi KA, Rippingale RJ, Morita M (1997) Arsenic transformations in short marine food chains studied by HPLC-ICP MS. Appl Organomet Chem 11:281–287

EPA (US Environmental Protection Agency) (2018) Groundwater and Drinking Water, National Primary Drinking water regulations. https://www.epa.gov/ground-water-and-drinking-water/national-primary-drinking-water-regulations#Inorganic (2018.05.28)

Even E, Masuda H, Shibata T, Nojima A, Chiba H, Yamanaka T (2017) Geochemical distribution and fate of arsenic in water and sediments of rivers from the Hokusetsu area, Japan. J Hydrol: Region Stud 9:34–47

Farooqi A, Masuda H, Firdous N (2007a) Toxic fluoride and arsenic contaminated groundwater in Lahore and Kasur districts, Punjab, Pakistan and possible contaminant. Environ Poll 145:837–849

Farooqi M, Masuda H, Kusakabe M, Naseem M, Firdous N (2007b) Deteriorating quality and large extent of arsenic and fluoride contaminated alluvial aquifers from East Punjab Pakistan and factors controlling contaminant's behavior. Geochem J 41:213–234

Fendorf S, Holly HA, van Geen A (2010) Spatial and temporal variations of groundwater arsenic in South and Southeast Asia. Science 328:1123–1127

Fernández-Martínez A, Cuello GJ, Johnson MR, Bardellli F, Póman-Ross G, Charlet L, Trrillas X (2008) Arsenate incorporation in gypsum probed by neutron, x-ray scattering and density functional theory modeling. J Phys Chem A 112: 5159–5166

Fujiwara S, Yamamoto K, Mimura K (2011) Dissolution processes of elements from subducting sediments into fluids: evidence from the chemical composition of the Sanbagawa pelitic schists. Geochem J 45:221–234

Ghosh D, Bhadury P, Bouth J (2014) Diversity of arsenite oxidizing bacterial communities in arsenic-rich deltaic aquifers in West Bengal, India. Front Microbiol 5, Article 602: doi: https://doi.org/10.3389/fmicb.2014.00602

Guo HM, Zhou YZ, Jia YF, Tang XH, Jiang YX, Lu H, Han SB, Zheng H, Cao YS, Zhao WG, Zhang D, Wei C, Norra S, Zhang FC (2016) Sulfur cycling-related biogeochemical processes of arsenic mobilization in the western Hetao basin, China: evidences from multiple isotope approaches. Environ Sci Technol 50:12650–12659

Gutiérrez-Caminero L, Weber B, Wurl J, Carrera-Muñoz M (2015) Tracing toxic elements sources using lead isotopes: an example from the San Antonio-El Triunfo mining district, Baja California Sur, México. Appl Geochem 59:23–32

Hasegawa H, Raman MA, Kitahara K, Itaya Y, Maki T, Ueda K (2010) Seasonal changes of arsenic speciation in lake waters in relation to eutrophication. Sci Total Environ 407:1418–1425

Hattori K, Arai S, Clarke DB (2002) Selenium, tellurium, arsenic and antimony contents of primary mantle sulfides. Can Mineral 40:637–650

Hattori K, Takahashi Y, Guillot S, Johanson B (2005) Occurrence of arsenic (V) in forearc mantle serpentinites based on X-ray absorption spectroscopy study. Geochim Cosmochim Acta 69:5585–5596

Health Canada (2017). Guidelines for Canadian drinking water quality—summary table. Water and Air Quality Bureau, Healthy Environments and Consumer Safety Branch, Health Canada, Ottawa, Ontario

Hecht H, Oguchi T (2017) Global evaluation of erosion rates in relation to tectonics. Prog Earth Planet Sci 4:40. https://doi.org/10.1186/s40645-017-0156-3

Imai N, Terashima S, Ohta M, Mikoshiba M, Okai T, Tachibana Y, Togashi S, Matsuhisa Y, Kanai Y, Ueoka A, Taniguchi M (2004) Geochemical map of Japan. Geological survey of Japan, p 209

Iskandar I, Koike K, Sendjaja P (2012) Identifying groundwater arsenic contamination mechanisms in relation to arsenic concentrations in water and host rocks. Environ Earth Sci 65:2015–2026

Islam FS, Gault AG, Boothman C, Polya DA, Charnock JM, Chatterjee D, Lloid JR (2004) Role of metal-reducing bacteria in arsenic release from Bengal delta sediments. Nature 430:68–71

Jin K, Kobayashi S, Takeuchi S, Tatsumi K, Miura K, Narita K (2012) Determination of As (III) and As (V) in river water of Toyohira, Sapporo and speciation of arsenic compounds in cyanobacteria by HPLC-ICPMS. Rep Hokkaido Inst Pub Health 62:35–41

JOGMEC (2012) Material flow of ore resources 2011: 44 Arsenic. http://mric.jogmec.go.jp/public/report/2011-07/As.pdf (Jan. 10, 2018)

Kawakami H, Nozaki H, Koga A (1956) Chemical study on Beppu Hoto spring (II) – trace elements of Peppu hot spring (II) distribution of arsenic. Nihon Kagaku Zasshi 77:1785–1789 Japanese

Kikawada K, Kawai S, Oi T (2006) Long term changes in the concentration of dissolved arsenic and its present supply in the Kusatsu hot springs Gunma Japan. Chikyukagaku (Geochemistry) 40:125–136 Japanese with English abstract

Kikawada K, Kyomen K, Oi T (2009) Behavior of arsenic in Yukawa River of the Kusatsu hot spring resource area, Gunma prefecture, Japan. J Hot Spring Science 59:81–87 Japanese with English abstract

Li Y, Audétat A (2012) Partitioning of V, Mn, Co, Ni, Cu, Zn, As, Mo, Ag, Sn, Sb, W, Au, Pb, and Bi between sulfide phases and hydrous basanite melt at upper mantle conditions. Earth Planet Sci Lett 355-356:327–340

Li Y, Audétat A (2015) Effects of temperature, silicate melt composition, and oxygen fugacity on the partitioning of V, Mn, Co, Ni, Cu, Zn, As, Mo, Ag, Sn, Sb, W, Au, Pb and Bi between sulfide phases and silicate melts. Geochim Cosmochim Acta 162:25–45

Majzlan J, Drahota P, Filippi M (2014) Parageneses and crystal chemistry of arsenic minerals. Rev Mineral Geochem 79:17–184

Mandal BK, Suzuki KT (2002) Arsenic round the world: a review. Talanta 58:201–235

Masuda H, Mitamura M, Farooqi AM, Muhanmad N, Owada M, Okazaki K, Seddique AA (2010) Geologic structure and geochemical characteristics of sediments of fluoride and arsenic contaminated groundwater aquifer in Kalalanwala and its vicinity, Punjab, Pakistan. Geochem J 44:489–505

Masuda H, Okabayashi K, Maeda S, Seddique AA, Mitamura M, Shinoda K (2013) Sequential chemical extraction of arsenic and related elements from the Holocene sediments of Sonargaon, Bangladesh, in relation to formation of arsenic-contaminated groundwater. Geochem J 47:651–661

Masuda H, Shinoda K, Okudaira T, Takahashi Y, Noguchi N (2012) Chlorite – source of arsenic groundwater pollution in the Holocene aquifer of Bangladesh. Geochem J 46:381–391

Masuda H, Yamatani Y, Okai M (2005) Transformation of arsenic compounds in modern intertidal sediments of Iriomote Island, Japan. J Geochem Explor 87: 73–81

Matschullat J (2000) Arsenic in geosphere – a review. Sci Total Environ 249:297–312

McDonough WF, Sun S-s (1995) The composition of the Earth. Chem Geol 120:223–253

McLaren RG, Naidu R, Smith J, Tiller KG (1998) Fraction and distribution of arsenic in soils contaminated by cattle dip. J Environ Qual 27:348–354

Minami H, Sato G, Watanuki K (1958) Concentrations of arsenic and lead of Tamagawa hot spring waters, Akita prefecture. Nihon Kagaku Zasshi 79: 860–865 Japanese

Mukhopadhyay R, Rozen BP, Pung LT, Silver S (2002) Microbial arsenic: from geocycles to genes and enzymes. FEM Microbiol Rev 26:311–325

NHMRC, NRMMC (2011) Australian drinking water guidelines paper 6 national water quality management strategy. National Health and Medical Research Council, National Resource Management Ministerial Council, Commonwealth of Australia, Canberra

Nickson R, McArthur JM, Ravencroft P, Burgess W, Ahmed KM (2000) Mechanism of arsenic release to groundwater, Bangladesh and West Bengal. Appl Geochem 15:403–413

Nicolli HB, Bundschuh J, Blanco MD, Tujchneider OC, Panarello HO, Dapena C, Rusansky JE (2012) Arsenic and associated trace-elements in groundwater from the Chaco-Panpean plain, Argentina: results from 100 years of research. Sci Total Environ 429:36–56

Nimick DA, Moore JN, Dalby CE, Savka MW (1998) The fate of geothermal arsenic in the Madison and Missouri Rivers, Montana and Wyoming. Water Resour Res 34:3051–3067

Nordstrom DK, Majzlan J, Königsberger E (2014) Thermodynamics properties for arsenic minerals and aqueous species. Rev Mineral Geochem 79:217–255

Norr PD Jr, Newson HE, Leeman WP, Ryan JG (1996) The role of hydrothermal fluids in the production of subduction zone magmas: evidence from siderophile and chalcophile trace elements and boron. Geochim Cosmochim Acta 60:587–611

Oguri T, Yoshinaga J, Tao H, Nakazato T (2014) Inorganic arsenic in the Japanese diet: daily intake and source. Archiv Environ Contam Toxic 66:100–112

Ohta A, Imai N, Terashima S, Tachibana Y, Ikehara K, Katayama H, Noda A (2010) Factors controlling regional spatial distribution of 53 elements in coastal sea sediments in northern Japan: comparison of geochemical data derived from stream and marine sediments. Appl Geochem 25:357–376

Okada H, Tada T, Chiba A, Nakazawa H, Mitsuhashi K, Ohara T (2002) Decontamination of geothermal water – removal of arsenic. Low Temp Sci 37:331–337 Japanese with English abstract

Oremland RS, Stolz JF (2003) The ecology of arsenic. Science 300:939–944

Pascua C, Charnock J, Polya DA, Sato T, Yokoyama S, Minato M (2005) Arsenic-bearing smectite from the geothermal environment. Mineral Mag 69:897–906

Paul D, Kazy SK, Gupta AK, Pal T, Sar P (2015) Diversity, metabolic properties and arsenic mobilization potential of indigenous bacteria in arsenic contaminated groundwater of West Bengal, India. PLoS One 10:e0118735. https://doi.org/10.1371/journal.pone.0118735

Pollizzotto ML, Harvey CF, Sutton S, Fendorf S (2005) Processes conductive to the release and transport of arsenic into aquifers of Bangladesh. Proc Nat Acad Sci US 102:18819–18823

Quéméneur M, Cébron A, Billard P, Battaglia-Brunet F, Garrido F, Leyval C, Joulian C (2010) Population structure and abundance of arsenite-oxidizing bacteria along an arsenic pollution gradient in waters of the upper Isle River Basin, France. Appl Environ Microbiol 76:4566–4570

Quichsall AN, Bostick BC, Sampson ML (2008) Linking organic matter deposition and iron mineral transformations to groundwater arsenic levels in the Mekong delta, Cambodia. Appl Geochem 23:3088–3098

Radloff KA, Manning AR, Mailloux B, Zheng Y, Rahman MM, Huq MR, Ahmed KM, van Geen A (2008) Considerations for conducting incubations to study the mechanisms of As release in reducing groundwater aquifers. Appl Geochem 23:3224–3235

Rango T, Vengosh A, Dwyer G, Bianchini G (2013) Mobilization of arsenic and other naturally occurring contaminants in groundwaters of the Main Ethiopian Rift. Water Res 47:5801–5818

Razo I, Carrizales L, Castro J, Díez-Barriga F, Monroy M (2004) Arsenic and heavy metal pollution of soil, water and sediments in a semi-arid climate mining area in Mexico. Water Air Soil Poll 152:129–152

Reimer KJ, Thompson JAJ (1988) Arsenic speciation in marine interstitial water. The occurrence of organoarsenicals Biogeochem 6:211–237

Ryan PC, Kim J, Wall AJ, Moen JC, Coremthal LG, Chow DR, Sullivan CM, Bright KS (2011) Ultra-mafic derived arsenic in a fractured bedrock aquifer. Appl Geochem 26:444–457

Saunders JA, Lee M-K, Shamsudduha M, Dhakal PUA, Chowdury MT, Ahmed KM (2008) Geochemistry and mineralogy of arsenic in (natural) anaerobic groundwaters. Appl Geochem 23:3205–3214

Savage KS, Ashley RP, Bird DK (2009) Geochemical evolution of a high arsenic, alkaline pit-lake in the Mother Lode Gold District, California. Econ Geol 104: 1171–1211

Savage KS, Tingle TN, O'Day PA, Waychunas GA, Bird DK (2000) Arsenic speciation in pyrite and secondary weathering phases, Mother Lode Gold District, Tuolumne County, California. Appl Geochem 15:1219–1244

Seddique AA, Masuda H, Mitamura M, Shinoda K, Yamanaka T, Nakaya S, Ahmed KM (2011) Mineralogy and geochemistry of shallow sediments of Sonargaon, Bangladesh and implications for arsenic dynamics: focusing on the role of organic matter. Appl Geochem 2:587–599

Shimada N (1996) Geochemical conditions enhancing the solubilization of arsenic into groundwater in Japan. Appl Organomet Chem 10:667–674

Shimada N (2009) The essence of problems on groundwater and soil pollutions caused by naturally occurring heavy metals and harmful elements: arsenic. Oyo Technical Report 29:31–59 Japanese with English abstract

Sigufusson B, Gislason SR, Meharg AA (2011) A field and reactive transport model study of arsenic in a basaltic rock aquifer. Appl Geochem 26:553–564

Sims KWW, Newsom HE, Gladney ES (1990) Chemical fractionation during formation of the Earth's core and continental crust: clues from As, Sb, W, and Mo. In: HE Newson and JH Jones (Editors). Origin of the earth. Oxford Univ Press, New York, N. Y., pp. 291–317

Smedley PL, Kinniburgh DG (2002) A review of source, behavior and distribution of arsenic in natural waters. Appl Geochem 17:517–568

Smedley PL, Zhang M, Zhang G, Luo Z (2003) Mobilization of arsenic and other trace elements in fluviolacustrine aquifers of the Huhhot Basin, Inner Mongolia. Appl Geochem 18:1453–1477

Stauffer RE, Thompson JM (1984) Arsenic and antimony in geothermal waters of Yellow Stone National Park, Wyoming, USA. Geochim Cosmochim Acta 48: 2547–2561

Takeuchi M, Kawahata H, Gupta LP, Kita N, Morishiba Y, Ono Y, Komai T (2007) Arsenic resistance and removal by marine and non-marine bacteria. J Biotechnol 127:434–442

Takeuchi M, Terada A, Nanba K, Kanai Y, Owaki M, Yoshida T, Kurolwa T, Nirei H, Komai T (2005) Distribution and fate of biologically formed organoarsenicals in coastal marine sediment. Appl Organomet Chem 19:945–951

Taylor SR, McLennan SM (1985) The continental crust: its composition and evolution. Blackwell, Oxford

van Geen A, Thoral S, Rose J, Garnier JM, Zheng Y, Bottero JY (2004) Decoupling of As and Fe release to Bangladesh groundwater under reducing conditions, part II: evidence from sediment incubations. Geochim Cosmochim Acta 68: 3475–3486

van Geen A, Zheng Y, Cheng Z, Aziz Z, Horneman A, Dhar RK, Mailloux B, Stute M, Weinman B, Goodbred S, Seddique AA, Hoque MA, Ahmed KM (2006) A transect of groundwater and sediment properties in Ariahazar, Bangladesh: further evidence of decoupling between As and Fe mobilization. Chem Geol 228:85–96

Vergasova LP, Krivovichev SV, Britvin SN, Burns PC, Ananiev VV (2004) Filatovite, K [(Al, Zn)$_2$(As, Si)$_2$O$_8$], a new mineral species from the Tolbachik volcano, Kamchatka peninsula, Russia. Eur J Mineral 16:533–536

Webster JG, Nordtrom K (2003) Chapter 4 geothermal arsenic the source, transport and fate of arsenic in geothermal systems. In: Welch AH, Stollenwerk KG (eds) Arsenic in groundwater. Kluwer Academic Publishers, Boston, pp 101–125

Webster-Brown JG, Lane V (2005) The environmental fate of geothermal arsenic in a major river system, New Zealand. In: Proceedings world geothermal congress 2005, Antalya, Turkey, 24–29 April

Webster-Brown JG, Lane V, Webster KS (2000) Arsenic in the Waikato River – an update. In: Proc the 22nd geothermal workshop, New Zealand

Wedepohl KH (1995) The composition of the continental-crust. Geochim Cosmochim Acta 59:1217–1232

WHO (World Health Organization) (2006) Guideline for drinking-water quality [electronic resource]: incorporating first addendum. Vol. 1, Recommendations. -3rd ed. WHO Press, Geneva, Switzerland, ISBN 92 4 154696 4. pp.515

Yamamura S, Amachi S (2014) Microbiology of inorganic arsenic: from metabolism to bioremediation. J Biosci Bioeng 118:1–9

You C-F, Castillo PR, Gieskes JM, Chan LH, Spivack AJ (1996) Trace element behavior in hydrothermal experiments: implications for fluid processes at shallow depths in subduction zones. Earth Planet Sci Lett 140:41–52

Zhu W, Young LY, Yee N, Serfes M, Rhine ED, Renfelder JR (2008) Sulfide-driven arsenic mobilization from arsenopyrite and black shale pyrite. Geochim Cosmochim Acta 72:5243–5250

8

The responses of the four main substitution mechanisms of H in olivine to H_2O activity at 1050 °C and 3 GPa

Peter M. E. Tollan[1,2], Rachel Smith[1], Hugh St.C. O'Neill[1*] and Jörg Hermann[1,2]

Abstract

The water solubility in olivine (C_{H_2O}) has been investigated at 1050 °C and 3 GPa as a function of water activity (a_{H_2O}) at subsolidus conditions in the piston-cylinder apparatus, with a_{H_2O} varied using H_2O–NaCl fluids. Four sets of experiments were conducted to constrain the effect of a_{H_2O} on the four main substitution mechanisms. The experiments were designed to grow olivine in situ and thus achieve global equilibrium (G-type), as opposed to hydroxylating olivine with a pre-existing point-defect structure and impurity content (M-type). Olivine grains from the experiments were analysed with polarised and unpolarised FTIR spectroscopy, and where necessary, the spectra have been deconvoluted to quantify the contribution of each substitution mechanism. Olivine buffered with magnesiowüstite produced absorbance bands at high wavenumbers ranging from 3566 to 3612 cm^{-1}. About 50% of the total absorbance was found parallel to the *a*-axis, 30% parallel to the *b*-axis and 20% parallel to the *c*-axis. The total absorbance and hence water concentration in olivine follows the relationship of $C_{H_2O} \propto a_{H_2O}^2$, indicating that the investigated defect must involve four H atoms substituting for one Si atom (labelled as [Si]). Forsterite buffered with enstatite produced an absorbance band exclusively aligned parallel the *c*-axis at 3160 cm^{-1}. The band position, polarisation and observed $C_{H_2O} \propto a_{H_2O}$ are consistent with two H substituting for one Mg (labelled as [Mg]). Ti-doped, enstatite-buffered olivine displays absorption bands, and polarisation typical of Ti-clinohumite point defects where two H on the Si-site are charge-balanced by one Ti on a Mg-site (labelled as [Ti]). This is further supported by $C_{H_2O} \propto a_{H_2O}$ and a 1:1 relationship of molar H_2O and TiO_2 in these experiments. Sc-doped, enstatite-buffered experiments display a main absorption band at 3355 cm^{-1} with $C_{H_2O} \propto a_{H_2O}^{0.5}$ and a positive correlation of Sc and H, indicating the coupled substitution of a trivalent cation plus a H for two Mg (labelled as [triv]). Our data demonstrate that extreme care has to be taken when inferences from experiments conducted at $a_{H_2O} = 1$ are applied to the mantle, where in most cases, a low a_{H_2O} persists. In particular, the higher exponent of the [Si] substitution mechanism means that the contribution of this hydrous defect to total water content will decrease more rapidly with decreasing a_{H_2O} than the contributions of the other substitution mechanisms. The experiments confirm previous results that the [Mg] mechanism holds an almost negligible amount of water under nearly all T-P-fO_2-fH_2O conditions that may be anticipated in nature. However, the small amounts of H_2O we find in substituting by this mechanism are similar in the experiments on forsterite doped with either Sc or Ti to those in the undoped forsterite at equivalent a_{H_2O} (all buffered by enstatite), confirming the assumption that, thermodynamically, C_{H_2O} substituting by each mechanism does not depend on the water concentration that substitutes by other mechanisms.

Keywords: Olivine, Nominally anhydrous minerals, Substitution mechanism, Point defect, Water in the mantle

* Correspondence: hugh.oneill@anu.edu.au
[1]Research School of Earth Sciences, The Australian National University, Building 142, Mills Road, Canberra, ACT 2601, Australia
Full list of author information is available at the end of the article

Introduction

Trace concentrations of hydrogen-bearing point defects can change the mechanical properties of olivine and the dominant phase of the Earth's upper mantle, producing a profound effect on mantle rheology (Demouchy et al. 2007; Karato et al. 1986; Mei and Kohlstedt 2000). Such defects also influence transport properties like electrical conductivity (Karato and Wang 2013). Fourier transform infrared spectroscopy (FTIR) shows that the H substitutes by bonding to O^{2-} anions, producing OH^- species according to the generalised reaction $O^{2-} + H_2O = 2OH^-$. The H incorporation thus depends on the fugacity of H_2O, justifying its colloquial designation as "water". Experimental studies on synthetic systems have identified four principal substitution mechanisms, which can be identified from their different O–H stretching modes, using FTIR spectroscopy. Following Kovacs et al. (2010), we use [Si] to denote the substitution mechanism whereby H is charge-balanced by silicon vacancies, [Mg] for H charge-balanced by Mg vacancies, [triv] for H co-substituting with a trivalent cation, and [Ti] for H co-substituting with titanium. The thermodynamic equilibria describing these four substitutions are:

$$[\mathrm{Si}] : 2\mathrm{H_2O} + \mathrm{Mg_2SiO_4} = \mathrm{Mg_2H_4O_4} + \mathrm{SiO_2} \quad (1)$$

$$[\mathrm{Mg}] : \mathrm{H_2O} + 0.5\ \mathrm{Mg_2SiO_4} + 0.5\ \mathrm{SiO_2} = \mathrm{MgH_2SiO_4} \quad (2)$$

$$[\mathrm{triv}] : 0.5\ \mathrm{H_2O} + \mathrm{SiO_2} + \mathrm{R^{3+}O_{1.5}} = \mathrm{R^{3+}HSiO_4} \quad (3)$$

$$[\mathrm{Ti}] : \mathrm{H_2O} + 0.5\ \mathrm{Mg_2SiO_4} + \mathrm{TiO_2} = \mathrm{MgTiH_2O_4} + 0.5\ \mathrm{SiO_2} \quad (4)$$

which give the following equilibrium constants for pure forsterite (that is, in the approximation that $a^{ol}_{Mg_2SiO_4}$ is unity):

$$K^{[Si]} = a^{ol}_{Mg_2H_4O_4} a_{SiO_2} (f(H_2O))^{-2} \quad (5)$$

$$K^{[Mg]} = a^{ol}_{MgH_2SiO_4} (a_{SiO_2})^{-0.5} (f(H_2O))^{-1} \quad (6)$$

$$K^{[triv]} = a^{ol}_{R^{3+}HSiO_4} (a_{SiO_2})^{-1} (a_{R^{3+}O_{1.5}})^{-1} (f(H_2O))^{-0.5} \quad (7)$$

$$K^{[Ti]} = a^{ol}_{MgTiH_2O_4} (a_{SiO_2})^{0.5} (a_{TiO_2})^{-1} (f(H_2O))^{-1} \quad (8)$$

The four substitution mechanisms therefore depend on the activity of silica (a_{SiO2}) in different ways, and in the case of the [triv] and [Ti] substitutions, the chemical potentials (effectively, the availability) of the relevant enabling components, $R^{3+}O_{1.5}$ and TiO_2, respectively. But the most significant implication is that the way that H is incorporated in olivine may vary with the fugacity of H_2O, depending on whether the H atoms needed to achieve charge balance are completely associated with the point defect by being bonded to the oxygen atoms surrounding it in specific locations, or are disordered over the lattice by being bonded to oxygen atoms without regard to location. For example, if the four H atoms of the [Si] mechanism are bonded to oxygen atoms surrounding the Si site vacancy to produce local charge balance (that is short-range order), as recently shown by Xue et al. (2017), then the activity of the $Mg_2H_4O_4$ component would be proportional to its mole fraction, such that $C^{[Si]}_{H_2O} \propto f(H_2O)^2$. Alternatively, if the H atoms were bonded to four oxygen atoms at random positions in the lattice without short-range order, the configurational entropy of the $Mg_2H_4O_4$ component would be correspondingly greater, with its activity being proportional to its mole fraction to the power of four (or five, if the configurational entropy due to the Si vacancy itself is included), such that $C^{[Si]}_{H_2O} \propto f(H_2O)^4$ or $f(H_2O)^5$. For the [triv] mechanism, there is only the one H atom per formula unit of $R^{3+}HSiO_4$, so if the position where the H bonds to an oxygen is determined by short-range order with respect to where the R^{3+} cation is substituting, the activity of the $R^{3+}HSiO_4$ component will again be proportional to its mole fraction, but with the result that $C^{[triv]}_{H_2O} \propto f(H_2O)^{0.5}$. The close association of H substituting by the [triv] mechanism with the R^{3+} cation is demonstrated by the correlation of the wavelength of the [triv] infrared absorption with the ionic radius of the R^{3+} cation (Berry et al. 2007a). If local charge balance proves to be the case, then even at one specific temperature and pressure, the way that "water" is incorporated into olivine should change with the activity of H_2O, a_{H_2O}, defined as $f(H_2O)/f(H_2O)^\circ$, where $f(H_2O)^\circ$ is the fugacity of pure H_2O at the T and P of interest. This is an important consideration when laboratory observations are to be extrapolated to olivine in natural environments such as the Earth's mantle. In the laboratory, experiments are typically conducted at high a_{H_2O} to maximise H_2O contents, which facilitates measurements, but in the Earth's mantle, a_{H_2O} is lowered by the presence of other components, and an upper limit to a_{H_2O} at a given T, P and composition is imposed by partial melting (Green et al. 2010). Whenever amphibole is present, a_{H_2O} is constrained by amphibole-pyroxene-olivine equilibria and is considerably lower than if a free aqueous fluid phase were present (Lamb and Popp 2009).

The aim of the experimental study presented here is to test the relationships between C_{H_2O} and a_{H_2O} for the four principal substitution mechanisms. We chose a representative condition of 3.0 GPa and 1050 °C for experimental convenience, with a_{H_2O} varied using H_2O–NaCl mixtures. Although the [triv] substitution in natural olivines is likely mostly due to Cr^{3+} and Fe^{3+} (e.g. Tollan et al. 2015), both these elements also occur in olivine in a 2+ valence state, Fe predominantly so, making the determination of the amounts of Fe^{3+} or Cr^{3+} from analysed Fe or Cr impractical. Yet, to understand the [triv] substitution mechanism, it is desirable to know the concentration of the R^{3+} cation in octahedral coordination. For this reason, we sought a redox-insensitive element with a substantial solubility in forsterite, and a large enough ionic radius to substitute into its octahedral sites only—unlike Al, which may also be important in promoting water solubility in olivine (Grant et al. 2007a). There is really only one choice, Sc^{3+}.

Methods/Experimental

Preparation of starting materials

For experiments investigating [Ti], [triv] and [Mg] hydrous defects, powdered synthetic Ti- and Sc-doped forsterite was prepared by a process of solution-gelation. Magnesium nitrate was dissolved in water slightly acidified by nitric acid, with Ti and Sc added to produce forsterite containing 2000 ppm of Ti and Sc, respectively, using ammonium bis(oxalato)oxotitanate(IV) hydrate or scandium oxide. Concentrations in this paper are given in ppm by weight, i.e., μg g^{-1}. Tetra-ethyl orthosilicate was added with ethanol, after which a few drops of ammonia initiated the gelation process, with the mixture then left overnight to allow complete precipitation. The precipitate was dehydrated by heating on a hot plate at low temperatures for 24 h followed by intensive heating, first over a Bunsen burner and then in a box furnace at 600 °C. Pure forsterite, produced by the same method (but without the addition of Sc and Ti), was then mixed with the Sc- and Ti-doped forsterite powders in various ratios to produce the starting compositions for the hydration experiments. Concentrations of Sc and Ti in recovered olivine crystals are given in Table 1. In order to buffer the silica activity, enstatite powder was also produced by sintering synthetic forsterite with SiO_2 in the appropriate stoichiometric proportions. For experiments investigating the [Si] defect, powdered San Carlos olivine was run with synthetic magnesiowüstite ($Fe_{0.7}Mg_{0.3}O$) to produce a low a_{SiO2}, which enhances water substitution by this mechanism.

Table 1 Initial compositions of the H_2O-NaCl fluids and dopant concentrations of recovered forsterite measured by LA-ICP-MS

Experiment	X_{H2O}	Ti or Sc (ppm)	Sd[a] (ppm)
[Si]–San Carlos olivine buffered by magnesiowüstite, Re/ReO_2 fO_2 buffer			
21	0.38	–	–
20	0.44	–	–
32	0.63	–	–
31	0.72	–	–
22	0.81	–	–
30	0.89	–	–
23	1	–	–
[Mg]–undoped forsterite buffered by enstatite, no fO_2 buffer			
84	0.56	–	–
85	1	–	–
[triv]–Sc forsterite buffered by enstatite, no fO_2 buffer			
36	1	586	61
81	1	317	18
82	1	179	11
83	1	950	108
89	0.82	1114	39
91	0.68	1159	8
40	0.50	1228	88
41	0.36	1031	35
75	0.24	1127	16
[Ti]–Ti forsterite buffered by enstatite, no fO_2 buffer			
62	1	307	17
34	1	253	11
37	1	176	19
79	1	88	6
42	1	330	47
76	0.81	292	15
69	0.7	252	23
70	0.57	197	16
77	0.41	164	26
80	0.22	162	24

[a]Standard deviation

Piston-cylinder experiments

One challenge of the experiments was to produce crystals of H_2O-bearing forsterite or olivine large enough for FTIR spectroscopy, but at sufficiently low temperature to avoid melting and excessive dissolution of silica in the fluid, which informed the choice of 1050 °C at 3 GPa. The crystals were grown from the starting materials in a piston-cylinder apparatus. Powders of pure or doped forsterite were packed inside Pt capsules, with a layer of enstatite powder to buffer a_{SiO2}, and oxygen fugacity was controlled by the intrinsic conditions of the assembly (estimated to be approximately equivalent to the fayalite-magnetite-quartz buffer); at which conditions, the only redox-sensitive element, Ti, should be essentially all Ti^{4+} (Mallmann and O'Neill 2009). The experiments using San Carlos olivine and magnesiowüstite were loaded into Au capsules, with oxygen fugacity internally buffered using layers of Re and ReO_2 powders (Pownceby and O'Neill 2000) to ensure a constant Fe^{2}

$^{+}/Fe^{3+}$ in the olivine and magnesiowüstite. In all experiments, a_{H_2O} was controlled by adding variable amounts of NaCl powder and distilled H_2O. Water was added with a microsyringe, and the amount added was monitored with a microbalance before and after welding to check for evaporative loss. The capsules were contained in MgO-graphite-glass-NaCl assemblies and run in an end-loaded Boyd-type piston-cylinder apparatus. Experiments were performed at 3.0 GPa and 1050 °C for 168 h with temperature monitored by a type B ($Pt_{0.7}Rh_{0.3}$-$Pt_{0.94}Rh_{0.06}$) thermocouple contained within mullite. The experiment was ended by switching off the power. The capsule was pierced and inspected for the presence of free fluid, which confirmed that fluid had not escaped during the run. The capsule was then cut with a scalpel and peeled open. Relatively large forsterite (or olivine) crystals were recovered by picking under an optical microscope.

All the forsterite crystals that were subsequently analysed grew during the experiments, and the finely ground San Carlos olivine completely recrystallised, which is necessary to achieve the equilibrium point-defect structure (Matveev et al. 2001). That all the crystals characterised in this study grew during their synthesis in equilibrium with the buffering assemblages and fluid phase is in contrast to the experimental approach that hydroxylates pre-existing crystals (e.g. Bai and Kohlstedt 1993; Zhao et al. 2004; Gaetani et al. 2014). In this latter approach, H moves into the crystal by solid-state diffusion, but the pre-existing point-defect structure of the olivine crystal is conserved. This phenomenon was exploited by Bai and Kohlstedt (1993) to study olivine pre-equilibrated at different oxygen fugacities at atmospheric pressure and then hydroxylated at one oxygen fugacity at a given temperature and pressure. In such experiments, the point-defect structure under which the water is incorporated is metastable and distinguishable to that expected at global equilibrium (Matveev et al. 2001). A spectacular example of conserving pre-existing point-defect structure is the study of Jollands et al. (2016) on the decoration of Ti^{3+} in forsterite to form the [triv] substitution, by hydroxylation under oxidizing conditions. Stabilizing Ti^{3+} in forsterite requires highly reducing conditions (Mallmann and O'Neill 2009); accordingly, Jollands et al. (2016) prepared their material using a CO–CO_2 gas mixture with 97% CO at 1500 °C and 1 bar. The hydroxylation was then carried out at oxygen fugacities up to twelve orders of magnitude higher, imposed by the Re–ReO_2 and Ag–Ag_2O buffers at 850 °C and 1.5 GPa (Jollands et al. 2016).

There is unnecessary controversy in the literature on the interpretation of experimental results on the incorporation of water in olivine due to the failure to distinguish clearly between the two types of experiments, those aimed at achieving global equilibrium under the conditions at which the water is incorporated and those aimed at hydroxylating existing point-defect structures and compositions, including the minor-element concentrations that determine the amounts of water associated with the [triv] substitution mechanisms. We suggest that to reduce the confusion, all experiments should be labelled "G-type" if global equilibrium is the target and "M-type" if a metastable equilibrium is sought. The experiments reported here are G-type. The interpretation of M-type experiments is likely to be the more difficult by far, because the metastable substitutions depend on the rates of diffusion of the point-defects and/or elements that enable them; in contrast, the global equilibrium that is the aim of a G-type experiment is independent of transport phenomena, by definition. The interpretation of M-type experiments may therefore be aided by H diffusion studies (e.g. Mackwell and Kohlstedt 1990; Demouchy and Mackwell 2006; Padron-Navarta et al. 2014). One potential problem is if the experimental products fall in the no man's land between the two types.

Fourier transform infrared spectroscopy

The forsterite and olivine crystals were analysed by FTIR spectroscopy, using a Bruker Tensor 27 infrared spectrometer with a liquid nitrogen cooled mercury cadmium telluride detector coupled to a Bruker Hyperion infrared microscope. The analysis chamber was continuously purged with dry air in order to minimise interference of the intrinsic forsterite absorbance band by atmospheric background. Individual crystals were placed on a 100 μm Cu wire grid and measured in transmission mode with polarised and unpolarised light. For polarised measurements, crystal orientation was deduced through comparison of the silicate overtone region (~1200–2200 cm^{-1}) with reference spectra for each principal orientation (Lemaire et al. 2004). Background analyses were taken at regular intervals, and both sample and background measurements were the average of 64 individual scans at a resolution of 4 cm^{-1}. After analysis, the acquired spectra were processed using the Bruker OPUS© software package. Any residual atmospheric contamination was minimised by application of the atmospheric compensation tool, followed by baseline correction using the "concave rubber band" software tool using 64 baseline points and four iterations. Occasionally, a further baseline correction was applied in order to remove broad absorbance bands due to fluid inclusions.

Quantification of data

The areas beneath the major absorbance bands from each experiment were calculated using the integration function of the OPUS software package, and the resulting absorbance value was normalised to a thickness of 1 cm. Crystal thickness was determined by exploiting the relationship with absorbance in the silica overtone region (1625–2150 cm^{-1}), whereby thickness (in microns) is calculated by dividing the integrated absorbance of the overtones (between 1625 and 2150 cm^{-1}) by the appropriate coefficient. For unpolarised measurements, the coefficient of Shen et al. (2014) was used (0.553). For polarised measurements, coefficients specific for each principal axis were determined by conducting unpolarised and polarised measurements on the same crystal and comparing the respective overtone absorbances. By doing this for multiple crystals, we established coefficients of 0.75, 0.50 and 0.56 for *a*, *b* and *c*-axes, respectively, which were then used in all subsequent thickness calculations for polarised data. Water contents for polarised measurements were calculated by summing the normalised absorbance of each major band along each of the three principal axes and dividing by the integral absorption coefficient. For each experiment design, only the absorbance due to the principal bands is reported (see Table 2 for the integration ranges used), guided by previous experimental and theoretical studies that have reported the association of different defects with specific band positions (e.g., Lemaire et al. 2004; Berry et al. 2005; Walker et al. 2007; Kovacs et al. 2010; Umemoto et al. 2011; Ingrin et al. 2013). Any additional less intense bands at distinctly different wavenumbers were not included in calculations of water, with the exception of bands attributed to Mg vacancies, which are reported separately. The reported uncertainty in the water contents is the standard deviation from measurements of many individual crystals, and is a measure of precision rather than accuracy, as it does not include systematic or semi-systematic uncertainties due to additional factors such as errors in published absorption coefficients, thickness calibrations and baseline corrections, which are likely to be significant, but are difficult to quantify. For the unpolarised absorbances, the measurements from at least 12 crystals from each experiment were averaged, divided by the same integral absorbance coefficient and then multiplied by three, following the method of Kovacs et al. (2008). Uncertainty in water contents following this method is dependent on the degree of polarisation of the defects, the absolute water contents and any unintentional bias in orientation due to crystal shape. Based on this and the comparison with Fig. 1 of Kovacs et al. (2008), an uncertainty of 10% was applied to all unpolarised data.

Spectral deconvolution

When appropriate, complex spectra were deconvoluted in order to isolate the absorbance of individual, overlapping bands. We used a similar method to that of Tollan et al. (2015): bands were fit to a Gaussian function by linear least squares regression, with the optimum half-width for each band held constant between spectra such that only band height was allowed to vary. The starting point for each model was the band positions from Ti-free [Mg] experiments. The positions of any additional bands were deduced through the regression procedure.

LA-ICP-MS

The concentrations of Ti and Sc were determined by laser ablation inductively coupled plasma mass spectrometry (LA-ICP-MS). The laser ablation system consists of a CompexPro 110F ArF excimer laser (193 nm wavelength) with a custom-built two-volume ablation cell for rapid aerosol extraction and washout. The laser was run with a frequency of 5 Hz in energy constant mode, which ensured a stable fluence at the sample surface of 3 to 4 J/cm^2. A focused spot with a diameter of 37 μm was used for sample and standard analyses. The cell was connected to an Agilent 7700x ICP-MS, and ablated material was transported to the ICP-MS in a stream of ultra-high purity He and Ar. Analyses were acquired as a time-resolved signal with 20 s of background followed by 40 s of counts on ablated material. NIST 610 and 612 glasses were used as calibration standards, and Si was used as internal standard.

Results and discussion

Each experiment yielded from three to 15 crystals large enough and sufficiently free of fluid inclusions to obtain quantifiable FTIR data. Polarised spectra from the three principal axes were recovered from each experiment, complemented by unpolarised spectra. Water concentrations calculated from both polarised spectra and unpolarised spectra following the method described by (Kovacs et al. 2008)) are in very good agreement for average water concentrations calculated from each experiment, ranging from 50 to 2500 ppm (Fig. 1), but particularly at lower water concentrations which are within the range typical of natural mantle olivine. This is consistent with the theory of Sambridge et al. (2008), as implemented by Kovacs et al. (2008), and illustrates the viability of using unpolarised data to calculate accurate water concentrations in olivine with a variety of defect populations. The concentration of H_2O associated with each species of hydrous point defect shows a clear relationship with the activity of water of the experiment (Fig. 2). Distinctly different polarised band positions and shapes can be identified from experiments investigating each of the different defect types.

Low SiO_2 activity experiments: [Si]

Olivines from experiments conducted at low silica activity (buffered by magnesiowüstite) are characterised by complex polarised spectra consisting of four major absorbance bands and six further minor absorbance bands (Fig. 3). Bands are moderately to strongly polarised, with very small FWHM (full width at half maximum). Parallel

Table 2 Calculated water concentrations in each of the three principal crystallographic orientations using polarised FTIR data and also total water calculated from unpolarised FTIR data using the method of Kovacs et al. (2008)

		E||a			E||b			E||c			Total			
Experiment	X_{H2O}	H_2O (ppm)	sd	*n*	H_2O (ppm)	sd	*n*	H_2O (ppm)	sd	*n*	H_2O (ppm)	sd	Unpolarised H_2O (ppm)	*n*
[Si] integration range 3650–3420 cm^{-1}														
21	0.379	67	23	11	39	24	7	44	7	6	150	33	163	19
20	0.438	164	51	11	55	21	5	49	30	14	268	55	307	22
32	0.63	418	30	12	241	31	12	139	36	12	798	43	676	21
31	0.717	472	40	17	281	28	14	218	22	11	970	49	918	35
22	0.814	609	101	15	381	147	10	248	81	16	1239	178	1147	26
30	0.892	719	124	13	430	88	15	287	75	13	1436	152	1302	34
23	1	1208	104	10	661	138	8	444	43	7	2314	173	2087	14
[Mg] integration range 3268–3107 cm^{-1}														
84	0.56							16	4	11	16	4		
85	1							24	3	15	24	3		
[triv] integration range 3375–3293 cm^{-1}														
40	1	79	5	3	40	5	6	48	3	7	167	7		
41	1	58	6	6	30	2	3	11	2	3	100	6		
75	1	34	5	6	21	6	3	20	2	7	75	8		
36	1	102	4	7	44	4	12	53	6	10	199	6	193	24
82	0.82	91	3	8	46	3	4	50	2	3	187	4	178	21
81	0.68	80	2	6	40	5	4	44	3	6	164	5	152	22
83	0.5	67	4	10	30	2	6	37	5	7	133	5	136	23
91	0.36	58	4	10	24	3	7	29	3	8	111	5	101	20
89	0.24	42	2	5	15	6	1	22	6	4	79	6	76	10
[Ti] integration range 3586–3495 cm^{-1}														
62	1	84	11	3	15	5	6	38	2	8	138	12	140	21
34	1	91	16	12	13	9	7	35	5	7	139	18	146	31
37	1	50	8	6	10	9	6	26	8	9	86	12	81	20
79	1	31	4	5	7	4	4	15	4	6	54	6	57	14
42	1	88	8	13	14	7	8	39	5	7	140	11	142	23
76	0.81	81	4	5	12	4	7	37	2	7	130	6	128	31
69	0.7	73	7	4	12	3	10	29	4	5	115	8	103	22
70	0.57	57	5	11	9	1	4	20	4	5	86	5	94	21
77	0.41	40	4	5	5	2	5	15	1	3	60	4	61	23
80	0.22	25	4	13	5	2	10	10	3	10	40	4	36	22

Only the principal absorption bands that dominate in each experiment were used, as indicated by the integration ranges. The data from the [triv] experiments have had the contribution from [Si] subtracted, as explained in the text. The expected standard deviation of the unpolarised data is ~10%, as discussed in Kovacs et al. (2008)

sd standard deviation in ppm

to E||a, the strongest bands are centred at 3612 cm^{-1} followed by 3578 cm^{-1}. The band at 3612 cm^{-1} shows strong asymmetry, with a prominent shoulder positioned at 3600 cm^{-1}, whilst the band at 3578 cm^{-1} shows minor asymmetry with a small shoulder positioned at 3550 cm^{-1}. Parallel to E||b, the strongest band is centred at 3550 cm^{-1}, followed by the band at 3578 cm^{-1} which has similar intensity as in the E||a direction. The band at 3612 cm^{-1} is also present, but at a much lower intensity than in the E||a direction. Parallel to E||c, only one major band is present, centred at 3566 cm^{-1}. This single band however is strongly asymmetric, with a double shoulder at 3545 and 3533 cm^{-1}. The band at 3612 cm^{-1} is again present, this time at even lower intensities than along either E||a or E||b. None of the other major bands from E||a or E||b are above the limit of detection.

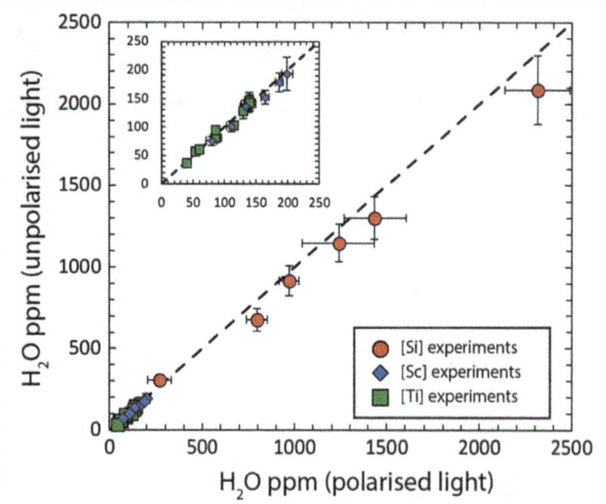

Fig. 1 Comparison of water concentrations for different point defects calculated using polarised light and unpolarised light, following the method of Kovacs et al. (2008). Note that the calculated water contents are based only on the principal absorbance bands from each experiment, as discussed in the "Methods" section. Integration ranges are as follows: [Si] 3650–3422 cm^{-1}; [Ti] 3586–3495 cm^{-1}; [triv] 3375–3293 cm^{-1}

Water associated to titanium: [Ti]

Two series of Ti-doped experiments were conducted: one set at varying a_{H_2O} but fixed bulk Ti concentration and another at fixed a_{H_2O} and varying bulk Ti concentration. Spectra from both types of experiment shared the same absorbance features. Forsterite from experiments doped with Ti reveals less complex spectra than those from [Si] experiments, with only two major bands and one minor band, which are present in each direction (Fig. 4). The two most prominent bands are centred at 3572 and 3525 cm^{-1}, consistent with previous studies (Berry et al. 2005; Kovacs et al. 2010; Padrón-Navarta et al. 2014). The 3572 cm^{-1} band is typically the most intense of the two, by a factor of approximately 2 and 1.5 in the E||a and E||b directions respectively. In the E||c direction, the 3572 cm^{-1} is more intense at high a_{H_2O}, but the ratio of the two peaks decreases to unity as a function of decreasing a_{H_2O}. Due to an overlapping contribution from [Si] bands with the 3572 cm^{-1} band (Padrón-Navarta et al. 2014), spectra from [Ti] experiments were deconvoluted as described in the "Methods" section, and the following discussion of [Ti] data is based on the results of this deconvolution (Fig. 7, Table 4). Titanium-doped forsterite covers a range of Ti contents, from 330 to 88 ppm (Table 1). Multiple

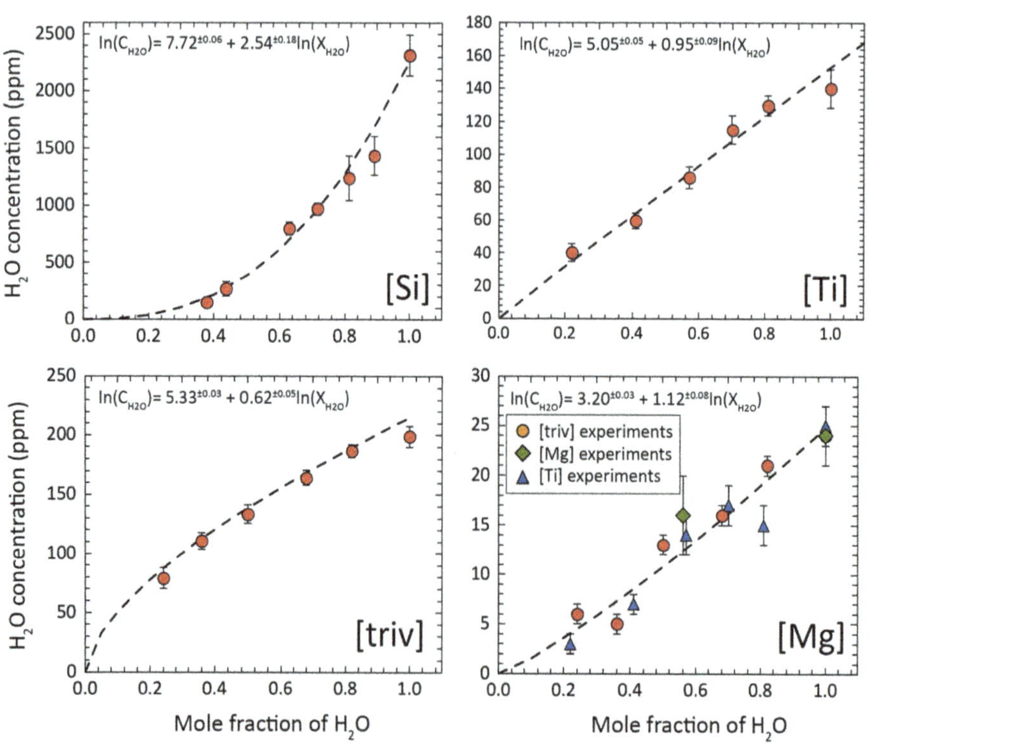

Fig. 2 Variation in water concentration with X_{H2O} in H_2O–NaCl fluids for the four different point defects. The calculated water contents are based only on the principal absorbance bands from each experiment, as discussed in the "Methods" section of the text. The best fit curves are from least squares non-linear regression with weighting of both variables according to the standard deviations in Tables 2 and 3

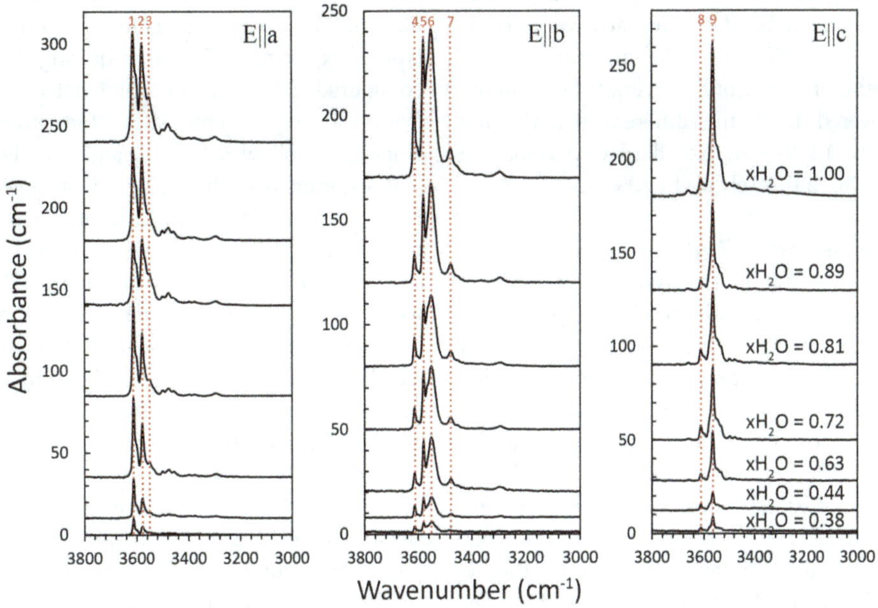

Fig. 3 Polarised spectra measured parallel to principal crystallographic axes for [Si] experiments, ordered from highest water activity (*top*) to lowest (*bottom*). Spectra are the average of multiple measurements, normalised to 1 cm and offset for clarity. The indicated band positions are (1) 3612 cm^{-1}, (2) 3578 cm^{-1}, (3) 3550 cm^{-1}, (4) 3612 cm^{-1}, (5) 3578 cm^{-1}, (6) 3550 cm^{-1}, (7) 3477 cm^{-1}, (8) 3612 cm^{-1} and (9) 3566 cm^{-1}

crystals from the same experiment had similar concentrations, with standard deviations of 5–15%.

Water associated to trivalent cations: [triv]

As with [Ti] experiments, two series were conducted: one with fixed a_{H_2O} and another with fixed Sc concentration, and likewise, spectra from both experiment types were consistent with each other. Scandium-doped forsterite have simple spectra which are consistent with the study of Berry et al. (2007a), with one major band with very small FWHM at 3355 cm^{-1}, which appears strongly in all three directions but most intensely along E||a and a minor band

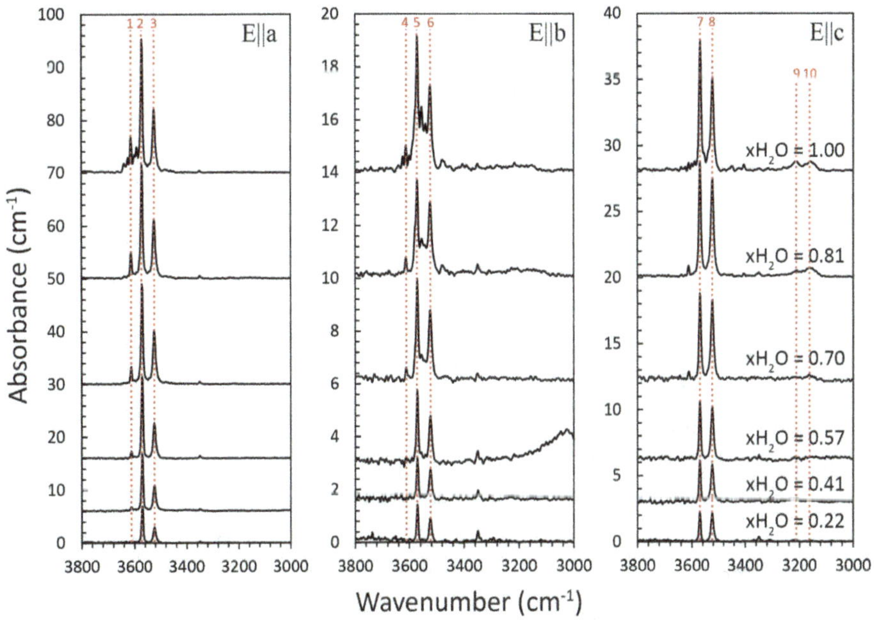

Fig. 4 Polarised spectra measured parallel to principal crystallographic axes for [Ti] experiments, ordered from highest water activity (*top*) to lowest (*bottom*). Spectra are the average of multiple measurements, normalised to 1 cm and offset for clarity. The indicated band positions are: (1) 3612 cm^{-1}, (2) 3572 cm^{-1}, (3) 3525 cm^{-1}, (4) 3612 cm^{-1}, (5) 3572 cm^{-1}, (6) 3525 cm^{-1}, (7) 3572 cm^{-1}, (8) 3525 cm^{-1}, (9) 3220 cm^{-1}, (10) 3160 cm^{-1}

at 3320 cm^{-1}, found principally along E||c (Fig. 5). A further minor, much broader band centred at 3160 cm^{-1} is similar to those found in [Ti] and [Mg] experiments and thus unrelated to the presence of Sc. Average Sc contents of forsterite recovered from the different experiments ranged from 179 to 1228 ppm. Standard deviations are typically less than 10%, but rarely up to 20%.

High SiO_2-activity experiments: [Mg]

Two experiments were run in the system $MgO–SiO_2–H_2O$ with excess enstatite and without Ti or Sc present to create hydrated Mg vacancies, to complement the results from both the [Ti] and [triv] experiments, which were also enstatite-buffered and produced the same absorbance bands in addition to those from the [Ti] and [triv] substitutions (Figs. 4 and 5). The absorption occurs at lower wavenumbers, with two broad bands typically present, one at 3220 cm^{-1} and another at 3160 cm^{-1}, with the former band generally more intense (an exception to this is discussed later). Neither of these bands was produced in [Si] experiments, which were conducted at much lower silica activity. Additional minor bands are observed in the [Mg] experiments, the most prominent of which is centred at 3567 cm^{-1} followed by a number of less intense bands between 3300 and 3500 cm^{-1}.

Band positions and point defect assignments

Olivine and forsterite from [Si], [Ti], [triv] and [Mg] experiments all have distinctive infrared spectra, indicating that the experiments were successful in producing different point defects. This is confirmed by the different polarisations of the principal bands for olivine/forsterite from each type of experiment (Fig. 6), requiring that O–H bonds be configured differently for each defect. Bands for olivine from [Si] experiments show the strongest absorbance along E||a, followed by E||b and E||c. For forsterite from [Ti] experiments, the order of strongest absorbance is E||a, E||c and then E||b, whilst for forsterite from [triv] experiments, E||a shows the strongest absorbance, with E||b and E||c the same within uncertainty. [Mg] bands meanwhile are only present along E||c. We have assigned the bands from each experiment to a different point defect stoichiometry based on the design and intent of each experiment and through comparison with similar experiments/spectra from the literature.

Bands for olivines from [Si] experiments (Fig. 3) have similar or identical positions with those in other low silica activity (wüstite- or magnesiowüstite-buffered) experimental studies (Aubaud et al. 2007; Lemaire et al. 2004; Padrón-Navarta et al. 2014; Withers and Hirschmann 2008), but also occur in experiments buffered at high silica activity conducted at significantly higher pressures (Smyth et al. 2006), with the highest intensity bands located at high wavenumbers (>3500 cm^{-1}). The precise location of the bands is occasionally offset by 1–8 cm^{-1}: for example, the band at 3550 cm^{-1} is found at slightly higher wavenumbers in other studies, likely due to differences in the configuration of Fe in M sites adjacent to the hydrated Si vacancy (Blanchard et al. 2017). The general complexity and number of bands is much

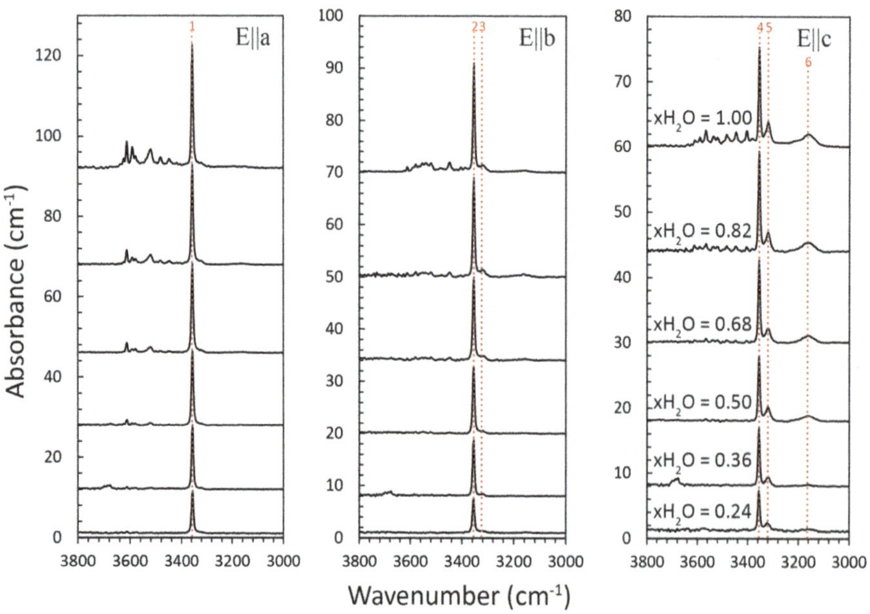

Fig. 5 Polarised spectra measured parallel to principal crystallographic axes for [triv] experiments (Sc-doped), ordered from highest water activity (*top*) to lowest (*bottom*). Spectra are the average of multiple measurements, normalised to 1 cm and offset for clarity. The indicated band positions are (1) 3355 cm^{-1}, (2) 3355 cm^{-1}, (3) 3320 cm^{-1}, (4) 3355 cm^{-1}, (5) 3320 cm^{-1} and (6) 3160 cm^{-1}

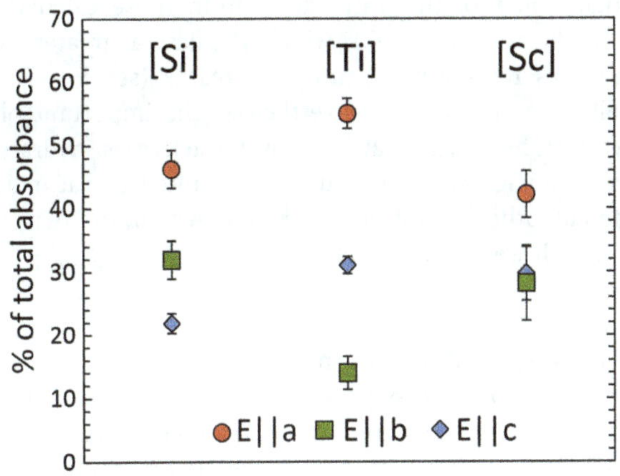

Fig. 6 Percentage that the absorbance from each orientation contributes to the total absorbance, for different point-defect types. The absorbance associated with [Mg] defects is only detectable along E||c

greater for defects in olivine from [Si] experiments than any other defect type investigated, indicating the involvement of multiple hydroxyl groups in a number of configurations, as explored through atomistic simulations (Balan et al. 2011; Umemoto et al. 2011; Walker et al. 2007). These observations are all consistent with hydrogen bonding to oxygen atoms surrounding silicon vacancies in the olivine structure, as denoted in Eq. 1 and supported by other studies investigating hydrous defect configurations (Ingrin et al. 2013; Blanchard et al. 2017).

Forsterite from the [Ti] experiments also displays absorbance at high wavenumbers, with bands at 3572 and 3524 cm^{-1} being the most intense (Fig. 4). These band positions are unique to [Ti] experiments, and are identical in wavenumber to previous experimental studies of Ti-doped olivine/forsterite (Berry et al. 2005; Kovacs et al. 2010; Padrón-Navarta et al. 2014). They are also the most commonly observed bands in natural upper mantle olivines sampled in spinel peridotite xenoliths, although their contribution to the total absorbance for a given sample varies significantly (Berry et al. 2005; Denis et al. 2013; Schmädicke et al. 2013). Precise band positions generally vary slightly in the literature, which can be explained in part due to the overlap with the [Si] bands at 3578–3580 and 3566 cm^{-1} (Padrón-Navarta et al. 2014). As with previous studies using Ti as a dopant, we assign these bands to formation of a titano-clinohumite point defect within the forsterite structure (Eq. 2), with the resultant generation of a silicon vacancy explaining the similar wavenumbers to bands from [Si] experiments.

The [triv] experiments produced forsterite with absorbance dominated by a single intense band at 3355 cm^{-1}, similar to the Sc-doped experiments of Berry et al. (2007a). This band, along with the minor band at 3320 cm^{-1}, are unique to the [triv] experiments and are assigned to coupled substitution of Sc^{3+} and H^{+}, with the latter providing local charge balance in octahedral site vacancies (Eq. 3). In natural olivines, bands in this wavenumber region occur sporadically, but can occasionally be the dominant defect type, particularly in olivines with very low Ti concentrations (Soustelle et al. 2013; Soustelle et al. 2010; Tollan et al. 2015). The precise position, polarisation and relative intensity of bands from natural olivines differ from those produced in our [triv] experiments, indicating that Sc is not a significant partner, as expected from the much lower concentrations of Sc in natural mantle olivine (typically 2–10 ppm). Ferric iron and Cr^{3+} are both much more abundant trivalent cations in natural olivine and are the likely candidates for the 3+ cation partnering H in this substitution mechanism (Tollan et al. 2015; Blanchard et al. 2017).

Two experiments were conducted specifically investigating [Mg] defects, which are generated most strongly under conditions of high silica activity (orthopyroxene/enstatite-buffered). However, identical bands were produced in all the other enstatite-buffered experiments, but not in the [Si] experiments, which were buffered at low a_{SiO_2} by magnesiowüstite. The bands associated to the [Mg] defect are located at lower wavenumbers, and display characteristically broad absorbance with large FWHM, but only along E||c. Two bands can be distinguished, one centred at 3160 cm^{-1} and another at 3220 cm^{-1}. Both of these bands are present in [Ti] and [Mg] experiments, with the band at 3220 cm^{-1} more intense. In [triv] experiments however, the band at 3160 cm^{-1} is more intense. Lemaire et al. (2004), Balan et al. (2011) and Umemoto et al. (2011) suggested that these band positions can be explained by protonation of vacant M1 and M2 sites. However, as pointed out by Blanchard et al. (2017), protonation of the M2 site was never explicitly modelled in these studies, and the variation in band position can be better explained by protonation of an M1 vacancy coupled with Fe^{2+} occupying either a neighbouring M1 or M2 site. Preferential ordering of Sc into one of the two kinds of octahedral sites would imply that vacancies concentrated in the other kind, with the H^{+} bonding to an O^{2-} surrounding the vacancy, to even out the perturbation of charge density. Based on site occupancy arguments in Blanchard et al. (2017), it is perhaps more likely that the greater intensity at 3160 cm^{-1} in [triv] experiments indicates that Sc is partitioning more favourably onto the larger M2 site, with the vacancy and its compensating H favouring M1, assuming that Sc is behaving similarly to Fe^{3+}. Similar or identical bands to these are commonly observed in other experimental studies and also in natural samples (e.g., Aubaud et al. 2007; Férot and Bolfan-Casanova 2012; Gaetani et al. 2014; Grant et al. 2007a, b). Band positions vary much more than bands from other defects, which may be due to the very broad and often low intensity of these bands, making it difficult to precisely locate the wavenumber of maximum intensity. In addition, partitioning of a wide range of trace and minor cations onto either or both octahedral sites could result in slightly different O–H bond lengths and an associated wavenumber shift.

Although our experimental strategy aims to emphasise only one of the four types of substitution mechanism in each group of experiments, it is important to recognise that absorption bands corresponding to some of the other substitution mechanisms are a minor feature of most of the spectra (Figs. 3, 4 and 5). A small band at 3350 cm^{-1} is often observed in [Si] and [Ti] experiments, most likely due to hydrogen bonding with trace amounts of Fe^{3+} (Berry et al. 2007a). However, the most significant additional band is the presence of relatively strong absorbance associated with silicon vacancies in [Ti] and [triv] experiments, despite the fact that they were run under enstatite-buffered conditions (high a_{SiO_2}). Whilst this observation combined with the general decrease in the ratio of water associated with Mg vacancies (i.e. $C_{H_2O}^{[Mg]}$) to that associated with Si vacancies $\left(C_{H_2O}^{[Si]}\right)$ with decreasing total water content (C_{H_2O}) in the [Ti] experiments (Table 3) is qualitatively consistent with the theoretical calculations of Walker et al. (2007), the actual values of this ratio differ from those calculated by Walker et al. (2007), which predicts a greater decrease as total water content increases (see Fig. 3 of Walker et al. (2007)). Nevertheless, the important observation here is that at lower water activities, hydrated Mg vacancies are favoured over hydrated Si vacancies when all other conditions are kept constant, the reasons for which we address next.

Table 3 Calculated water contents in ppm for the [Mg] bands (hydrated Mg vacancies), based on the integration range 3268–3107 cm^{-1}, from [Ti] and [triv] experiments. Also reported is the water associated with silicon vacancies in [Ti] experiments, calculated from band deconvolution. See the text for further details

Experiment	X_{H2O}	[Mg] H_2O (ppm)	sd	[Si] H_2O (ppm)
[Ti]				
62	1			30
34	1			29
37	1			26
79	1			32
42	1	25	2	25
76	0.81	15	2	15
69	0.7	17	2	10
70	0.57	14	2	5
77	0.41	7	1	2
80	0.22	3	1	1
[triv]				
36	1	24	1	
82	0.82	21	1	
81	0.68	16	1	
83	0.5	13	1	
91	0.36	5	1	
89	0.24	6	1	

Relationships with water activity, a_{H_2O}

Significant controversy prevails regarding the issue of relating measured absorptions to water contents. Several theoretical and experimental studies have demonstrated that the absorption coefficient should vary as a function of the vibrational energy characteristic of the configuration of H in a given defect (e.g., Kovacs et al. 2010; Balan et al. 2011; Ingrin et al. 2014; Blanchard et al. 2017). Hence, there should be a negative correlation between the infrared wavenumber and absorption coefficient. At present, however, whilst computational studies are useful in defining the expected trend, absolute values are not reliable. Experimental studies on the other hand produce more accurate values, but have so far explored a relatively small number of appropriate defects. Amongst the available experimental studies, the value of 28,450 L mol^{-1} cm^{-2} from Bell et al. (2003) has been most widely applied. Kovacs et al. (2010) produced an identical value for olivine bearing [Ti] and [triv] defects, although their absorption coefficients for [Si] and [Mg] defects were compromised by the occurrence of a previously unknown defect involving boron which comprised a significant proportion of the water in the "Pakistani" olivine used in their experiments (Ingrin et al. 2014). More notably, Withers et al. (2012) determined a strikingly different value of 45,200 L mol^{-1} cm^{-2} for synthetic olivines with most of the absorbance in their infrared spectra occurring at similarly high wavenumbers to the spectra in the study of Bell et al. (2003), albeit with a different defect ([Si] as oppose to [Ti]). This study utilised elastic recoil detection analysis as the matrix-independent method of absolute H_2O concentration determination, as opposed to nuclear reaction analysis which was used by Bell et al. (2003). The reason for the discrepancy between absorption coefficients is currently unclear, but since the absolute determination of water content is of minor significance for this study, we apply the more commonly used value (Bell et al. 2003), 28,450 L mol^{-1} cm^{-2}, with the caveat that calculated values may be systematically wrong as a result, something we address further in the discussion. Furthermore, provided that only one substitution mechanism is being considered, use of a different calibration would change C_{H_2O} proportionately, and the relationship to a_{H_2O} would not be affected.

The water contents, C_{H_2O}, associated with the [Si], [Mg] and [triv] were fitted by weighted least squares regression with uncertainties in both C_{H_2O} and X_{H_2O} as given in Table 2, to the equation:

$$\ln(C_{H_2O}) = a + b * \ln(X_{H_2O}) \quad (9)$$

where C_{H_2O} is the concentration of water in ppm (μg g^{-1}) for a particular point defect: a and b are constants and X_{H_2O} is the mole fraction of H_2O in the H_2O–NaCl fluids loaded into the capsule of the experiment, and is related to the activity of H_2O by $a_{H_2O} = X_{H_2O}\gamma_{H_2O}$. For the purposes of this immediate discussion, we assume that γ_{H2O}, the activity coefficient, is unity (ideal mixing). This assumption will be revisited below. With this assumption, the constant b is the exponent from Eqs. 5, 6, 7 and 8, and the value derived from the regression provides a means of testing the stoichiometry of the different point defects.

Exponents for [Si], [triv] and [Mg] experiments are 2.5 ± 0.2, 0.6 ± 0.1 and 1.1 ± 0.1, which are close to but consistently higher than the predicted exponents of 2, 0.5 and 1. Such similar values are a strong indication that the substitution stoichiometries given by equilibria 5 to 7 are substantially correct. The data for the [Ti] experiments present the problem that there is substantial overlap of bands associated with [Si] with those that appear in the [Ti] experiments in the wavenumber region 3650–3450 cm^{-1}. This is confirmed by [Mg] experiments, which were performed under identical conditions but without being doped with Ti. Spectra from these forsterite crystals show absorption bands at 3578 cm^{-1} along E||a and 3567 cm^{-1} along E||c, bands which are most prominently observed in experiments conducted at lower silica activity and higher pressure (Lemaire et al. 2004; Mosenfelder et al. 2011). In order to obtain the correct exponent for water associated with Ti, spectra were deconvoluted into 7–8 bands in an effort to isolate the overlapping, defect-specific bands (Fig. 7, Table 4). Following deconvolution, the areas corresponding to both the principal [Ti] bands and subsidiary bands associated with [Si] were calculated (Tables 2 and 3). Regression of these data gives exponents of [Ti] 0.95 ± 0.09, while for [Si], we obtain 2.6 ± 0.4, in good agreement with the value of 2.5 ± 0.2 calculated for olivine from the [Si] experiments. The agreement is evidence that we have successfully deconvoluted the absorbances associated with the [Si] and [Ti] defects from their measured sum.

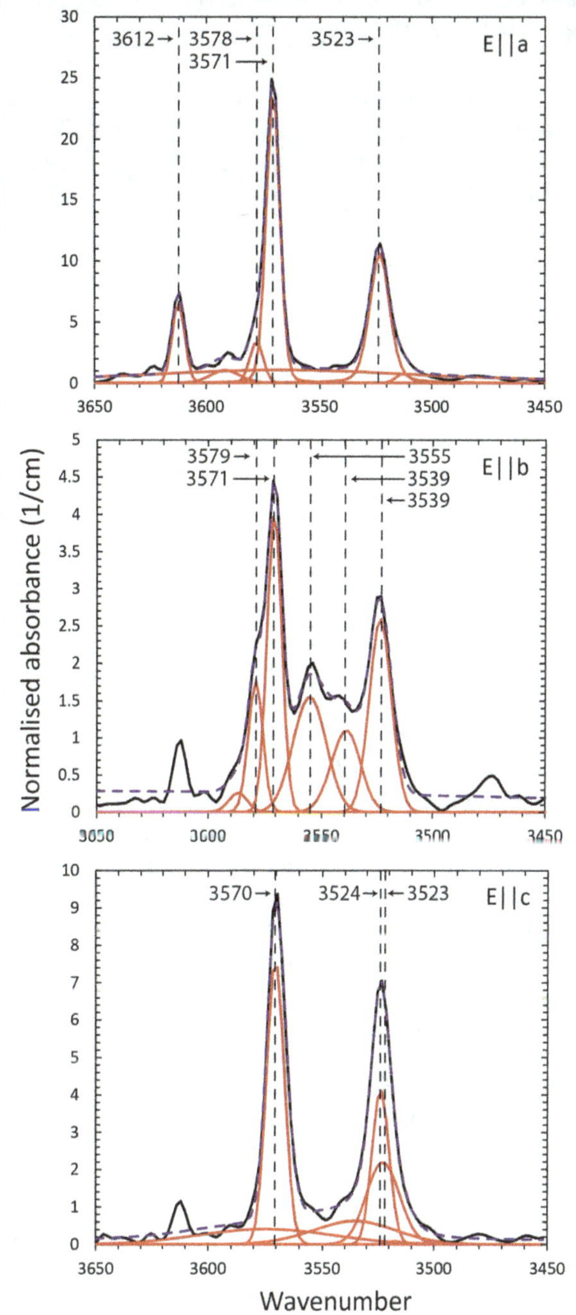

Fig. 7 Examples of deconvoluted spectra along E||a, E||b and E||c for [Ti] experiments. The deconvolution is necessary to isolate the absorbance due to [Ti] from that due to [Si]

There is a tendency for the exponents to be somewhat higher than those expected from the stoichiometry of reactions 1 to 4. This is explicable if there is a negative departure from ideal mixing in H_2O–NaCl fluids. The data can be reconciled to the expected values of the exponents if we describe the H_2O–NaCl binary at 1050 °C and 3 GPa by a regular solution model, such that RT ln $\gamma_{H_2O} = W_{H_2O\text{-}NaCl}(1-X_{H_2O})^2$, with an interaction parameter $W_{H_2O\text{-}NaCl}$ of ~−5 kJ/mol. This is not in agreement with the experimental study of Aranovich and Newton (1996), who found that while a_{H_2O} was nearly proportional to X_{H2O} in NaCl–H_2O solutions at 0.2 GPa, with increasing pressure, a_{H_2O} –X_{H_2O} relations tended towards $a_{H_2O} \propto (X_{H_2O})^2$. If this were the relationship for our

Table 4 Calculated water contents associated with each band deconvoluted from the measured spectra of [Ti] experiments. The method used is explained in the text

		E‖a				E‖b						E‖c		
		Wavenumber of band centre (cm^{-1})				Wavenumber of band centre (cm^{-1})						Wavenumber of band centre (cm^{-1})		
		H_2O (ppm)				H_2O (ppm)						H_2O (ppm)		
Experiment	X_{H2O}	3612	3578	3571	3523	3587	3579	3571	3555	3539	3524	3570	3524	3523
62	1	9.7	5.4	35.4	27.0	0.1	3.9	8.9	6.8	4.1	8.1	16.7	9.3	11.7
34	1	10.3	5.6	41.1	26.8	0.6	3.2	7.4	6.1	4.0	7.2	16.2	8.7	9.0
37	1	8.2	4.9	20.6	15.9	0.6	3.4	5.9	6.0	3.5	5.3	12.5	6.9	5.9
79	1	12.3	5.3	14.4	8.5	0.3	3.9	4.6	6.5	3.8	3.6	10.6	4.5	0.0
42	1	6.1	2.7	37.9	28.2	1.3	2.2	8.3	5.6	2.4	7.8	14.1	10.5	6.6
76	0.81	5.0	3.1	33.2	26.6	0.0	2.0	5.8	3.1	1.5	6.9	12.7	10.3	10.5
69	0.7	3.1	2.6	29.6	24.2	0.0	1.2	6.6	2.3	1.0	7.0	10.1	8.2	9.4
70	0.57	1.3	1.1	25.3	16.6	0.2	0.5	5.1	1.1	0.5	4.9	7.9	6.1	5.4
77	0.41	0.6	0.7	17.5	11.6	0.1	0.2	2.9	0.3	0.2	3.0	5.4	4.0	4.1
80	0.22	0.1	0.4	11.4	7.3	0.1	0.1	2.6	0.2	0.0	2.4	3.7	3.1	3.1

experiments, then, although the relative differences between exponent terms for different defects would remain the same, the absolute values would differ, casting doubt on our identifications of their stoichiometries. However, we note that our experiments were conducted at a higher temperature and pressure than the range investigated by Aranovich and Newton (1996), which extends only to 900 °C and 1.5 GPa. It should also be noted that we have not allowed for other components, such as SiO_2, dissolving in H_2O–NaCl fluids, which would lower a_{H_2O} compared to the simple binary.

Composition and water incorporation

The association between olivine composition and water incorporation remains a subject of debate in the literature, particularly the role of certain trace elements, such as Ti and trivalent species such as Fe^{3+}, Al and Cr^{3+}. Experimental studies have shown clear evidence for coupled substitution of hydrogen with these species in simple systems (Berry et al. 2005; Berry et al. 2007a; Faul et al. 2016; Férot and Bolfan-Casanova 2012; Grant et al. 2007a, b; Kovacs et al. 2010), supported by atomistic simulations (Balan et al. 2011; Blanchard et al. 2017; Walker et al. 2007). Furthermore, the FTIR spectra produced by these studies reproduce well the dominant absorbance bands of natural olivine crystals (Berry et al. 2005; Schmädicke et al. 2013; Tollan et al. 2015). However, there are few studies reporting good correlations between trace element composition and water content in natural olivine crystals, leading some to conclude that such incorporation mechanisms are of limited importance (Gaetani et al. 2014; Withers and Hirschmann 2008).

Our study confirms a clear relationship between water content and the presence of Ti and Sc. Of particular significance are the correlations between Ti concentration, water concentration and water activity in [Ti] experiments at both fixed water activity and fixed bulk Ti concentration (Figs. 2 and 8). This indicates that Ti and H are being incorporated together as part of the same point defect, and that the formation of this defect is the dominant method of incorporating both Ti and H in the system studied. Note that Ti concentrations overlap the range found in natural mantle olivines (De Hoog et al. 2010). Extrapolating the trend of H_2O concentration vs. Ti concentration to anhydrous conditions gives a Ti

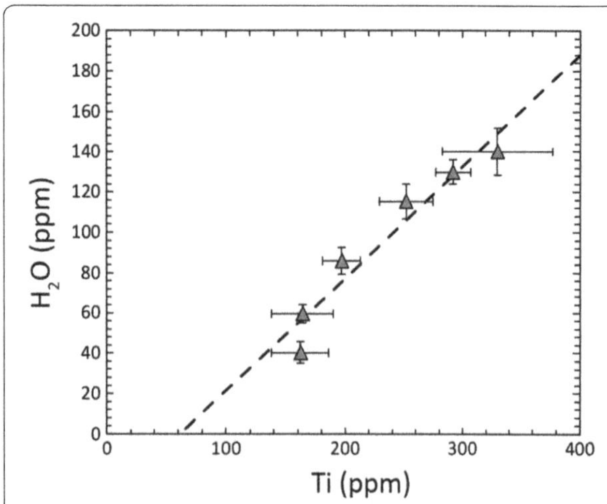

Fig. 8 Variation of H_2O concentration (C_{H_2O}) with Ti concentration for [Ti] experiments conducted at variable a_{H_2O}. The *dashed line* is a best fit through the deconvoluted data

concentration of ~60 ppm (Fig. 8), which is close to the maximum solubility of Ti in olivine under anhydrous conditions at the same temperature and pressure, as estimated from the experimental study of Hermann et al. (2005). This illustrates how Ti solubility in olivine is dramatically increased at high water activities. Converting H_2O and Ti to molar concentrations allows a further opportunity to assess the relationship between these two species (Fig. 9). The stoichiometry of Ti-clinohumite, $MgTi(OH)_2O_2$, requires two moles of OH (or one mole of H_2O) for every mole of Ti. Taking [Ti] experiments with a constant water activity of 1 but variable Ti concentration in olivine, we find that the trend of the data is very close to the 1:1 line (moles of Ti to moles of H_2O), consistent with this stoichiometry. Furthermore, this indicates that at water, saturated conditions ($a_{H_2O} = 1$) at 3 GPa and 1050 °C, essentially all Ti substitutes by the Ti-clinohumite point-defect mechanism, as opposed to tetrahedral Ti as in the Mg_2TiO_4 substitution (Berry et al. 2007b). The latter substitution becomes more important at higher temperatures and lower pressures (Hermann et al. 2005) and with increasing Fe content (O'Neill 1998). Finally, the good agreement between molar concentrations of Ti and H_2O is a strong indication that the absorption coefficient of Bell et al. (2003) is reliable for this particular defect. The Bell value was determined using olivines with identical spectra to our [Ti] experiments. Likewise, the identical value obtained from [Ti] experiments by Kovacs et al. (2010) is also validated.

Similar trends can also be observed for the [triv] experiments, where water activity was held constant and bulk Sc concentration varied. In this case, there is a good positive correlation between Sc concentration and H_2O concentration; however, when converted to molar concentrations, there is a deviation of approximately one third away from the expected Sc:H_2O molar ratio, such that there is more water than expected, given the Sc concentration. This deviation is consistent with the results of the study by Berry et al. (2007a), implying that this is not isolated just to Sc but to all R^{3+} cations (Fig. 10). We suggest that the simplest explanation for this is that the absorption coefficient used is not appropriate for this particular defect and should be increased by one third from the values reported by Bell et al. (2003) and Kovacs et al. (2010) and close to the value reported by Withers et al. (2012), in order to compensate for the excess water calculated here and by Berry et al. (2007a). This is also consistent with predictions from theoretical calculations for this type of defect (Blanchard et al. 2017), although the absolute value obtained in that study is somewhat unlikely. Since this defect is relatively common in natural olivines and thus a major contributor to mantle water storage (Tollan et al. 2015), an experimental study specifically designed to determine the absorption coefficient for this defect is required.

A further result that contrasts with the [Ti] experiments is that there is no correlation between the very

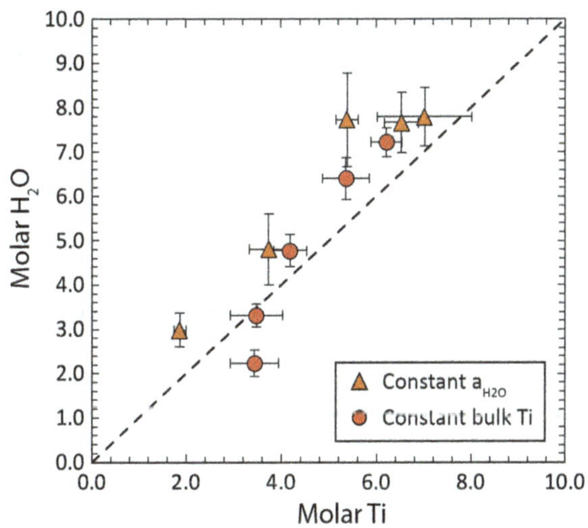

Fig. 9 Variation of molar H_2O concentration with molar Ti concentration for [Ti] experiments conducted both at both constant a_{H_2O} but variable at Ti concentration and variable in a_{H_2O} and constant in bulk Ti concentration. The *dashed line* represents 1:1 molar concentrations

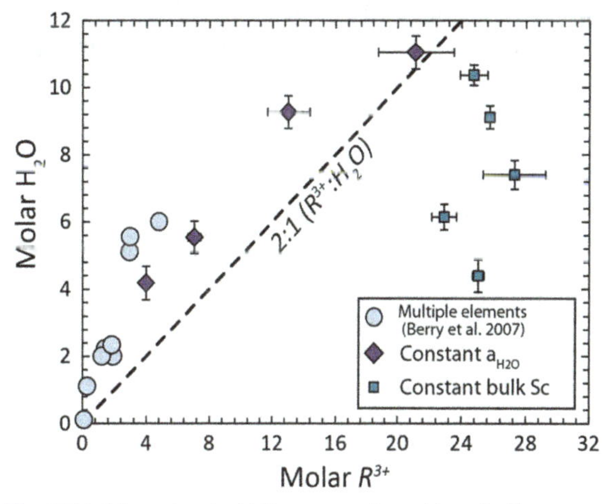

Fig. 10 Variation of molar H_2O concentration with molar Sc concentration for [triv] experiments conducted at constant water activity but variable in Sc concentration and at fixed bulk Sc concentration. Note that the latter experiments show no correlation between forsterite water concentration and Sc concentration, for reasons discussed in the text. For comparison, we have also plotted data from Berry et al. (2007a), which were similar experiments but at 1400 °C and 1.5 GPa and performed with a variety of trivalent cations in addition to Sc (Al, V, Ga, Y, In, Gd, Dy, Tm and Lu). Both sets of data confirm a consistent increase of C_{H_2O} with the availability of an R^{3+} cation at fixed water activity, but the deviation from the expected stoichiometry for the substitution may indicate that the absorption coefficient used in this study to calculate C_{H_2O} is in error; the value obtained by Withers et al. (2012) may be more appropriate (see text for further discussion)

high concentrations of Sc and H_2O concentration in the series at variable water activity, suggesting that another charge-balancing mechanism for Sc is operating in addition to the hydrous defect proposed here. Under anhydrous conditions, considerable Sc in olivine can be charge-balanced by an octahedral site vacancy (Spandler and O'Neill 2010) or, alternatively, Na^+ could be incorporated from the salt-water solution and charge-balance Sc instead of H^+. The latter would be increasingly important at lower water activities, where the $NaCl/H_2O$ ratio of the fluid increases substantially. But Grant and Wood (2010) found little evidence for coupled substitution of Na and Sc, so this reaction seems unlikely. It is possible that other charge-balancing species (such as Li) were present as contaminants in the salt; this would require further work to demonstrate.

Comparisons with other studies

Matveev et al. (2001) conducted experiments on natural olivines from Mount Porndon (initially Fo_{90-91}) at a range of water activities, controlled by diluting H_2O with CO_2 at 1300 °C and 2 GPa. Although their experiments were designed to be M-type, the olivine partially recrystallised to some extent in all runs, and FTIR spectra were reported only for the recrystallised material, using the marked change in Fo content (to $\sim Fo_{96-97}$) as a discriminant. Their results are therefore for G-type experiments, which explain why they were able to obtain completely different FTIR spectra in orthopyroxene-buffered experiments from magnesiowüstite-buffered experiments, as in this study. They found a factor of 1.3 decrease in water concentration in olivine over the range in X_{H_2O} from 1.0 to 0.3, in four orthopyroxene-buffered experiments. Inspection of their spectra shows that the [triv] defect dominates the total absorbance, and we can therefore directly compare their results to ours for this defect, but with the caveat that the trivalent cations associated with the [triv] substitution in the natural olivines are likely Fe^{3+} or/and Cr^{3+} (cf. Tollan et al. 2015) rather than Sc. Considerable Fe^{3+} in their olivines is expected given the relatively high oxygen fugacities of their experiments, achieved using the Re–ReO_2 buffer. The exponent for this defect derived from their data (following the same data-fitting procedure as for this study) is 0.58 ± 0.21, which is identical within uncertainty to the exponent calculated for the [triv] substitution in this study, supporting our hypothesis that the configurational entropy of the [triv] substitution is independent of the identity of the trivalent cation. The [triv] substitution was absent in the spectra from their magnesiowüstite-buffered experiments, although these were conducted at the same oxygen fugacity (Re–ReO_2); this observation confirms the role of high a_{SiO_2} in promoting this substitution, according to reaction (3). Their attribution of the [Si] substitution mechanism to the FTIR spectra of natural olivines published by Miller et al. (1987) was a mistake due to confusing the [Si] substitution in their experiments with the [Ti] substitution in the natural crystals. In the experiments, the Ti in the starting crystals of Mount Porndon olivine, which is low anyway (~20 ppm in an example analysed by Eggins et al. 1998), would have been lost during recrystallisation, while the infrared fingerprint of the [Ti] substitution was only recognised later (Berry et al. 2005).

Gaetani et al. (2014) performed what they assumed were M-type experiments on San Carlos olivine, but with re-equilibration of fO_2-sensitive defects. Only the orthopyroxene-buffered condition was investigated. They used mixtures of H_2O and CO_2 to vary a_{H_2O} but explored only three conditions $(X_{H_2O} = 1, 0.855, 0.185)$. There is a factor of ~2.5 difference between the lowest and highest water concentrations. Given that their published FTIR spectra show the [Ti], [triv] and [Mg] substitutions, this is in good agreement with our results. Gaetani et al. (2014) also demonstrate a decrease in intensity of the [triv] peaks between experiments at high and low fO_2, which suggests that the [triv] substitution was mainly associated with Fe^{3+}, which re-equilibrated during the hydroxylation. Interestingly, in M-type experiments, such reduction may be mediated by the fugacity of H_2 if the reaction occurs without diffusion of other elements:

$$Fe^{3+}HSiO_4 + 0.5\,H_2 = Fe^{2+}H_2SiO_4 \qquad (10)$$

Here, the component $Fe^{2+}H_2SiO_4$ represents essentially the same substitution mechanism in ferromagnesian olivines that we refer to as [Mg] in forsterite (Eq. 2). This reaction suggests that reduction of the Fe^{3+} [triv] substitution would increase the [Mg] substitution in M-type experiments, but as the olivines used by Gaetani et al. (2014) differed in composition, and the partitioning of $C_{H_2O}^{Total}$ amongst the three different substitution mechanisms was not reported, this hypothesis cannot yet be tested.

Yang et al. (2014) also used different ratios of H_2O and CO_2, finding a difference of a factor of 2 between values of C_{H_2O} in olivines from experiments with pure H_2O and those with H_2O–CO_2, but changing X_{CO2} from 0.22–0.50 apparently had no effect on the water solubility. This observation disagrees with those of Matveev et al. (2001) and Gaetani et al. (2014). Otsuka and Karato (2011) studied the effect of variable a_{H_2O} on water in San Carlos olivine in M-type experiments at 1000 °C and 5 GPa by using two three-phase assemblages involving a hydrous phase in the system MgO–SiO_2–H_2O, which produces known a_{H_2O} in the absence of a fluid phase, plus one experiment with H_2O fluid $(a_{H_2O}\,\tilde{}\,1)$. Thus, three values of a_{H_2O} were investigated. They observed that $C_{H_2O}^{Total} \propto a_{H_2O}$ (i.e. the exponent was 1), but their published polarised infrared spectra show that the hydroxyl in their run products is held by a combination of both the [Si] and the [triv] mechanisms. Deconvolution of the spectra would be needed to see if $C_{H_2O}^{[Si]} \propto a_{H_2O}{}^2$ and $C_{H_2O}^{[triv]} \propto a_{H_2O}{}^{0.5}$. The proportion of the [Si] substitution did increase with a_{H_2O}, in agreement at least qualitatively with the results of this study. This interpretation is also supported by the reported changes in the anisotropy of the absorption with a_{H_2O}.

Assignment of infrared absorption bands to substitution mechanisms

Our data also provides new evidence on the interpretation of infrared bands at wavenumbers greater than 3450 cm^{-1}, which typically dominate the spectra of natural olivines. There has been debate in the literature concerning whether silicon vacancies or magnesium vacancies are the main host of water in olivine. Zhao et al. (2004), Mosenfelder et al. (2006), Kohlstedt (2006), Bali et al. (2008) and Otsuka and Karato (2011) suggest that water associated with Mg vacancies is the most important incorporation mechanism and is responsible for the generation of high wavenumber bands, based on relationships between water solubility and fH_2O in experimental studies conducted over a broad range of conditions. A similar assignment has also been made by Smyth et al. (2006) through X-ray site occupancy refinement measurements on very water-rich samples equilibrated at high pressure and high a_{H_2O}.

On the other hand, Matveev et al. (2001), Lemaire et al. (2004), Berry et al. (2005, 2007a), Walker et al. (2007), Kovacs et al. (2010), Balan et al. (2011), Umemoto et al. (2011) and Ingrin et al. (2013) concluded that high wavenumber bands are instead due to hydrated Si vacancies, based on experimental studies at a range of silica activities, atomistic simulations of different hydroxyl configurations and the response of absorption bands to heating from liquid nitrogen temperatures. Furthermore, Blanchard et al. (2009), Umemoto et al. (2011) and Ingrin et al. (2013) pointed out that the use of the empirical relationship between OH frequencies and O–O bond distances by Smyth et al. (2006) is not valid for predicting site occupancy of hydroxyl groups in nominally anhydrous phases, since such a correlation was established for hydrous phases. Most recently, Xue et al. (2017) used ^1H NMR to determine the H positions in forsterite with 0.5 wt% H_2O, synthesised at 1200 °C and 12 GPa. They concluded that "this study has provided unambiguous evidence supporting that hydrogen is incorporated in forsterite at relatively high pressure dominantly as (4H)Si defects, with (2H)M1 defects playing only a very minor role".

By showing again that the appropriate compositional variations yield forsterite (or olivine) with different O–H infrared spectra, the results presented here confirm these latter interpretations of the identities of the substitution mechanisms of H_2O in olivine. The fact that the concentrations of hydroxyl, C_{H_2O}, in these different substitutions depend on a_{H_2O} in different ways that accord with the expected stoichiometry of the mechanisms (Eqs. 5, 6, 7 and 8) introduces a new line of supporting evidence for the identification of the substitution mechanisms, as does the variation of $C_{H_2O}^{[Ti]}$ with the concentration of Ti, and $C_{H_2O}^{[triv]}$ with the concentration of Sc (Fig. 10). The substitution mechanism associated with Mg site vacancies, [Mg], is quantitatively minor even at high a_{SiO_2} (buffered by enstatite) and unobservable at low a_{SiO_2} (buffered by magnesiowüstite), thereby refuting assertions that this mechanism is the main one by which H_2O substitutes into olivine. Nevertheless, the results on the [Mg] substitution are of considerable theoretical interest, because they show that the concentration of H_2O associated with this mechanism is the same at a given a_{H_2O} in the Sc- and Ti-doped forsterites as in the undoped forsterite, despite the fact that the total concentrations of H_2O $\left(C_{H_2O}^{Total}\right)$ in the latter experiments are much higher (Fig. 2d). This confirms that the concentrations of H_2O by the different mechanisms are additive, although from thermodynamic principals it is hard to conceive how this could be otherwise. Therefore, the total concentration of water in forsterite or olivine must be described by an equation of the type:

$$C_{H_2O}^{Total} = C_{H_2O}^{[Si]} + C_{H_2O}^{[Mg]} + C_{H_2O}^{[Ti]} + \sum_{R^{3+}} C_{H_2O}^{[triv]} \quad (11)$$

where the R^{3+} cations of importance in natural olivines are likely Fe^{3+} and Cr^{3+} (Tollan et al. 2015) and perhaps Al (Grant et al. 2007a).

For each substitution mechanism X, there is a separate equilibrium constant $K^{[X]}$ that relates $C_{H_2O}^{[X]}$ to $f(H_2O)$ and a_{SiO_2} in different ways (Eqs. 5, 6, 7 and 8), and each $K^{[X]}$ is a function of temperature and pressure. In addition, the [Ti] and [triv] substitutions depend on the concentrations of Ti and relevant R^{3+} cations, hence the activities of TiO_2 and the $R^{3+}O_{1.5}$ components in a system at global equilibrium. The $R^{3+}O_{1.5}$ components important in natural olivines are likely mainly $Fe^{3+}O_{1.5}$ and $Cr^{3+}O_{1.5}$ (Tollan et al. 2015; Blanchard et al. 2017), which are both redox-sensitive and depend on fO_2. It might be tempting to develop a single constitutive equation for the solubility of water in olivine as a function of these variables, but the convenience of this should be balanced against the guarantee that it would produce the wrong answers (Ingrin et al. 2013). A sufficient equation necessitates the parameterisation of each term on the right-hand side of Eq. 11 separately as a function of their relevant variables.

Implications for the assessment of water incorporation in mantle olivine

Not addressed in this study are the effects of temperature and pressure on the individual substitution mechanisms, but it would be extraordinary if the solubility of water in the four different mechanisms did not respond very differently to these variables. From the thermodynamic perspective, the effect of pressure acts in three ways at a given a_{H_2O} and a_{SiO_2}: (1) the increase of $f(H_2O)°$ with pressure; (2) the effect that $f(H_2O)°$ has through the stoichiometries of the reactions (reactions 1 to 4 and Eqs. 5, 6, 7 and 8); and (3) the effect on the equilibrium constants (Eqs. 5, 6, 7 and 8). As regards this third factor, the effect of pressure at constant $f(H_2O)$ (NB not constant $f(H_2O)°$) is:

$$\Delta \ln K = \int_{P1}^{P_2} \Delta \overline{V}_{solids}(P, T)dP \tag{12}$$

where $\Delta \overline{V}_{solids}$ is the change in partial molar volume of the solid components. For the example of the [Si] mechanism (reaction 1), $\Delta \overline{V}_{solids}$ is given by:

$$\Delta \overline{V}^{[Si]}_{solids} = \overline{V}^{olivine}_{Mg_2H_4O_4} - V^{olivine}_{Mg_2SiO_4} - V_{SiO_2} \tag{13}$$

Here, $\overline{V}^{olivine}_{Mg_2H_4O_4}$ is the partial molar volume of the $Mg_2H_4O_4$ component in the olivine, and V_{SiO_2} is the effective molar volume of the system, as determined by the coexisting phases. The standard states are the conventional ones of the pure components at the P, T of interest for the solid phases, and pure H_2O as an ideal gas at T and 1 bar. The value of $\overline{V}^{olivine}_{Mg_2H_4O_4}$ (298,1) may be calculated from the data in Smyth et al. (2006) as 45.2 cm^3/mol, by recalculating the H_2O contents of hydrous forsterites given in Table 2 to mole fractions of the $Mg_2H_4O_4$ component, and extrapolating linearly to this end-member. Hence, with V_{SiO_2} (298,1) = 19.0 cm^3/mol (from Mg_2SiO_4 + $Mg_2Si_2O_6$) and $V^{olivine}_{Mg_2SiO_4}$ (298,1) = 43.7 cm^3/mol, all from Holland and Powell (2011), we calculate $\Delta \overline{V}_{solids^{[Si]}}(298, 1) = 20.5\, cm^3/mol$. However, what is really needed is to compare the response of the [Si] substitution to pressure with those of [Mg], [Ti] and [triv], which requires the partial molar volumes of the components MgH_2SiO_4 and $MgTiH_2O_4$, and the relevant $R^{3+}HSiO_4$ components, which are at present unknown.

In contrast, the differential effect of the change in $f(H_2O)°$ with pressure comes simply from the stoichiometry of the reaction: increasing pressure will favour the [Si] substitution, because $C_{H_2O} \propto f(H_2O)^2$. That the [Si] mechanism is indeed the dominant mechanism in experiments with a_{H_2O} ~1 at P >> 3 GPa, even at high a_{SiO_2} (buffered by orthopyroxene) is evident from several experimental studies (Kohlstedt et al. 1996; Mosenfelder et al. 2006; Withers and Hirschmann 2008). However, decreasing a_{H_2O} at any pressure and temperature will decrease the proportion of [Si] relative to [Mg], [Ti] and especially [triv]. Higher temperatures should favour the mechanisms with higher configurational entropy, if differences in vibrational entropies are of subsidiary importance (Walker et al. 2007). With this assumption, the relative amounts of water associated with the different defects should increase with increasing temperature in the reverse order from the effect of pressure at a_{H_2O} ~1: [triv] > [Mg] ~ [Ti] > [Si] (Eqs. 1, 2, 3 and 4), other variables remaining constant.

We reiterate the point made by Ingrin et al. (2013), who emphasised that water incorporation into olivine cannot be described by a single solubility law. The existence of the four main solubility mechanisms with their different controlling stoichiometries mandates that the relationship between C_{H_2O} and the intensive thermodynamic variables of temperature, pressure, fH_2O and compositional variables requires four separate terms, one for each mechanism. Quantifying water substitution in olivine must also allow for the availability of Ti and the trivalent cations, particularly Fe^{3+}, whose concentration is sensitive to fO_2. Varying a_{SiO_2} also affects substitution mechanisms in different ways. Obviously, increasing a_{SiO_2} increases Mg vacancies but decreases Si vacancies and vice versa, but increasing a_{SiO_2} should also promote the [triv] substitution but decrease [Ti] (Eqs. 3 and 4). The values of a_{SiO_2} as buffered at low or high values by magnesiowüstite or pyroxene, respectively, also both change with P and T. The effect of the major element compositional variable Mg/Fe^{2+} on each mechanism also requires systematic study.

The fact that there are three ways by which increasing pressure at constant given a_{H_2O} and relative a_{SiO_2} changes $C^{[X]}_{H_2O}$ for each of the four substitution mechanisms means that it is difficult to deduce the nature of the substitution mechanism from the change of $C^{[X]}_{H_2O}$ with pressure; it would be even more difficult to deduce how water dissolves in olivine with the change of $C^{Total}_{H_2O}$ with pressure.

Interactions between the four defect mechanisms should also be taken into account in the interpretation of the water contents of natural olivines and their FTIR spectra. The speciation of water in olivine observed in simple-system experiments may not be preserved in natural olivines because of rearrangements on cooling and/or decompression. Where this occurs without mass transfer of components into or out of the olivine by diffusion, such rearrangements may be rapid. To give one example, it is possible that water may transfer from the [triv] mechanism to the [Ti] mechanism according to the reaction:

$$4/3R^{3+}HSiO_4 + 2/3\,Mg_2TiO_4 = R^{3+}{}_{4/3}SiO_4 + 2/3\,MgTiH_2O_4 + 1/3\,Mg_2SiO_4 \quad (14)$$

where all components are in olivine. The $R^{3+}_{4/3}SiO_4$ component is the anhydrous substitution of R^{3+} into olivine, charge-balanced by octahedral site vacancies (Evans et al. 2008), while the Mg_2TiO_4 component is the anhydrous substitution of Ti for Si in olivine studied by Hermann et al. (2005). Therefore, the observation that [Ti] is the dominant mechanism for water incorporation in olivine from San Carlos xenoliths (Berry et al. 2005) may reflect internal re-equilibration during cooling to some extent as yet unknown. Loss of water from olivine, which has often been inferred from both phenocryst olivines in magmas during eruption, or from olivines in xenoliths during their exhumation (e.g. Tollan et al. 2015 and references therein), may also affect speciation. Water loss changes a_{SiO_2} (inter alia), hence the point-defect structure of the olivine, in different ways according to the different substitution mechanisms (Eqs. 1, 2, 3 and 4). The effects of this are further complicated by the different rates of diffusion of H associated with the different substitution mechanisms (Padrón-Navarta et al. 2014).

Conclusions

This study presents the results of G-type experiments at one condition of temperature and pressure, namely 1050 °C and 3 GPa, which confirm the identities of the four main substitution mechanisms by which water is likely to be incorporated into mainstream mantle olivines. The results demonstrate the role that a_{H_2O} plays in determining which mechanisms are significant at the experimental condition, but from the stoichiometries of the substitution mechanisms, the same principle should apply at other pressures and temperatures. Evaluating water solubilities is not straightforward because it depends on the mechanisms by which water is incorporated (the type of hydrous defect), which in turn is dictated by other factors, most notably composition and silica activity. The consequence of this is that under otherwise identical conditions, higher water activities will favour hydrated silicon vacancies ([Si]) over other defects. Conversely, at low water activities, the ratio of hydrated silicon vacancies to other hydrous defects will be much lower. This has profound impacts for our understanding of water storage in the mantle and interpretation of measurements of natural olivine from different mantle domains. For example, the absence of the infrared absorption bands associated with the [Si] mechanism does not necessarily indicate high silica activity but could be due to low water activities, as expected in most mantle environments (Bali et al. 2008; Lamb and Popp 2009). This is highlighted by our observation that the [Si] substitution features prominently in enstatite-buffered experiments at high pressures and high a_{H2O}, even dominating total absorbance (Smyth et al. 2006). In order to understand the role of water in the mantle, careful consideration needs to be given to several factors that have hitherto been underappreciated, including the limitations on a_{H_2O} and the role of minor components, especially Ti and the more abundant R^{3+} cations, Fe^{3+} and Cr^{3+}.

Acknowledgements
We thank Dean Scott and Dave Clark for their tireless efforts maintaining the running of the experimental laboratories, Jung Park for his assistance with the LA-ICP-MS measurements and both Mike Jollands and Alberto Padrón-Navarta for their assistance with the FTIR and many valuable and fruitful discussions. We thank the two anonymous reviewers and Shun-ichiro Karato for his review and editorial handling.

Funding
We gratefully acknowledge the Australian Research Council (ARC) support through DP110103134 to JH and HON, and FL130100066 to HON, which partly supported PT during the final stages of this project. RS acknowledges an Australian Postgraduate Award.

Authors' contributions
The project was conceived by HON and JH, with the experiments and analyses conducted by RS. PT processed and modelled the data. PT and HON fit the data to the thermodynamic model. PT and HON wrote the manuscript with help from JH. All authors read and approved the final manuscript.

Competing interests
The authors declare that they have no competing interests.

Author details
[1]Research School of Earth Sciences, The Australian National University, Building 142, Mills Road, Canberra, ACT 2601, Australia. [2]Institute of Geological Sciences, Universität Bern, Bern 3012, Switzerland.

References
Aranovich LY, Newton RC (1996) H_2O activity in concentrated NaCl solutions at high pressures and temperatures measured by the brucite-pericalse equilibrium. Contrib Mineral Petrol 125:200–212

Aubaud C et al (2007) Intercalibration of FTIR and SIMS for hydrogen measurements in glasses and nominally anhydrous minerals. Am Mineral 92(5–6):811–828

Bai Q, Kohlstedt DL (1993) Effects of chemical environment on the solubility and incorporation mechanism for hydrogen in olivine. Phys Chem Miner 19:460–471

Balan E, Ingrin J, Delattre S, Kovács I, Blanchard M (2011) Theoretical infrared spectrum of OH-defects in forsterite. Eur J Mineral 23(3):285–292

Bali E, Bolfan-Casanova N, Koga KT (2008) Pressure and temperature dependence of H solubility in forsterite: an implication to water activity in the Earth interior. Earth Planet Sci Lett 268(3–4):354–363

Bell DR, Rossman GR, Maldener J, Endisch D, Rauch F (2003) Hydroxide in olivine: a quantitative determination of the absolute amount and calibration of the IR spectrum. J Geophys Res 108

Berry AJ, Hermann J, O'Neill HSC, Foran GJ (2005) Fingerprinting the water site in mantle olivine. Geology 33(11):869

Berry AJ, O'Neill HSC, Hermann J, Scott DR (2007a) The infrared signature of water associated with trivalent cations in olivine. Earth Planet Sci Lett 261(1–2):134–142

Berry AJ, Walker AM, Hermann J, O'Neill HSC, Foran GJ, Gale JD (2007b) Titanium substitution mechanisms in forsterite. Chem Geol 242(1–2):176–186

Blanchard M, Balan E, Wright K (2009) Incorporation of water in iron-free ringwoodite: a first-principles study. Am Mineral 94(1):83–89

Blanchard M, Ingrin J, Balan E, Kovács I, Withers AC (2017) Effect of iron and trivalent cations on OH defects in olivine. Am Mineral 102(2):302–311

De Hoog JCM, Gall L, Cornell DH (2010) Trace-element geochemistry of mantle olivine and application to mantle petrogenesis and geothermobarometry. Chem Geol 270(1–4):196–215

Demouchy S, Mackwell S (2006) Mechanisms of hydrogen incorporation and diffusion in iron-bearing olivine. Phys Chem Miner 33:347_355

Demouchy S, Mackwell SJ, Kohlstedt DL (2007) Influence of hydrogen on Fe–Mg interdiffusion in (Mg, Fe)O and implications for Earth's lower mantle. Contrib Mineral Petrol 154(3):279–289

Denis CMM, Demouchy S, Shaw CSJ (2013) Evidence of dehydration in peridotites from Eifel Volcanic Field and estimates of the rate of magma ascent. J Volcanol Geotherm Res 258:85–99

Eggins SM, Rudnick RL, McDonough WF (1998) The composition of peridotites and their minerals: a laser-ablation ICP-MS study. Earth Planet Sci Lett 154:53–71

Evans TM, O'Neill HSC, Tuff J (2008) The influence of melt composition on the partitioning of REEs, Y, Sc, Zr and Al between forsterite and melt in the system CMAS. Geochim Cosmochim Acta 72:5708–5721

Faul UH, Cline CJ, David EC, Berry AJ, Jackson I (2016) Titanium-hydroxyl defect-controlled rheology of the Earth's upper mantle. Earth Planet Sci Lett 452: 227–237

Férot A, Bolfan-Casanova N (2012) Water storage capacity in olivine and pyroxene to 14 GPa: implications for the water content of the Earth's upper mantle and nature of seismic discontinuities. Earth Planet Sci Lett 349–350:218–230

Gaetani GA et al (2014) Hydration of mantle olivine under variable water and oxygen fugacity conditions. Contrib Mineral Petrol 167(2):965

Grant KJ, Wood BJ (2010) Experimental study of the incorporation of Li, Sc, Al and other trace elements into olivine. Geochim Cosmochim Acta 74(8):2412–2428

Grant KJ, Kohn SC, Brooker RA (2007a) The partitioning of water between olivine, orthopyroxene and melt synthesised in the system albite-forsterite-H_20. Earth and Planet Sci 260: 227–241.

Grant KJ, Brooker RA, Kohn SC, Wood BJ (2007b) The effect of oxygen fugacity on hydroxyl concentrations and speciation in olivine: implications for water solubility in the upper mantle. Earth Planet Sci Lett 261(1–2):217–229

Green DH, Hibberson WO, Kovacs I, Rosenthal A (2010) Water and its influence on the lithosphere-asthenosphere boundary. Nature 467(7314):448–451

Hermann J, O'Neill HSC, Berry AJ (2005) Titanium solubility in olivine in the system TiO_2-MgO-SiO_2: no evidence for an ultra-deep origin of Ti-bearing olivine. Contrib Mineral Petrol 148:746–760

Holland TJB, Powell R (2011) An improved and extended internally consistent thermodynamic dataset for phases of petrological interest, involving a new equation of state for solids. J Metamorph Geol 29:333–383

Ingrin J et al (2013) Low-temperature evolution of OH bands in synthetic forsterite, implication for the nature of H defects at high pressure. Phys Chem Miner 40(6):499–510

Ingrin J, Kovács I, Deloule E, Balan E, Blanchard M, Kohn SC, Hermann J (2014) Identification of hydrogen defects linked to boron substitution in synthetic forsterite and natural olivine. Am Mineral 99:2138–2141

Jollands MC, Padrón-Navarta JA, Hermann J, O'Neill HSC (2016) Hydrogen diffusion in Ti-doped forsterite and the preservation of metastable point defects. Am Mineral 101(7–8):1571–1583

Karato S-I, Wang D (2013) Electrical conductivity of minerals and rocks. In: Karato S-I (ed) Physics and Chemistry of the Deep Earth. Wiley-Blackwell, New York, pp 145–182

Karato S-I, Paterson MS, FitzGerald JD (1986) Rheology of synthetic olivine aggregates: Influence of grain size and water. J Geophys Res 91(B8):8151

Kohlstedt DL (2006) The role of water in high-temperature rock deformation. Rev Mineral Geochem 62:377–396

Kohlstedt DL, Keppler H, Rubie DC (1996) Solubility of water in the α, β and γ phases of (Mg, Fe)2SiO_4. Contrib Mineral Petrol 123:345–357

Kovacs I et al (2008) Quantitative absorbance spectroscopy with unpolarized light: Part II. Experimental evaluation and development of a protocol for quantitative analysis of mineral IR spectra. Am Mineral 93(5–6):765–778

Kovacs I, O'Neill HSC, Hermann J, Hauri EH (2010) Site-specific infrared O-H absorption coefficients for water substitution into olivine. Am Mineral 95(2–3):292–299

Lamb WM, Popp RK (2009) Amphibole equilibria in mantle rocks: determining values of mantle aH_2O and implications for mantle H_2O contents. Am Mineral 94(1):41–52

Lemaire C, Kohn SC, Brooker RA (2004) The effect of silica activity on the incorporation mechanisms of water in synthetic forsterite: a polarised infrared spectroscopic study. Contrib Mineral Petrol 147(1):48–57

Mackwell SJ, Kohlstedt DL (1990) Diffusion of hydrogen in olivine: implications for water in the mantle. J Geophys Res 95:5079–5088

Mallmann G, O'Neill HSC (2009) The crystal/melt partitioning of V during mantle melting as a function of oxygen fugacity compared with some other elements (Al, P, Ca, Sc, Ti, Cr, Fe, Ga, Y, Zr and Nb). J Petrol 50(9):1765–1794

Matveev S, O'Neill HSC, Ballhaus C, Taylor WR, Green DH (2001) Effect of silica activity on OH^- IR spectra of olivine: implications for low-$aSiO_2$ mantle metasomatism. J Petrol 42(4):721–729

Mei S, Kohlstedt DL (2000) Influence of water on plastic deformation of olivine aggregates: 1. Diffusion creep regime. J Geophys Res Solid Earth 105(B9): 21457–21469

Miller GH, Rossman GR, Harlow GE (1987) The natural occurrence of hydroxide in olivine. Phys Chem Miner 14:461–472

Mosenfelder JL, Deligne NI, Asmiow PD, Rossman GR (2006) Hydrogen incorporation in olivine from 2-12 GPa. Am Mineral 91:285–294

Mosenfelder JL et al (2011) Analysis of hydrogen in olivine by SIMS: evaluation of standards and protocol. Am Mineral 96(11–12):1725–1741

O'Neill HSC (1998) Partitioning of Fe and Mn between ilmenite and olivine at 1100 °C: constraints on the thermodynamic mixing properties of (Fe, Mn)TiO_3 ilmenite solid solutions. Contrib Mineral Petrol 133:284–296

Otsuka K, Karato S-I (2011) Control of the water fugacity at high pressures and temperatures: applications to the incorporation mechanisms of water in olivine. Phys Earth Planet Inter 189(1–2):27–33

Padrón-Navarta JA, Hermann J, O'Neill HSC (2014) Site-specific hydrogen diffusion rates in forsterite. Earth Planet Sci Lett 392:100–112

Pownceby MI, O'Neill HSC (2000) Thermodynamic data from redox reactions at high temperatures. VI. Thermodynamic properties of CoO–MnO solid solutions from emf measurements. Contrib Mineral Petrol 140(1):28–39

Sambridge M, Gerald JF, Kovacs I, O'Neill HSC, Hermann J (2008) Quantitative absorbance spectroscopy with unpolarized light: Part I. Physical and mathematical development. Am Mineral 93(5–6):751–764

Schmädicke E, Gose J, Witt-Eickschen G, Bratz H (2013) Olivine from spinel peridotite xenoliths: hydroxyl incorporation and mineral composition. Am Mineral 98(10):1870–1880

Shen T, Hermann J, Zhang L, Padrón-Navarta JA, Chen J (2014) FTIR spectroscopy of Ti-chondrodite, Ti-clinohumite, and olivine in deeply subducted serpentinites and implications for the deep water cycle. Contrib Mineral Petrol 167:992–1009

Smyth JR, Frost DJ, Nestola F, Holl CM, Bromiley G (2006) Olivine hydration in the deep upper mantle: effects of temperature and silica activity. Geophys Res Lett 33(15)

Soustelle V, Tommasi A, Demouchy S, Ionov DA (2010) Deformation and fluid-rock interaction in the supra-subduction mantle: microstructures and water contents in peridotite xenoliths from the Avacha Volcano, Kamchatka. J Petrol 51(1–2):363–394

Soustelle V, Tommasi A, Demouchy S, Franz L (2013) Melt-rock interactions, deformation, hydration and seismic properties in the sub-arc lithospheric mantle inferred from xenoliths from seamounts near Lihir, Papua New Guinea. Tectonophysics 608:330–345

Spandler C, O'Neill HSC (2010) Diffusion and partition coefficients of minor and trace elements in San Carlos olivine at 1,300 °C with some geochemical implications. Contrib Mineral Petrol 159(6):791–818

Tollan PME, O'Neill HSC, Hermann J, Benedictus A, Arculus RJ (2015) Frozen melt–rock reaction in a peridotite xenolith from sub-arc mantle recorded by diffusion of trace elements and water in olivine. Earth Planet Sci Lett 422:169–181

Umemoto K, Wentzcovitch RM, Hirschmann MM, Kohlstedt DL, Withers AC (2011) A first-principles investigation of hydrous defects and IR frequencies in forsterite: The case for Si vacancies. Am Mineral 96(10):1475–1479

Walker AM, Hermann J, Berry AJ, O'Neill HSC (2007) Three water sites in upper mantle olivine and the role of titanium in the water weakening mechanism. J Geophys Res 112(B5)

Withers AC, Hirschmann MM (2008) Influence of temperature, composition, silica activity and oxygen fugacity on the H_2O storage capacity of olivine at 8 GPa. Contrib Mineral Petrol 156(5):595–605

Withers AC, Bureau H, Raepsaet C, Hirschmann MM (2012) Calibration of infrared spectroscopy by elastic recoil detection analysis of H in synthetic olivine. Chem Geol 334:92–98

Xue X, Kanzaki M, Turner D, Loroch D (2017) Hydrogen incorporation mechanisms in forsterite: New insights from ^{1}H and ^{29}Si NMR spectroscopy and first-principles calculation. Am Mineral 102:519–536

Yang X, Liu D, Xia Q (2014) CO_2-induced small water solubility in olivine and implications for properties of the shallow mantle. Earth Planet Sci Lett 403: 37–47

Zhao Y-H, Ginsberg SB, Kohlstedt DL (2004) Solubility of hydrogen in olivine: dependence on temperature and iron content. Contrib Mineral Petrol 147: 155–161

9

Evaluation of geochemical records as a paleoenvironmental proxy in the hypercalcified demosponge *Astrosclera willeyana*

Ryuji Asami[1*], Akira Kinjo[2], Daiki Ohshiro[2], Tohru Naruse[3], Masaru Mizuyama[4], Ryu Uemura[5], Ryuichi Shinjo[2], Yuji Ise[6], Yoshihisa Fujita[7] and Takashi Sakamaki[8]

Abstract

The geochemistry of the calcium carbonates of marine organisms is an excellent proxy for reconstruction of the paleoceanographic history. However, previous studies of hypercalcified demosponges (sclerosponges) are considerably fewer than those of corals, foraminifers, and bivalves. Here, we investigated stable oxygen ($\delta^{18}O$) and carbon ($\delta^{13}C$) isotopes and minor and trace element (Mg, Sr, Ba, Pb, and U) to Ca ratios of 36 living sclerosponges (*Astrosclera willeyana*) collected from Kume Island in the Ryukyu Islands, southwestern Japan, to evaluate the utility of geochemistry as a paleoenvironmental proxy. The $\delta^{18}O$, $\delta^{13}C$, and Sr/Ca deviations of the coevally precipitated skeleton were extremely small and almost constant at all skeletal portions, strongly suggesting that within-skeletal variations in the chemical components are negligibly small for non-symbiotic sclerosponges. Mean $\delta^{18}O$, $\delta^{13}C$, and Sr/Ca values (N = 36), falling within a quite narrow range, showed no significant evidence for intraspecific (inter-specimen) variations in the sclerosponges. The sclerosponges $\delta^{18}O$ and $\delta^{13}C$ were consistent with those of the aragonites precipitated in isotopic equilibrium with seawater at the growth site. The sclerosponge Sr/Ca was close to that of inorganically precipitated aragonite, and the estimated partition coefficient of 1.1 (almost unity) is identical to previously reported values for different species (*Ceratoporella nicholsoni*). Consequently, these results suggest that *A. willeyana* sclerosponge, having little vital effects on the geochemistry, is a robust indicator of paleocean environments (seawater $\delta^{18}O$, temperature, and dissolved inorganic carbon $\delta^{13}C$). Further, our evaluation study documents that sclerosponges living in deeper ocean environments can support the reconstruction of spatial and vertical paleoceanographic changes in conjunction with coral proxy records. The sclerosponge U/Ca showed little within-skeletal and intraspecific variations, but the heterogeneity and individual difference of the Mg/Ca, Ba/Ca, and Pb/Ca were relatively large, the reasons of which still remain unresolved.

Keywords: Sclerosponge, *Astrosclera willeyana*, Oxygen and carbon isotopic composition, Minor and trace elements, Isotope equilibrium, Paleoenvironmental proxy

* Correspondence: ryuji.asami.b5@tohoku.ac.jp
[1]Institute of Geology and Paleontology, Graduate School of Science, Tohoku University, Aobayama, Sendai 980-8578, Japan
Full list of author information is available at the end of the article

Introduction

The hypercalcified demosponges ("sclerosponge" hereinafter), members of the earliest branching metazoan taxon (Porifera), have been dominant reef-building marine organisms throughout the Phanerozoic (e.g., Vacelet 1985; Reitner and Engeser 1987; Wood 1990; Reitner 1992). The taxon *Astrosclera* is present in the fossil record from the Late Triassic (Vacelet 1979, 1985). In modern coral reefs, the sclerosponges can be found in cryptic niches of low-light environments, unlike zooxanthellate scleractinian corals (e.g., Wörheide 1998). Sclerosponges, commonly mushroom-shaped in form, deposit a $CaCO_3$ skeleton at a very slow growth rate of mostly less than about 2 mm/year and can grow for up to several hundred years (e.g., Benavides and Druffel 1986; Böhm et al. 1996; Reitner and Gautret 1996; Wörheide et al. 1997; Wörheide 1998; Swart et al. 2002; Fallon et al. 2003; Grottoli et al. 2010).

To elucidate the processes underpinning climate change and global warming, environmental proxy archives such as trees, sediments, and corals for documenting air and sea surface temperature variation over the last millennia are needed (PAGES 2 k Consortium 2013; Abram et al. 2016). The stable oxygen ($\delta^{18}O$) and carbon ($\delta^{13}C$) isotope records of sclerosponges are a useful proxy for reconstructing seawater temperature and salinity (e.g., Wörheide 1998; Böhm et al. 2000; Moore et al. 2000; Rosenheim et al. 2005a) and dissolved inorganic carbon $\delta^{13}C$ of seawater (e.g., Böhm et al. 1996; Swart et al. 2010) for the past. Results of previous studies have shown that, like corals, the $\delta^{18}O$ and $\delta^{13}C$ values of sclerosponges have respectively reflected interannual climate variability (Swart et al. 2002; Grottoli 2006; Grottoli et al. 2010; Wu and Grottoli 2010) and increasing fossil fuel CO_2 in seawater (Böhm et al. 2002; Swart et al. 2010). Moreover, other minor and trace elements and isotopes can be a proxy for monitoring ocean environments and chemistry (e.g., Lazareth et al. 2000; Fallon et al. 2005; Rosenheim et al. 2005b; Ohmori et al. 2014). Some studies showed that sclerosponge strontium/calcium (Sr/Ca) can offer a potential for paleotemperature determination of seawater (e.g., Rosenheim et al. 2004; Waite et al. 2018).

Importantly, sclerosponges can live in environments with less sunlight (ex. reef caves and overhangs) and at greater water depths of up to several hundred meters in the Indo-Pacific tropical regions (e.g., Hartman 1980) and in the Atlantic and Caribbean regions (e.g., Druffel and Benavides 1986; Böhm et al. 1996), unlike zooxanthellate corals and giant clams living in shallow coral reefs. For that reason, their skeletal chemistry can be useful for elucidating vertical variations of physical and chemical parameters of seawater and generating their several-century-long time series that overlap with and extend beyond the instrumental climate data. Nevertheless, sclerosponge-based reconstruction studies are extremely few (especially in the North Pacific) relative to those of other marine $CaCO_3$ organisms such as corals, foraminifers, and mollusks. Furthermore, evaluation and calibration studies of sclerosponges for fidelity as a paleoenvironmental proxy are few and they were only based on time series analyses along the growth direction (Böhm et al. 2000; Rosenheim et al. 2005b, 2009). Therefore, distribution (homogeneity/heterogeneity) and intraspecific difference of chemical composition in the sclerosponge skeleton should be evaluated for further robust application to paleoenvironmental reconstruction.

We investigated 36 specimens of living sclerosponges collected in Kume Island, southwestern part of Japan in the North Pacific to clarify (1) within-skeletal variations in the $\delta^{18}O$, $\delta^{13}C$, and elements/Ca ratios (magnesium/Ca (Mg/Ca), Sr/Ca, barium/Ca (Ba/Ca), lead/Ca (Pb/Ca), and uranium/Ca (U/Ca)) and (2) their intraspecific variations (individual difference = inter-specimen variations). In addition, to evaluate the fidelity as a paleoenvironmental proxy, we compared the isotopic values of the sclerosponge samples to theoretically calculated values of aragonites precipitated close to isotope equilibrium with seawater at the living site and estimated partition coefficients for elements/Ca between seawater and aragonitic skeleton. Knowledge acquired through the results of our evaluation study can contribute to the expansion of applications of multi-proxy reconstructions for climate and ocean environmental changes and can provide insight into a better understanding of calcium carbonate biomineralization for marine organisms.

Materials and methods

Study site and environment

The study site (26°20′12–17′′ N, 126°43′40–49′′ E) is located off the western coast of Kume Island in the Ryukyu Islands, southwestern Japan (Fig. 1). The climate of this region is subtropical, with atmospheric temperatures of 17–29 °C and an annual mean of 23 °C (Japan Meteorological Agency 2016). The average annual precipitation is ca. 2100 mm/year. The warm Kuroshio current (the western North Pacific current) flows into the East China Sea and passes northeastward along the Ryukyu Islands. It allows the development of coral reefs with a high bio-diversity at Kume Island (e.g., Naruse et al. 2012).

In situ seawater temperature and salinity at the study site in water depths of 34–40 m were measured every 1 h from the 5th of October 2015 through the 13th of November 2016 using data loggers (Infinity CTW; JFE Advantech Co., Ltd. and HOBO Tidbit v2; Onset Computer Corp.). The observation data were calibrated with actual temperature and salinity measurements of seawater collected at the study site using a calibration thermometer and a conductivity meter (CM-41X: DKK-

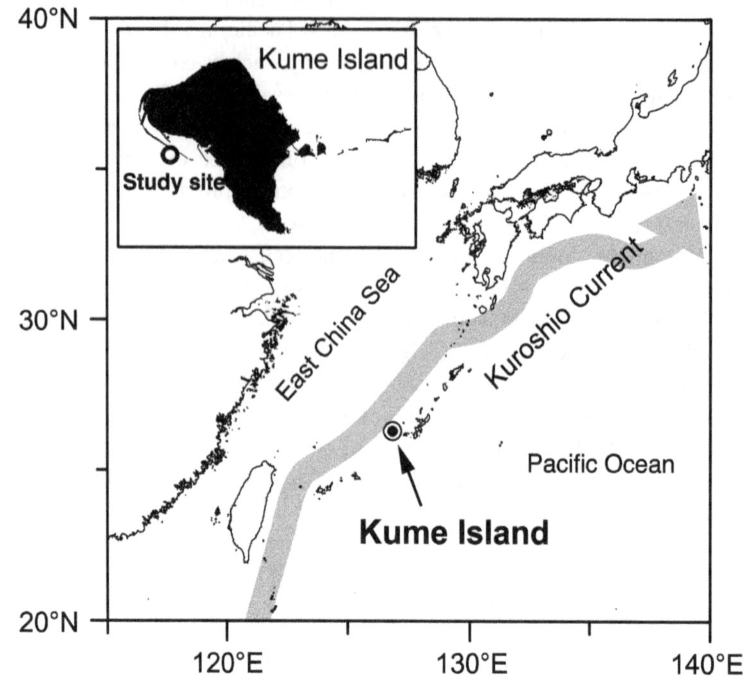

Fig. 1 Maps showing Kume Island and study site locations

TOA Corp.). The in situ data for 2015–2016 are fundamentally consistent with monthly average data at a depth of 30 m, available from the Japan Oceanographic Data Center (JODC) for 1906–2003 (26°–27° N, 126°–127° E; http://www.jodc.go.jp/) (Fig. 2). The seawater temperature and salinity vary respectively, as 20 °C–31 °C and 33.6–35.0 (psu) with annual means of 25.4 °C and 34.6 (psu).

Samples

The hypercalcified demosponges (*Astrosclera willeyana* Lister 1900) examined in this study were growing on the ceiling or wall in dark submarine caves, crevasses, or overhung of reef outer edge. In October 2015 and November 2016, 36 living sclerosponge samples were collected by scuba diving. Specifically, 6 samples (KMJ-40), 4 samples (KMJ-34), and 26 samples (KMJ-38) were collected respectively at water depths of 40 m, 34 m, and 38 m (Table 1 and Fig. 3). The surfaces of samples are mostly light-to-dark brown and/or orange. The samples, which were from 5.4 mm × 5.0 mm × 2.5 mm to 24.9 mm × 18.0 mm × 11.6 mm, were rinsed three times ultrasonically with milli-Q water (resistance = 18.2 mΩ; Millipore Corp.) and were then dried in a desiccator.

To check the mineralogy and preservation state of the skeleton, five selected sclerosponges (KMJ-38-14, KMJ-38-20, KMJ-38-22, KMJ-38-25, and KMJ-38-26) were examined using X-ray diffraction (XRD) analysis and scanning electron microscopy (SEM) (Fig. 4). At the University of the Ryukyus, XRD analysis was conducted using a multipurpose XRD device (RINT Ultima/PC; Rigaku Corp.) to test for the presence of calcite. Powdered samples were mounted on well-cleaned glass slides with HCl, ethanol, and milli-Q water and were analyzed using Cu Kα radiation (40 kV, 30 mA) by scanning from 20 to 37° at a 2θ angle at 0.02° steps. SEM imaging of the skeleton was conducted using a microscope (TM3030; Hitachi High-Tech. Corp.).

Stable isotope analyses

For stable isotope analyses, powdered subsamples (1–20 subsamples per sclerosponge) were taken from the skeleton surface along the X-axis (= long axis) and Y-axis (= short axis) directions (see Fig. 4), to evaluate within-skeleton variations in the isotope values. The top of the sample is regarded as the coordinate origin ($X = 0$, $Y = 0$). A subsample (Φ, 1 mm; depth, ca. 1.5 mm) was taken at intervals of 1–4 mm from the coevally precipitated skeleton using a pre-cleaned diamond drill. The common skeletal growth rates of *A. willeyana* scleroponges are estimated to be about 0.2–1.9 mm/year (e.g., Wörheide 1998; Moore et al. 2000; Fallon and Guilderson 2005; Fallon et al. 2005; Grottoli et al. 2010). The tissue layer in *A. willeyana* can occupy upwards of about 5 mm and the time between initial deposition and secondary infilling of aragonite skeleton appears to be several years (Fallon et al. 2005; Allison et al. 2012). Furthermore, the skeletal density of *A. willeyana* increases gradually with depth. That in the surface living zone is about half that of inner structures, which is indicative of continual thickening of

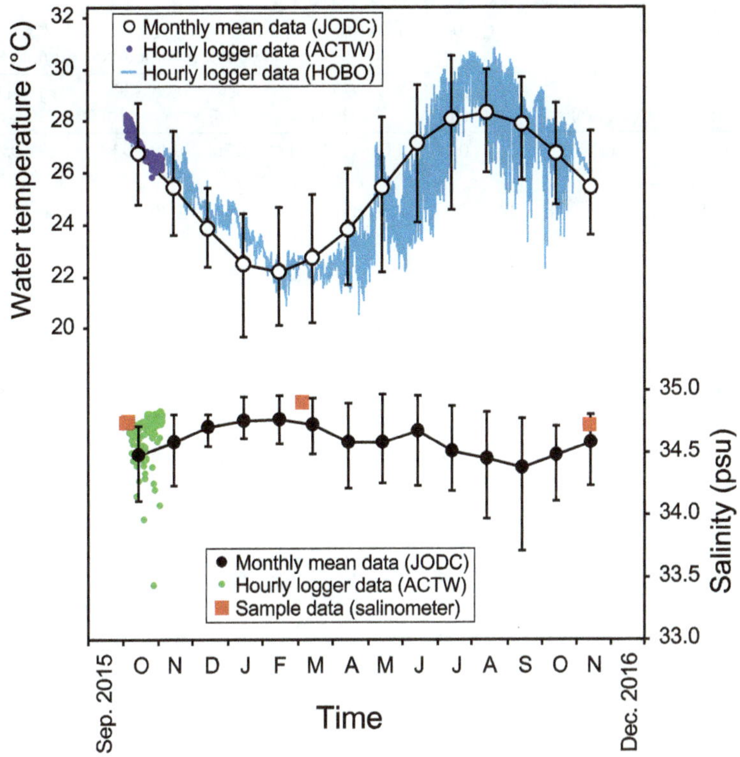

Fig. 2 Monthly mean climatology and logging data of seawater temperature and salinity at the study site. Data are derived from the Japan Oceanographic Data Center (JODC) for 1906–2003 (26°–27° N, 126°–127° E; http://www.jodc.go.jp/) and in situ observations made during October 2015 to November 2016 using loggers in this study. Error bars represent monthly maximum and minimum values in the past. Salinity of seawater samples collected at the site was also measured for cross-validation

aragonite throughout the living tissue layer (Fallon et al. 2005). Considering such characteristics of skeletal growth, the subsampling portion (Φ, ~ 1 mm; depth, ~ 1.5 mm) probably corresponds to an increment during about a few years, assumed to be the years 2013–2016 in this study.

Stable oxygen ($\delta^{18}O$) and carbon ($\delta^{13}C$) isotope analyses of calcium carbonate (0.14–0.16 mg) were conducted using a continuous flow isotope ratio mass spectrometer attached to a Gasbench II and a GC-PAL auto-sampler (Delta V Advantage; ThermoFisher Scientific Inc.) at the University of the Ryukyus, following the method presented by Asami et al. (2015). The seawater $\delta^{18}O$ was measured using a cavity ring-down spectroscopy (CRDS L2130-i; Picarro Inc.) at the University of the Ryukyus, following the method reported by Uemura et al. (2012). Isotopic ratios were reported in the conventional δ notation relative to Vienna Pee Dee Belemnite (VPDB) for carbonate and Vienna Standard Mean Ocean Water (VSMOW) for seawater. External precision (1σ) throughout the entire analysis was ± 0.07‰ ($\delta^{18}O$) and ± 0.05‰ ($\delta^{13}C$) for carbonate and ± 0.08‰ ($\delta^{18}O$) for seawater. Accuracy of the measurements of sclerosponge samples was evaluated based on replicates of GSJ/AIST JCp-1 (aragonite; Okai et al. 2004), yielding average $\delta^{18}O$ and $\delta^{13}C$ values of − 4.65 ± 0.07‰ and − 1.59 ± 0.06‰ (N = 42), which show excellent agreement with previously reported values of − 4.71‰ and − 1.63‰ (Nakayama et al. 2008), respectively.

Elements/Ca analyses

Elements/Ca ratios (Mg/Ca, Sr/Ca, Ba/Ca, Pb/Ca, U/Ca) were analyzed using an inductively coupled plasma-mass spectrometer (XSeries II, Thermo Fisher Scientific) at the University of the Ryukyus, basically following the method reported by Asami et al. (2009, 2013). Each sclerosponge powder sample (~ 0.15 mg) was dissolved in 4.5 mL of 0.5 mol/L high-purity HNO_3 diluted with ultrapure Milli-Q water. Each seawater sample of 125 μL was diluted 80 times with high-purity HNO_3 and ultrapure Milli-Q water. Internal standard elements (Sc, Y, and Yb) were added to all of the solutions to produce equal concentrations to control matrix effect and to correct for instrumental noise. Solutions were analyzed for ^{24}Mg, ^{43}Ca, ^{44}Ca, ^{45}Sc, ^{87}Sr, ^{88}Sr, ^{89}Y, ^{137}Ba, ^{138}Ba, ^{172}Yb, ^{208}Pb, and ^{238}U. Measurements were conducted in triplicate. Calibrations of the five gravimetric standard solutions yielded high correlation coefficients of $r > 0.99998$ for Ca, Mg, and Sr and of $r > 0.99995$ for Ba, Pb, and U. A reference solution, gravimetrically matched to the Ca concentration of the average sclerosponge sample

Table 1 Details of sclerosponge samples and analytical results of geochemistry

Sample ID	Collection Date	Water Depth (m)	Size			N	δ13C (‰)		δ18O (‰)		Mg/Ca (mmol/mol)		Sr/Ca (mmol/mol)		Ba/Ca (μmol/mol)		Pb/Ca (μmol/mol)		U/Ca (μmol/mol)	
			X (mm)	Y (mm)	H (mm)		Avg.	SD (1σ)	Avg.	SD (1σ)	Avg.	SD (1σ)	Avg.	SD (1σ)	Avg.	SD (1σ)	Avg.	SD (1σ)	Avg.	SD (1σ)
KMJ-40-1	6-Oct-15	40	9.4	8.8	6.0	9	3.49	0.07	-1.08	0.07	1.45	0.36	9.42	0.17	0.59	0.70	0.45	0.28	3.26	0.04
KMJ-40-2			11.4	7.9	4.8	8	3.40	0.05	-1.16	0.10	1.49	0.64	9.36	0.11	1.06	0.64	0.51	0.17	3.23	0.03
KMJ-40-3			12.3	11.8	5.5	9	3.43	0.05	-1.00	0.17	2.11	0.61	9.41	0.11	1.39	0.76	0.71	0.17	3.13	0.06
KMJ-40-4			19.0	15.5	8.9	14	3.42	0.06	-1.00	0.11	1.59	0.25	9.52	0.15	1.90	1.02	0.63	0.15	3.25	0.02
KMJ-40-5			24.9	18.0	11.6	13	3.47	0.08	-1.01	0.13	1.85	0.36	9.42	0.16	2.74	0.77	N/A		3.24	0.06
KMJ-40-6			21.0	18.6	12.6	14	3.51	0.08	-0.75	0.12	1.33	0.38	9.58	0.19	2.97	0.69	N/A		3.21	0.07
KMJ-34-1	7-Oct-15	34	10.3	10.2	5.9	13	3.51	0.05	-0.98	0.07	1.45	0.47	9.63	0.13	1.96	0.44	0.71	0.11	3.17	0.04
KMJ-34-2			12.4	11.3	6.0	12	3.45	0.09	-0.88	0.06	1.26	0.23	9.76	0.09	2.06	0.40	0.81	0.13	3.12	0.04
KMJ-34-3			13.9	10.8	6.8	14	3.42	0.12	-1.02	0.09	1.47	0.39	9.52	0.09	2.04	0.44	0.74	0.12	3.17	0.07
KMJ-34-4			15.2	13.8	11.4	17	3.44	0.07	-1.07	0.12	1.61	0.37	9.43	0.14	1.86	0.43	0.73	0.12	3.27	0.05
ALL (KMJ-34 and -40): 95% CL							**3.454 ± 0.024**		**-0.995 ± 0.072**		**1.560 ± 0.160**		**9.503 ± 0.077**		**1.857 ± 0.451**		**0.663 ± 0.086**		**3.205 ± 0.033**	
KMJ-38-1	13-Nov-16	38	14.7	13.2	8.8	16	3.41	0.05	-0.98	0.06	1.31	0.32	9.15	0.17	1.44	0.94	N/A		3.27	0.14
KMJ-38-2			17.7	12.5	10.5	18	3.40	0.05	-1.01	0.16	1.29	0.34	9.27	0.28	1.47	1.11	N/A		3.24	0.12
KMJ-38-3			5.4	5.0	2.5	1	3.37		-1.17		2.14		9.26		N/A		N/A		3.36	
KMJ-38-4			6.0	5.7	2.0	1	3.48		-0.95		1.40		9.29		N/A		0.08		3.27	
KMJ-38-5			6.0	6.0	3.0	1	3.44		-0.99		1.45		9.55		N/A		0.53		3.42	
KMJ-38-6			7.2	6.9	3.5	5	3.36	0.02	-0.89	0.10	1.78	0.51	9.43	0.28	0.59	1.07	N/A		3.29	0.10
KMJ-38-7			7.8	7.1	3.0	5	3.38	0.03	-0.95	0.08	1.67	0.17	9.45	0.16	0.75	0.99	N/A		3.40	0.15
KMJ-38-8			7.7	7.2	4.0	5	3.34	0.05	-1.02	0.09	1.33	0.20	9.35	0.12	0.31	0.29	N/A		3.54	0.08
KMJ-38-9			9.7	8.7	4.0	5	3.38	0.04	-1.00	0.09	0.95	0.10	9.74	0.08	N/A		2.32	1.64	3.35	0.06
KMJ-38-10			10.0	9.9	6.0	5	3.41	0.05	-0.93	0.03	1.07	0.12	9.80	0.15	0.43	0.97	0.98	0.75	3.34	0.14
KMJ-38-11			11.0	10.3	7.0	15	3.40	0.06	-1.09	0.11	0.90	0.06	9.33	0.09	0.12	0.33	2.53	1.16	3.37	0.10
KMJ-38-12			11.4	8.9	5.1	12	3.42	0.05	-0.99	0.08	0.96	0.28	9.41	0.08	0.09	0.31	3.53	1.37	3.30	0.07
KMJ-38-13			12.8	8.9	7.3	11	3.44	0.05	-0.94	0.06	1.03	0.18	9.51	0.09	0.09	0.20	4.55	2.49	3.48	0.07
KMJ-38-14			12.3	10.8	7.0	14	3.37	0.04	-0.90	0.05	1.08	0.12	9.73	0.40	2.41	0.85	3.20	1.36	3.11	0.07
KMJ-38-			12.7	11.8	6.4	13	3.40	0.03	-1.06	0.09	1.14	0.22	9.44	0.25	1.59	1.14	3.82	2.70	3.18	0.06

Table 1 Details of sclerosponge samples and analytical results of geochemistry *(Continued)*

Sample ID	Collection Date	Water Depth (m)	Size			*N*	$\delta^{13}C$ (‰)		$\delta^{18}O$ (‰)		Mg/Ca (mmol/mol)		Sr/Ca (mmol/mol)		Ba/Ca (μmol/mol)		Pb/Ca (μmol/mol)		U/Ca (μmol/mol)	
			X (mm)	Y (mm)	H (mm)		Avg.	SD (1σ)	Avg.	SD (1σ)	Avg.	SD (1σ)	Avg.	SD (1σ)	Avg.	SD (1σ)	Avg.	SD (1σ)	Avg.	SD (1σ)
15																				
KMJ-38-16			15.9	15.5	8.8	16	3.39	0.06	-1.17	0.05	1.02	0.21	9.28	0.09	1.20	0.82	3.01	1.55	3.22	0.06
KMJ-38-17			17.1	15.1	8.8	14	3.41	0.05	-1.07	0.11	1.19	0.31	9.56	0.15	0.84	0.95	N/A	3.80	3.21	0.07
KMJ-38-18			15.6	12.8	8.0	12	3.34	0.06	-1.12	0.12	1.09	0.16	9.36	0.25	1.86	1.03	2.24	0.87	3.16	0.05
KMJ-38-19			14.8	13.6	9.0	15	3.37	0.06	-0.94	0.07	1.22	0.49	9.30	0.17	2.24	0.45	1.50	0.44	3.22	0.05
KMJ-38-20			18.2	15.9	9.5	19	3.38	0.04	-1.01	0.09	1.04	0.14	9.38	0.09	1.01	0.71	N/A	3.49	3.24	0.04
KMJ-38-21			20.0	16.8	10.0	20	3.39	0.05	-1.00	0.07	1.10	0.12	9.41	0.25	1.46	0.86	3.19	1.64	3.20	0.05
KMJ-38-22			18.5	18.2	9.0	15	3.41	0.03	-1.02	0.07	1.10	0.27	9.32	0.19	1.31	0.90	2.37	1.37	3.20	0.07
KMJ-38-23			20.5	15.8	9.0	14	3.35	0.06	-0.95	0.09	1.23	0.20	9.74	0.31	1.75	0.92	1.43	0.61	3.20	0.05
KMJ-38-24			21.6	16.2	10.0	17	3.37	0.06	-0.96	0.11	1.15	0.07	9.43	0.18	2.13	0.44	0.79	0.18	3.17	0.04
KMJ-38-25			19.8	19.7	10.2	18	3.39	0.05	-1.02	0.11	0.99	0.14	9.38	0.20	2.15	0.47	0.77	0.39	3.22	0.06
KMJ-38-26			21.6	19.1	11.8	20	3.39	0.06	-1.01	0.08	0.86	0.09	9.24	0.18	1.55	0.79	0.72	0.12	3.25	0.06
ALL (KMJ-38): 95% CL							**3.392 ± 0.012**		**-1.006 ± 0.029**		**1.211 ± 0.114**		**9.428 ± 0.067**		**1.217 ± 0.312**		**2.086 ± 0.611**		**3.277 ± 0.041**	

solutions, was measured at intervals of three samples to correct for instrumental drift. Based on replicate measurements of the solution of aragonite reference material JCp-1 coral (Okai et al. 2002), external precisions for Mg/Ca, Sr/Ca, Ba/Ca, Pb/Ca, and U/Ca determinations were better than 0.43%, 0.31%, 1.1%, 4.0%, and 0.82% relative standard deviations ($N = 150$) and the measurement values of samples were calibrated using the widely accepted values of Hathorne et al. (2013).

Results and discussion

Skeletal structures and preservation

The results of XRD analysis indicate that the skeleton samples of *Astrosclera willeyana* consist exclusively of aragonite (> 99.9%), with no detection of calcite. The SEM images of skeletal samples also reveal no evidence of the existence of calcite (Fig. 5). Skeletal aspects of our sclerosponge samples closely resemble those found in earlier studies (e.g., Wörheide 1998; Jackson et al. 2010; Gilis et al. 2013), clearly showing a typical macrostructure of reticulate skeletal organization with branches composed of aragonitic spherulitic structural elements (Ayling 1982; Reitner 1992; Allison et al. 2012). Pristine aragonite crystals were observed without evidence of diagenetic alteration such as overgrowth of inorganically precipitated secondary cements, traces of dissolution, or bioerosion. Consequently, these results indicate that *A. willeyana* samples used for this study can retain pristine skeletons with original values of isotope composition and elemental concentrations.

Within-skeleton variations

Oxygen and carbon isotopes

Biological carbonates are known to precipitate out of ^{13}C and ^{18}O equilibrium with seawater attributable to kinetic and metabolic isotope effects (e.g., McConnaughey 1998, 2003). For a photosynthetic scleractinian aragonitic coral with symbiotic algae, McConnaughey (1989) showed that the $\delta^{13}C$ and $\delta^{18}O$ variations (1σ) of the coeval skeletal increments were significantly different between sunshine side samples (± 0.17‰ and ± 0.04‰, respectively) and shade side samples (± 0.47‰ and ± 0.49‰, respectively). Furthermore, a strong positive correlation ($r = 0.90$) between coral $\delta^{13}C$ and $\delta^{18}O$ was found for the shade side. These results suggest that the isotopic composition is different depending on the skeletal portion of $CaCO_3$ in photosynthetic corals.

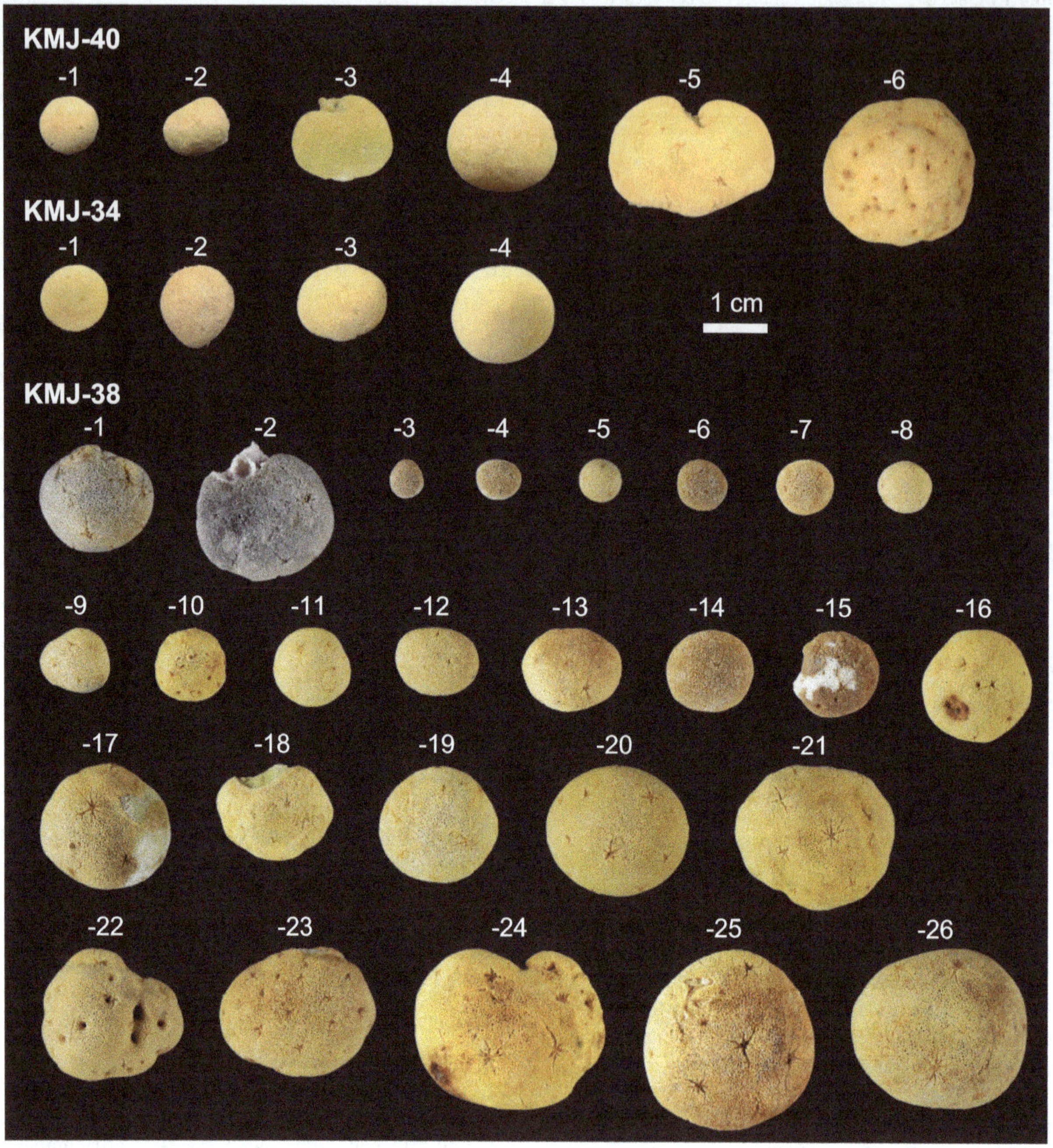

Fig. 3 Photographs of sclerosponge samples

The average and standard deviation (1σ) of $\delta^{13}C$ and $\delta^{18}O$ in an *A. willeyana* specimen were estimated from values of sub-samples ($N < 20$) collected at the coevally precipitated skeleton (Table 1; Fig. 4). The $\delta^{13}C$ and $\delta^{18}O$ records are almost similar among KMJ-40, KMJ-34, and KMJ-38. The $\delta^{13}C$ and $\delta^{18}O$ variations (standard deviations, 1σ) of the coeval skeleton were quite small: from ± 0.02 to ± 0.12‰ and from ± 0.03 to ± 0.17‰, respectively. To identify within-skeletal variations of all samples, the respective deviations relative to the average (differences between isotopic values and the average) in an individual specimen are depicted in Fig. 6. The whole standard deviation (2σ) of the isotopic differences was small (± 0.11‰ for $\delta^{13}C$ and ± 0.19‰ for $\delta^{18}O$) and almost constant at all skeletal portions, with no significant variations and trends (Fig. 6). These results show that the within-skeletal variations in the isotopic composition are negligibly slight for the species *A. willeyana*.

The distribution of correlation coefficients among respective samples in within-skeleton showed little

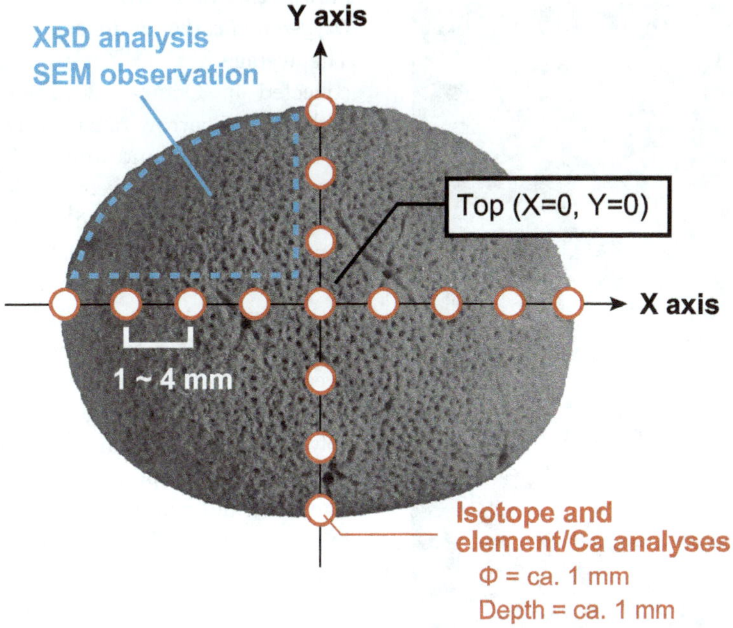

Fig. 4 Example showing the sub-sampling method conducted on a sclerosponge specimen. Subsamples for geochemical analyses (red) and for XRD analysis and SEM observation (blue) were collected

systematic relationships and trends (Fig. 7). Only 5 sclerosponge samples showed positive relations between the $\delta^{13}C$ and $\delta^{18}O$ with high *r* values of 0.6–0.8, but these are expected to be arbitrary in consideration of the very slight within-skeletal variation of ± 0.03–0.06‰ for $\delta^{13}C$ and ± 0.08–0.11‰ for $\delta^{18}O$ and considering analytical errors (Table 1). It is considered that the homogeneity of the isotopic composition in the coevally skeleton for *A. willeyana* is due to little kinetic fractionation effect because the species has no symbiont and a very slow growth rate of the $CaCO_3$ precipitation.

Elements/Ca ratios

Despite $\delta^{13}C$ and $\delta^{18}O$ records, only a few studies have reported minor and trace elements/Ca ratios of sclerosponges (e.g., Fallon et al. 2005; Rosenheim et al. 2005b; Allison et al. 2012). The Sr/Ca and U/Ca records in the coevally precipitated skeletons of *A. willeyana* samples are almost similar among KMJ-40, KMJ-34, and KMJ-38 (Table 1). Overall, the Mg/Ca, Sr/Ca, and U/Ca variations (1σ) of the coeval skeleton were small, ranging respectively from ± 0.06 to ± 0.64 mmol/mol, ± 0.08 to ± 0.40 mmol/mol, and from ± 0.02 to ± 0.15 μmol/mol. The sclerosponge Ba/Ca and Pb/Ca had large fluctuations with respective standard deviations of ± 0.20–1.14 μmol/mol and ± 0.11–3.80 μmol/mol. The deviations (2σ) relative to the average for an individual were almost constant at all skeletal portions (± 0.55 mmol/mol for Mg/Ca, ± 0.37 mmol/mol for Sr/Ca, ± 1.45 μmol/mol for Ba/Ca, ± 2.45 μmol/mol for Pb/Ca, and ± 0.13 μmol/mol for U/Ca), with no significant variations and trends (Fig. 6). These results show that the elements are homogeneously distributed in coeval skeleton of *A. willeyana* sclerosponge, whereas the relative standard deviations of Sr/Ca (3.9%) and U/Ca (4.2%) were quite smaller than those of Mg/Ca (45%), Ba/Ca (97%), and Pb/Ca (138%), which can be partially caused by the difference in elemental concentrations and respective analytical errors. Therefore, our results suggest that the within-skeletal variations of Sr/Ca and U/Ca in *A. willeyana* sclerosponge are so slight as to be ignored.

The distribution of correlation coefficients among respective samples in within-skeleton showed little systematic relationships and trends, showing that most of the *r* values fell within a range from − 0.4 to 0.4 (Fig. 7). The relationships mostly explain only < 20% of the covariance among respective geochemical composition, thereby indicating that the within-skeletal geochemistry for *A. willeyana* can be less influenced by kinetic fractionation effect. It is known that sclerosponges are composed of two key crystal structures in the skeleton at a micro-scale resolution: fused aragonitic spherulites (formed intracellularly and fused together at the surface) and epitaxial backfill (deposited extracellularly at the base of the sponge tissue). High-resolution spatial analyses showed that elements/Ca ratios were highly variable in and among these structures (Allison et al. 2012), probably indicating a heterogeneity of chemical composition in the sclerosponge skeleton.

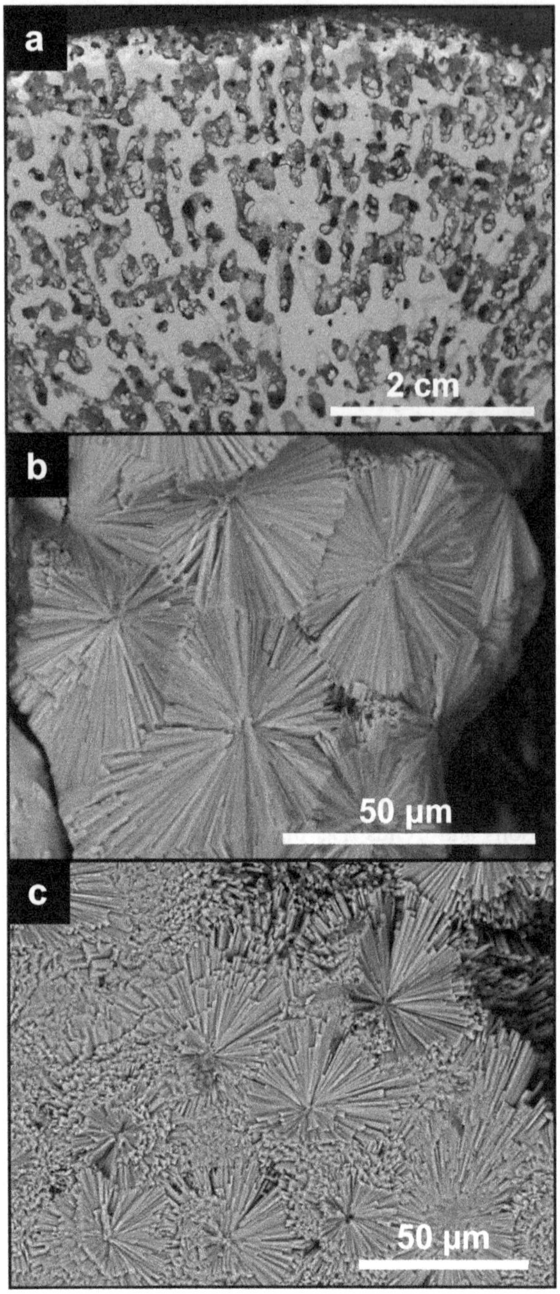

Fig. 5 Scanning electron micrograph images of sclerosponges. A well-preserved pristine aragonite skeleton is visible (**a**, **b** KMJ-38-14; **c** KMJ-38-25)

On the other hand, our data showed that within-skeletal Sr/Ca and U/Ca of *A. willeyana* sclerosponges were quite small and constant (Table 1 and Fig. 6). These suggest that the chemical components in the sclerosponge skeleton are commonly homogeneous at millimeter-scale resolution and annually resolved reconstruction of ocean environmental changes can be applied.

Intraspecific variations

Oxygen and carbon isotopes

The average $\delta^{13}C$ ($\delta^{18}O$) values of *A. willeyana* samples collected in October 2015 and November 2016 fell within quite narrow ranges from 3.40 to 3.51‰ (from − 0.75 to − 1.16‰) and from 3.35 to 3.48‰ (from − 0.89 to − 1.17‰), respectively (Table 1 and Fig. 8). The respective averages of the $\delta^{13}C$ and $\delta^{18}O$ with 95% confidence levels were 3.45 ± 0.02‰ and − 0.99 ± 0.07‰ for KMJ-40 and KMJ-34 samples (N = 10) and 3.39 ± 0.01‰ and − 1.01 ± 0.03‰ for KMJ-38 samples (N = 26) (Table 1). No significant difference was found between the two population averages for $\delta^{18}O$ (0.01 ± 0.08‰). The $\delta^{13}C$ average for KMJ-38 samples (presumably 2014–2016) was slightly lower than that for KMJ-40 and KMJ-34 samples (presumably 2013–2015) with a significant difference of 0.06 ± 0.03‰, which may reflect the Suess Effect widely accepted as a result of the burning of ^{12}C-enriched fossil fuels (Keeling et al. 1979; Quay et al. 1992). The effect has decreased $\delta^{13}C$ of dissolved inorganic carbon (DIC) in seawater since the mid-20th century because of the accelerated increase of anthropogenically derived CO_2 with isotopically low $\delta^{13}C$ values into the ocean from the atmosphere, results of which are recorded in corals and sclerosponges (e.g., Böhm et al. 2002; Asami et al. 2005; Swart et al. 2010). The difference is roughly consistent with decreasing $\delta^{13}C$ rates of − 0.03‰/year for 2000–2016 (atmospheric CO_2 at Hawaii: derived from Scripps CO_2 Program initiated in 1956 by Keeling CD) and − 0.07‰/year for 2000–2010 (coral data in the Ryukyu Islands; Asami et al. unpublished data).

Cross plots of $\delta^{13}C$ and $\delta^{18}O$ records of our 36 sclerosponge samples converged respectively to the averages of 3.41‰ and − 1.00‰ with very slight standard deviations of ± 0.04‰ and ± 0.08‰ (Fig. 8). The relationship between average $\delta^{13}C$ and $\delta^{18}O$ values of all samples yielded a very weak correlation coefficient of 0.27 (p > 0.01, Table 2), which is unlike those found in photosynthetic marine organisms. Therefore, one can reasonably infer that the intraspecific $\delta^{13}C$ and $\delta^{18}O$ variations (individual difference) for the species *A. willeyana* are quite small and hardly influenced by vital effects because the sclerosponges grow without the need for light and symbiotic algae.

Elements/Ca ratios

The average Sr/Ca (U/Ca) ratios of *A. willeyana* samples collected in October 2015 and November 2016 fell within quite narrow ranges from 9.36 to 9.76 mmol/mol (from 3.12 to 3.27 μmol/mol) and from 9.15 to 9.80 mmol/mol (from 3.11 to 3.54 μmol/mol), respectively (Table 1 and Fig. 8). The respective Sr/Ca and U/Ca averages with 95% confidence levels were 9.50 ± 0.08

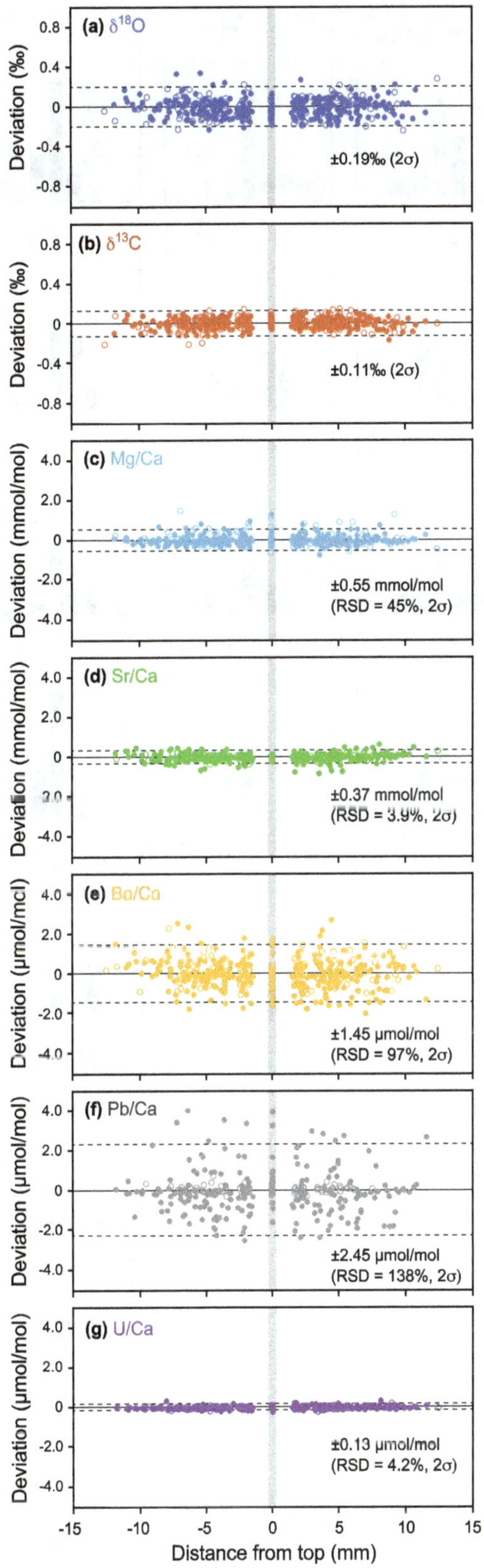

Fig. 6 Within-skeleton variations in **a** $\delta^{18}O$ and **b** $\delta^{13}C$ values and **c**–**g** elements/Ca ratios of sclerosponges. Deviations to the average in an individual sample are shown. Horizontal lines and dotted lines respectively represent zero and standard deviation (2σ)

mmol/mol and 3.21 ± 0.03 µmol/mol for KMJ-40 and KMJ-34 samples (N = 10) and 9.43 ± 0.07 mmol/mol and 3.28 ± 0.04 µmol/mol for KMJ-38 samples (N = 26) (Table 1). No significant difference was found between the two population averages for Sr/Ca (0.08 ± 0.10 mmol/mol) and U/Ca (0.07 ± 0.05 µmol/mol). Furthermore, the Sr/Ca and U/Ca records are in agreement with those of the same species from other regions in the Pacific Ocean (Table 3). Consequently, these results strongly suggest that intraspecific variations are negligibly small for Sr/Ca and U/Ca of *A. willeyana*.

The mean values of within-skeletal Mg/Ca, Ba/Ca, and Pb/Ca ratios fall within 1.33–2.11 mmol/mol, 0.59–2.97 µmol/mol, and 0.45–0.81 µmol/mol for KMJ-40 and KMJ-34 samples (collected in 2015) and 0.86–2.14 mmol/mol, 0.09–2.41 µmol/mol, and 0.08–4.55 µmol/mol for KMJ-38 samples (collected in 2016), respectively (Table 1 and Fig. 8). The respective Mg/Ca and Ba/Ca averages with 95% confidence levels were 1.56 ± 0.16 mmol/mol and 1.86 ± 0.45 µmol/mol for KMJ-40 and KMJ-34 and 1.21 ± 0.11 mmol/mol and 1.22 ± 0.31 µmol/mol for KMJ-38 (Table 1). The differences between the two population averages for Mg/Ca (0.35 ± 0.20 mmol/mol) and for Ba/Ca (0.64 ± 0.55 µmol/mol) were statistically significant but not so large (Fig. 8). The respective Pb/Ca averages with 95% confidence levels were 0.66 ± 0.09 µmol/mol for KMJ-40 and KMJ-34 and 2.09 ± 0.61 µmol/mol for KMJ-38, the difference of which is significant (1.42 ± 0.62 µmol/mol). The intraspecific variations for Mg/Ca, Ba/Ca, and Pb/Ca may be due to an individual difference in biological controls on skeletal partitioning of the trace elements.

For Mg/Ca, the mechanisms of magnesium incorporation into aragonite are very complicated: loosely bound in the crystal lattice, bound to metal sites of organic compounds, and/or adsorbed onto crystal surface (Amiel et al. 1973). The intraspecific Mg/Ca variations may be due to individual differences in organic phases and the metal binding capacity of sclerosponge tissue layers. For Ba/Ca, the mode of incorporation of barium into the sclerosponge skeleton remains unknown (e.g., Rosenheim et al. 2005b; Allison et al. 2012). Sclerosponges contain a large amount of bacteria living within their tissue (Wörheide 1998). One possibility to explain the intraspecific Mg/Ca and Ba/Ca variations found in this study is that the physiological processes associated with these bacteria may have an effect on the incorporation of Mg^{2+} and Ba^{2+} for calcification at a micro-environmental level (Rosenheim et al. 2005b). Further, the intraspecific variation indicates that

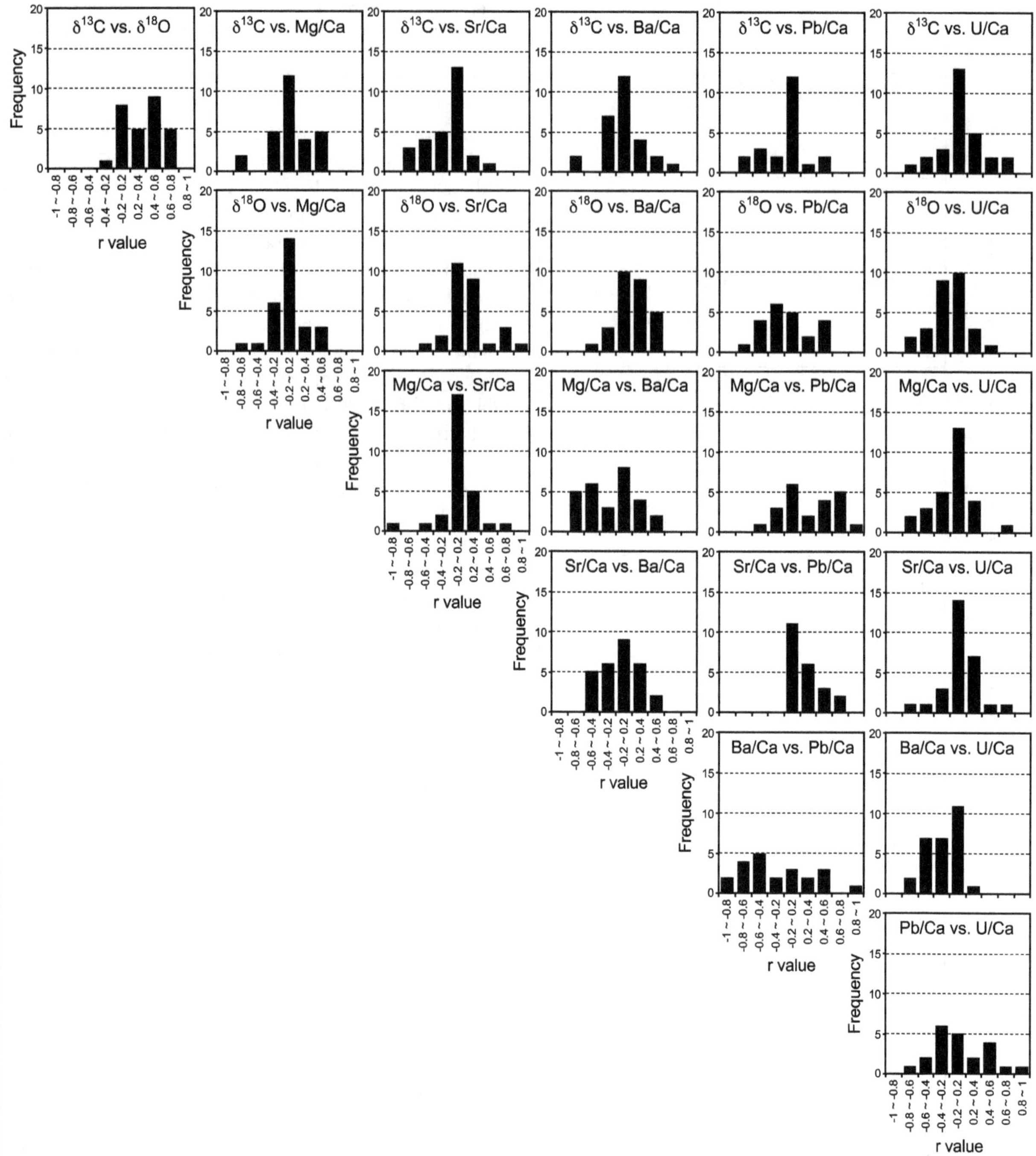

Fig. 7 Frequency distributions (histograms) of correlation coefficients among geochemical records of sclerosponges. The *Y*-axis represents the number of individuals (sample number: *N*). The *r* values are divided into nine classes in the *X*-axis. A trend of the respective relations of two within-skeleton geochemical components is visualized. High frequency with *r* values of − 0.4–0.4 indicates a little significant relationship in within-skeletal geochemistry

the efficiency for transporting elements into the calcification site may be different among individuals because of the large discrepancy in the cationic radii of Mg^{2+} and Ba^{2+} relative to Ca^{2+} unlike Sr^{2+} (Gaetani and Cohen 2006).

The Pb/Ca profiles along a growth direction shows historical changes associated with environmental lead pollution (e.g., Lazareth et al. 2000; Swart et al. 2002; Ohmori et al. 2014). Skeletal Pb/Ca in a *Acanthochaetetes wellsi* (high-Mg calcite) collected from the same locality as our study (Kume Island, Japan) varied roughly from 200 to 1000 nmol/mol for 2000–2007 (Ohmori et al. 2014) and that in a *Ceratoporella nicholsoni* (aragonite) from the

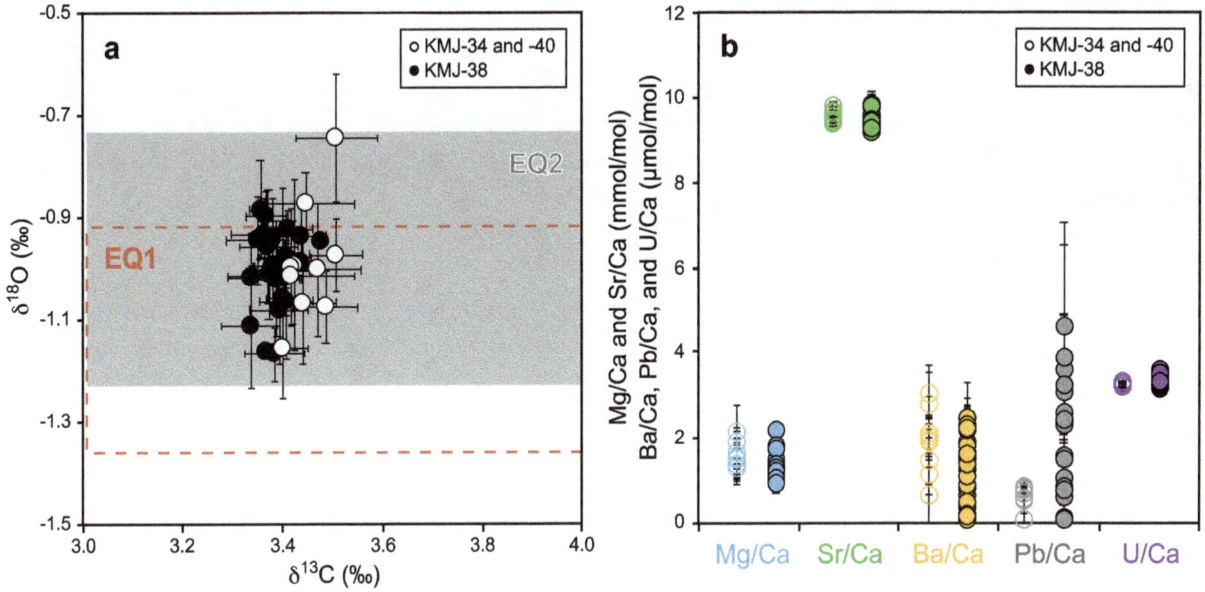

Fig. 8 Scatter plot of **a** δ^{18}O and δ^{13}C and **b** elements/Ca ratios of sclerosponges. Data are shown as averages with 1σ standard deviations in individual samples, which show intraspecific variations in sclerosponge geochemistry. Red (EQ1) and gray (EQ2) areas are shown respectively as estimated ranges of isotopically equilibrium aragonite precipitated at the study site using a relation between seawater δ^{18}O and salinity by Oba (1990) and Abe et al. (2009)

Caribbean Sea varied roughly from 90 to 120 nmol/mol for 1999–2001 (Rosenheim et al. 2005b). Considering the Pb/Ca averages of our samples (1648 ± 421 nmol/mol), interspecific variations are large. Sclerosponges incorporate ~ 10 times higher Pb than corals because pumping and filtering particulates, which absorb heavy metal species, are the only source of nutrition for sclerosponges (e.g., Hartman and Goreau 1966; Willenz and Hartman 1989). Therefore, it is inferred that the amounts, size, and/or species of filtered particles may be considerably different among individuals and species.

Differences in the geochemistry among micro-scale skeletal structures may reflect variations in the composition of the precipitation fluid, the incorporation of ions into the skeleton, and/or kinetic effects (Gabitov et al. 2008). In most cases, apparent relationships among elements/Ca and isotope data of all samples were not founded, yielding a very weak correlation coefficient ranging from − 0.38 to 0.29 (Table 2). This may indicate that the millimeter scale analyses carried out in this study smoothed the heterogeneity of skeletal geochemistry, which results in a smoothing of environmental signals to no small extent through the living tissue layer. Meanwhile, there significantly exist weak correlations between Sr/Ca and δ^{18}O (r = 0.47), Mg/Ca and Pb/Ca (r = − 0.56), and Ba/Ca and U/Ca (r = − 0.69), but their relationships can explain only 22–48% in the respective variance. The former may reflect a difference in skeletal growth rate among the individual samples because both sclerosponge Sr/Ca and δ^{18}O strongly correlated negatively with seawater temperature. At this point, no clear reason underlying the latter two can be inferred. However, the negative relation between Mg/Ca and Pb/Ca found in this study can be explained by an indication that sclerosponges would acquire more Mg^{2+} nutritionally with a decrease in pumping of ambient water and particulates with adsorbed Pb species (Willenz and Hartman 1989; Rosenheim et al. 2005b). The significant negative correlation of the comparatively high r value of − 0.69 between Ba/Ca and U/Ca in

Table 2 Correlations between average values of sclerosponge geochemistry*

	δ^{18}O	Mg/Ca	Sr/Ca	Ba/Ca	Pb/Ca	U/Ca
δ^{13}C	*0.27*	*0.22*	*0.17*	*0.22*	*-0.35*	*-0.11*
δ^{18}O		*-0.06*	**0.47**	*0.29*	*-0.04*	*-0.06*
Mg/Ca			*-0.11*	*0.18*	**-0.56**	*0.02*
Sr/Ca				*0.16*	*-0.01*	*-0.12*
Ba/Ca					-0.38	**-0.69**
Pb/Ca						*0.19*

*Coefficent values in italics and bold fall outside (p > 0.01) and inside (p < 0.01) the 99% confidence intervals, respectively

Table 3 Average elemental ratios and partition coefficients for aragonite sclerosponges

	Species	Site	Depth (m)	Mg/Ca (mmol/mol)	Sr/Ca (mmol/mol)	Ba/Ca (μmol/mol)	Pb/Ca (nmol/mol)	U/Ca (μmol/mol)
This study								
Avg. (95% CL)	*A. willeyana*	Kume Is., Japan	30-40	1.31 ± 0.11	9.45 ± 0.05	1.42 ± 0.26	1648 ± 421	3.26 ± 0.03
$D_{Me,\ Arag}$				3.62-4.27 x 10^{-4}	1.09-1.11	0.007-0.017		2.28-2.35
Rosenheim et al. (2005)								
Avg. (±1σ)	*C. nicholsoni*	Jamaica	22	0.85-0.92	9.45-9.89	3.60-4.14	8.13-91.9	3.20-3.37
$D_{Me,\ Arag}$				1.81 x 10^{-4}	1.10	0.098		
Average (±1σ)	*C. nicholsoni*	Bahamas	143	0.77-0.90	10.15-10.74	3.16-3.66	52.10-289	3.07-3.28
$D_{Me,\ Arag}$				1.6 x 10^{-4}	1.17	0.86	171.85	2.98
Fallon et al. (2005)								
Avg. (1980-1994)	*A. willeyana*	Otta Island, Micronesia	25	0.9-1.2	9.5-12.0	2.7-4.1		3.1-3.6
Allison et al. (2012)								
Avg. (-2004, several yrs)	*A. willeyana*	Nomuka Ika, Tonga	9.5	0.3-1.1	8.5-12.5	2.5-8.0		

our sclerosponge samples possibly reflects a significant positive relation between Ba/Ca and sample size (a parameter presumably related to biological and skeletal growth rate) (Fig. 9) and the significant negative relationship between U/Ca and seawater carbonate ion concentration (relating to pH and aragonite saturation degree) for deep-sea azooxanthellate corals (Anagnostou et al. 2011).

Compared with equilibrium aragonite for stable isotopes

To evaluate the fidelity of *A. willeyana* as a paleoenvironmental proxy, we compared our data with the $\delta^{13}C$ and $\delta^{18}O$ values of aragonites that are inorganically precipitated in isotopic equilibrium with seawater at the sclerosponge growth site: "equilibrium aragonite." The $\delta^{18}O$ values were calculated using an equation ($1000 \times \ln \alpha = 17.88(10^3/T) - 31.14$) derived from laboratory synthesis experiments of aragonite at water temperatures of 0–40 °C (Kim et al. 2007). Therein, α is the fractionation factor between aragonite and water; T is in degrees Kelvin. Although showing good consistency with the climatological data products (Fig. 2), the in situ observational data of this study (October 2015 to November 2016) are temporally limited relative to the growth periods (a few years) of investigated skeletal portions. Based on skeletal growth characteristics for *A. willeyana* reported from earlier works (see the "Materials and methods" section), for this study, the investigated skeletal increments of our sclerosponges were

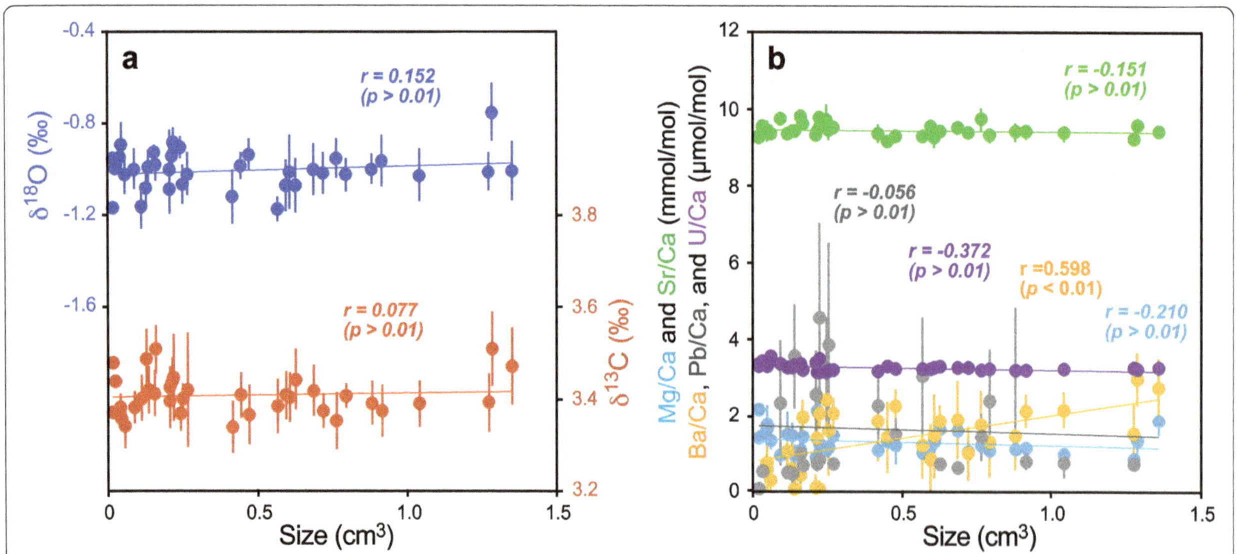

Fig. 9 Relationship between sample size and **a** $\delta^{18}O$ and $\delta^{13}C$ and **b** elements/Ca ratios of sclerosponges

assumed to be precipitated for 2013–2016. Consequently, the annual mean seawater temperature (25.4 °C) and the variation (± 0.9 °C, 2σ) for 2013–2016 were estimated from JODC and Integrated Global Ocean Services System gridded data (26°–27° N, 126°–127° E: Reynolds et al. 2002). The seawater $\delta^{18}O$ values of 0.26 ± 0.05‰ or 0.43 ± 0.08‰ were estimated respectively from JODC salinity data using a relation for the East China Sea – Kuroshio Current region (Oba 1990) (case 1) or for Ishigaki Island in the southern Ryukyus (Abe et al. 2009) (case 2). These estimations are consistent with $\delta^{18}O$ values (0.18‰ in November 2009, 0.25–0.29‰ in October 2015, 0.35‰ in March 2016) of seawater samples collected at water depths of 34–40 m in this study area. Consequently, the $\delta^{18}O$ values of equilibrium aragonites precipitated at the study site for 2013–2016 are expected to fall within a range of – 0.92‰ to – 1.37‰ (case 1: shown as EQ1 in Fig. 8) or from – 0.73‰ to – 1.23‰ (case 2: shown as EQ2 in Fig. 8) at 95% confidence intervals.

The $\delta^{13}C$ values of the equilibrium aragonites were estimated using DIC $\delta^{13}C$ of seawater, the carbon-isotope fractionation factors among CO_2 species (Zhang et al. 1995), and the aragonite–HCO_3– enrichment factor (2.7 ± 0.6‰; Romanek et al. 1992). Because of the lack of data on DIC $\delta^{13}C$ of seawater around the study site, the annual $\delta^{13}C$ value of DIC at a water depth of 40 m is assumed to be the medium value (0.68‰) between November and August estimated using two significant relations established west of Okinawa Island (r^2 = 0.810; Takayanagi et al. 2012) and southeast of Okinawa Island (r^2 = 0.997; Suzuki et al. 2009). Consequently, the $\delta^{13}C$ values of the equilibrium aragonites are expected to be roughly 3.0–4.2‰ (Fig. 8). It is noteworthy that the estimated $\delta^{13}C$ values are approximate because of the limitations and uncertainty of the dataset.

The skeletal $\delta^{13}C$ and $\delta^{18}O$ values in our 36 sclerosponge samples almost fell within the probable ranges (case 1 and case 2) of the equilibrium aragonites precipitated at the living site (Fig. 8). This result strongly shows that the $\delta^{13}C$ and $\delta^{18}O$ of *A. willeyana* is not be influenced isotopically by vital physiological effects. Furthermore, no strong correlation exists between the isotope composition and the sclerosponge sample size (r = 0.076 for $\delta^{13}C$ and r = 0.150 for $\delta^{18}O$, $p > 0.01$) (Fig. 9), unlike symbiont corals for which the isotopic variations (the degree of disequilibrium) are related to the skeletal growth rate (e.g., McConnaughey 1989; Felis et al. 2003). $\delta^{13}C$ and $\delta^{18}O$ values of corals with symbiotic algae are well known to differ greatly from those of equilibrium aragonite because of kinetic and metabolic fractionation effects (e.g., Aharon 1991; McConnaughey 2003). We therefore suggest that the fractionation effects have little influence on the isotopic composition of *A. willeyana* because the skeletal growth is expected to constant at a very slow rate. Our results are supported strongly by evidence from earlier studies showing that $\delta^{18}O$ values of living sclerosponges (seven *C. nicholsoni* from the Caribbean Sea and a *A. willeyana* from the Great Barrier Reef) are in good agreement with published values from synthetic aragonite and near-equilibrium values of mollusks and foraminifers (Böhm et al. 2000) and that sclerosponges (three *C. nicholsoni* from the Caribbean Sea and a *A. wellsi* from New Caledonia) precipitate $CaCO_3$ skeletons close to isotopic equilibrium for $\delta^{13}C$ values (Böhm et al. 1996).

Partition coefficients of elements/Ca

Partition coefficients, apparent distribution coefficients ($D_{Me,\ Arag}$), were estimated using element/Ca ratios between seawater and aragonitic skeleton of sclerosponge, except for Pb/Ca (Table 3). As it is known that the $D_{Sr,\ Arag}$ of inorganic aragonite depends on water temperature (e.g., Kinsman and Holland 1969; DeCarlo et al. 2015), sclerosponge Sr/Ca can be useful for a proxy of seawater temperature (e.g., Rosenheim et al. 2004; Haase-Schramm et al. 2005; Waite et al. 2018). The $D_{Sr,\ Arag}$ from our *A. willeyana* samples was 1.09–1.11 (almost unity), which is consistent with that of *C. nicholsoni* from the Caribbean Sea (Rosenheim et al. 2005b) (Table 3) and experimentally precipitated abiogenic aragonite (Gaetani and Cohen 2006; DeCarlo et al. 2015). The average Sr/Ca of ~ 9.5 mmol/mol in our sclerosponges was identical to an estimation of inorganic aragonite of 9.42–9.50 mmol/mol at 25.4 ± 0.9 °C calculated by using the experimentally established relationship (Gaetani and Cohen 2006) and almost consistent with previously reported values for the same species in the Pacific Ocean (Fallon et al. 2005; Grottoli et al. 2010; Allison et al. 2012). Furthermore, high-resolution Sr/Ca records of aragonitic skeleton for *A. willeyana* (Fallon et al. 2005) and *C. nicholsoni* (Rosenheim et al. 2004) using LA-ICP-MS have been shown to be seasonal variations, yielding significant calibrations for seawater temperature. Therefore, these lines of evidence suggest that *A. willeyana* sclerosponge Sr/Ca can be a robust indicator of seawater temperature.

The value of $D_{Mg,\ Arag}$ estimated from our samples shows that *A. willeyana* sclerosponge Mg/Ca ratios are several orders (10^{-3} to 10^{-4}) of magnitude smaller than those of seawater (Table 3). The value is about twice that of the species *C. nicholsoni* (Rosenheim et al. 2005b), but about one order of magnitude less than experimentally precipitated inorganic aragonite (Gaetani and Cohen 2006). The $D_{Ba,\ Arag}$ estimated from our samples shows that Ba/Ca ratios in *A. willeyana* sclerosponge are variable and about two orders of magnitude less than those of seawater. The $D_{Ba,\ Arag}$ value estimated in this study is about several orders (10^{-1} to 10^{-2}) of magnitude less than those for the species *C. nicholsoni* (Rosenheim et al. 2005b) and inorganic aragonite (Gaetani and Cohen 2006; Mavromatis et al. 2018). Rosenheim et al. (2005b) also reported that *C. nicholsoni* sclerosponges showed large variations in their $D_{Ba,\ Arag}$ values of about 0.1–0.9 (Table 3). Such a small $D_{Ba,\ Arag}$ value of < 1 for sclerosponge samples can be reasonably explained by the lower incorporation of an incompatible trace metal because of the large difference of ~ 25% in ionic radius between Ba^{2+} and Ca^{2+}, as was true also for other aragonite materials such as synthetic aragonite (Mavromatis et al. 2018), speleothems (e.g., Wassenburg et al. 2016), bivalves (e.g., Poulain et al. 2015), and fish otoliths (e.g., Bath et al. 2000) in contrast to corals, which were greater than unity (e.g., Gonneea et al. 2017). However, such a different value of ~ 0.01 to ~ 0.9 (Table 3) indicates that physiological processes for the incorporation of Ba^{2+} into aragonitic sclerosponges for calcification differ greatly not only among individuals but also among species. Perhaps for this reason, significant correlations exist between the Ba/Ca and the sclerosponge sample size found in this study ($r = \sim 0.6$, $p < 0.01$) (Fig. 9). Furthermore, results of a study of element partitioning during aragonite precipitation suggest that the calcifying fluid from which coral skeleton precipitates is depleted in Mg and Ba, unlike Sr. Results also suggest that seasonal fluctuations exist in the mass fraction of aragonite precipitated from the fluid (precipitation efficiency) (Gaetani and Cohen 2006), which can qualitatively explain our results showing lower $D_{Mg,\ Arag}$ and $D_{Ba,\ Arag}$ values and a value of $D_{Sr,\ Arag}$ that is almost equal to unity for *A. willeyana* sclerosponge. Quantitative verification of these results requires additional experimental work.

For U/Ca, the average value (~ 3.3 µmol/mol) of our *A. willeyana* sclerosponges was consistent with those of the same species (Fallon et al. 2005) and with *C. nicholsoni* (Rosenheim et al. 2005b) (Table 3). The value of $D_{U,\ Arag}$ (2.3–2.4) estimated in this study was slightly lower than that (2.98) reported for *C. nicholsoni* (Rosenheim et al. 2005b) and much higher than the experimentally determined value (0.31 ± 0.06) of inorganic aragonite precipitation at 25.5 °C (DeCarlo et al. 2015). Degradation of tissue layers and/or bacteria can influence the pH and carbonate species system of the calcifying fluid, thereby altering the U/Ca ratios in sclerosponge skeleton at different levels, which possibly caused differences in U/Ca among sclerosponge species. Although coralline aragonite U/Ca records are apparently correlated with temperature (e.g., Min et al. 1995; Shen and Dunbar 1995), several recent studies demonstrated that the U/Ca ratios of azooxanthellate deep sea coral (Anagnostou et al. 2011) and experimentally precipitated aragonite from seawater (DeCarlo et al. 2015) have significant negative correlation with carbonate ion concentrations because the aqueous chemistry of uranium is influenced by the carbonate ion, which forms complexes with the uranyl ion (Langmuir 1978). The sclerosponge U/Ca is about 2–3 times higher than that of zooxanthellate corals, which may be related to lowering of pH and carbonate ion concentrations in the calcifying fluid. These lines of implication might bring higher $D_{U,\ Arag}$ values in aragonitic sclerosponges than that of synthetic aragonite. Although the complicated mechanism for the incorporation of uranyl complexes into calcium carbonate has not been resolved, the slight intraspecific variations (± 0.03 µmol/mol, the relative variation is 0.9%) of sclerosponge U/Ca for the *A. willeyana* species examined in this study (Fig. 8) support the possibility of its use as a proxy for ocean environments.

Conclusions

Results of this study are important scientific findings for the hypercalcified demosponge *Astrosclera willeyana*; (1) within-skeletal variation is quite small for $\delta^{13}C$, $\delta^{18}O$, Sr/Ca, and U/Ca (namely, little heterogeneity in the coevally precipitated skeleton); (2) intraspecific variations are quite small for $\delta^{13}C$, $\delta^{18}O$, Sr/Ca, and U/Ca (namely, little individual difference); (3) the aragonite skeleton is precipitated in or very close to isotopic equilibrium with seawater at the living site; (4) the isotopic composition of the skeleton is scarcely affected by kinetic and biological effects; (5) the partition coefficient of Sr/Ca is consistent with that for different species and close to inorganic aragonite.

We believe that sclerosponge U/Ca may be a paleoenvironmental proxy although there are still unknowns relating to uranium into the aragonitic skeleton. Relatively large variations and individual differences in the sclerosponge Mg/Ca, Ba/Ca, and Pb/Ca were shown in this study. To clarify the reasons for controlling factors of incorporation (distribution) of the elements into the sclerosponge aragonite skeleton, further investigations such as ultra-high resolution analyses and culture experiments are needed. Importantly, we suggest that *A. willeyana* $\delta^{18}O$ ($\delta^{13}C$) is very suitable as an environmental proxy to reconstruct seawater temperature and $\delta^{18}O$ (DIC $\delta^{13}C$) for the past. Coupled determination of Sr/Ca and $\delta^{18}O$ in a sclerosponge can potentially enable the construction of independent time series for both seawater temperature and $\delta^{18}O$ (≒ salinity) at near-annual resolution by using a millimeter-scale sampling protocol for widely used and convenient methods. Furthermore, in combination with the geochemistry of marine calcified organisms living in shallow waters, such as corals and giant clams, long-lived sclerosponge-based reconstruction of deeper ocean environments can generate a long time series of spatial and vertical changes in coral reef environments for the past few hundred years, extending beyond the observation data. This study also presents the important implication that even if only a few skeletal fragments and a few sclerosponge samples are collected, one can reconstruct annual average paleoceanographic conditions using well-preserved fossils that have not been altered diagenetically. Regarded from these viewpoints, our results can support contributions toward broad applications not only to climatology and oceanography but also to geology and biogeochemistry.

Abbreviations

JODC: Japan Oceanographic Data Center; VPDB: Vienna Standard Pee Dee Belemnite; VSMOW: Vienna Standard Mean Ocean Water

Acknowledgements

We are grateful to a chief editor (Prof. H. Kawahata) and two anonymous reviewers for their helpful and valuable suggestions and comments on the manuscript. We also thank M. Kondo (Okinawa Institute of Science and Underwater Technology) for help with field surveys and Dr. H. Fujimura, M. Chinen, and Y. Akamine (University of the Ryukyus) for help with experiments.

Authors' contributions

RA designed the study, analyzed the samples and data, and wrote the manuscript. TN, MM, YI, YF, and TS collaborated with RA in the fieldwork and sampling. AK, DO, RU, and RS collaborated with RA in the chemical analyses. The authors read and approved the manuscript.

Funding

This research was supported by the Japan Society for the Promotion of Science KAKENHI (Grant-in-Aid for Scientific Research) Grant numbers 26550012, 26707028, and 18K18522 to RA and 15H01729 to RU, and the NEXT Program Grant Number GR083 to TS.

Competing interests

The authors declare that they have no competing interests.

Author details

[1]Institute of Geology and Paleontology, Graduate School of Science, Tohoku University, Aobayama, Sendai 980-8578, Japan. [2]Department of Earth Science, Faculty of Science, University of the Ryukyus, 1 Senbaru, Nishihara, Okinawa 903-0213, Japan. [3]Tropical Biosphere Research Center, Iriomote Station, University of the Ryukyus, 870 Uehara, Taketomi, Okinawa 907-1541, Japan. [4]Department of Chemistry, Biology, and Marine Science, Faculty of Science, University of the Ryukyus, 1 Senbaru, Nishihara, Okinawa 903-0213, Japan. [5]Division of Earth and Environmental Sciences, Graduate School of Environmental Studies, Nagoya University, Furo-cho, Chikusa-ku, Nagoya 464-8601, Japan. [6]Centre for Marine and Coastal Studies (CEMACS), Universiti Sains Malaysia, 11800 USM, Penang, Malaysia. [7]Okinawa Prefectural University of Arts, 1-4 Shuri-Tounokura, Naha, Okinawa 903-8602, Japan. [8]Department of Civil and Environmental Engineering, Graduate School of Engineering, Tohoku University, 6-6-06 Aramaki-Aza-Aoba, Sendai 980-8579, Japan.

References

Abe O, Agata S, Morimoto M, Abe M, Yoshimura K, Hiyama T, Yoshida N (2009) A 6.5-year continuous record of sea surface salinity and seawater isotopic composition at Harbour of Ishigaki Island, southwest Japan. Isot Environ Health Stud 45:247–258. https://doi.org/10.1080/10256010903083847

Abram NJ, HV MG, Tierney JE, Evans MN, NP MK, Kaufman DS, the PAGES 2 k Consortium (2016) Early onset of industrial-era warming across the oceans and continents. Nature 536:411–418. https://doi.org/10.1038/nature19082

Aharon P (1991) Recorders of reef environment histories: stable isotopes in corals, giant clams, and calcareous algae. Coral Reefs 10:71–90 https://doi.org/10.1007/BF00571826

Allison N, Tudhope AW, EIMF (2012) The skeletal geochemistry of the sclerosponge *Astrosclera willeyana*: implications for biomineralisation processes and paleoenvironmental reconstruction. Palaeogeogr Palaeoclimatol Palaeoecol 313–314:70–77. https://doi.org/10.1016/j.palaeo.2011.10.009

Amiel AJ, Friedman GM, Miller DS (1973) Distribution and nature of incorporation of trace elements in modern aragonitic corals. Sedimentology 20:47–64 https://doi.org/10.1111/j.1365-3091.1973.tb01606.x

Anagnostou E, Sherrell RM, Gagnon A, LaVigne M, Field MP, McDonough WF (2011) Seawater nutrient and carbonate ion concentrations recorded as P/Ca, Ba/Ca, and U/Ca in the deep-sea coral *Desmophyllum dianthus*. Geochim Cosmochim Acta 75:2529–2543. https://doi.org/10.1016/j.gca.2011.02.019

Asami R, Felis T, Deschamps P, Hanawa K, Iryu Y, Bard E, Durand N, Murayama M (2009) Evidence for tropical South Pacific climate change during the Younger Dryas and the Bølling–Allerød from geochemical records of fossil Tahiti corals. Earth Planet Sci Lett 288:96–107. https://doi.org/10.1016/j.epsl.2009.09.011

Asami R, Iryu Y, Hanawa K, Miwa T, Holden P, Shinjo R, Paulay G (2013) MIS 7 interglacial sea-surface temperature and salinity reconstruction from a southwestern subtropical Pacific coral. Quat Res 80:575–585 https://doi.org/10.1016/j.yqres.2013.09.002

Asami R, Konishi M, Tanaka K, Uemura R, Furukawa M, Shinjo R (2015) Late Holocene coral reef environment recorded in Tridacnidae shells from archaeological sites in Okinawa-jima, subtropical southwestern Japan. Island Arc 24:61–72. https://doi.org/10.1111/iar.12076

Asami R, Yamada T, Iryu Y, Quinn TM, Meyer CP, Paulay G (2005) Interannual and decadal variability of the western Pacific sea surface condition for the years 1787–2000: reconstruction based on stable isotope record from a Guam coral. J Geophys Res 110:C05018. https://doi.org/10.1029/2004JC002555

Ayling A (1982) Redescription of *Astrosclera willeyana* Lister 1900 (Ceratoporellida, Demospongiae), a new record from the Great Barrier Reef. Mem Nat Mus Victoria **43**:99–103 https://doi.org/10.24199/j.mmv.1982.43.06

Bath GE, Thorrold SR, Jones CM, Campana SE, McLaren JW, Lam JWH (2000) Strontium and barium uptake in aragonitic otoliths of marine fish. Geochim Cosmochim Acta 64:1705–1714 https://doi.org/10.1016/S0016-7037(99)00419-6

Benavides LM, Druffel ERM (1986) Sclerosponge growth rate as determined by ^{210}Pd and Δ^{14}C chronologies. Coral Reefs 4:221–224. https://doi.org/10.1007/BF00298080

Böhm F, Haase-Schramm A, Eisenhauer A, Dullo W C, Joachimski MM, Lehnert H, Reitner J (2002) Evidence for preindustrial variations in the marine surface water carbonate system from coralline sponges. Geochem Geophys Geosyst 3:1019. https://doi.org/10.1029/2001GC000264

Böhm F, Joachimski MM, Dullo W-C, Eisenhauer A, Lehnert H, Reitner J, Wörheide G (2000) Oxygen isotope fractionation in marine aragonite of coralline sponges. Geochim Cosmochim Acta 64:1695–1703. https://doi.org/10.1016/S0016-7037(99)00408-1

Böhm F, Joachimski MM, Lehnert H, Morgenroth G, Kretschmer W, Vacelet J, Dullo W-C (1996) Carbon isotope records from extant Caribbean and South Pacific sponges: evolution of δ^{13}C in surface water DIC. Earth Planet Sci Lett 139:291–303. https://doi.org/10.1016/0012-821X(96)00006-4

DeCarlo TM, Gaetani GA, Holcomb M, Cohen AL (2015) Experimental determination of factors controlling U/Ca of aragonite precipitated from seawater: implications for interpreting coral skeleton. Geochim Cosmochim Acta 162:151–165 https://doi.org/10.1016/j.gca.2015.04.016

Druffel ERM, Benavides LM (1986) Input of excess CO_2 to the surface ocean based on $^{13}C/^{12}C$ ratios in a banded Jamaican sclerosponge. Nature 321:58–61. https://doi.org/10.1038/321058a0

Fallon SJ, Guilderson TP (2005) Extracting growth rates from the nonlaminated coralline sponge *Astrosclera willeyana* using bomb radiocarbon. Limnol Oceanogr Methods 3:455–461 https://doi.org/10.4319/lom.2005.3.455

Fallon SJ, Guilderson TP, Caldeira K (2003) Carbon isotope constraints on vertical mixing and air-sea CO_2 exchange. Geophys Res Lett 30:2289. https://doi.org/10.1029/2003GL018049

Fallon SJ, McCulloch MT, Guilderson TP (2005) Interpreting environmental signals from the coralline sponge *Astrosclera willeyana*. Palaeogeogr Palaeoclimatol Palaeoecol 228:58–69. https://doi.org/10.1016/j.palaeo.2005.03.053

Felis T, Pätzold J, Loya Y (2003) Mean oxygen-isotope signatures in *Porites* spp. corals: inter-colony variability and correction for extension-rate effects. Coral Reefs 22:328–336 https://doi.org/10.1007/s00338-003-0324-3

Gabitov RI, Gaetani GA, Watson EB, Cohen AL, Ehrlich HL (2008) Experimental determination of growth rate effect on U^{6+} and Mg^{2+} partitioning between aragonite and fluid at elevated U^{6+} concentration. Geochim Cosmochim Acta 72:4058–4068. https://doi.org/10.1016/j.gca.2008.05.047

Gaetani GA, Cohen AL (2006) Element partitioning during precipitation of aragonite from seawater: a framework for understanding paleoproxies. Geochim Cosmochim Acta 70:4617–4634. https://doi.org/10.1016/j.gca.2006.07.008

Gilis M, Grauby O, Willenz P, Dubois P, Heresanu V, Baronnet A (2013) Biomineralization in living hypercalcified demosponges: toward a shared mechanism? J Struct Biol 183:441–454 https://doi.org/10.1016/j.jsb.2013.05.018

Gonneea ME, Cohen AL, DeCarlo TM, Charette MA (2017) Relationship between water and aragonite barium concentrations in aquaria reared juvenile corals. Geochim Cosmochim Acta 209:123–134 https://doi.org/10.1016/j.gca.2017.04.006

Grottoli AG (2006) Monthly resolved stable oxygen isotope record in a Palauan sclerosponge *Acanthocheatetes wellsi* for the period of 1977–2001. In Proc 10th Int Coral Reef Symp 2004:572–579.

Grottoli AG, Adkins JF, Panero WR, Reaman DM, Moots K (2010) Growth rates, stable oxygen isotopes ($\delta^{18}O$), and strontium (Sr/Ca) composition in two species of Pacific sclerosponges (*Acanthocheatetes wellsi* and *Astrosclera willeyana*) with $\delta^{18}O$ calibration and application to paleoceanography. J Geophys Res 115:C06008. https://doi.org/10.1029/2009JC005586

Haase-Schramm A, Böhm F, Eisenhauer A, Garbe-Schönberg D, Dullo W-C, Reitner J (2005) Annual to interannual temperature variability in the Caribbean during the Maunder sunspot minimum. Paleoceanography 20:PA4015. https://doi.org/10.1029/2005PA001137

Hartman WD (1980) Ecology of recent sclerosponges. In: Hartman WD, Wendt JW, Wiedenmayer F (eds.) Living and Fossil Sponges. Sedimenta, Miami, pp 253–255.

Hartman WD, Goreau TF (1966) *Ceratoporella*, a living sponge with stromatoporoid affinities. Am Zool 6:563–564

Hathorne EC et al (2013) Interlaboratory study for coral Sr/Ca and other element/Ca ratio measurements. Geochem Geophys Geosyst 14:3730–3750. https://doi.org/10.1002/ggge.20230

Jackson DJ, Thiel V, Wörheide G (2010) An evolutionary fast-track to biocalcification. Geobiology 8:191–196 https://doi.org/10.1111/j.1472-4669.2010.00236.x

Keeling CD, Mook WG, Tans PP (1979) Recent trends in the $^{13}C/^{12}C$ ratio of atmospheric carbon dioxide. Nature 277:121–123. https://doi.org/10.1038/277121a0

Kim S-T, O'Neil JR, Hillaire-Marcel C, Mucci A (2007) Oxygen isotope fractionation between synthetic aragonite and water: influence of temperature and Mg^{2+} concentration. Geochim Cosmochim Acta 71:4704–4715. https://doi.org/10.1016/j.gca.2007.04.019

Kinsman DJ, Holland HD (1969) The co-precipitation of cations: IV. The co-precipitation of Sr^{2+} with aragonite between 16–96 °C. Geochim Cosmochim Acta 33:1–17 https://doi.org/10.1016/0016-7037(69)90089-1

Langmuir D (1978) Uranium solution-mineral equilibria at low temperatures with applications to sedimentary one deposits. Geochim Cosmochim Acta 42: 547–569 https://doi.org/10.1016/0016-7037(78)90001-7

Lazareth C, Willenz P, Navez J, Keppens E, Dehairs F, Andre L (2000) Sclerosponges as a new potential recorder of environmental changes: lead in *Ceratoporella nicholsoni*. Geology 28:515–518, doi:10.1130/0091-7613(2000)28 < 515:SAANPR>2.0.CO;2

Lister JJ (1900) *Astrosclera willeyana*, the type of a new family of sponges. Zool Res 4:461–482

Mavromatis V, Goetschl KE, Grengg C, Konrad F, Purgstaller B, Dietzel M (2018) Barium partitioning in calcite and aragonite as a function of growth rate. Geochim Cosmochim Acta 237:65–78 https://doi.org/10.1016/j.gca.2018.06.018

McConnaughey TA (1989) ^{13}C and ^{18}O isotopic disequilibrium in biological carbonates: I. Patterns. Geochim Cosmochim Acta 53:151–162. https://doi.org/10.1016/0016-7037(89)90282-2

McConnaughey TA (2003) Sub-equilibrium oxygen-18 and carbon-13 levels in biological carbonates: carbonate and kinetic models. Coral Reefs 22:316–327 https://doi.org/10.1007/s00338-003-0325-2

Min GR, Edwards RL, Taylor FW, Recy J, Gallup CD, Beck JW (1995) Annual cycles of U/Ca in coral skeletons and U/Ca thermometry. Geochim Cosmochim Acta 59:2025–2042 https://doi.org/10.1016/0016-7037(95)00124-7

Moore MD, Charles CD, Rubenstone JL, Fairbanks RG (2000) U/Th-dated sclerosponges from the Indonesian Seaway record subsurface adjustments to west Pacific winds. Paleoceanography 15:404–416. https://doi.org/10.1029/1999PA000396

Nakayama H, Iijima H, Nakamura N, Kayanne H (2008) Carbon and oxygen stable isotope ratios of GSJ carbonate reference materials (JCp-1 and JCt-1). Bull Geol Surv Japan 59:461–466 (in Japanese with English abstract)

Naruse T, Chan TY, Tan HH, Ahyong ST, Reimer JD (2012) Scientific results of the Kumejima marine biodiversity expedition—KUMEJIMA 2009. Zootaxa 3367:1–280

Oba T (1990) Paleoceanographic information obtained by the isotopic measurement of individual foraminiferal specimens. In: Wang P, Lao Q, He Q (eds) Asian Marine Geology. China Ocean Press, Beijing, pp 169–180

Ohmori K, Watanabe T, Tanimizu M, Shirai K (2014) Lead concentration and isotopic composition in the Pacific sclerosponge (*Acanthochaetetes wellsi*) reflects environmental lead pollution. Geology 42:287–290. https://doi.org/10.1130/G34316.1

Okai T, Suzuki A, Kawahata H, Terashima S, Imai N (2002) Preparation of a new Geological Survey of Japan geochemical reference material: coral JCp-1. Geostand Newslett 26:95–99 https://doi.org/10.1111/j.1751-908X.2002.tb00627.x

Okai T, Suzuki A, Terashima S, Inoue M, Nohara M, Kawahata H, Imai N (2004) Collaborative analysis of GSJ/AIST geochemical reference materials JCp-1(Coral) and JCt-1(Giant Clam). Chikyu Kagaku (Geochemistry) 38:281–286 https://doi.org/10.14934/chikyukagaku.38.281 (in Japanese with English abstract)

PAGES 2 k Consortium (2013) Continental-scale temperature variability during the past two millennia. Nat Geosci 6:339–346. https://doi.org/10.1038/ngeo1797

Poulain C, Gillikin DP, Thebault J, Munaron JM, Bohn M, Robert R, Paulet Y-M, Lorrain A (2015) An evaluation of Mg/Ca, Sr/Ca, and Ba/Ca ratios as environmental proxies in aragonite bivalve shells. Chem Geol 396:42–50 https://doi.org/10.1016/j.chemgeo.2014.12.019

Quay PD, Tilbrook B, Wong CS (1992) Oceanic uptake of fossil fuel CO_2: carbon-13 evidence. Science 256:74–79. https://doi.org/10.1126/science.256.5053.74

Reitner J (1992) "Coralline Spongien". Der Versuch einer phylogenetisch-taxonomischen Analyse. Coralline sponges an attempt of a phylogenetic-taxonomic analysis. Berliner Geowiss. Abh. Reihe E Palaeobiol 1:1–352

Reitner J, Engeser TS (1987) Skeletal structures and habitats of recent and fossil *Acanthochaetetes* (subclass Tetractinomorpha, Demospongiae, Porifera). Coral Reefs 6:13–18. https://doi.org/10.1007/BF00302207

Reitner J, Gautret P (1996) Skeletal formation in the modern but ultraconservative chaetetid sponge *Spirastrella* (*Acanthochaetetes*) *wellsi* (Demospongiae, Porifera). Facies 34:193–208. https://doi.org/10.1007/BF02546164

Reynolds RW, Rayner NA, Smith TM, Stokes DC, Wang W (2002) An improved in situ and satellite SST analysis for climate. J Clim 15:1609–1625 https://doi.org/10.1175/1520-0442(2002)015<1609:AIISAS>2.0.CO;2

Romanek CS, Grossman EL, Morse JW (1992) Carbon isotopic fractionation in synthetic aragonite and calcite: effects of temperature and precipitation rate. Geochim Cosmochim Acta 56:419–430 https://doi.org/10.1016/0016-7037(92)90142-6

Rosenheim BE, Swart PK, Thorrold SR (2005b) Minor and trace elements in sclerosponge *Ceratoporella nicholsoni*: biogenic aragonite near the inorganic endmember? Palaeogeogr Palaeoclimatol Palaeoecol 228:109–129 https://doi.org/10.1016/j.palaeo.2005.03.055

Rosenheim BE, Swart PK, Thorrold SR, Eisenhauer A, Willenz P (2005a) Salinity change in the subtropical Atlantic: secular increase and teleconnections to the North Atlantic Oscillation. Geophys Res Lett 32:L02603. https://doi.org/10.1029/2004GL021499

Rosenheim BE, Swart PK, Thorrold SR, Willenz P, Berry L, Latkoczy C (2004) High-resolution Sr/Ca records in sclerosponges calibrated to temperature in situ. Geology 32(2):145–148 https://doi.org/10.1130/G20117.1

Rosenheim BE, Swart PK, Willenz P (2009) Calibration of sclerosponge oxygen isotope records to temperature using high-resolution $\delta^{18}O$ data. Geochim Cosmochim Acta 73:5308–5319. https://doi.org/10.1016/j.gca.2009.05.047

Shen GT, Dunbar RB (1995) Environmental controls on uranium in reef corals. Geochim Cosmochim Acta 59:2009–2024 https://doi.org/10.1016/0016-7037(95)00123-9

Suzuki A, Kawamura N, Itaki T, Katayama H, Murakami S, Usami T, Kuroyanagi A (2009) Geochemical analyses on seawater samples collected during the GH08 cruise in the eastern offshore of Okinawa Island. In: Arak K (ed) Marine Geological and Geophysical Studies around Okinawa Islands eastern off of Okinawa Island, Preliminary Reports on Researches in the 2008 Fiscal Year, GSJ Interim Rep 46. Geol Surv Jpn, Natl Inst Adv Ind Sci and Technol, Tsukuba, pp 86–92

Swart PK, Greer L, Rosenheim BE, Moses CS, Waite AJ, Winter A, Dodge RE, Helmle K (2010) The ^{13}C Suess effect in scleractinian corals mirror changes in the anthropogenic CO_2 inventory of the surface oceans. Geophys Res Lett 37:L05604. https://doi.org/10.1029/2009GL041397

Swart PK, Thorrold S, Rosenheim B, Eisenhauer A, Harrison CGA, Grammer M, Latkoczy C (2002) Intra-annual variation in the stable oxygen and carbon and trace element composition of sclerosponges. Paleoceanography 17:1045. https://doi.org/10.1029/2000PA000622

Takayanagi H, Asami R, Abe O, Miyajima T, Kitagawa H, Iryu Y (2012) Carbon- and oxygen-isotope compositions of a deep-water modern brachiopod *Campagea japonica* collected off Aguni-jima, Central Ryukyu Islands, southwestern Japan. Geochem J 46:77–87 https://doi.org/10.2343/geochemj.1.0153

Uemura R, Yonezawa N, Yoshimura K, Asami R, Kadena H, Yamada K, Yoshida N (2012) Factors controlling isotopic composition of precipitation on Okinawa Island, Japan: implications for paleoclimate reconstruction in the East Asian Monsoon region. J Hydrol 475:314–322 https://doi.org/10.1016/j.jhydrol.2012.10.014

Vacelet J (1979) Description et affinit.s d'une .ponge actuelle. In: Levi C, Boury-Esnault N (eds) Biologie des Spongiaires - sponge biology. Colloques Internationaux du Centre National de la Recherche Scientifique, CNRS, Paris, pp 483–493

Vacelet J (1985) Coralline sponges and the evolution of Porifera. In: Conway Morris S, George JD, Gibson R, Platt HM (eds) The Origins and Relationships of Lower Invertebrates. Clarendon Press, Oxford, pp 1–13

Waite AJ, Swart PK, Rosenheim BE, Rosenberg AD (2018) Improved calibration of the Sr/Ca-temperature relationship in the sclerosponge *Ceratoporella nicholsoni*: Re-evaluating Sr/Ca derived records of post-industrial era warming. Chem Geol 488:56–61 https://doi.org/10.1016/j.chemgeo.2018.03.005

Wassenburg JA, Scholz D, Jochum KP, Cheng H, Oster J, Immenhauser A, Richter DK, Hager T, Jamieson RA, Baldini JUL, Hoffmann D, Breitenbach SFM (2016) Determination of aragonite trace element distribution coefficients from speleothem calcite-aragonite transitions. Geochim Cosmochim Acta 190:347–367 https://doi.org/10.1016/j.gca.2016.06.036

Willenz P, Hartman WD (1989) Micromorphology and ultrastructure of Caribbean sclerosponges. Mar Biol 103:387–401 https://doi.org/10.1007/BF00397274

Wood R (1990) Reef-building sponges. Am Sci 78:224–235

Wörheide G (1998) The reef cave dwelling ultraconservative coralline demosponge *Astrosclera willeyana* Lister 1900 from the Indo-Pacific. Micromorphology, ultrastructure, biocalcification, isotope record, taxonomy, biogeography, phylogeny. Facies 38:1–88. https://doi.org/10.1007/BF02537358

Wörheide G, Reitner J, Gautret P (1997) Comparison of biocalcification processes in the two coralline demosponges *Astrosclera willeyana* Lister 1900 and *"Acanthochaetetes" wellsi* Hartman and Goreau 1975. In: Lessios HA, Macintyre IG (eds) Proc 8th Int Coral Reef Symp. Smithsonian Tropical Research Institute, Panama, pp 1427–1432

Wu HC, Grottoli AG (2010) Stable oxygen isotope records of corals and a sclerosponge in the Western Pacific warm pool. Coral Reefs 29:413–418. https://doi.org/10.1007/s00338-009-0576-7

Zhang J, Quay PD, Wilbur DO (1995) Carbon isotope fractionation during gas-water exchange and dissolution of CO_2. Geochim Cosmochim Acta 59:107–114 https://doi.org/10.1016/0016-7037(95)91550-D

10

Hydrogen and carbon isotope systematics in hydrogenotrophic methanogenesis under H_2-limited and H_2-enriched conditions: implications for the origin of methane and its isotopic diagnosis

Tomoyo Okumura[1,2], Shinsuke Kawagucci[1,2,3*], Yayoi Saito[2,4], Yohei Matsui[2,3], Ken Takai[1,2,3] and Hiroyuki Imachi[2,3]

Abstract

Hydrogen and carbon isotope systematics of H_2O–H_2–CO_2–CH_4 in hydrogenotrophic methanogenesis and their relation to H_2 availability were investigated. Two H_2-syntrophic cocultures of fermentatively hydrogenogenic bacteria and hydrogenotrophic methanogens under conditions of $<10^2$ Pa-H_2 and two pure cultures of hydrogenotrophic methanogens under conditions of $\sim10^5$ Pa-H_2 were tested. Carbon isotope fractionation between CH_4 and CO_2 during hydrogenotrophic methanogenesis was correlated with pH_2, as indicated in previous studies. The hydrogen isotope ratio of CH_4 produced during rapid growth of the thermophilic methanogen *Methanothermococcus okinawensis* under high pH_2 conditions ($\sim10^5$ Pa) was affected by the isotopic composition of H_2, as concluded in a previous study of *Methanothermobacter thermautotrophicus*. This "δD_{H_2} effect" is a possible cause of the diversity of previously reported values for hydrogen isotope fractionation between CH_4 and H_2O examined in H_2-enriched culture experiments. Hydrogen isotope fractionation between CH_4 and H_2O, defined by $(1000 + \delta D_{CH_4})/(1000 + \delta D_{H_2O})$, during hydrogenotrophic methanogenesis of the H_2-syntrophic cocultures was in the range 0.67–0.69. The hydrogen isotope fractionation of our H_2-syntrophic dataset overlaps with those obtained not only from low-pH_2 experiments reported so far but also from natural samples of "young" methane reservoirs (0.66–0.74). Conversely, such hydrogen isotope fractionation is not consistent with that of "aged" methane in geological samples (≥0.79), which has been regarded as methane produced via hydrogenotrophic methanogenesis from the carbon isotope fractionation. As a possible process inducing the inconsistency in hydrogen isotope signatures between experiments and geological samples, we hypothesize that the hydrogen isotope signature of CH_4 imprinted at the time of methanogenesis, as in the experiments and natural young methane, may be altered by diagenetic hydrogen isotope exchange between extracellular CH_4 and H_2O through reversible reactions of the microbial methanogenic pathway in methanogenic region and/or geological methane reservoirs.

Keywords: Methane, Hydrogenotrophic methanogenesis, Hydrogen isotope ratio, Carbon isotope ratio, H_2 availability, Culture experiments

* Correspondence: kawagucci@jamstec.go.jp
[1]Laboratory of Ocean-Earth Life Evolution Research (OELE), Japan Agency for Marine-Earth Science and Technology (JAMSTEC), 2-15 Natsushima-cho, Yokosuka 237-0061, Japan
[2]Department of Subsurface Geobiological Analysis and Research (D-SUGAR), Japan Agency for Marine-Earth Science and Technology (JAMSTEC), 2-15 Natsushima-cho, Yokosuka 237-0061, Japan
Full list of author information is available at the end of the article

Background

Methane is a greenhouse gas, a primary product of methanogenic archaea, a useful fuel for human activity, and the simplest organic molecular. The geochemical origins of methane in environmental samples have been investigated in terms of global warming (e.g., Patra et al. 2011), limits of the deep biosphere (e.g., Inagaki et al. 2015), exploration and development of energy resources (e.g., Kvenvolden 1988), and prebiotic chemical evolution on the early Earth (e.g., McCollom 2013). The geochemical origin of methane can be classified into four types in terms of the carbon source (inorganic vs. organic) and the generation process (chemical vs. microbial): abiotic, thermogenic, aceticlastic and methylotrophic, and hydrogenotrophic methane (e.g., Kawagucci et al. 2013a). Because hydrogenotrophic methanogenesis is one of the most important energy metabolisms in living ecosystems in the ancient Earth and habitable extraterrestrial environments (McCollom 1999; Takai et al. 2004; Hsu et al. 2015), distinguishing hydrogenotrophic methanogenesis from the other methanogenic processes is a key to discuss whether life is present or absent there.

The stable isotope ratios of carbon ($^{13}C/^{12}C$) and hydrogen (D/H) in methane have been used as geochemical tracers to deduce its origin (e.g., Schoell 1980; 1988; Whiticar et al. 1986; Whiticar 1999). Although the usefulness of the stable isotope tracers for methane origins have been long debated (Martini et al. 1996; Waldron et al. 1999; Tang et al. 2000), the stable isotope diagnosis becomes more accurate by enhancing the understanding of stable isotope systematics with respect to each of the methanogenic processes in addition to subsequent alteration of the imprinted isotope signature. For this purpose, carbon isotope fractionation between CH_4 and CO_2 in hydrogenotrophic methanogenesis has been well studied by laboratory incubations (e.g., Valentine et al. 2004; Penning et al. 2005; Takai et al. 2008). The magnitude of the carbon isotope fractionation is thought to depend on the thermodynamic state of the methanogenic environment: a greater ^{13}C-depletion in the produced CH_4 occurs at a lower H_2 availability. This relationship has been explained by "differential reversibility" in the multistep pathway of methanogenesis from H_2 and CO_2 (Valentine et al. 2004): higher reversibility exhibits at a lower H_2 availability.

In contrast to the carbon isotopic characteristics, the hydrogen isotope characteristics of hydrogenotrophic methanogenesis remain less understood. In general, the hydrogen isotope ratio of produced CH_4 is related solely to that of H_2O (e.g., Daniels et al. 1980). Based on this finding, numerous studies have examined the CH_4–H_2O hydrogen isotope fractionation both in experiments using various methanogen cultures (Balabane et al. 1987; Sugimoto and Wada 1995; Valentine et al. 2004; Yoshioka et al. 2008; Hattori et al. 2012; Stolper et al. 2015; Wang et al. 2015) and by observations on methane-bearing environments (e.g., Nakai et al. 1974; Whiticar et al. 1986; Burke 1993; Waldron et al. 1999). Large variation of the CH_4–H_2O hydrogen isotope fractionation has been reported in previous studies (Additional file 1), and the possible relationship between the fractionation and the H_2 availability in the methanogenic environment has been discussed (Burke 1993; Yoshioka et al. 2008; Hattori et al. 2012; Stolper et al. 2015; Wang et al. 2015). On the other hand, Kawagucci et al. (2014) recently demonstrated that the hydrogen isotope ratio of substrate H_2 also affects the hydrogen isotope ratio of the produced CH_4 during the rapid growth of a thermophilic methanogen, *Methanothermobactor thermautotrophicus* strain ΔH, in H_2-enriched batch culture. If this "δD_{H2} effect" occurs ubiquitously in other hydrogenotrophic methanogens and under other growing conditions such as low H_2 availability, it may be necessary to reconsider the geochemical implications of the traditional interpretation based on the CH_4–H_2O hydrogen isotope ratio fractionation.

In this study, we therefore investigated the hydrogen and carbon isotope systematics of H_2O–H_2–CO_2–CH_4 during the growth of different hydrogenotrophic methanogens under various growth conditions and the effect of H_2 availability. We tested two different species of methanogens under syntrophic growth conditions with fermentatively hydrogenogenic bacteria that attained very low H_2 availability (corresponding to $<10^2$ Pa-H_2 in the headspace gas) and two different species of hydrogenotrophic methanogens under H_2 enriched growth conditions (~10^5 Pa-H_2 in the headspace gas). This is the first attempt to monitor both hydrogen and carbon isotope systematics in low-pH_2 cultures, although the hydrogen and carbon isotope systematics have been investigated separately (Penning et al. 2005; Yoshioka et al. 2008) (Additional file 1). Compilation of the hydrogen and carbon isotope fractionations from the experimental results and environmental observations reveals three provinces in the dataset and provides new insights into stable isotope diagnosis of the origin and fate of environmental methane.

Methods

Organism details and medium compositions

Table 1 provides information on the four culture sets and the experimental conditions used in this study. To assess the growth of methanogens under low H_2-availability conditions at different temperatures, two species of butyrate-oxidizing hydrogenogenic fermentative bacteria were cocultured with two species of methanogens as syntrophic consortia at 55 °C and 25 °C (thermophilic and mesophilic cocultures). The partial pressure of H_2 for the cocultures was lower than 10^2 Pa-H_2 during the

Table 1 Experimental and organismal information examined in this study

Culture type	T (°C)	Head space H_2 (Pa)	CH_4 substrate	Organism information	
				Syntrophic bacteria	Methanogen
Thermophilic coculture	55	$<6.0 \times 10^1$	Butyrate	*Syntrophothermus lipocalidus* (DSMZ 12681)	*Methanothermobacter thermautotrophicus* strain ΔH (JCM10044)
Mesophilic coculture[a]	25	$<1.0 \times 10^1$	Butyrate	*Syntrophomonas* sp.	*Methanobacterium* sp. strain MO-MB1
Thermophilic pure culture	60	$<2.1 \times 10^5$	H_2/CO_2	–	*Methanothermococcus okinawensis* strain IH1 (JCM11175)
Mesophilic pure culture	30	$<2.2 \times 10^5$	H_2/CO_2	–	*Methanobacterium* sp. strain MO-MB1

[a]The mesophilic coculture included a contaminant bacterium, *Geosporobacter* sp. (Saito et al. 2014)

incubation (see Results). In addition to the coculture sets, two hydrogenotrophic methanogen species were grown as pure culture under high H_2-availability conditions at 60 °C and 30 °C (thermophilic and mesophilic pure cultures). The initial partial pressure of H_2 for the pure culture sets was $2.0–2.2 \times 10^5$ Pa-H_2. The metabolic reactions in the culture sets are as follows:

$$4H_2 + CO_2 \rightarrow CH_4 + 2H_2O$$
(R1: Hydrogenotrophic methanogenesis),
$$CH_3CH_2CH_2COO^- + 2H_2O \rightarrow 2CH_3COO^- + H^+ + 2H_2$$
(R2: Butyrate oxidation),
$$2CH_3CH_2CH_2COO^- + 2H_2O + CO_2 \rightarrow 4CH_3COO^- + 2H^+ + CH_4$$
(R3: Consortium).

For the thermophilic coculture experimental set (hereafter, TC), we purchased a pure coculture of *Syntrophothermus lipocalidus* strain TGB-C1 and *Mtb. thermautotrophicus* strain ΔH (DSM 12681) from Deutsche Sammlung von Mikroorganismen und Zellkulturen (DSMZ; Braunschweig, Germany). *Mtb. thermautotrophicus* is one of the best-studied methanogens in terms of its genetic, biochemical, and biogeochemical characteristics (e.g., Smith et al. 1997; Kawagucci et al. 2014). *S. lipocalidus* was originally isolated from a granular sludge in a thermophilic upflow anaerobic sludge blanket reactor (Sekiguchi et al. 2000). The syntrophism between these microbes was previously studied by Sekiguchi et al. (2000). In addition, the CH_4–H_2O hydrogen isotope fractionation for this syntrophic consortium has been previously investigated (Yoshioka et al. 2008). The syntrophic consortium was cultivated in the medium as follows (L^{-1}): 0.54 g NH_4Cl, 0.14 g KH_2PO_4, 0.20 g $MgCl_2$–$6H_2O$, 0.15 g $CaCl_2$–$2H_2O$, 25 g NaCl, 2.0 g $NaHCO_3$, 0.5-mL resazurin solution (1 g L^{-1}), 1-mL vitamin solution, and 1-mL trace element solution. The composition of the vitamin solution was as follows (L^{-1}): 4.88 mg biotin, 8.82 mg folic acid, 4.12 mg pyridoxine–HCl, 6.74 mg thiamine–HCl, 7.52 mg riboflavin, 2.44 mg nicotinamide, 9.54 mg calcium pantothenate, 27.1 mg vitamin B12, 2.74 mg 4-aminobenzic acid, and 4.12 mg lipoic acid. The trace element solution contained (L^{-1}) the following composition: 1.27 g $FeCl_2$–$4H_2O$, 0.20 g $MnCl_2$–$4H_2O$, 0.13 g $CoCl_2$, 0.14 g $ZnCl_2$, 1.3 mg $CuCl_2$–$2H_2O$, 13.3 mg $AlCl_3$, 6.2 mg H_3BO_3, 24.2 mg Na_2MoO_4–$2H_2O$, 13.0 mg $NiCl_2$, 1.7 mg Na_2SeO_3, and 3.3 mg Na_2WO_4–$2H_2O$. Batches of the culture in 120-mL serum bottles, each containing 40 mL of medium, were flushed with mixture gas of N_2/CO_2 (80:20, v/v) for 5 min and sealed with butyl rubber septums and aluminum crimp seals. After autoclaving at 121 °C for 20 min, the headspace was pressurized with N_2 gas up to 150 kPa. After gas injection, 0.5 mL of the filter-sterilized reducing solution was added to the bottles. The reducing solution contained (L^{-1}) 6.25 g Na_2S–$9H_2O$ and 6.25 g cysteine-HCl. Butyrate solution sterilized in an autoclave was added to the culture bottles at a final concentration of 20 mmol/L immediately before the inoculation. The amount of inoculum for each TC batch was 0.55 mL. Batches of TC were incubated at 55 °C in the dark without shaking.

For the mesophilic coculture experimental set (hereafter, MC), we used a butyrate-oxidizing enrichment culture from our laboratory. The culture was obtained from subseafloor sediments collected off the Shimokita Peninsula, Japan (Imachi et al. 2011; Saito et al. 2014). The culture is a highly purified enrichment culture that constitutes only three types of microorganism: a mesophilic hydrogenotrophic methanogen, *Methanobacterium* sp. strain MO-MB1; a mesophilic butyrate-oxidizing hydrogenogenic bacterium, *Syntrophomonas* sp.; and a heterotrophic anaerobic bacterium, *Geosporobacter* sp. The procedure for culture batch preparation for MC was almost the same as that for TC with some differences, as follows. The base medium composition for MC was as follows (L^{-1}): 0.54 g NH_4Cl, 0.10 g KH_2PO_4, 4.0 g $MgCl_2$–$6H_2O$, 1.0 g $CaCl_2$–$2H_2O$, 25 g NaCl, 2.0 g $NaHCO_3$, 0.5-mL resazurin solution (1 g L^{-1}), 1-mL vitamin solution, and 1-mL trace element solution. The butyrate solution was added to the bottles at a final concentration of 10 mmol/L. The amount of inoculum for each MC batch was 1.5 mL. Batches of MC were incubated at 25 °C without shaking in the dark. The preliminary experiment showed that the growth of the consortium was optimal at 25 °C with minimum growth of the contaminant bacterium *Geosporobacter* sp. (Saito et al. 2014). We also confirmed that butyrate oxidation occurred together with methane production stoichiometrically, which is explained by the eq. R1, 2, and 3 (see Additional file 2).

For the thermophilic pure culture experimental set (hereafter, TP), we used a thermophilic hydrogenotrophic methanogen, *Methanothermococcus okinawensis* strain IH1 (JCM 11175), from the culture collection of JAMSTEC. *Mtc. okinawensis* was isolated from a deep-sea hydrothermal vent chimney at the Iheya Ridge, Okinawa Trough (Takai et al. 2002). The base medium composition for TP was as follows (L^{-1}): 0.09 g KH_2PO_4, 0.09 g K_2HPO_4, 3.0 g $MgCl_2$–$6H_2O$, 4.0 g $MgSO_4$–$7H_2O$, 0.8 g $CaCl_2$, 0.33 g KCl, 30 g NaCl, 0.25 g NH_4Cl, 10.0 mg $NiCl_2$–$6H_2O$, 0.24 mg Na_2MoO_4–$2H_2O$, 4.9 mg $NaSeO_4$, 10.0 mg $Fe_2(SO_4)_3$, 0.5-mL resazurin solution (1 g L^{-1}), 10-mL trace element solution, and 2.5 g $NaHCO_3$. The trace element solution contained (L^{-1}): 1.0 g NaCl, 0.5 g $MnSO_4$–H_2O, 0.18 g $CoSO_4$–$2H_2O$, 0.1 g $CaCl_2$–$2H_2O$, 0.18 g $ZnSO_4$–$7H_2O$, 0.01 g $CuSO_4$–$5H_2O$, 0.02 g $KAl(SO_4)_2$–$12H_2O$, 0.01 g H_3BO_3, 0.025 g $NiCl_2$–$6H_2O$, and 0.3 mg Na_2SeO_3–$5H_2O$. Batches of the culture in 160-mL pressure-resistant serum bottles, each containing 20 mL of medium, were flushed with N_2 for 5 min, sealed with butyl rubber septums and aluminum crimp seals, and autoclaved at 121 °C for 20 min. The headspace for the TP comprised 100 kPa N_2, 100 kPa CO_2, and 200 kPa H_2. The medium was reduced by the addition of autoclaved 5 % (w/v) Na_2S–$9H_2O$ solution to a final concentration of 2.1 mmol/L. The amount of inoculum for each TP batch was 0.1 mL. Batches of TP were incubated at 60 °C without shaking in the dark.

For the mesophilic pure culture experimental set (hereafter, MP), we used *Methanobacterium* sp. MO-MB1 isolated from subseafloor sediment in our laboratory (Imachi et al. 2011). The base liquid medium composition for MP was almost the same as that for MC, except that butyrate solution was not added. The MP batches were prepared using almost the same procedure as for the TP batches, except that the headspace gas composition was 160 kPa N_2, 40 kPa CO_2, and 200 kPa H_2. The amount of inoculum for each TP batch was 0.1 mL. Batches of MP were incubated at 30 °C without shaking in the dark.

General notation of isotope ratios

The term "hydrogen" in this paper includes both deuterium (D or ^2H) and protium (H or ^1H). The relative abundance of D to H or ^{13}C to ^{12}C was noted *R*:

$$R = [D]/[H] \text{ or } [^{13}C]/[^{12}C] \quad (1)$$

We used delta (δ) notation to express the relative difference between the isotope ratio of a sample and a standard in the permil scale:

$$\delta = \left[(R_{sample}/R_{standard}) - 1\right] \times 1000\ (‰). \quad (2)$$

The δD and $\delta^{13}C$ values were calculated with respect to the international standards VSMOW and VPDB, respectively. The isotope fractionation factor, α, is generally defined as the division between the isotope ratios of two molecules as follows:

$$\alpha_{A\text{-}B} = R_A/R_B, \quad (3)$$

$$\alpha_{A\text{-}B} = (1000 + \delta_A)/(1000 + \delta_B) \quad (4)$$

In this study, each C or H is superscribed with α to indicate carbon or hydrogen isotope fractionation, respectively, for convenience. The fractionation factors at isotopic equilibrium between $H_2 - H_2O$ ($\alpha^{eq}{}_{H_2\text{-}H_2O}$), $CH_4 - H_2O$ ($\alpha^{eq}{}_{CH_4\text{-}H_2O}$), and $CH_4 - CO_2$ ($\alpha^{eq}{}_{CH_4\text{-}CO_2}$), and their temperature dependency (Bardo and Wolfsberg 1976; Horita and Wesolowski 1994; Horibe and Craig 1995; Horita 2001) are used.

Experimental settings of isotope ratios

To investigate the hydrogen isotope systematics during methanogen growth, the initial δDH_2O and δDH_2 values in the batch cultures were adjusted by adding D_2O and DH molecules (Table 2). For the coculture experiments, the initial δD_{H_2O} values were adjusted to three levels: nonD_2O-labeled (−62‰ to −58‰), moderately D_2O-labeled (+1406‰ to +1639‰), and highly D_2O-labeled (+3286‰ to +3787‰) (see the next section Sampling and analyses for δDH_2O determination). When the H_2–H_2O equilibrium is attained with these δD_{H_2O} values, the equilibrated δD_{H_2} values ($\delta D^{eq}{}_{H_2}$) will be −755‰ to −709‰ (nonD_2O-labeled), −369‰ to −189‰ (moderately D_2O-labeled), and +120‰ to +479‰ (highly D_2O-labeled) (see Additional file 2), which are calculated from the temperature-dependent $\alpha^{eq}{}_{H_2\text{-}H_2O}$ values (0.309 and 0.260 at 55 °C and 25 °C: Horibe and Craig 1995). Duplicate batches were examined for each of the δD_{H_2O} levels in both coculture experiments (batches A1–A6 for TC and batches B1–B6 for MC). Precultures of the coculture experiments were conducted using the same medium in terms of δD_{H_2O} levels in order to reduce carry-over δD signatures from the precultures.

For the pure culture experiments, the initial δD_{H_2O} values were also adjusted to three levels: nonD_2O-labeled (−59‰ to −43‰), moderately D_2O-labeled (δD_{H_2O} = +1097‰ to +1427‰), and highly D_2O-labeled (δD_{H_2O} = +2317‰ to +3142‰). In addition, the initial δD_{H_2} values were adjusted with respect to the theoretical δD_{H_2} values expected from the H_2–H_2O equilibrium, i.e., we prepared initial δD_{H_2} values that were higher (DH-rich), lower (D_2O-rich), or almost identical to the equilibrated δD_{H_2} value at the incubation temperatures

Table 2 Initial stable hydrogen isotopic conditions for the hydrogenotrophic methanogenesis experiments

Batches	Isotope labeling		Initial value		$\delta^{eq}D_{H_2-H_2O}$
	D_2O	DH	δD_{H_2O}	δD_{H_2}	
Thermophilic coculture experimental set (TC, at 55 °C)					
A1	–	–	−58	–	–
A2	–	–	−62	–	–
A3	+	–	+1624	–	–
A4	+	–	+1639	–	–
A5	++	–	+3771	–	–
A6	++	–	+3787	–	–
Mesophilic coculture experimental set (MC, at 25 °C)					
B1	–	–	−62	–	–
B2	–	–	−58	–	–
B3	+	–	+1406	–	–
B4	+	–	+1413	–	–
B5	++	–	+3662	–	–
B6	++	–	+3286	–	–
Thermophilic pure culture experimental set (TP, at 60 °C)					
C1	–	–	−59	−358	DH-rich
C2	–	+	−53	+330	DH-rich
C3	+	–	+1097	−287	Equilibrated
C4	+	++	+1210	+1886	DH-rich
C5	++	–	+2574	−350	D_2O-rich
C6	++	+	+2317	+70	Equilibrated
Mesophilic pure culture experimental set (MP, at 30 °C)					
D1	–	–	−54	−344	DH-rich
D2	+	+	+1427	+145	DH-rich
D3	+	+	+1424	+22	DH-rich
D4	+	–	+1369	−350	Equilibrated
D5	+	–	+1382	−349	Equilibrated
D6	++	–	+3096	−351	D_2O-rich
D7	++	+	+3142	+85	Equilibrated

(Table 2). Moreover, we predicted that the initial δDH_2 value of the batch during methanogen growth would approach a δD_{H_2} value isotopically equilibrated with the medium H_2O (Valentine et al. 2004; Kawagucci et al. 2014) because it is known that the functions of hydrogenases promote isotope exchange (Vignais 2005; Campbell et al. 2009; Yang et al. 2012; Walter et al. 2012). Six batches for TP (C1–C6) and seven batches for MP (D1–D7) were examined (Table 2). In the same manner as the coculture experiments, precultures for the pure culture experiments were conducted with media containing the same δD_{H_2O} and δD_{H_2} levels in order to reduce carry-over δD signatures from the precultures.

Sampling and analyses

To determine the partial pressures (pH_2 and pCH_4) and isotope ratios (δD_{CH_4}, δD_{H_2}, $\delta^{13}C_{CH_4}$, $\delta^{13}C_{CO_2}$), the headspace gas of each batch was collected during incubation. The sampling frequency was determined on the basis of pCH_4 monitoring in preculture to include both the exponential and stationary growth phases of the methanogens. For the TC and MC, 8 mL of the headspace gas was subsampled into an evacuated 8-mL glass vial at each sampling time to quantify pH_2 and δD_{H_2} at low H_2 level. In contrast, for the TP and MP, 0.5 mL of the headspace gas was subsampled into an 8-mL glass vial filled with ultrapure grade helium (purity is >99.9999 %: Iwatani Gasnetwork Corporation, Osaka, Japan) at 100 kPa for convenience for the isotope ratio analyses. In an attempt to assess the change of δD_{H_2O} during the growth, 2 mL of the medium fluid was collected immediately after the inoculation and at the end of the experiment, filtered with 0.2-μm pore-size filters (ADVANTEC, Tokyo, Japan), and measured within 12 h of sampling. Because of slight changes before and after the incubation (see Additional file 2), we used the δD_{H_2O} value at the end of incubation to evaluate the isotope systematics (Additional file 2). For only TP, cell density was determined using 0.1 mL of medium subsample by a direct cell count with DAPI (4',6-diamidino-2-phenylindole), to evaluate the growth rate and the cell-specific H_2 consumption rate (Kawagucci et al. 2014 and references therein). The cell densities of the other three experimental sets (TC, MC, and MP) could not be determined using this procedure because the cells did not diffuse in the medium but were tightly attached to the surface of the culture bottle.

The partial pressures of H_2 and CH_4 were determined by gas chromatography using a helium ionization detector (GC-HID) with an in-house standard comprising 100 ppm H_2 and CH_4 over a He matrix. Overall errors for pH_2 and pCH_4 analyses were expected to be within 10 %.

The stable hydrogen and carbon isotope ratios for CH_4 were determined by continuous-flow isotope ratio mass spectrometry (CF-IRMS) with an on-line gas preparation and introduction system connected with a mass spectrometer, MAT253 (Thermo Fisher Scientific, Bremen, Germany), based on a previous study (Umezawa et al. 2009). Details of the system and analytical procedure are as follows. A helium-purged purification line made of stainless-steel tubing and including several two-position valves (VICI Precision Sampling, Inc. Louisianna, USA) with chemical and cold traps was used as the on-line gas preparation and introduction system. Ultra-pure-grade helium (purity >99.9999 %: Iwatani Gasnetwork Corporation) was used with further purification by a Molecular Sieve 5A column at −196 °C (liquid-N_2 bath). The sample gas including CH_4 was introduced via a gas-tight syringe (PRESSURE-LOK® series, VICI Precision Sampling, Inc., Louisiana, USA) into a 30 mL/min (~ + 0.2 MPa) helium stream, named the "precon stream," of the purification line. CH_4 in the sample gas in the precon stream was separated from CO_2 and H_2O by a stainless-steel tubing coil trap held at −110 °C (ethanol/liquid-N_2 bath) and a chemical trap filled with magnesium perchlorate ($Mg(ClO_4)_2$; Merck KGaA, Darmstadt, Germany) and Ascarite II (sodium-hydroxide-coated silica; Thomas Scientific, Swedesboro, New Jersey, USA). Subsequently, CH_4 was condensed on a stainless-steel tubing trap filled with HayeSep-D porous polymer (60/80 mesh, Hayes Separations Inc., Texas, USA) held at −130 °C (ethanol/liquid-N_2 bath) mounted on a two-position six-port valve. After the turn of the valve position to introduce another helium stream set at 1.0 mL/min, named the "GC stream," into the Hayesep-D trap, the condensed CH_4 was released at >80 °C (hot water bath). The CH_4 on the GC stream was again condensed on a capillary trap made of PoraPLOT Q (20 cm long, 0.32 mm i.d.) held at −196 °C (liquid-N_2 bath) for cryofocus and finally released at room temperature. After complete separation of CH_4 from the other molecules by a HP-PLOT Molesieve capillary column (30 m long, 0.32 mm i.d.) at 40 °C, the effluent CH_4 went to the pyrolysis or combustion units (Thermo Fisher Scientific, Massachusetts, USA) to be converted into H_2 or CO_2, respectively. During the analyses, the pyrolysis and combustion units were maintained at 1440 °C and 960 °C, respectively. The pyrolysis unit was conditioned twice a month by repeated injection of 0.2 mL of pure CH_4 to form a graphite coat on the inner wall of the tubing for quantitative conversion of the sample CH_4 to H_2. The CH_4-derived H_2 and CO_2 were finally introduced via an open-split interface, GC Combustion Interface III (Thermo Fisher Scientific, Massachusetts, USA), into MAT253. Hydrogen or carbon isotope ratios were obtained through simultaneous monitoring of H_2^+ isotopologues at $m/z = 2$ and 3 or CO_2^+ isotopologues at $m/z = 44$, 45, and 46, respectively. Stable hydrogen isotope ratio for H_2 was also determined with MAT253 by CF-IRMS in another publication (Kawagucci et al. 2010). Stable carbon isotope ratio for the CO_2 was determined with an isotope-ratio mass spectrometer, DELTA Plus Advantage (Thermo Fisher Scientific), connected to a universal on-line gas preparation and introduction system, GASBENCH II (Thermo Fisher Scientific). Data acquisition for all the IRMS analyses was based on the ISODAT software package (Thermo Fisher Scientific). The amount of H_3^+ ions generated in the ion source, the so-called H_3^+ factor, in the analysis of hydrogen isotope ratios was determined on ISOTAD and was <4.5 (ppm/mV) during the study. The analytical precisions for the $\delta^{13}C_{CO_2}$, $\delta^{13}C_{CH_4}$, δD_{H_2}, and δD_{CH_4} values were estimated by repeated analyses of standard gases to be within 0.5‰, 0.3‰, 10‰, and 5‰, respectively. The determined δ values were calibrated with commercial and/or in-house standard gases as follows: −40.56‰ and −3.57‰ for $\delta^{13}C_{CO_2}$, −74.01‰ and −39.03‰ for $\delta^{13}C_{CH_4}$, −758.22‰ and −11.09‰ for δD_{H_2}, and −185.9‰ for δD_{CH_4}. To minimize analytical errors, amplitudes of ion beams were matched between samples and standards by regulating amounts of the analyses introduced into IRMS. The δD_{H_2O} of the mediums were analyzed using a liquid water isotope analyzer (Los Gatos Research, Inc.) with an analytical precision of 0.5‰. The determined δD_{H_2O} values were calibrated with three commercial standard waters (−154.3‰, −96.4‰, and −9.5‰). Negligible memory effects were confirmed on the analyses that alternated between D-labeled and nonD-labeled samples. For the D-enriched batches, which had δD values far from the calibration ranges (particularly δD_{H_2O}), the analytical accuracy was presumably worse and the calculated $\alpha^H_{CH_4-H_2O}$ values from the batches with D-enriched H_2O should be carefully considered.

Results

All the experimental results are illustrated in Fig. 1 and listed in Additional file 2. The results of each experiment are summarized separately below for each culture type.

Results of Thermophilic Coculture (TC)

The pH_2 values of the TC batches were 4.0–6.8 Pa immediately after the inoculation, increased during the early period of growth, and decreased toward the initial level at the later period (Fig. 1a). The highest pH_2 of each TC batch was somewhat different but within the same order of values (28.6–64.9 Pa). The timings of the pH_2 peaks were slightly variable between the batches (5–8 days). Note that the observed pH_2 values were close to the threshold at which hydrogenotrophic methanogens cease H_2 consumption (10^0–10^1 Pa; Lovley 1985; Thauer et al. 2008), suggesting that our experiment covered the lower end of H_2 availability for microbial methanogenesis. The pCH_4 generally showed an exponential increment with the time of growth and reached 5.5–6.2 kPa (Fig. 1b). The increment of pCH_4 accompanied or followed that of pH_2 in each batch (5–8 days). This implies that the growth of hydrogenotrophic methanogens is triggered by the preceding metabolic function (H_2 production) and growth of the hydrogenogenic bacterial counterpart.

The δD_{H_2} values of TC batches were determined only when the pH_2 values were sufficiently high for the δD_{H_2}

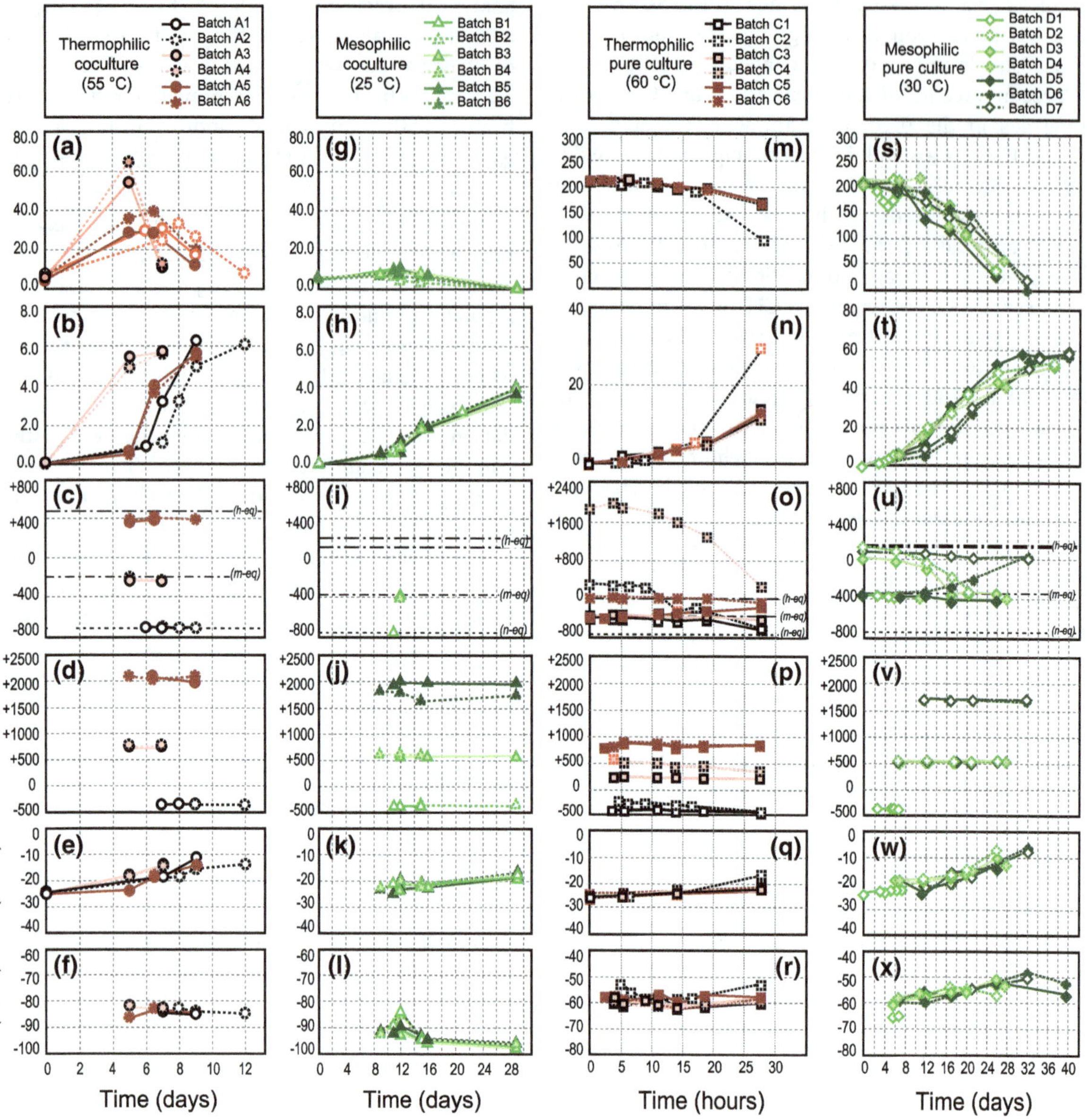

Fig. 1 Temporal changes in H_2, CH_4, and CO_2 in four culture experimental sets. The experimental sets for the thermophilic coculture (**a**–**f**; TC), the mesophilic coculture (**g**–**l**; MC), the thermophilic pure culture (**m**–**r**; TP), and the mesophilic pure culture (**s**–**x**; MP) consisted of experimental batches A1–A6, B1–B6, C1–C6, and D1–D7, respectively. The pH_2 values (panels **a**, **g**, **m**, and **s**), the pCH_4 values (panels **b**, **h**, **n**, and **t**), the δD_{H_2} values (panels **c**, **i**, **o**, and **v**), the δD_{CH_4} values (panels **d**, **j**, **p**, and **v**), the $\delta^{13}C_{CO_2}$ values (panels **e**, **k**, **q**, and **w**), and the $\delta^{13}C_{CH_4}$ values (panels **f**, **l**, **r**, and **x**) were determined for a certain interval of growth from immediately after inoculation to the stationary phase of growth. The symbols are common within each experimental set (TC, MC, TP, and MP)

analysis (pH_2 of >10 Pa) (Fig. 1c). The δD_{H_2} values were virtually identical within batches that had almost identical initial δD_{H_2O} values regardless of pH_2 and time (Fig. 1c). The δD_{H_2} values ranged from −714‰ to −698‰, −231‰ to −213‰, and +374‰ to +413‰ for batches with nonlabeled δD_{H_2O} (A1 and A2), moderately labeled δD_{H_2O} (A3 and A4), and highly labeled δD_{H_2O} (A5 and A6), respectively (Additional file 2). All of these δD_{H_2} values were close to the theoretical δD_{H_2} value at the isotopic equilibrium between H_2 and H_2O ($\delta D^{eq}{}_{H_2}$: see Additional file 2). Our observation is consistent with previous studies examining biologically produced H_2 that also revealed δD_{H_2} values close to the equilibrium (Walter et al. 2012; Yang et al. 2012). Note that the measured δD_{H_2O}–δD_{H_2} relationship showed increased deviation from the theoretically expected relationship in batches with higher δD_{H_2O} values (deviations of <26‰ for A1 and A2, <42‰ for A3 and A4, and <100‰ for A5 and A6). The deviations were beyond the analytical precision of <10‰. Moreover, the deviations were possibly derived from the worse accuracy for the excessively

D-enriched δD_{H_2O} measurement resulting from the over range of calibration by using the commercial standards having relatively lower isotopic values (see Methods) than the δD_{H_2O} values of the experimental batches (Table 2).

The δD_{CH_4} values ranged from −356‰ to −366‰ in A1 and A2, +720‰ to +722‰ in A3 and A4, and +1984‰ to +2092‰ in A5 and A6 (Additional file 2). The δD_{CH_4} values were similar within batches that had similar initial δD_{H_2O} values regardless of pH_2 and time (Fig. 1d); this was also observed for the δD_{H_2} values (Fig. 1c). The $\alpha^H{}_{CH_4-H_2O}$ values calculated from the measured δD_{CH_4} and δD_{H_2O} values were in the ranges 0.673–0.683, 0.655–0.656, and 0.637–0.659 for the nonlabeled, moderately labeled, and highly labeled δD_{H_2O} batches, respectively (Fig. 2 and Additional file 2). There appeared to be a negative correlation between the $\alpha^H{}_{CH_4-H_2O}$ and δD_{H_2O} values. However, when the δD_{H_2O} value calculated from the measured δD_{H_2} value and $\alpha^{eq}{}_{H_2-H_2O}$ was used instead of the measured δD_{H_2O} value, the resulting $\alpha^H{}_{CH_4-H_2O}$ values fell in a narrower range (0.668–0.697) regardless of the D_2O labels (Fig. 2 and Additional file 2). This $\alpha^H{}_{CH_4-H_2O}$ range is similar to the $\alpha^H{}_{CH_4-H_2O}$ values obtained from the nonlabeled batches using the measured δD_{H_2O} value (0.673–0.683: A1 and A2, Additional file 2). These facts imply that the measured δD_{H_2O} values in the D-enriched batches (e.g., > +1000‰) would be overestimates of the actual values, probably due to the unreliable analytical accuracy in

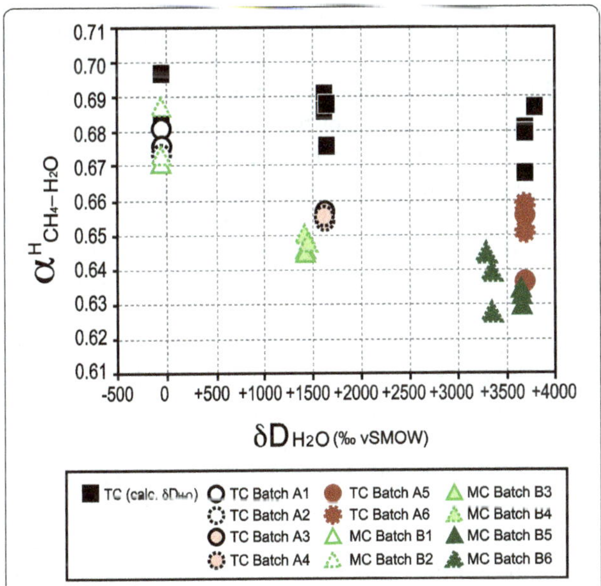

Fig. 2 Relations between δD_{H_2O} and $\alpha^H{}_{CH_4-H_2O}$ values in the two coculture experiments (TC and MC). For TC batches, the δD_{H_2O} (and $\alpha^H{}_{CH_4-H_2O}$) values calculated from the measured δD_{H_2} values and the H_2–H_2O isotope fractionation factor at the temperature are also illustrated for comparison

these cases that caused the underestimation of the $\alpha^H{}_{CH_4-H_2O}$ value.

The $\delta^{13}C_{CO_2}$ value for each of the TC batches increased with the time of growth from −23.5‰ to −11.5‰ (Fig. 1e). The $\delta^{13}C_{CH_4}$ values were nearly constant throughout growth, between −86.3‰ and −82.0‰ (Fig. 1f). The resulting $\alpha^C{}_{CH_4-CO_2}$ values generally decreased with time from 0.936 to 0.925 (Fig. 3 and Additional file 2). The $\alpha^C{}_{CH_4-CO_2}$ decreases appeared to be synchronized with the pH_2 decrease (Fig. 4a and Additional file 2), which is harmonious to the pH_2 $-\alpha^C{}_{CH_4-CO_2}$ relationship previously reported (Valentine et al. 2004; Penning et al. 2005; Takai et al. 2008).

Results of Mesophilic Cocultures (MC)

The pH_2 values of the MC batches showed trends similar to those of the TC batches: an initial increase during the early phase of growth and a subsequent decrease to the initial level during the later period of growth (Fig. 1g). The highest pH_2 values of each MC batch were 6.8–10.7 Pa, approximately one-fifth lower than those of the TC batches. The pH_2 peaks occurred around 9–12 days after inoculation. The pCH_4 showed an exponential increase after the pH_2 peak and rose to 3.4–3.9 kPa at the end of incubation (29 days, Fig. 1h).

The δD_{H_2} values of MC batches were obtained from only three samples; B1-11 h, B3-12 h, and B4-12 h, because the pH_2 values of the other samples were below the detection limits of δD_{H_2} measurement (Fig. 1i). The δDH_2 values of MC batches (−746‰ for B1-11 h, −357‰ for B3-12 h, and −394‰ for B4-12 h) were similar to those of H_2 isotopically equilibrated with the medium H_2O, as for the TC batches.

The δD_{CH_4} values of the MC batches were virtually identical through the growth in each of the batches except B6 (Fig. 1j). The δD_{CH_4} values ranged from −371‰ to −354‰ in B1 and B2, +560‰ to +589‰ in B3 and B4, +1934‰ to +1959‰ in B5, and +1608 to +1802‰ in B6, whereas reason(s) of the large variation in B6 remain unclear. The measured δD_{CH_4} and δD_{H_2O} values resulted in $\alpha^H{}_{CH_4-H_2O}$ values of 0.671–0.687, 0.645–0.650, and 0.628–0.645 for the nonlabeled, moderately labeled, and highly labeled δD_{H_2O} batches, respectively (Fig. 2 and Additional file 2). As for the TC batches, the MC batches showed a negative correlation between $\alpha^H{}_{CH_4-H_2O}$ and δD_{H_2O} (Fig. 2). The $\alpha^H{}_{CH_4-H_2O}$ values for the nonD_2O-labeled batches were similar between TC and MC (0.673–0.683 and 0.671–0.687, respectively) in spite of the different species, growth temperatures, and pH_2 conditions (Fig. 4b and Additional file 2).

The $\delta^{13}C_{CO_2}$ values of the MC batches increased with the time of growth from −24.7‰ to −17.2‰ (Fig. 1k). The $\delta^{13}C_{CH_4}$ values of the MC batches increased until

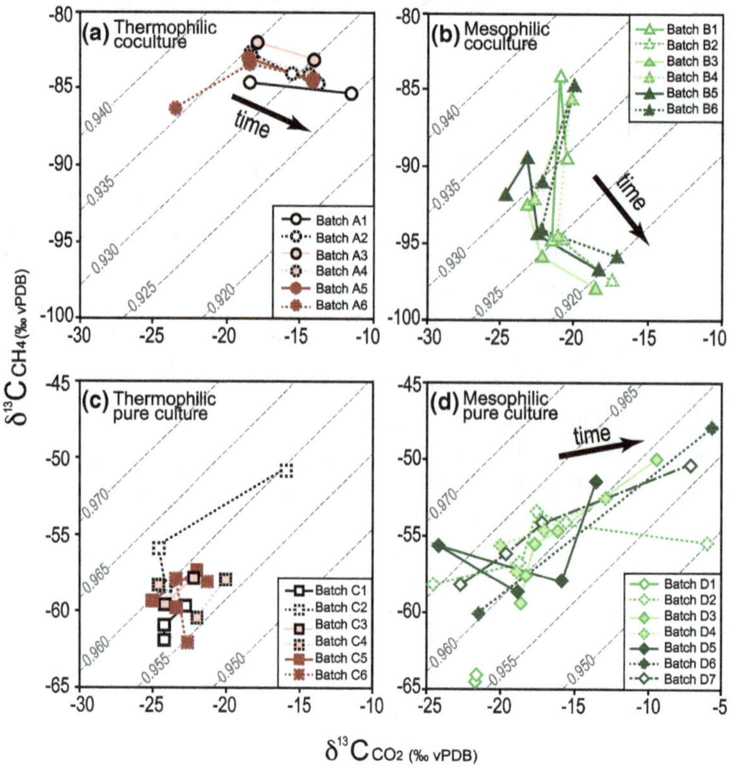

Fig. 3 Relations between $\delta^{13}C_{CO_2}$ and $\delta^{13}C_{CH_4}$ values during growth in each of the four experiment sets (**a**–**d**). Symbols are the same as those in Fig. 1. The black arrow in each panel shows orders of the values along with the time of growth. Diagonal dashed lines indicate the theoretical isotope fractionation factors between CO_2 and CH_4

12 days and subsequently decreased through time. The $\delta^{13}C_{CH4}$ values of the MC batches ranged from −97.9‰ to −84.3‰ (Fig. 1l). The resulting $\alpha^{C}{}_{CH_4\text{-}CO_2}$ values of the MC batches generally decreased through time from 0.935 to 0.919 (Fig. 3b and Additional file 2). The $\alpha^{C}{}_{CH_4\text{-}CO_2}$ range of the MC batches overlapped with but was partly lower than those of the TC batches (Fig. 4a). The last samples of each batch showed the lowest $\alpha^{C}{}_{CH_4\text{-}CO_2}$ values (~0.920) and pH_2 (≤1 Pa), which is harmonious with the positive $pH_2\text{-}\alpha^{C}{}_{CH_4\text{-}CO_2}$ relationship.

Results of Thermophilic Pure cultures (TP)

The sampling and analysis for the TP batches were conducted with hourly frequency because of the rapid growth of *Mtc. okinawensis* under high pH_2 conditions (Fig. 1). The pH_2 of the TP batches decreased with time from 210 kPa to 94 kPa for one batch, C2, but to ~160 kPa in the other five batches (Fig. 1m and Additional file 2). The pCH_4 increased exponentially to 30 kPa in batch C2, but to ~12 kPa in the other five batches (Fig. 1n). The H_2 decrement and CH_4 increment roughly matched the stoichiometry in the net reaction of hydrogenotrophic methanogenesis (R1). The cell densities were ~4×10^5 cells/mL at the time of inoculation and exponentially increased to ~5×10^7 cells/mL within 6–15 h in all the batches (Additional file 2). Specific growth rates during the exponential growth phase were estimated to be $0.19 \pm 0.05\ h^{-1}$ ($n = 6$). Cell-specific H_2 consumption rates during the exponential growth phase were estimated from the specific growth rate, pCH_4, and the methanogenic stoichiometry (R1) to be 0.14–1.6 fmol-H_2 $cell^{-1}\ s^{-1}$ ($n = 6$). These values are within cell-specific H_2 consumption rates evaluated in incubation of *Mtb. Thermautotrophicus* under 10^5 Pa-H_2 (0.1–58 fmol-H_2 $cell^{-1}\ s^{-1}$), in which the δD_{H_2} effect was exhibited (Kawagucci et al. 2014).

The temporal changes in the δD_{H_2} values showed different directions in the batches that had different initial isotope compositions of H_2 and H_2O (Fig. 1o and Table 2). For the batches that initially had isotopically unequilibrated conditions between H_2 and H_2O (batches C1, C2, C4, and C5), the δD_{H_2} values unidirectionally shifted, either upward or downward, from the initial values toward the equilibrated values (Figs. 5a–c). In the cases of batches that initially had isotopically equilibrated conditions between H_2 and H_2O (batches C3 and C6), the δD_{H_2} values changed little during growth (Fig. 5b and c). The difference between the theoretically expected δD_{H_2} values (equilibrated with δD_{H_2O}) and the

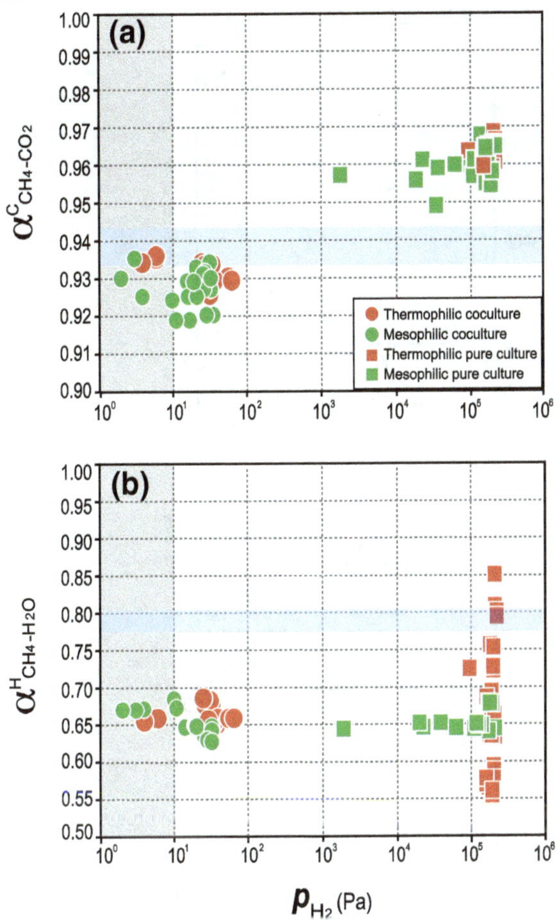

Fig. 4 The isotope fractionations of carbon (**a**) and hydrogen (**b**) with pH_2. The horizontal axis uses a logarithmic scale. Datasets are derived from the thermophilic coculture (TC), the mesophilic coculture (MC), the thermophilic pure culture (TP), and the mesophilic pure culture (MP) in this study

measured δD_{H_2} values of the last samples were within 100‰ except for the batch with the highly DH-labeled H_2 (C4, Table 2) (~600‰).

The δD_{CH_4} values of two batches with highly DH-labeled H_2 (Table 2) decreased with the time of growth from −191‰ to −416‰ in C2 and +585‰ to +360‰ in C4 (Figs. 1p, 5a–c, and Additional file 2). The δD_{CH_4} value of the slightly DH-rich batch (C1) varied in a smaller range (−375‰ to −414‰) than the values of batches C2 and C4. The δD_{CH_4} value of a D_2O-rich batch (C5) and batches C3 and C6 also showed slight changes, which were initially in H_2–H_2O isotopic equilibrium, (Additional file 2). The measured δD_{CH_4} and δD_{H_2O} values resulted in $\alpha^H{}_{CH_4-H_2O}$ values of 0.558–0.850 (Figs. 4b, 5a–c and Additional file 2).

The $\delta^{13}C_{CO_2}$ values for each of the TP batches slightly increased with time of growth from −25.7‰ to −16.8‰ (Fig. 1q and Additional file 2). The $\delta^{13}C_{CH_4}$ values ranged from −62.0‰ to −52.1‰ (Fig. 1r and Additional file 2). The $\alpha^C{}_{CH_4-CO_2}$ values ranged from 0.960 and 0.968 through growth (Fig. 3 and Additional file 2), which were higher than the values of the TC and MC batches (Fig. 4a).

Results of Mesophilic Pure cultures (MP)

The pH_2 values of MP batches decreased with time of growth from ~200 to 1.8–37 kPa (Fig. 1s). The pCH_4 values showed an exponential increase during the early stages of growth and finally plateaued after >30 days of incubation (Fig. 1t). The pCH_4 increment of the MP batches was more than one order of magnitude slower than that of the TP batches. Although the cell-specific H_2 consumption rate cannot be evaluated due to the lack of cell density information, the pCH_4 increment rate implies an H_2 consumption rate that is more than one order of magnitude slower than that of the TP batches (i.e., <~0.1 fmol-H_2 cell^{-1} s^{-1} for MP batches).

Temporal changes of δD_{H_2} values in the MP batches were variable between batches that had different initial isotope compositions of H_2 and H_2O (Fig. 1u and Table 2), as observed in the TP batches. The δD_{H_2} values of batches that initially had isotopically unequilibrated conditions between H_2 and H_2O (D1, D2, D3, and D6) (Table 2) shifted toward the equilibrated values (Fig. 1u). The δD_{H_2} values of batches with initially in isotopic equilibrium between H_2 and H_2O (D4, D5, and D7) showed little change during growth. The differences between the theoretically expected δD_{H_2} values (equilibrated with δD_{H_2O}) and the measured δD_{H_2} values of the last samples were less than 100‰ (Additional file 2).

The δD_{CH_4} values of the MP batches were at around −358‰ in D1, ranged from +536‰ to +581‰ in D2–D5, and were between +1663‰ and +1728‰ in D6 and D7 (Fig. 5d–f and Additional file 2), respectively. The δD_{CH_4} values were highly variable among all the batches but showed small variations between batches that had the same initial δD_{H_2O} values regardless of the time of growth (Fig. 1v). This result suggests that the δD_{CH_4} values depend mostly on the δD_{H_2O} values. The δD_{CH_4} and δD_{H_2O} values resulted in $\alpha^H{}_{CH_4-H_2O}$ values of 0.677–0.679 in D1, 0.647–0.654 in D2–D5, and 0.642–0.657 in D6 and D7 (Fig. 4b and Additional file 2).

The $\delta^{13}C_{CO_2}$ values of the MP batches increased with time of growth from −24.9‰ to −5.7‰ (Fig. 1w). The $\delta^{13}C_{CH_4}$ values also increased from −67.0‰ to −48.0‰ with time (Fig. 1x). The resulting $\alpha^C{}_{CH_4-CO_2}$ values ranged from 0.955 and 0.968 (Fig. 3 and Additional file 2). The $\alpha^C{}_{CH_4-CO_2}$ values in both TP and MP were similar to each other, although the experiments were conducted using different methanogen species with different growth and metabolic kinetics and under different temperature conditions

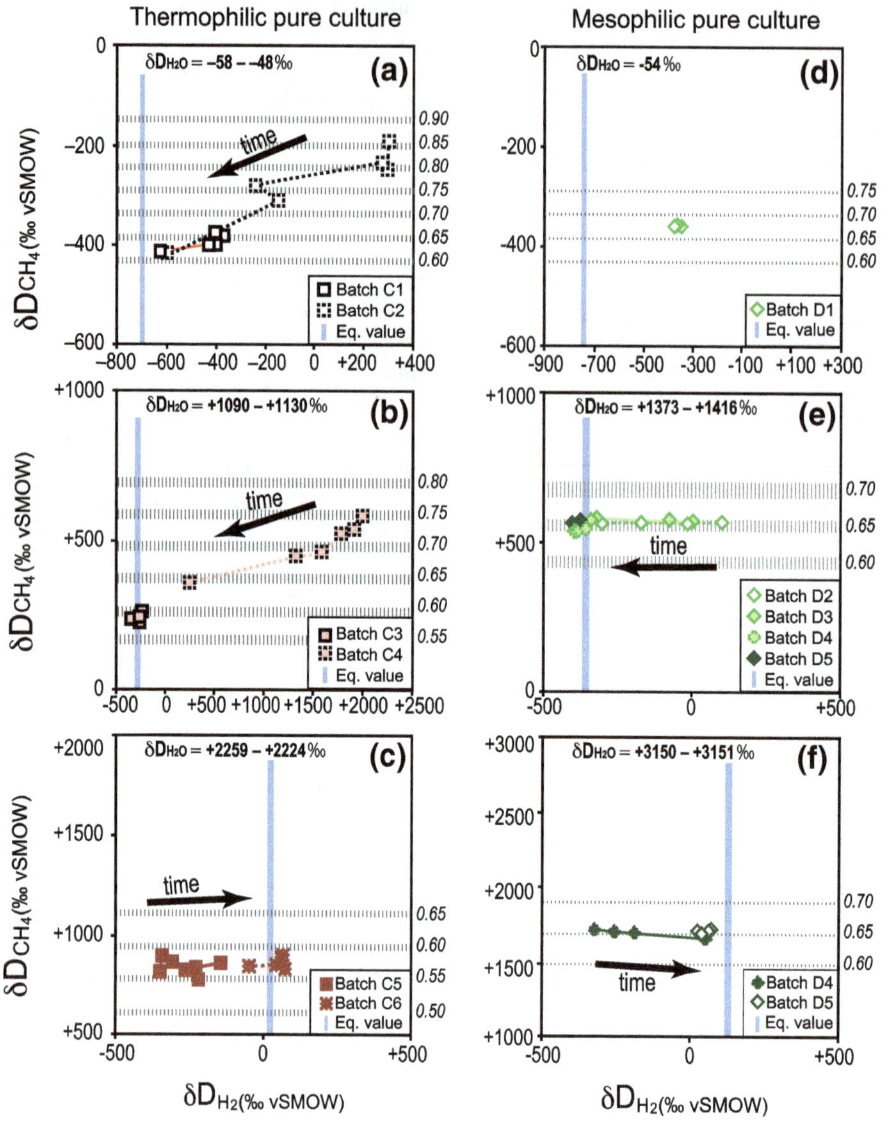

Fig. 5 Relations between δD_{H_2} and δD_{CH_4} in the two pure culture experiment sets (TP and MP). The results were divided into three subsets within each of the experimental sets according to the δD_{H_2O} label conditions: nonD_2O-labeled (panels **a** and **d**); moderately D_2O-labeled (panels **b** and **e**); and highly D_2O-labeled conditions (panels **c** and **f**). The black arrow in each panel marks the trend in the temporal changes of the values. The horizontal dashed lines in each panel indicate the theoretical fractionation factors between CH_4 and H_2O. The vertical gray bar in each panel indicates the equilibrated δD_{H_2} value calculated according to Horibe and Craig (1995)

(Fig. 4a). This result suggests that the stable carbon isotopic fractionation of these hydrogenotrophic methanogens is regulated little with these factors.

Discussion

Carbon isotope fractionation of pure culture and coculture experiments

Comparing the four experimental sets with each other, the $\alpha^C{}_{CH_4-CO_2}$ values were markedly different between the cocultures (0.919–0.936: TC and MC) and the pure cultures (0.949–0.968: TP and MP) (Fig. 4a) in spite of same methanogen used in mesophilic cultures (MC and MP). The $\alpha^C{}_{CH_4-CO_2}$ values in our incubations sometimes overlap but are not always consistent with those at carbon isotope equilibrium (Fig. 4a; 0.933–0.943 at 25–60 °C: Horita 2001). All the $\alpha^C{}_{CH_4-CO_2}$ values obtained in this study (0.919–0.968) fall into the $\alpha^C{}_{CH_4-CO_2}$ range previously determined by experiments between 0.913 (Penning et al. 2005) and 0.989 (Takai et al. 2008) (Fig. 6). The $\alpha^C{}_{CH_4-CO_2}$ values show a positive correlation with the pH_2 condition (Fig. 4a), as determined in previous studies (Valentine et al. 2004; Penning et al. 2005; Takai et al. 2008). These facts allows us to regard the $\alpha^C{}_{CH_4-CO_2}$ value as a proxy for H_2 availability during hydrogenotrophic methanogenesis. As

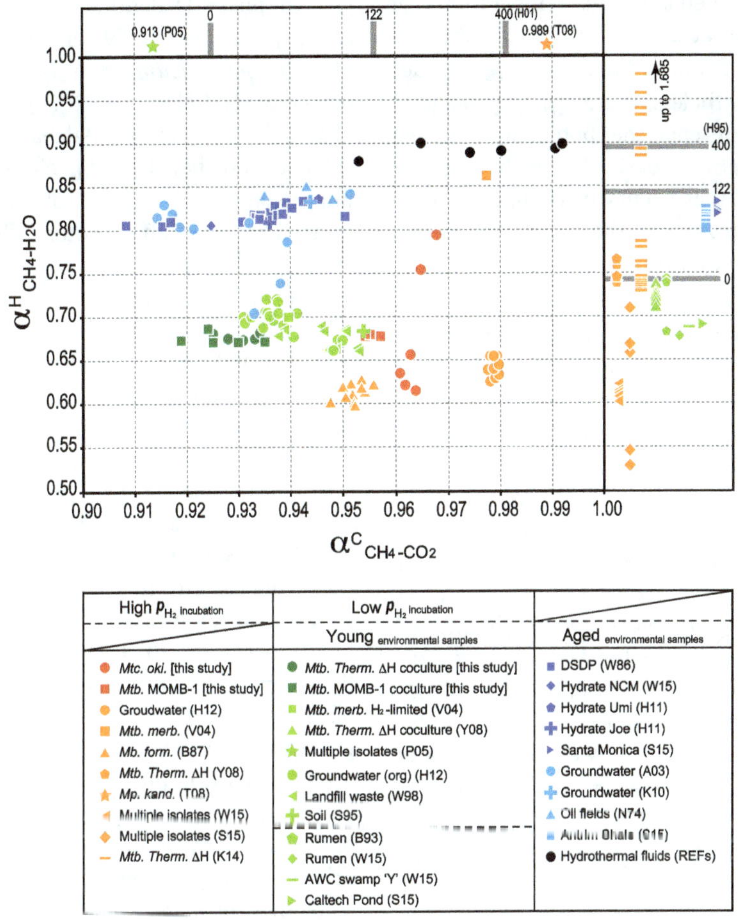

Fig. 6 Compilation of the isotope systematics of hydrogen ($\alpha^{H}_{CH_4-H_2O}$) and carbon ($\alpha^{C}_{CH_4-CO_2}$) associated with in hydrogenotrophic methanogenesis. Symbol colors of reddish, greenish, and bluish respectively represent datasets of laboratory incubations under high pH_2 conditions, incubations under low pH_2 condition and observations of "young" methane reservoirs in nature, and observations of "aged" methane reservoirs. Horizontal and vertical gray bars respectively represent the fractionation factors at the equilibriums of hydrogen ($\alpha^{eq}_{CH_4-H_2O}$) and carbon ($\alpha^{eq}_{CH_4-CO_2}$) at the indicated shown temperatures (0 °C, 122 °C, and 400 °C). All the data sources are listed in the main text (see Discussion)

previous and current incubations examining the $\alpha^{C}_{CH_4-CO_2}$ values have covered a full spectrum of pH_2 conditions from the methanogenic threshold (10^{-1}–10^{1} Pa: e.g., the MC batches in this study) to unrealistic enrichment (10^{7} Pa: Takai et al. 2008), the potential range of $\alpha^{C}_{CH_4-CO_2}$ variation corresponding to pH_2 change has been fully covered.

Hydrogen isotope systematics of syntrophic consortium incubation

The cocultures experiments (TC and MC) exhibited $\alpha^{H}_{CH_4-H_2O}$ values of 0.63–0.69 through the growth phase shift, although the experiments were conducted under different temperatures and with different syntrophic couples (Fig. 6). The $\alpha^{H}_{CH_4-H_2O}$ values of 0.67–0.69 from the nonD_2O-labeled batches represent the most reliable range, as discussed above (Fig. 2). A slightly higher $\alpha^{H}_{CH_4-H_2O}$ range (0.71–0.74) was obtained in previous coculture experiments with the same hydrogenotrophic methanogenic consortium, *S. lipocalidus* and *Mtb. thermautotrophicus* at 55 °C (Yoshioka et al. 2008) (Fig. 6). In addition to these syntrophic consortiums reconstructed in the laboratory, similar $\alpha^{H}_{CH_4-H_2O}$ values have been reported from experiments using potential H_2-syntrophic communities in natural environments such as in soil (0.68) (Sugimoto and Wada 1995), groundwater (0.66–0.73) (Hattori et al. 2012), and landfill waste (0.66–0.69) (Waldron et al. 1998) (Fig. 6). Note that there was considerable variation in $\alpha^{H}_{CH_4-H_2O}$ (0.66–0.74) in these studies compared to the typical analytical errors of δD_{H_2O} and δD_{CH_4} values ($\leq \sim 10‰$), corresponding to the $\alpha^{H}_{CH_4-H_2O}$ deviation of $\leq \sim 0.01$. The reason why such variation of stable hydrogen isotope fractionation occurs in various potentially H_2-syntrophic communities is currently unknown. The hydrogen isotope equilibrium at methanogenic temperatures of our culture (0.775–0.809 at 25 °C – 60 °C: Horibe and Craig 1995) does not account for the $\alpha^H{}_{CH_4\text{-}H_2O}$ values observed (Fig. 4b). Nevertheless, the $\alpha^H{}_{CH_4\text{-}H_2O}$ range of 0.66–0.74 probably represents the potential hydrogen isotopic signature of methane derived from hydrogenotrophic methanogenesis that occurs in H_2-limited habitats at methanogenic rates within the experimental time scale (typically shorter than a year).

Hydrogen isotope systematics of pure culture incubation

Stable hydrogen isotope fractionation in hydrogenotrophic methanogenesis ($\alpha^H{}_{CH_4\text{-}H_2O}$) under H_2-enriched conditions (>10^4 Pa-H_2) has been examined using pure cultures of methanogens (Additional file 1): 0.55–0.86 for *Mtb. marburgensis* (Valentine et al. 2004; Stolper et al. 2015); 0.60–1.69 for *Mtb. thermautotrophicus* (Yoshioka et al. 2008; Kawagucci et al. 2014; Wang et al. 2015); 0.528 for *Methanosarcina barkeri* (Stolper et al. 2015); 0.59–0.63 for *Methanobacterium formicicum* (Balabane et al. 1987); 0.64–0.68 for *Methanobacterium* sp. MO-MB1 (MP in this study); and 0.56–0.85 for *Mtc. okinawensis* (TP in this study) (Fig. 6). In addition to the pure cultures, natural groundwater was also incubated with 10^5 Pa-H_2 supply that yielded $\alpha^H{}_{CH_4\text{-}H_2O}$ values of 0.63–0.66 (Hattori et al. 2012). The variation of $\alpha^H{}_{CH_4\text{-}H_2O}$ values from the H_2-enriched condition (0.55–1.69) is much greater than that from the H_2-limited cultures (0.66–0.74) (Fig. 4b). The high variability of $\alpha^H{}_{CH_4\text{-}H_2O}$ values under high pH_2 conditions makes it difficult to estimate the relationship between the $\alpha^H{}_{CH_4\text{-}H_2O}$ values and pH_2 conditions during hydrogenotrophic methanogenesis (Burke 1993; Sugimoto and Wada 1995; Valentine et al. 2004; Yoshioka et al. 2008; Hattori et al. 2012; Stolper et al. 2015; Wang et al. 2015). Instead, the large $\alpha^H{}_{CH_4\text{-}H_2O}$ variation seems to be regulated by some other factors than pH_2.

One possible factor is the δD_{H_2} effect: the stable isotope signature of the substrate H_2 affects that of the product CH_4 during methane production by rapidly growing hydrogenotrophic methanogens (Kawagucci et al. 2014). In fact, TP batches in this study clearly showed a positive correlation between $\alpha^H{}_{CH_4\text{-}H_2O}$ and δD_{H_2} values during the early growth phase (Fig. 5), as also observed in a previous study of *Mtb. thermautotrophicus* (Kawagucci et al. 2014). In addition, the δD_{H_2} effect has been detected during reexamination of experimental datasets published in the literatures. For example, the $\alpha^H{}_{CH_4\text{-}H_2O}$ variation was observed during the growth of *Mtb. marburgensis* in a flow-through culture fed with D-enriched H_2 (δD_{H_2} = −187‰) and slightly D-depleted H_2O (δD_{H_2O} = −93‰), and the highest $\alpha^H{}_{CH_4\text{-}H_2O}$ value of 0.86 was obtained from the early growth phase (Valentine et al. 2004). The lowermost $\alpha^H{}_{CH_4\text{-}H_2O}$ value of 0.59 was obtained during the growth of *Mtb. formicicum* in a batch culture fed with D-depleted H_2 (δD_{H_2} = −728‰) and slightly D-enriched H_2O (δD_{H_2O} = +39‰) (Balabane et al. 1987). These results suggest that the δD_{H_2} effect may have a significant impact on the $\alpha^H{}_{CH_4\text{-}H_2O}$ values under high pH_2 conditions. The δD_{H_2} effect potentially leads to a great variation of $\alpha^H{}_{CH_4\text{-}H_2O}$ values as a result of the contribution from δD_{H_2} variation.

In contrast, it was notable that little δD_{H_2} effect was exhibited in MP batches, which cultivated a hydrogenotrophic methanogen under high pH_2 conditions (Fig. 5a–c). This finding implies that the δD_{H_2} effect is not directly associated with the pH_2 conditions of growth. The hydrogen isotope fractionation in hydrogenotrophic methanogenesis is thought to be related to the methanogenic rate (Valentine et al. 2004; Stolper et al. 2015). As the MP batches reached the stationary growth phase after several weeks, which is slower than the typical growth rates of other pure culture methanogens at >10^4 Pa-H_2 with favorable growth conditions (timescales of hours to days: Additional file 1), the δDH_2 effect may instead be related to the metabolic and growth kinetics of methanogens.

As discussed in previous studies (Burke 1993; Sugimoto and Wada 1995; Kawagucci et al. 2014), the incorporation of intracellularly produced H^+ from the H_2 into CH_4 may drive the δD_{H_2} effect. From this viewpoint, a previous study that cultivated *Mtb. thermautotrophicus* under high-pH_2 conditions attempted to yield a quantitative understanding of the magnitude of the δD_{H_2} effect (Kawagucci et al. 2014). We can know only two quantitative pieces of information relating to the δD_{H_2} effect: the cell-specific H_2 consumption rate (i.e., the intracellular H^+ production rate) of 10^{-1}–10^2 fmol/cell/s and the quantity of intracellular H_2O, which is 10^1–10^2 fmol-H_2O. If the H_2 consumption rate is comparable with that trans-membrane H_2O and H^+ flux, which remains unknown, the intracellular δD_{H_2O} value should be affected by H_2-derived H_2O (H^+) that results in exhibition of the δD_{H_2} effect. Moreover, if the trans-membrane H_2O flux is constant regardless of the cell growth situation, the slower H_2 consumption rate would lead to a smaller δD_{H_2} effect. It would be reasonable for a significant δD_{H_2} effect to be apparent in the TP batches and *Mtb. thermautotrophicus* batches reported previously (Kawagucci et al. 2014), in which the H_2 consumption rate was rapid (10^{-1}–10^2 fmol/cell/s), whereas little δD_{H_2} effect was detected in the MP batches, in which the H_2 consumption rate was slow (presumably < ~10^{-1} fmol/cell/s). This finding seems consistent with the decrement in magnitude of the δD_{H_2} effect along with the growth phase shift from exponential growth to the stationary phase (Kawagucci et al. 2014).

Comparison between methanogenic experiments and environmental methane

To compare the experimental results with natural environments, we here review hydrogen and carbon isotope fractionations that have been obtained from environmental observations in addition to the incubations (Additional file 1). As the timescale of methanogenesis is a probable key factor controlling hydrogen isotope fractionation, as aforementioned, the compiled dataset of natural isotope fractionations is classified by considering the timescales of natural methane reservoirs (Fig. 6). Methane in the bovine rumen, one of the natural environments in which methane generation occurs as rapidly as in laboratory incubations, showed the $\alpha^{H}_{CH_4-H_2O}$ range between 0.68 and 0.74 (Burke 1993; Wang et al. 2015). The pH$_2$ condition of the rumen was estimated at ~10^2 Pa (Burke 1993; Wang et al. 2015), higher than the values exhibited in our cocultures (<65 Pa: Fig. 1) and the methanogenic threshold (<10^1 Pa: Lovley 1985; Thauer et al. 2008). Terrestrial sediments in freshwater systems of Caltech pond and Swamp Y (Stolper et al. 2015; Wang et al. 2015) possessed $\alpha^{H}_{CH_4-H_2O}$ values of ~0.69. The $\alpha^{H}_{CH_4-H_2O}$ values in these environmental samples from relatively "young" methane reservoirs (0.68–0.74) are consistent with those obtained from the regulated laboratory cocultures of hydrogenotrophic methanogens (≤0.74). Note that aceticlastic and/or methylotrophic methanogenesis may also occur in the wild environment.

Hydrogen and carbon isotope fractionations have also been investigated in geological methane reservoirs: natural gas fields (Nakai et al. 1974; Whiticar et al. 1986); continental groundwater at the Elk Valley coalbed methane field (Aravena et al. 2003); terrestrial groundwater associated with the accretionary prism in south-east Japan (Kimura et al. 2010); deep-sea sediments (Whiticar et al. 1986); and subseafloor methane hydrate deposits from the Northern Cascadia Margin (Wang et al. 2015), the Umitaka Spur, and the Joetsu Knoll of the Japan Sea (Hachikubo et al. 2011) (Fig. 6). We hereafter methane in geological reservoirs is named as "aged" methane, for comparison with "young" methane in laboratory cultures and natural environments, as aforementioned. Almost all of these aged samples show a $\alpha^{C}_{CH_4-CO_2}$ range of 0.91–0.95, suggesting that hydrogenotrophic methanogenesis under H$_2$-limited conditions is plausible as the geochemical origin of these methane. Almost all of the $\alpha^{H}_{CH_4-H_2O}$ values of the aged methane samples fall within a narrow range of 0.79–0.84 (Fig. 6). The $\alpha^{H}_{CH_4-H_2O}$ values are not consistent with the values obtained from the H$_2$-limited experiments and young environmental samples (≤0.74; Fig. 6) in spite of the commonly low $\alpha^{C}_{CH_4-CO_2}$ values (≤0.95).

Possible mechanisms exhibiting variable isotope fractionation

We define three major provinces in the compilation of isotope fractionations (Fig. 6) and infer the metabolic mechanisms functioning to yield each of the provinces as follows (Fig. 7). The first province has an $\alpha^{C}_{CH_4-CO_2}$ range higher than 0.95 with highly variable $\alpha^{H}_{CH_4-H_2O}$ values (Fig. 6). This province has occurred in >10^4 Pa-H$_2$ culture but never observed in any natural samples from biologically functionable temperature environments (≤122 °C). This finding is reasonable because such high-pH$_2$ environments are rare in nature, except for several H$_2$-enriched geofluid systems (Charlou et al. 2002; Proskurowski et al. 2006; Ishibashi et al. 2014). The high $\alpha^{C}_{CH_4-CO_2}$ value in >10^4 Pa-H$_2$ cultures has been thought to be a result of an almost straightforward progress of carbon reduction from CO$_2$ to CH$_4$ in a multistep methanogenic pathway (Valentine et al. 2004; Penning et al. 2005; Fig. 7a and b). The large $\alpha^{H}_{CH_4-H_2O}$ variation probably results from the δD_{H_2} effect (see discussion above), and the magnitude of the δD_{H_2} effect appears to be linked to the H$_2$ consumption rate (Fig. 7a and b). The δD_{H_2} shift, probably due to H$_2$–H$_2$O isotope exchange toward the isotope equilibrium by reversible function of hydrogenase, suggests that H$_2$ reproduction functioned even during the almost straightforward methanogenesis (Fig. 7b).

The second province has an $\alpha^{C}_{CH_4-CO_2}$ range lower than 0.95 and an $\alpha^{H}_{CH_4-H_2O}$ range of 0.67–0.74 (Fig. 6). This province has been found in both laboratory cultures of <10^4 Pa-H$_2$ conditions and also in natural reservoirs of relatively young methane such as the bovine rumen and terrestrial sediments. The low $\alpha^{C}_{CH_4-CO_2}$ values are thought to result from the extent of reversibility of the methanogenic pathway under low-pH$_2$ conditions (Valentine et al. 2004; Fig. 7c). The narrow $\alpha^{H}_{CH_4-H_2O}$ range is probably caused by a combination of the higher reversibility and the slower H$_2$ consumption rate. The slow H$_2$ consumption would alter the intracellular δD_{H_2O} value little and leave it the same as the environmental (extracellular) δD_{H_2O} value (Fig. 7c). The repeated formation and cleavage of the C–H bonds of methane precursors through the reversible methanogenic pathway lead to hydrogen isotope equilibrium between the precursor and intracellular water. Note that the $\alpha^{H}_{CH_4-H_2O}$ values of this province (0.67–0.74) are out of the $\alpha^{eq}_{CH_4-H_2O}$ range between 0 °C and 400 °C (Horibe and Craig 1995). This finding suggests that some kinetic processes are also involved in the hydrogen isotope fractionation (Fig. 7c).

The third province has an $\alpha^{C}_{CH_4-CO_2}$ range lower than 0.95 and an $\alpha^{H}_{CH_4-H_2O}$ range higher than 0.80 (Fig. 6). The third province overlaps with the second

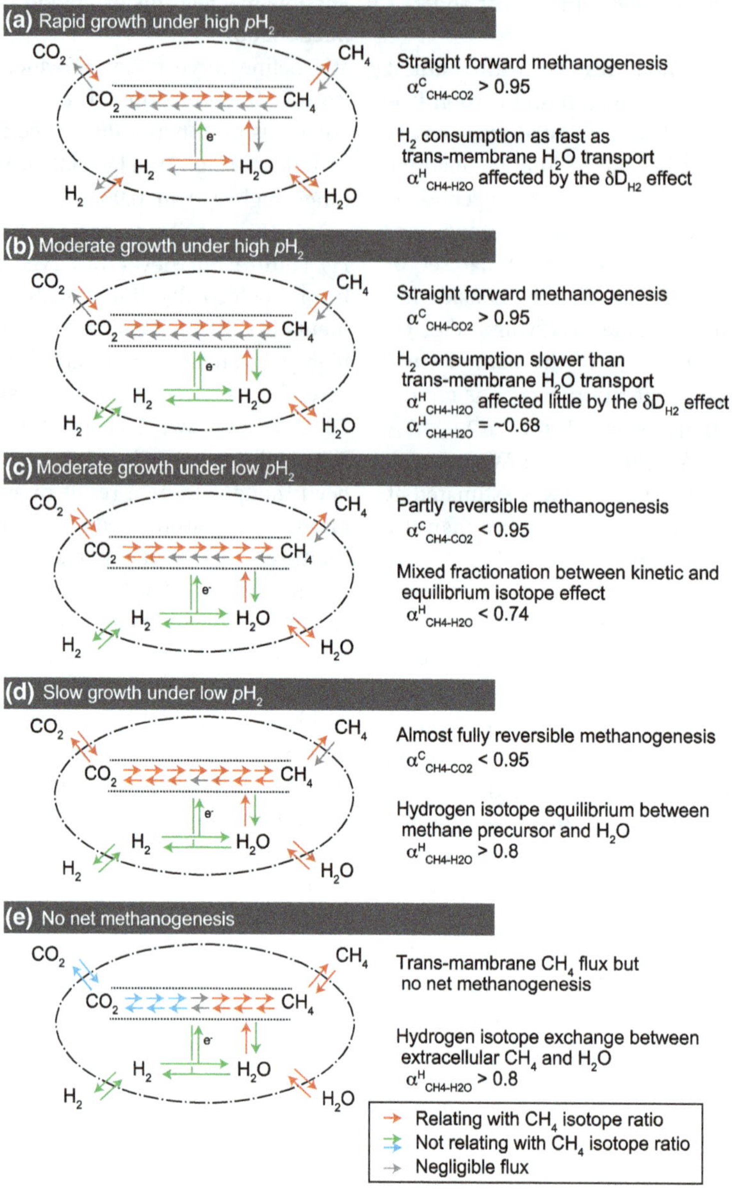

Fig. 7 Cartoons of intra- and trans-membrane fluxes of the molecules in hydrogenotrophic methanogenesis relating to hydrogen and carbon isotope systematics. Each five panel represents methanogenic characteristics with respect to growth rate and pH_2 condition for the growth. Red arrows represent significant molecular fluxes relating with CH_4 isotope ratios. Light green and blue arrows represent significant molecular fluxes not relating with CH_4 isotope ratios. Grey arrows represent negligible molecular fluxes. Dash-dotted circles are cell membrane, and arrows enclosed by dotted lines represent multistep methanogenic reaction (numbers of the arrows have no meaning). A term "H_2O" in cartoon represents both H_2O and H^+. See main text for details

province in $\alpha^{C}{}_{CH_4-CO_2}$ values but differs in the $\alpha^{H}{}_{CH_4-H_2O}$ value. This province has been observed typically in geological reservoirs of aged methane such as natural gas and seafloor methane hydrate. It was thought that methanogenesis at an extremely slow rate could produce the third province because of isotopic equilibrium between H_2O, H_2, CO_2, and methane precursors through a highly reversible methanogenic pathway (Stolper et al. 2015; Wang et al. 2015; Fig. 7d). Unfortunately, we cannot examine this possibility experimentally because microbial methanogenesis at a rate as slow as geological timescale is not reproducible by experiments. In addition, we are aware of the $\alpha^{C}{}_{CH_4-CO_2}$ variation in spite of the narrow $\alpha^{H}{}_{CH_4-H_2O}$ range within the third province (Fig. 6). The $\alpha^{C}{}_{CH_4-CO_2}$ variation may be inconsistent with this explanation because methanogenesis through the geological timescale is expected to occur under constantly low pH_2 conditions and slow growth rate, resulting in the constantly low $\alpha^C{}_{CH_4-CO_2}$ value of approximately 0.92.

We here hypothesize that the higher $\alpha^H{}_{CH_4-H_2O}$ values of the third province may be caused by a post-methanogenesis process over geological timescales (Fig. 7e) rather than the methanogenesis through geological timescale (Fig. 7d). This process is diagenetic hydrogen isotope exchange between extracellular CH_4 and H_2O promoted by the reversible function of the multistep methanogenic pathway in methanogenic region and/or methane reservoirs after the migration (Fig. 7e). Multiple lines of circumstantial evidence based on biological and geochemical viewpoints can be cited to put forward the possibility of the diagenetic hydrogen isotope exchange as following. First, the microbial methanogenic pathway is also able to operate reversibly for methane consumption; e.g., it is known that anaerobic methanotrophic archaea can oxidize CH_4 via potential reverse methanogenesis (Knittel and Boetius 2009) and the methanogens can perform trace methane oxidation even during active methanogenesis (Moran et al. 2005). In addition, trans-membrane exchange of CH_4 would be not negligible in geological timescale. Thus, the reversible enzymatic reactions, in particular at the later steps of the methanogenesis, i.e., cleavage and reproduction of C–H bonds of CH_4, could result in promoting the hydrogen isotope exchange between extracellular CH_4 and H_2O (Fig. 7e). Second, the $\alpha^H{}_{CH_4-H_2O}$ values in the third province (0.79–0.86) overlap with the $\alpha^{eq}{}_{CH_4-H_2O}$ values in the known biologically-functional temperature range (0.758–0.843 at 0–122 °C: Horibe and Craig 1995) (Fig. 6). Third, a one-year hydrothermal experiment at 323 °C (Reeves et al. 2012) revealed negligible hydrogen isotope exchange between CH_4 and H_2O. This finding suggests that hydrogen isotope exchange is extremely sluggish in the habitable temperature range without effective catalysts such as enzymes in the reversible methanogenic pathway. Fourth, the third province is only detected in the aged methane samples obtained from both marine and terrestrial environments (Fig. 6) while relatively young methane samples such as from the rumen and laboratory experiments are never located in the third province. This age-isotope relationship is also found in a "clumped" isotope composition, i.e., the abundance of ^{13}C–D bonds in CH_4, which may yield information about the temperature at which the C–H bonds were formed or last equilibrated (Stolper et al. 2014, 2015; Ono et al. 2014; Wang et al. 2015). The clumped isotope composition so far observed in aged methane reservoirs indicates the CH_4–H_2O equilibrated signature, whereas clumped signatures that cannot be explained by expected formation temperature have been found frequently in young biogenic methane from natural reservoirs and laboratory incubations (Stolper et al. 2014, 2015; Wang et al. 2015). Fifth, the diagenetic hydrogen isotope exchange is not in conflict with the broad $\alpha^C{}_{CH_4-CO_2}$ range of the third province. If a methane reservoir incorporates multiple methane sources, each of which has distinct isotope signatures, the diagenetic hydrogen isotope exchange could unify the $\alpha^H{}_{CH_4-H_2O}$ values regardless of the sources but leave the $\alpha^C{}_{CH_4-CO_2}$ value as a mixed signature because of incomplete molecular equilibrium between CO_2 and CH_4 (Fig. 7e). Such diagenetic unification of the hydrogen isotope ratio while maintaining the $\alpha^C{}_{CH_4-CO_2}$ variation is evident in deep-sea, high-temperature hydrothermal fluids (Fig. 6: Proskurowski et al. 2006; McCollom 2008; Kawagucci et al. 2013a; 2013b; Wang et al. 2015). Methane in the hydrothermal fluids so far observed in various geological settings shows a broad $\alpha^C{}_{CH_4-CO_2}$ range (0.953–0.992) but a narrow $\alpha^H{}_{CH_4-H_2O}$ range (0.88–0.90), which are close to the $\alpha^{eq}{}_{CH_4-H_2O}$ values at the endmember fluid temperatures (~0.87 at 300 °C–400 °C: Horibe and Craig 1995).

These pieces of circumstantial evidence put forward that the hydrogen isotope signature of CH_4 imprinted at the time of hydrogenotrophic methanogenesis ($\alpha^H{}_{CH_4-H_2O}$ = 0.66–0.74), belonging to second province, can be altered by the reversible reactions of the multistep methanogenic pathway, which leads to CH_4–H_2O isotopic equilibrium at habitable temperatures ($\alpha^H{}_{CH_4-H_2O}$ > 0.76) belonging to the third province. If this is true, the hydrogen isotope ratio of methane in geological samples would be useless as a tracer to deduce the origin of methane. However, it is also possible to criticize this hypothesis: the diagenetic isotope exchange proposed here is not required when methanogenesis through geological time actually exhibits isotope fractionation in the third province, as aforementioned. Furthermore, an in vitro biochemical study of methyl coenzyme M reductase activity, which is an enzyme that catalyzes the last step of methanogenesis, showed that little hydrogen isotope exchange between CH_4 and H_2O occurred during the enzyme reaction for several tens of minutes (Scheller et al. 2013). It is still uncertain whether diagenetic hydrogen isotope exchange can be promoted by methanogens or not and how rapid the isotope exchange is if it occurs. To solve these uncertainties, experiments using methanogenic cells concentrated in mediums with isotopically disequilibrated CH_4 and H_2O under thermodynamic states favorable to each of methane generation and methane consumption will be conducted.

That work will provide experimental justification of the possible post-methanogenesis isotope diagenesis hypothesis.

Conclusions

This study confirmed five facts as follows.

1. Carbon isotope fractionation between CH_4 and CO_2 during hydrogenotrophic methanogenesis is correlated with pH_2, as suggested in previous studies (Valentine et al. 2004; Penning et al. 2005; Takai et al. 2008).
2. The hydrogen isotope ratio of CH_4 produced by a thermophilic methanogen, *Mtc. okinawensis*, under high pH_2 conditions (~10^5 Pa-H_2) is affected by the isotope ratio of H_2, as pointed out in a previous incubation of *Mtb. thermautotrophicus* (Kawagucci et al. 2014). This pattern is named as the δD_{H_2} effect. This effect also appears to account for the diverse hydrogen isotope fractionation between CH_4 and H_2O previously observed in H_2-enriched culture incubations (Balabane et al. 1987; Valentine et al. 2004; Stolper et al. 2015; Wang et al. 2015).
3. A mesophilic methanogen, *Methanobacterium* sp. MO-MB1, showed little δD_{H_2} effect even under high-pH_2 conditions (~10^5 Pa). This result suggests that methanogenic and growth rates rather than the pH_2 condition are significant factors controlling the δD_{H_2} effect.
4. The hydrogen isotope fractionation between CH_4 and H_2O in hydrogenotrophic methanogenesis under low pH_2 conditions ($<10^2$ Pa) observed in experiments is in the range of 0.66–0.74.
5. Hydrogen isotope fractionation exhibited in laboratory incubations under low-pH_2 conditions is consistent with that observed in "young" methane reservoirs but inconsistent with that observed in "aged" methane reservoirs.

In conclusion, we propose that diagenetic hydrogen isotope exchange between extracellular CH_4 and H_2O catalyzed by a reversible methanogenic pathway of methanogenic populations alters the hydrogen isotope signature of CH_4 imprinted at the time of generation.

Additional files

Additional file 1: Compilation of hydrogen and carbon isotope systematics from incubation and observation. Description of data: Type of ecosystem, name of ecosystem, temperature of methanogen growth (Celsius), approximate timescale for growth, fractionation factors of the carbon isotope ratio between CH_4 and CO_2 ($\alpha^C_{CH_4-CO_2}$), fractionation factors of the hydrogen isotope ratio between CH_4 and H_2O ($\alpha^H_{CH_4-H_2O}$), and references.

Additional file 2: All results. Description of data: Batch culture names, sampling times (h; hours or d; days), hydrogen concentration (Pa), methane concentration (kPa), hydrogen isotope ratio of H_2O, H_2, and CH_4 (‰ vs. VSMOW), carbon isotope ratio of CO_2 and CH_4 (‰ vs. VPDB), values of equilibrated hydrogen isotope ratio of H_2 ($\delta^{eq}D_{H_2}$) between H_2 and H_2O, fractionation factors of the carbon isotope ratio between CH_4 and CO_2 ($\alpha^C_{CH_4-CO_2}$), and fractionation factors of the hydrogen isotope ratio between CH_4 and H_2O ($\alpha^H_{CH_4-H_2O}$). Cell numbers were counted only in the thermophilic pure culture batches.

Abbreviations

CF-IRMS: continuous-flow isotope ratio mass spectrometry; DSMZ: Deutsche Sammlung von Mikroorganismen und Zellkulturen; Eq: isotope equilibrium; GC-HID: gas chromatography using a helium ionization detector; MC: Mesophilic coculture experimental set; MP: Mesophilic pure culture experimental set; Mtb: *Methanothermobacter*; Mtc: *Methanothermococcus*; S: *Syntrophothermus*; TC: Thermophilic coculture experimental set; TP: Thermophilic pure culture experimental set; VPDB: Vienna Pee Dee Belemnite; VSMOW: Vienna standard mean ocean water.

Competing interests

The authors declared that they have no competing interests.

Authors' contributions

SK proposed the topic, conceived, and designed the study. TO performed all the experiments and measurements. YS and HI contributed to incubation and interpretation of the TC, MC, and MP experiments. YM analyzed the carbon isotopes of CO_2 and CH_4 and assisted with their interpretation. TO and SK compiled the results and drafted the manuscript. KT collaborated with the corresponding author in the design of the study and the construction of the manuscript. All authors read and approved the final manuscript.

Authors' information

TO is a JSPS research fellow for young scientist adopted at the Laboratory of Ocean-Earth Life Evolution Research (OELE), Japan Agency for Marine-Earth Scienceand Technology (JAMSTEC). SK is a researcher at OELE, Department of Subsurface Geobiological Analysis and Research (D-SUGAR), and Research and Development (R&D) Center for Submarine Resources, JAMSTEC. During this study, YS was a graduate school student at the Department of Environmental Systems Engineering, Nagaoka University of Technology. HI is a senior researcher at the D-SUGAR and R&D Center for Submarine Resources, JAMSTEC. YM is an engineer at the R&D Center for Submarine Resources, JAMSTEC. KT is a principal scientist at D-SUGAR, OELE, and the R&D Center for Submarine Resources, JAMSTEC.

Acknowledgements

We wish to thank Uta Konno, Keiko Tanaka, Eiji Tasumi, and Sanae Sakai for their assistance on microbial incubation and chemical analysis. The authors would like to thank Enago (www.enago.jp) for the English language review. Funds for this project were partly provided by the Ministry of Education, Culture, Sports, Science, and Technology, Japan as Grant-in-aids for Young Scientists (A) (25701004 and 15H05468), Scientific Research (A) (15H02419), Scientific Research (C) (22560811), and the JSPS research fellowship for young scientist (26–10363). This project was also supported by Cross-Ministerial Strategic Innovation promotion Program, organized by the Cabinet Office of Japan, for the Development of New-Generation Research Protocol for Submarine Resources.

Author details

[1]Laboratory of Ocean-Earth Life Evolution Research (OELE), Japan Agency for Marine-Earth Science and Technology (JAMSTEC), 2-15 Natsushima-cho, Yokosuka 237-0061, Japan. [2]Department of Subsurface Geobiological Analysis and Research (D-SUGAR), Japan Agency for Marine-Earth Science and Technology (JAMSTEC), 2-15 Natsushima-cho, Yokosuka 237-0061, Japan. [3]Research and Development Center for Submarine Resources, Japan Agency for Marine-Earth Science and Technology (JAMSTEC), 2-15 Natsushima-cho, Yokosuka 237-0061, Japan. [4]Department of Environmental Systems Engineering, Nagaoka University of Technology, 1603-1 Kamitomiokamachi, Nagaoka 940-2188, Japan.

References

Aravena R, Harrison SM, Abercrombie H, Rudolph D (2003) Origin of methane in the Elk Valley coalfield, southeastern British Columbia, Canada. Chem Geol 195:219–227

Balabane M, Galimov E, Hermann M, Letolle R (1987) Hydrogen and carbon isotope fractionation during experimental production of bacterial methane. Org Geochem 11:115–119

Bardo RD, Wolfsberg M (1976) A theoretical calculation of the equilibrium constant for the isotopic exchange reaction between H_2O and HD. J Physic Chem 80:1068–1071

Burke RA (1993) Possible influence of hydrogen concentration on microbial methane stable hydrogen isotopic composition. Chemosph 26:55–67

Campbell BJ, Li C, Sessions AL, Valentine DL (2009) Hydrogen isotopic fractionation in lipid biosynthesis by H_2-consuming *Desulfobacterium autotrophicum*. Geochim Cosmochim Acta 73:2744–2757

Charlou JL, Donval JP, Fouquet Y, Jean-Baptiste P, Holm N (2002) Geochemistry of high H2 and CH4 vent fluids issuing from ultramafic rocks at the Rainbow hydrothermal field (36°14′N, MAR). Chem Geol 191:345–359

Daniels L, Fulton G, Spencer RW, Ormejohnson WH (1980) Origin of hydrogen in methane produced by *Methanobacterium Thermoautotrophicum*. J Bacteriol 141:694–698

Hachikubo A, Tomaru H, Matsumoto R (2011) Molecular and isotopic characteristics of hydrocarbons in sediments and gas hydrate at eastern margin of Japan Sea. In Proceedings of the 7th Conference on Gas Hydrates (ICGH 2011), Edinburgh, Scotland, United Kingdom, 17–21 July 2011.

Hattori S, Nashimoto H, Kimura H, Koba K, Yamada K, Shimizu M, Watanabe H, Yoh M, Yoshida N (2012) Hydrogen and carbon isotope fractionation by thermophilic hydrogenotrophic methanogens from a deep aquifer under coculture with fermenters. Geochem J 46:193–200

Horibe Y, Craig H (1995) D/H fractionation in the system methane-hydrogen-water. Geochim Cosmochim Acta 59:5209–5217

Horita J (2001) Carbon isotope exchange in the system CO_2-CH_4 at elevated temperatures. Geochim Cosmochim Acta 65:1907–1919

Horita J, Wesolowski DJ (1994) Liquid–vapor fractionation of oxygen and hydrogen isotopes of water from the freezing to the critical temperature. Geochim Cosmochim Acta 58:3425–3437

Hsu HW, Postberg F, Sekine Y, Shibuya T, Kempf S, Horanyi M, Juhasz A, Altobelli N, Suzuki K, Masaki Y, Kuwatani T, Tachibana S, Sirono S, Moragas-Klostermeyer G, Srama R (2015) Ongoing hydrothermal activities within Enceladus. Nature 519:207–210

Imachi H, Aoi K, Tasumi E, Saito Y, Yamanaka Y, Saito Y, Yamaguchi T, Tomaru H, Takeuchi R, Morono Y, Inagaki F, Takai K (2011) Cultivation of methanogenic community from subseafloor sediments using a continuous-flow bioreactor. The ISME J 5:1913–1925

Inagaki F, Hinrichs K-U, Kubo Y, Bowles MW, Heuer VB, Hong W-L, Hoshino A, Ijiri A, Imachi M, Ito M, Kaneko M, Lever MA, Lin C-H, Murayama M, Ohkouchi N, Ono S, Park Y-S, Phillips SC, Prieto-Mollar X, Purley M, Riedinger N, Sanada Y, Sauvage J, Snyder G, Susilawati R, Yakano Y, Tasumi E, Terada T, Tomaru H, Trembath-Reichert E, Wang DT, Yamada Y (2015) Exploring deep microbial life in coal-bearing sediment down to ~2.5 km below the ocean floor. Science 349:420–424

Ishibashi J, Noguchi T, Toki T, Miyabe S, Yamagami S, Onishi Y, Yamanaka T, Yokoyama Y, Omori E, Takahashi Y, Hatada K, Nakaguchi Y, Yoshizaki M, Konno U, Shibuya T, Takai K, Inagaki F, Kawagucci S (2014) Diversity of fluid geochemistry affected by processes during fluid upwelling in active hydrothermal fields in the Izena Hole, the middle Okinawa Trough back-arc basin. Geochem J 48:1–13

Kawagucci S, Toki T, Ishibachi J, Takai K, Ito M, Oomori T, Gamo T (2010) Isotopic variation of molecular hydrogen in 20 °C–375 °C hydrothermal fluids as detected by a new analytical method. J Geophysic Res-Biogeosci 115:G03021

Kawagucci S, Ueno Y, Takai K, Toki T, Ito M, Inoue K, Makabe A, Yoshida N, Muramatsu Y, Takahata N, Sano Y, Narita T, Teranishi G, Obata H, Nakagawa S, Nunoura T, Gamo T (2013a) Geochemical origin of hydrothermal fluid methane in sediment-associated fields and its relevance to the geographical distribution of whole hydrothermal circulation. Chem Geol 339:213–225

Kawagucci S, Miyazaki J, Nakajima R, Nozaki T, Takaya Y, Kato Y, Shibuya T, Konno U, Nakaguchi Y, Hatada K, Hirayama H, Fujikura K, Furushima Y, Yamamoto H, Watsuji T, Ishibashi J, Takai K (2013b) Post-drilling changes in fluid discharge pattern, mineral deposition, and fluid chemistry in the Iheya North hydrothermal field, Okinawa Trough. Geochem Geophys Geosyst 14:4774–4790

Kawagucci S, Kobayashi M, Hattori S, Yamada K, Ueno Y, Takai K, Yoshida N (2014) Hydrogen isotope systematics among H_2–H_2O–CH_4 during the growth of the hydrogenotrophic methanogen Methanothermobacter thermautotrophicus strain ΔH. Geochim Cosmochim Acta 142:601–614

Kimura H, Nashimoto H, Shimazu M, Hattori S, Yamada K, Koba K, Yoshida N, Kato K (2010) Microbial methane production in deep aquifer associated with the accretionary prism in Southwest Japan. The ISME J 4:531–541

Knittel K, Boetius A (2009) Anaerobic oxidation of methane: progress with unknown process. Annu Rev Microbiol 63:311–333

Kvenvolden KA (1988) Methane hydrate—a major reservoir of carbon in the shallow geosphere. Chem Geol 71:41–51

Lovley DR (1985) Minimum threshold for hydrogen metabolism in methanogenic bacteria. Appl Environ Microbiol 49:1530–1531

Martini AM, Budai JM, Walter LM, Schoell M (1996) Microbial generation of economic accumulations of methane within a shallow organic-rich shale. Nature 383:155–158

McCollom TM (1999) Methanogenesis as a potential source of chemical energy for primary biomass production by autotrophic organisms in hydrothermal systems on Europa. J Geophys Res 104:30729–30742

McCollom TM (2008) Observation, experimental, and theoretical constraints on carbon cycling in Mid-Ocean Ridge hydrothermal systems. In: Lowell RP, Seewald JS, Metaxas A, Perfit MR (eds) Magma to Microbe: Modeling hydrothermal processes at ocean spreading centers. Geophysical Monograph Series. American Geophysical Union, Washington, D. C., pp 193–213

McCollom TM (2013) Laboratory simulations of abiotic hydrocarbon formation in Earth's deep subsurface. Rev Mineral Geochem 75:467–494

Moran JJ, House CH, Freeman KH, Ferry JG (2005) Trace methane oxidation studied in several Euryarchaeota under diverse conditions. Archaea 1:303–309

Nakai N, Yoshida Y, Ando N (1974) Isotopic studies on oil and natural gas fields in Japan. Chikyukagaku (in Japanese with English abstract) 7(8):87–98

Ono S, Wang DT, Gruen DS, Sharwood-Lollar B, Zahniser MS, McManus BJ, Nelson DD (2014) Measurement of a doubly substituted methane isotpologue, $^{13}CH_3D$, by tunable infrared laser direct absorption spectroscopy. Anal Chem 86(13):6487–6494

Patra PK, Houweling S, Krol M, Bousquet P, Belikov D, Bergmann D, Bian H, Cameron-Smith P, Chipperfield MP, Corbin K, Fortems-Cheiney A, Fraser A, Gloor E, Hess P, Ito A, Kawa SR, Law RM, Loh Z, Maksyutov S, Meng L, Palmer PI, Prinn RG, Rigby M, Saito R, Wilson C (2011) TransCom model simulations of CH_4 and related species: linking transport, surface flux and chemical loss with CH_4 variability in the troposphere and lower stratosphere. Atmos Chem Physic 11:12813–12837

Penning H, Plugge CM, Galand PE, Conrad R (2005) Variation of carbon isotope fractionation in hydrogenotrophic methanogenic microbial cultures and environmental samples at different energy status. Global Change Biol 11: 2103–2113

Proskurowski G, Lilley MD, Kelley DS, Olson EJ (2006) Low temperature volatile production at the Lost City hydrothermal field, evidence from a hydrogen stable isotope geothermometer. Chem Geol 229:331–343

Reeves E, Seewald JS, Sylva SP (2012) Hydrogen isotope exchange between n-alkanes and water under hydrothermal conditions. Geochim Cosmochim Acta 77:582–599

Saito Y, Sakai S, Saito Y, Aoki M, Miyazaki M, Tasumi E, Yamaguchi T, Inagaki F, Takai K, Imach H (2014) Cultivation of anaerobic, syntrophic heterotrophic bacteria with hydrogenotrophic methanogens from deep subseafloor sediments off Shimokita, Japan. In: Proceedings of International Symposium on Microbial Ecology (ISME15). Seoul, Korea, 24–29 August 2014.

Scheller S, Goenrich M, Thauer RK, Jaun B (2013) Methyl-coenzyme M reductase from methanogenic archaea: Isotope effects on the formation and anaerobic oxidation of methane. J Am Chem Soci 135:14975–14984

Schoell M (1980) The hydrogen and carbon isotopic composition of methane from natural gases of various origins. Geochim Cosmochim Acta 44:649–661

Schoell M (1988) Multiple origins of methane in the Earth. Chem Geol 71:1–10

Sekiguchi Y, Kamagata Y, Nakamura K, Ohashi A, Harada H (2000) *Syntrophothermus lipocalidus* gen. nov., sp. Nov., a novel thermophilic, syntrophic, fatty-acid-oxidizing anaerobe which utilizes isobutyrate. Int J Syst Evol Microbiol 50:771–779

Smith DR, Doucette-Stamm LA, Deloughery C, Lee HM, Dubois J, Aldredge T, Bashirzadeh R, Blakely D, Cook R, Gilbert K, Harrison D, Hoang L, Keagle P, Lumm W, Pothier B, Qiu DY, Spadafora R, Vicaire R, Wang Y, Wierzbowski J, Gibson R, Jiwani N, Caruso A, Bush D, Safer H, Patwell D, Prabhakar S, McDougall S, Shimer G, Goyal A, Pietrokovski S, Church GM, Daniels CJ, Mao JI, Rice P, Nolling J, Reeve JN (1997) Complete genome sequence of

Methanobacterium thermoautotrophicum ΔH: Functional analysis and comparative genomics. J Bacteriol 179:7135–7155
Stolper DA, Lawson M, Davis CL, Ferreia AA, Santos Neto EV, Ellis GS, Lewan MD, Martini AM, Tang Y, Schoell M, Sessions AL, Eiler JM (2014) Formation temperatures of thermogenic and biogenic methane. Science 344:1500–1503
Stolper DA, Martini AM, Clog M, Douglas PM, Shusta SS, Valentine DL, Sessions AL, Eiler JM (2015) Distinguishing and understanding thermogenic and biogenic sources of methane using multiply substituted isotopologues. Geochim Cosmochim Acta 161:219–247
Sugimoto A, Wada E (1995) Hydrogen isotopic composition of bacterial methane - CO_2/H_2 reduction and acetate fermentation. Geochim Cosmochim Acta 59: 1329–1337
Takai K, Inoue A, Horikoshi K (2002) *Methenothermococcus okinawensis* sp. nov., a thermophilic, methane-producing archaeon isolated from a Western Pacific deep-sea hydrothermal vent system. Int J Syst Evol Microbiol 52:1089–1095
Takai K, Gamo T, Tsunogai U, Nakayama N, Hirayama H, Nealson KH, Horikoshi K (2004) Geochemical and microbiological evidence for a hydrogen-based, hyperthermophilic subsurface lithoautotrophic microbial ecosystem (HyperSLiME) beneath as active deep-sea hydrothermal field. Extremophiles 8:269–282
Takai K, Nakamura K, Toki T, Tsunogai U, Miyazaki M, Miyazaki J, Hirayama H, Nakagawa S, Nunoura T, Horikoshi K (2008) Cell proliferation at 122 °C and isotopically heavy CH_4 production by a hyperthermophilic methanogen under high-pressure cultivation. Proc Natl Acad Sci U S A 105:10949–10954
Tang Y, Perry JK, Henden PD, Schoell M (2000) Mathematical modeling of stable carbon isotope ratios in natural gases. Geochim Cosmochim Acta 64:2673–2687
Thauer RK, Kaster A-K, Seedorf H, Buckel W, Hedderich R (2008) Methanogenic archaea: ecologically relevant differences in energy conservation. Nature Reviews in Microbiology 6:579–591
Umezawa T, Aoki S, Nakazawa T, Morimoto S (2009) A high-precision measurement system for carbon and hydrogen isotopic ratios of atmospheric methane and its application to air samples collected in the Western Pacific region. J Meteorol Soc Japan 87:365–379
Valentine DL, Chidthaisong A, Rice A, Reeburgh WS, Tyler SC (2004) Carbon and hydrogen isotope fractionation by moderately thermophilic methanogens. Geochim Cosmochim Acta 68:1571–1590
Vignais PM (2005) H/D exchange reactions and mechanistic aspects of the hydrogenases. Coord Chem Rev 249:1677–1690
Waldron S, Waston-Craik IA, Hall AJ, Fallick AE (1998) The carbon and hydrogen stable isotope composition of bacteriogenic methane: A laboratory study using a landfill inoculum. Geomicrobiol J 15:157–169
Waldron S, Lansdown JM, Scott EM, Fallick AE, Hall AJ (1999) The global influence of the hydrogen isotope composition of water on that of bacteriogenic methane from shallow freshwater environments. Geochim Cosmochim Acta 63:2237–2245
Walter S, Laukenmann S, Stams AJM, Vollmer MK, Glexner G, Roeckmann T (2012) The stable isotopic signature of biologically produced molecular hydrogen (H_2). Biogeosciences 9:4115–4123
Wang DT, Gruen DS, Sherwood-Lollar B, Hinrichs K-U, Stewart LC, Holden JF, Hristov AN, Pohlman JW, Morrill PL, Konneke M, Delwiche KB, Reeves EP, Sutcliffe CN, Ritter DJ, Seewald JS, McIntosh JC, Hemond HF, Kubo MD, Cardace D, Hoeler TM, Ono S (2015) Nonequilibrium clumped isotope signals in microbial methane. Science 348:428–431. doi:10.1126/science.aaa4326
Whiticar MJ (1999) Carbon and hydrogen isotope systematics of bacterial formation and oxidation of methane. Chem Geol 161:291–314
Whiticar MJ, Faber E, Schoell M (1986) Biogenic methane formation in marine and freshwater environments: CO_2 reduction vs. acetate fermentation—Isotope evidence. Geochim Cosmochim Acta 50:693–709
Yang H, Gandhi H, Shi L, Kreuzer HW, Ostrom NE, Hegg EL (2012) Using gas chromatography/isotope ratio mass spectrometry to determine the fractionation factor for H2 production by hydrogenase. Rapid Commun Mass Spectrom 26:61–68
Yoshioka H, Sakata S, Kamagata Y (2008) Hydrogen isotope fractionation by *Methanothermobacter thermoautotrophicus* in coculture and pure culture conditions. Geochim Cosmochim Acta 72:2687–2694

11

Direct comparison of site-specific and basin-scale denudation rate estimation by *in situ* cosmogenic nuclides: an example from the Abukuma Mountains, Japan

Atsunori Nakamura[1,2], Yusuke Yokoyama[1,2,3*], Kazuyo Shiroya[4], Yosuke Miyairi[1] and Hiroyuki Matsuzaki[5]

Abstract

Accurate determination of denudation rates is important in understanding Earth surface system dynamics. *In situ* produced cosmogenic nuclides, such as ^{10}Be and ^{26}Al, provide a valuable insight but are not entirely free from complications. In this study, denudation rates are determined for the Abukuma Mountains, Japan, using both site-specific and basin-scale methods. Considered with density measured in the field, distinct and systematic differences between the two methods are identified. Site-specific rates calculated from depth profiles of cosmogenic nuclides (^{10}Be and ^{26}Al) at topographic highs indicate a rate of 67 to 85 mm/kyr, whereas basin-scale averaged denudation rates derived from the concentration of cosmogenic nuclides in fluvial sediments show 114 to 180 mm/kyr. This is the first comparison of these two commonly used methods in the same region in Japan, where the entire study area is characterized by well-developed saprolite. These results indicate that differential denudation rates between topographic highs and valleys reflect increasing local topographic relief of the study area. Comparison between rates derived from depth profiles and those applicable to the entire basin is important for understanding landscape development.

Keywords: Denudation rate; Erosion rate; *In situ* cosmogenic nuclide; ^{10}Be and ^{26}Al; Depth profile; Density

Background

Site-specific and basin-scale denudation rates

Quantifying denudation rates is important across a diverse range of disciplines in geology, geomorphology, and paleoclimatology (Yokoyama et al. 2005; Yokoyama and Esat 2011). Denudation is often associated with sediment generation, which exposes fresh rock to CO_2-consuming weathering reactions, and with creating soil, changing landforms, and moving mass from continents to oceans (Berner et al. 1983; Matsumoto and Yokoyama 2013). While numerous studies underline the significance and complexity of relationships between denudation, tectonics, and climatic processes (e.g., Granger and Riebe 2007 and references therein; Shiroya et al. 2013), determination of long-term denudation rates over the timescales of soil formation and transport has, until recently, remained difficult. *In situ* cosmogenic nuclide concentrations such as ^{10}Be and ^{26}Al are directly related to denudation rate, allowing quantification on long-term timescales (Gosse and Phillips 2001). Initial application of cosmogenic nuclides to denudation rate determination was limited to areas of widely distributed, exposed rock, often at high latitudes or in arid environments (e.g., Nishiizumi et al. 1986; Cockburn et al. 1999). Later work employed these methods in more humid regions, where soil covers much of the land surface (e.g., Heimsath et al. 2001a).

There are two common methods for determining denudation rates using *in situ* cosmogenic nuclides in humid regions. One relies on fluvial sediments to obtain basin-scale averaged rates (Brown et al. 1995; Bierman and Steig 1996; Granger et al. 1996), and the other uses depth profiles obtained from topographic highs to determine site-specific rates (Siame et al. 2004; Matsushi et al. 2006). Combination of these two methods has the

* Correspondence: yokoyama@aori.u-tokyo.ac.jp
[1]Atmosphere and Ocean Research Institute, University of Tokyo, 5-1-5 Kashiwanoha, Kashiwa, Chiba 277-8564, Japan
[2]Department of Earth and Planetary Science, Graduate School of Science, University of Tokyo, 7-3-1 Hongo, Bunkyo-ku, Tokyo 113-0033, Japan
Full list of author information is available at the end of the article

potential for providing greater insight into the complex surface processes and landscape development in any particular region (Brown et al. 1995). In this study, we present a direct comparison between denudation rates determined from these two methods in the same lithological region, where the entire study area is characterized by well-developed saprolite.

The regional topographic relief increases if denudation rates on topographic highs are less than basin-scale averaged rates. In contrast, a reduction in relief occurs if the rates on topographic highs exceed basin-scale rates. Comparison of denudation rates from soil-mantled upland obtained from single samples, rather than a depth profile, with basin-scale averaged rates has been previously performed in the Oregon Coast Range (Heimsath et al. 2001a), Australian southeastern highlands (Heimsath et al. 2001b, 2006), and German Black Forest (Meyer et al. 2010). Comparison between summit bedrock and basin-averaged denudation rates has also been performed in the Negev Desert (Clapp et al. 2000), southern Africa (Bierman and Caffee 2001), and the Appalachians (Hancock and Kirwan 2007; Portenga et al. 2013). The advantage of a depth profile of *in situ* cosmogenic nuclides is that it has the potential to provide information about equilibrium conditions, which cannot be obtained from a single sample.

Denudation rates from *in situ* cosmogenic nuclides

Measuring *in situ* cosmogenic nuclides in river-born sediments allows investigation of denudation rates on drainage basin scales (Brown et al. 1995; Bierman and Steig 1996; Granger et al. 1996; Riebe et al. 2000; Moon et al. 2011; Siame et al. 2011). If exposure to cosmic rays is sufficiently long with respect to the denudation rate, the accumulation of *in situ*-produced ^{10}Be and ^{26}Al reaches a secular equilibrium, providing information on the maximum denudation rate (Lal 1991). Because rivers carry and mix the sediment produced within a given drainage basin, denudation rates deduced from river-born quartz were identified as ideal material for quantifying denudation rates of an entire watershed (Brown et al. 1995; Bierman and Steig 1996; Granger et al. 1996).

Although the principle is simple, the culmination of a series of assumptions and approximations may compound some uncertainty. Stream sediment is not always representative of the entire watershed, especially within smaller watersheds prone to episodic mass wasting (Niemi et al. 2005; Granger and Riebe 2007). Storage and burial of river sediments would also disturb the initial concentration of cosmogenic nuclides by radioactive decay, which creates a case where the measured nuclide concentration underestimates the steady-state nuclide concentration (Wittmann et al. 2011). This methodology has been tested for small (1 to 10 km^2; Granger et al. 1996) and large (10^2 to 10^5 km^2; Schaller et al. 2001) drainage basins, as well as in different geologic and climatic settings (Portenga and Bierman 2011 and references therein). Many regional-scale cosmogenic studies indicate that environmental parameters, such as precipitation, temperature, altitude, basin slope, and tectonics, can influence denudation rates (Riebe et al. 2000; von Blanckenburg 2005; Henck et al. 2011; Moon et al. 2011, Siame et al. 2011). However, quantitative assessment of the contributions from independent environmental parameters is still under debate, due in part to the lack of direct comparison between basin scale and site-specific denudation rates.

Depth profiles of *in situ* cosmogenic nuclides allow determination of site-specific denudation rates in humid regions without the complex assumptions required for basin-scale studies (Siame et al. 2004; Matsushi et al. 2006). Topographic highs such as mountain summits, hilltops, and ridges are appropriate sites for depth profiling due to the minimized influence of along-slope mass transport (Shiroya et al. 2010). Another advantage stemming from the usage of depth profiles is that information regarding nuclide equilibrium is also obtained (see the following). Cumulative concentration of a particular nuclide at an outcrop is a function of the denudation rate (D) and exposure age (T), so that measurement of one nuclide (typically ^{10}Be) at the surface cannot uniquely identify D and T. The use of depth profiles has recently been introduced to alleviate this problem (Siame et al. 2004; Braucher et al. 2009). Depth profiles have been used to determine the ages of fluvial terraces (Hein et al. 2009; Hidy et al. 2010; Rixhon et al. 2011).

In this study, we present denudation rates for a granitic saprolite in the Abukuma Mountains, Japan, using *in situ* cosmogenic nuclides measured in both depth profile and fluvial sediments. A known saprolite density is required to determine the denudation rate using cosmogenic nuclides. However, previous studies have often treated density as a free parameter (Siame et al. 2004; Rixhon et al. 2011). Therefore, we also conduct actual field measurements of the saprolite density. Together with ^{26}Al/^{10}Be plots of fluvial sediment, these data indicate that there are distinct variations in denudation rates obtained from the topographic highs and from basin-scale approaches.

Study area and sampling strategy

The Abukuma Mountains, a major granitic terrain in Japan, formed as an uplifted peneplain (e.g., Koike 1968) (Figure 1). The mountains consist of low-relief hills of granitic saprolite at an elevation of 400 to 600 m above sea level covered by soil and vegetation. Freshly exposed bedrock and gully development are not observed. Because the mountains are not steep, landslides are also

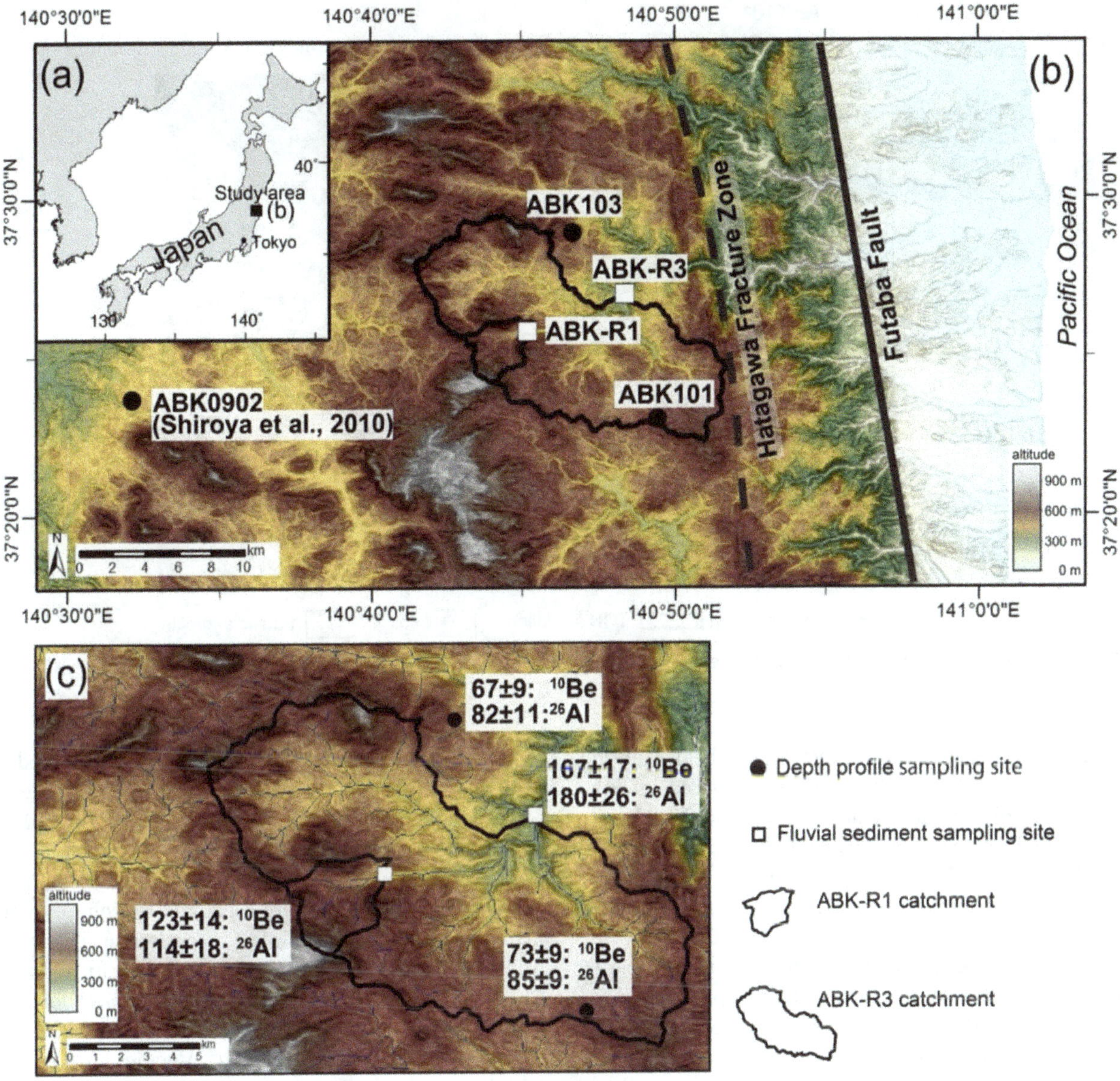

Figure 1 The Abukuma Mountains, a major granitic terrain in Japan, formed as an uplifted peneplain. (a) Location of the study area. **(b)** Digital elevation map of the sampling sites. **(c)** Close-up of the catchments. Elevation data is obtained from ASTER GDEM (http://gdem.ersdac.jspacesystems.or.jp/). Black dots indicate the sampling sites for depth profiles. Open squares indicate the sampling sites for fluvial sediments. Catchments of the sites are also indicated as solid lines. Numbers described are denudation rates (mm/kyr) obtained from this study.

not reported in this region (Shimizu and Oyagi 1988). The Abukuma granitic terrain is petrologically separated by the N-S trending Hatagawa fracture zone and the Futaba fault (Figure 1). Recent activity of a west-dipping reverse fault with a dip slip rate of 50 to 100 mm/kyr has been reported for Futaba fault (Fukushimaken 1999). Annual mean precipitation is 1,465 mm with a peak in September, and annual mean temperature is 10.3°C (calculated from 1981 to 2010 at Kawauchi, Japan; http://www.jma.go.jp/jma/indexe.html).

Depth profile sampling sites, ABK101 (N 37.384, E 140.824) and ABK103 (N 37.482, E 140.778), are roadside outcrops covered by approximately 20 cm of black soil and are located near the topographic highs in the eastern part of the Abukuma Mountains at an altitude of 620 and 540 m, respectively (Figures 1 and 2, Table 1). These are rounded, gently sloping, and deeply weathered (>3 m deep), which indicates no significant influence from catastrophic mass wasting. The saprolite exhibits original granitic textures, also indicating that no significant disturbance or mixing has occurred during *in situ* weathering. The entire study area is characterized by soil of relatively uniform thickness that is underlain by well-developed saprolite to appreciable depth. Samples for cosmogenic

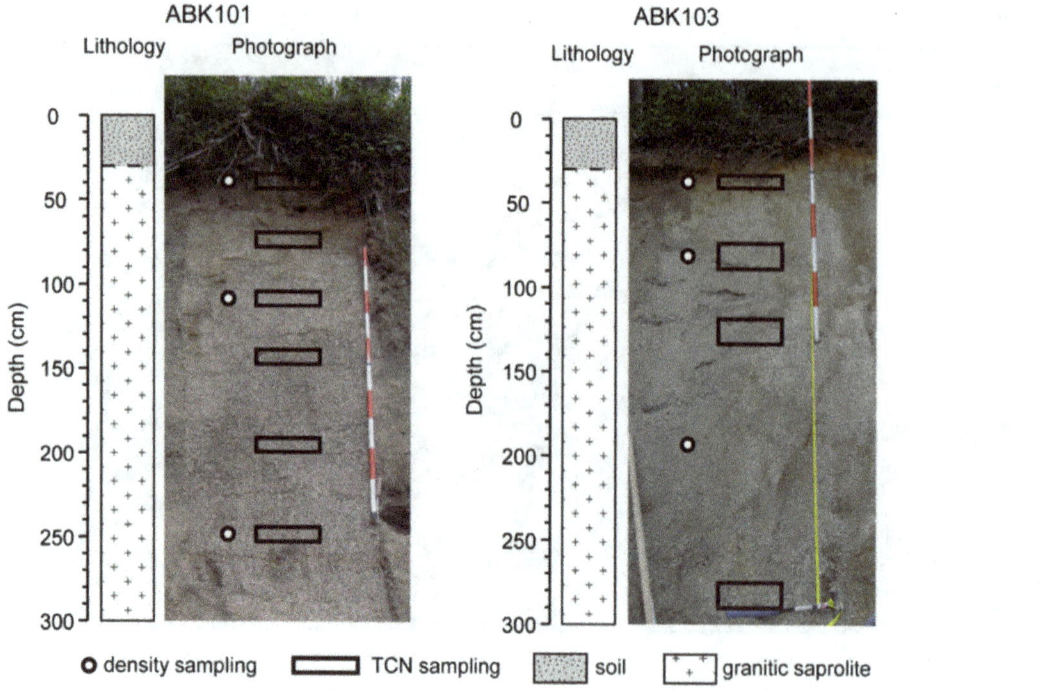

Figure 2 Columnar sections and photographic views of sampling sites.

nuclide measurement were collected from below the soil horizon. Plastic tubes (5 cm long) were placed into the granitic saprolite to determine its wet and dry density (Figure 3). Here, we use mean wet bulk density at the sampling site for calculation because granitic saprolite contains water, in contrast to fresh rock. ABK-R1 (N 37.431, E 140.754) and ABK-R3 (N 37.450, E 140.807) are the fluvial sites, located in the Takase River (Figure 1, Table 1). The catchment areas of the fluvial sampling sites are between ABK101 and ABK103 and do not include the Hatagawa fracture zone.

Methods

Samples for depth profiles were crushed to a grain size of 0.35 to 1 mm, and quartz grains were subsequently separated by chemical etching following the method of Kohl and Nishiizumi (1992). River sediments were sieved to 0.35 to 1 mm size fractions for analysis. The purified quartz samples (approximately 25 g) were dissolved with HF and spiked with Be carrier (approximately 0.3 mg). Aliquots for Be and Al were separated by column chromatography and then oxidized for accelerator mass spectrometry (AMS) measurements at the University of Tokyo. The standards used were 07KNSTD (KNB5-1) for ^{10}Be (^{10}Be/^{9}Be ratio 2.709×10^{-11}; Nishiizumi et al. 2007) and KNSTD (KNA4-2) for ^{26}Al (^{26}Al/^{27}Al ratio 3.029×10^{-11}; Nishiizumi 2004). The decrease in cosmogenic nuclide concentrations with depth has previously been modeled using a chi-square (χ^2) inversion method (Siame et al. 2004; Braucher et al. 2009). For a surface experiencing

Table 1 Site parameters and geometric properties of the drainage basins sampled for cosmogenic measurements

Site	^{10}Be production rate (atoms/g/yr)			^{26}Al production rate (atoms/g/yr)			Latitude °N	Longitude °E	Altitude (m)	Mean basin altitude (m)[c]	Mean slope (deg)
	Spallation[a]	Slow muon[b]	Fast muon[b]	Spallation[a]	Slow muon[b]	Fast muon[b]					
ABK101	6.90	0.016	0.045	46.64	1.1	0.093	37.384	140.824	620	-	-
ABK103	6.49	0.015	0.044	43.82	1.1	0.092	37.482	140.778	540	-	-
ABK-R1	7.07	0.016	0.045	47.77	1.1	0.094	37.431	140.754	463	649	13.9
ABK-R3	6.58	0.015	0.044	44.43	1.1	0.092	37.450	140.807	339	551	12.0

[a]Production rates by high-energy spallation are calculated following Stone (2000) for altitudinal and latitudinal scaling. This is the most commonly used scaling scheme in the existing literature. The scaling factor is a function of atmospheric pressure and geographic latitude; [b]production rates by slow and fast muons are calculated according to Braucher et al. (2011); [c]mean altitudes of the catchments are calculated by ArcGIS for production rates of fluvial sediments.

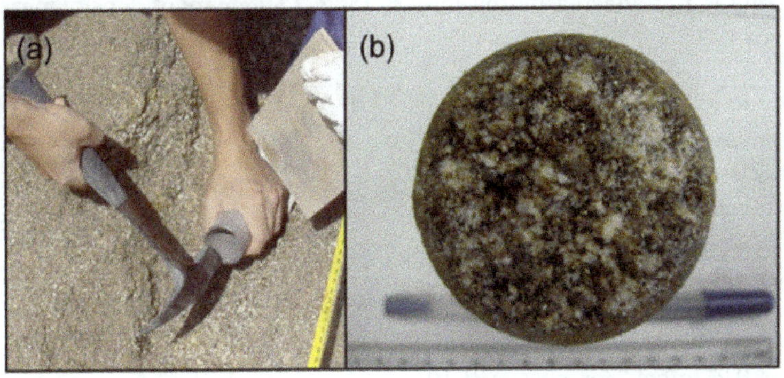

Figure 3 The sampled saprolite. (a) Plastic tubes were placed into the granitic saprolite. **(b)** The sampled saprolite shows the original granite textures, indicating that no disturbance occurred during sampling.

denudation at rate ε (cm/year) during a time interval dt, the concentration of cosmogenic nuclides C (atoms/g) with time t (year), and depth x (cm) is described as

$$\frac{\partial C}{\partial t} = P_0 e^{\frac{-\rho x}{\Lambda}} + \varepsilon \frac{\partial C}{\partial x} - C\lambda \tag{1}$$

where P_0 (atoms/g/year) is the surface production rate, Λ (g/cm^2) is the attenuation length, ρ (g/cm^3) is the density, and λ is the radioactive decay constant (year^{-1}) for ^{10}Be (5.10×10^{-7}) (Nishiizumi et al. 2007) and ^{26}Al (9.83×10^{-7}) (Nishiizumi 2004). Considering the three reaction mechanisms (neutrons, stopping, and fast muons) and assuming constant rates of denudation, Equation 1 is solved as follows using the initial condition at time $t = 0$ that $C = 0$.

$$\begin{aligned} C(x,t) = & \frac{P_{spal}}{\frac{\rho\varepsilon}{\Lambda_{spal}} + \lambda} e^{\frac{-\rho x}{\Lambda_{spal}}} \left(1 - e^{-\left(\lambda + \frac{\rho\varepsilon}{\Lambda_{spal}}\right)t}\right) \\ & + \frac{P_{stop}}{\frac{\rho\varepsilon}{\Lambda_{stop}} + \lambda} e^{\frac{-\rho x}{\Lambda_{stop}}} \left(1 - e^{-\left(\lambda + \frac{\rho\varepsilon}{\Lambda_{stop}}\right)t}\right) \\ & + \frac{P_{fast}}{\frac{\rho\varepsilon}{\Lambda_{fast}} + \lambda} e^{\frac{-\rho x}{\Lambda_{fast}}} \left(1 - e^{-\left(\lambda + \frac{\rho\varepsilon}{\Lambda_{fast}}\right)t}\right) \end{aligned} \tag{2}$$

Here, P_{spal}, P_{stop}, and P_{fast} are surface production rates for the three production mechanisms, and Λ_{spal}, Λ_{stop}, and Λ_{fast} are the effective attenuation lengths, respectively, of those three rates. Production rates for neutrons are scaled using the method of Stone (2000) based on the reference ^{10}Be production rate of 4.49 (atoms/g/year) with ^{26}Al/^{10}Be = 6.75 (Balco et al. 2008). Production rates for muons are calculated following Braucher et al. (2011). This study uses the values: $\Lambda_{spal} = 160$ (g/cm^2), (Lal 1991; Brown et al., 1995), $\Lambda_{stop} = 1{,}500$ (g/cm^2), $\Lambda_{fast} = 4{,}320$ (g/cm^2) (Heisinger et al. 2002a, b; Braucher et al. 2011).

The χ^2 inversion model (Siame et al. 2004) for best fitting of the model is as follows:

$$\chi^2 = \sum_{i=1}^{N} \left(\frac{C_i - C(x_i, \varepsilon, t)}{\sigma_i} \right)^2 \tag{3}$$

Here, C_i and σ_i are the measured cosmogenic nuclide concentration and its standard deviation at the depth x_i, respectively, and $C(x_i, \varepsilon, t)$ is the theoretical concentration computed using Equation 3 with a given time-denudation (ε, t) pair. A nuclide concentration at a variety of depths below the surface can provide both D and T information. Not only does this approach allow determination of ages (Hein et al. 2009; Hidy et al. 2010; Rixhon et al. 2011) but also allows testing of the equilibrium condition. On topographic highs, where cosmic ray exposure is infinitely long, exposure time calculated by the χ^2 method is expected to indicate equilibrium. If mass movement has significantly influenced the outcrop, χ^2 inversion would exhibit lower, nonequilibrium values. Therefore, this method is useful for confirming the existence of the equilibrium condition, which is required to accurately calculate denudation rates. Uncertainties of denudation rates include both uncertainty of density and uncertainty of χ^2 inversion, determined as $\chi^2 \le \chi^2_{min} + 1$ following the method of Rixhon et al. (2011).

Basin-scale denudation rates from river-born quartz are also determined following the method of Granger et al. (1996), where the terms are the same as those listed previously.

$$\varepsilon = \frac{P_{spal}\Lambda_{spal} + P_{stop}\Lambda_{stop} + P_{fast}\Lambda_{fast}}{C\rho} \tag{4}$$

Mean altitudes of the catchments calculated by ArcGIS with ASTER GDEM elevation data (http://www.gdem.aster.ersdac.or.jp/) are used for production rates of fluvial sediments (Table 1). The effects of topographic shielding are insignificant and therefore not included because the production rate is not affected by the low-relief geometry of the study area.

Results

χ^2 fitting for the depth profiles

Measured wet bulk densities ranged between 1.59 and 1.68 g/cm^3 with no systematic depth trend (Figure 4, Table 2). Similar stability of the dry bulk density indicates constant water content. Therefore, we used the average wet bulk density of each sampling site for calculation of cosmogenic nuclide depth profiles. The values are 1.66 ± 0.07 and 1.61 ± 0.07 g/cm^3 for sites ABK101 and ABK103, respectively.

Figure 5 shows that measured ^{10}Be and ^{26}Al concentrations decrease exponentially with depth (Table 3). Low χ^2 values are distributed parallel to the axes of exposure age (Figure 6). Therefore, χ^2 fitting indicates that steady-state concentrations have been reached. Denudation rates at site ABK101 are 73 ± 9 and 85 ± 9 mm/kyr for ^{10}Be and ^{26}Al, respectively, whereas 67 ± 9 and 82 ± 11 mm/kyr were measured for ^{10}Be and ^{26}Al, respectively, at site ABK103 (Figures 5 and 6). The uncertainty reflects uncertainties in both measurements of density and χ^2 inversion. Although the modeled curves are not perfectly fit to the measured concentration, two independent lines of evidence indicate that the derived denudation rates are reliable. First, rates calculated from both ^{10}Be and ^{26}Al are consistent, and second, both sampling sites exhibit almost identical denudation rates. Thus, we interpret that the cosmogenic nuclides data accurately represent local denudation rates. It should be noted that relatively large χ^2 values are observed for ^{10}Be profiles compared with ^{26}Al profiles at site ABK103 (Figure 6). This might due to relatively high ^{10}Be/^{9}Be ratios for the process blanks (Table 3) compared with those used in recent publications (e.g., Larsen et al. 2014).

Table 2 Summary of density measurements

Sample	Depth (m)	Wet density (g/cm^3)	1σ error	Dry density (g/cm^3)	1σ error
ABK101-8d	39.0	1.64	0.07	1.41	0.06
ABK101-12d	108.3	1.68	0.07	1.47	0.06
ABK101-20d	248.5	1.65	0.07	1.43	0.06
ABK103-1d	37.2	1.59	0.07	1.43	0.06
ABK103-3d	81.7	1.61	0.07	1.45	0.06
ABK103-8d	193.2	1.62	0.07	1.45	0.06

Denudation rates from fluvial sediments

Basin-scale denudation rates are calculated from the concentration of cosmogenic nuclides using Equation 4. For river sediments, we used 1.63 ± 0.07 g/cm^3 as the density of the basin, which is the average value of sites ABK101 and ABK103. This assumption is justified because these two values are consistent within their uncertainties. Furthermore, well-developed saprolite is widespread and representative at the scale of the entire catchment, which allows us to calculate the denudation rates for the basins using the measured density at the sites of the depth profiles. The catchment-averaged denudation rate of site ABK-R1 is 123 ± 14 and 114 ± 18 mm/kyr, as obtained from ^{10}Be and ^{26}Al, respectively. ABK-R3 exhibits rates of 167 ± 17 and 180 ± 26 mm/kyr from ^{10}Be and ^{26}Al, respectively. In most previous studies, only a single nuclide, typically ^{10}Be, is measured for determination

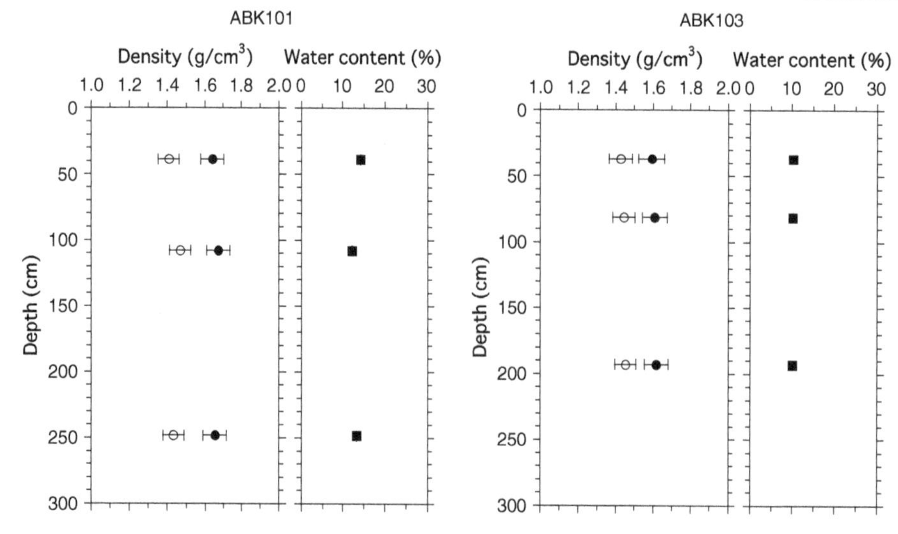

Figure 4 Dry density (open circles), wet density (filled circles), and water content (squares) of the samples.

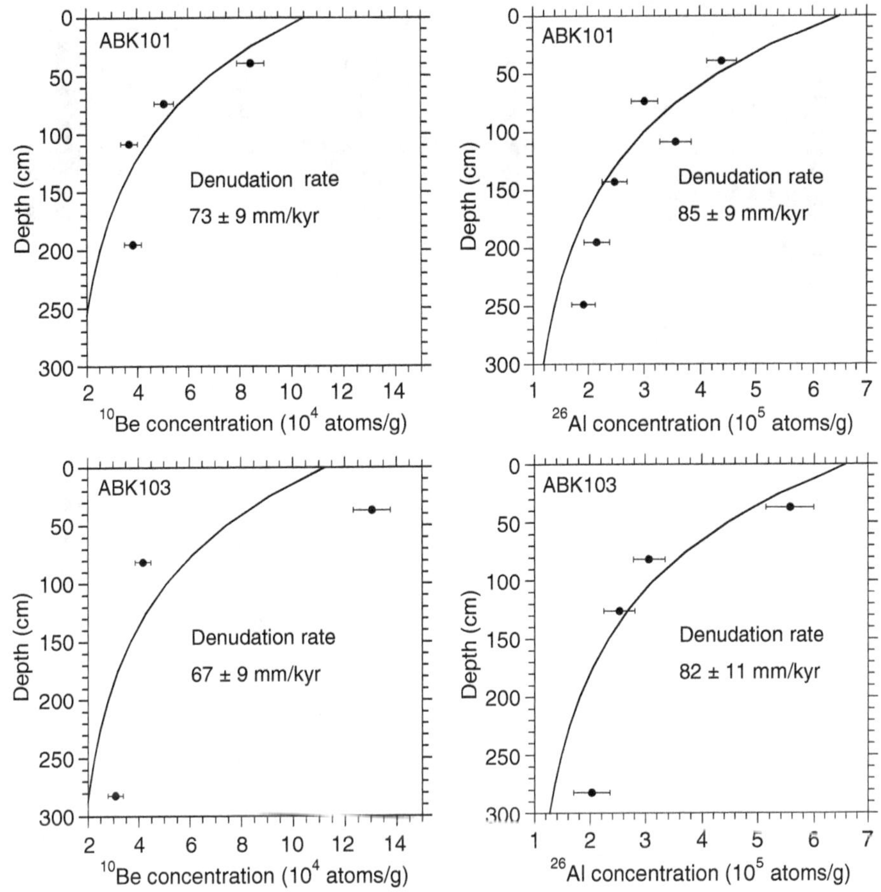

Figure 5 Depth profile of cosmogenic nuclides (circles) with model-best-fit (curve) concentrations based on steady denudation (Equation 2). Error bars show analytical uncertainties (1σ).

Table 3 Summary of sample data from the Abukuma Mountains

Sample	Depth (cm)	Dissolved quartz (mg)	$^{10}Be/^{9}Be$ (10^{-14})	1σ error (10^{-14})	^{10}Be concentration (10^4 atoms/g)	1σ error (10^4 atoms/g)	$^{26}Al/^{27}Al$ (10^{-13})	1σ error (10^{-13})	^{26}Al concentration (10^5 atoms/g)	1σ error (10^5 atoms/g)
ABK101-8	39.0	25,028.70	10.63	0.66	8.43	0.53	2.37	0.14	4.38	0.26
ABK101-10	73.6	25,020.16	6.35	0.48	5.05	0.38	1.73	0.14	3.02	0.24
ABK101-12	108.3	24,973.98	4.59	0.41	3.68	0.33	1.70	0.13	3.57	0.28
ABK101-14	142.9	25,009.59	-	-	-	-	1.32	0.12	2.47	0.23
ABK101-17	194.9	25,012.77	4.81	0.41	3.82	0.33	1.09	0.12	2.15	0.23
ABK101-20	248.5	24,978.36	-	-	-	-	0.99	0.11	1.91	0.21
ABK103-1	37.2	25,007.06	16.63	0.92	13.05	0.72	2.40	0.18	5.57	0.42
ABK103-3	81.7	25,003.43	5.24	0.39	4.17	0.31	1.34	0.12	3.07	0.28
ABK103-5	126.3	24,997.33	-	-	-	-	1.29	0.15	2.53	0.29
ABK103-12	282.4	22,166.50	3.44	0.33	3.08	0.29	0.87	0.14	2.03	0.33
ABK-R1	-	29,839.39	10.42	0.53	6.96	0.36	2.98	0.30	5.29	0.52
ABK-R3	-	24,123.85	5.83	0.26	4.80	0.22	2.01	0.17	3.16	0.26

Background values of ^{10}Be and ^{26}Al are subtracted using the value of each process blank, which is treated in the same batch. ^{10}Be ratios of process blanks for depth profile and river sediment are $2.63 \pm 0.20 \times 10^{-14}$ and $2.13 \pm 0.12 \times 10^{-14}$, respectively. These values are indistinguishable from the value of $^{10}Be/^{9}Be$ of added commercial 100 ppm Be carrier (Wako Co., Ltd.; 2×10^{-14}), and the long-term average value of the process blank (2×10^{-14}). ^{26}Al ratios of process blanks for depth profile and river sediment are $5.28 \pm 0.74 \times 10^{-15}$ and $7.12 \pm 7.12 \times 10^{-16}$, respectively. Counting errors (1σ) of standards and process blanks measured with AMS are propagated to the counting uncertainty on the sample ratios. We do not have Be data for ABK101-14, −20, and ABK103-5 because sample treatment was unsuccessful.

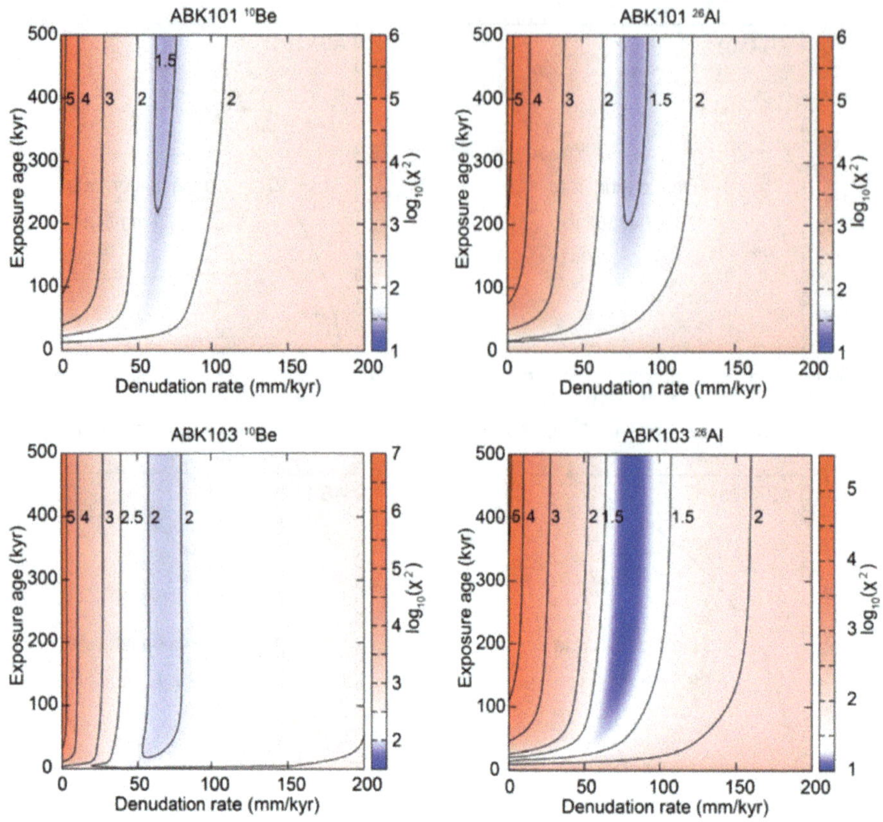

Figure 6 Contour plots of the log(χ^2) based on Equation 2. Blue areas indicate where equilibrium conditions are reasonable.

of catchment-averaged denudation rate. However, denudation rates derived from a single nuclide may be biased by complex exposure histories (Wittmann et al. 2011). In order to test the steady-state denudation model, the data are plotted on an 'erosion-island' graph (Lal 1991) (Figure 7). The plots show close agreement between denudation rates based on ^{10}Be and ^{26}Al, within error limits, inside the 'erosion-island' area, supporting that these samples do not record complex exposure histories. Although complex histories of sediment transportation cannot be ruled out completely because of the large uncertainties, the combination of ^{10}Be and ^{26}Al results indicates negligible Myr-scale sediment burial. The large uncertainties are more likely the result of low cosmogenic nuclide concentration, resulting from the fast denudation rate taking place in low elevation area, which leads low nuclide production.

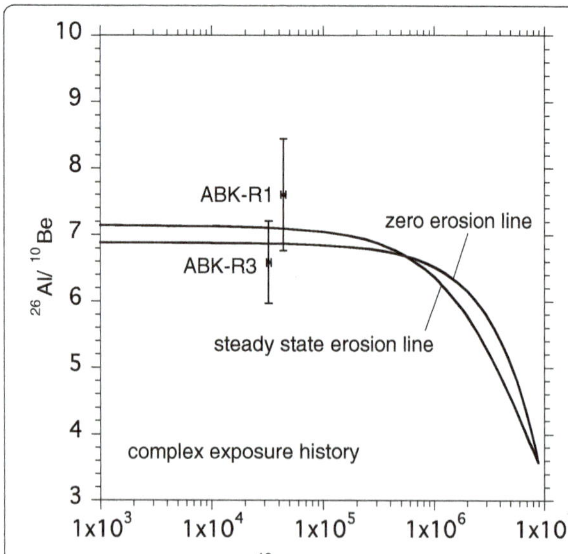

Figure 7 Plot of ^{26}Al/^{10}Be ratios versus ^{10}Be concentration of fluvial samples with 1σ errors. ^{10}Be concentrations are normalized for values at sea level and high latitude. Zero denudation and steady-state denudation lines are shown as solid lines. ^{26}Al/^{10}Be ratios indicate that these samples do not have complex exposure histories detected by the ^{26}Al/^{10}Be system.

Discussion

Density dependence of nuclide concentration profiles

Denudation rates calculated from Equations 2 and 4 are highly sensitive to density variation, with large changes resulting from reasonable density adjustments. For example, substitution of 2.00 g/cm^3 for the measured value of 1.66 g/cm^3 changes the ^{10}Be denudation rate of ABK101 from 73 to 54 mm/kyr, a 25% change resulting from only a 0.34 g/cm^3 shift in density. This underscores the importance of accurate density measurements, as has been previously reported by several authors.

Braucher et al. (2009) suggested that the density used by Hidy et al. (2005) was too high and advocated consideration of density as a free parameter for denudation and time. Hidy et al. (2010) measured density by inserting a pliable plastic bag into the sampling hole and filling it with a measured quantity of water to determine the volume of the excavated mass of rock. They state that density measured in the field should be considered a maximum value because excavation of a very coarse boulder to gravel-sized fillings may lead to systematic error. They therefore allowed the density to vary randomly using Monte Carlo simulations. Rixhon et al. (2011) treated density as a free parameter, adjusting it within a 1.8 to 2.4 g/cm^3 range.

Although these studies have attempted to identify density variations, accurate measurement of density for terrace sediments is complicated by the presence of boulders. In contrast, we successfully measured density by employing the newly developed technique of using constant volume plastic tubes for granitic saprolite. Time-dependent variation in density (Rodés et al. 2011) is negligible in the study area because of the homogeneous distribution of granitic saprolite with little differential weathering. Measured density exhibits a constant value with depth, indicating that chemical weathering gradients are not significant for the time scale of interest. In addition, granitic saprolite is widely and homogenously distributed at scales larger than outcrop size (>3 m deep, >10 m wide). Therefore, denudation rates determined in our study are indicative of those following saprolite formation. Note that the density of soil above the saprolite was not measured because plant roots blocked the sampling tube. Nevertheless, density of soil does not affect our interpretation because its thickness is only 20 cm and its density, typically regarded as 1.4–1.8 g/cm^3 (Siame et al. 2011; Heimsath et al. 2012), is indistinguishable from that of the saprolite.

We reanalyzed the denudation rates reported by Shiroya et al. (2010) in the western part of the Abukuma Mountains using actual measured density. They determined denudation rates from the depth profile of a quartz vein in granitic saprolite along a mountain ridge assuming 2.2 g/cm^3 for the density of wet granitic saprolite. The reported denudation rates were 61 + 6/−4 and 62 + 4/−3 mm/kyr based on ^{10}Be and ^{26}Al depth profiles, respectively. In the current study, we applied the actual density of the rock at this site (1.76 ± 0.03 g/cm^3) and parameters described in Shiroya et al. (2010) for recalculation, which resulted in rates of 86 ± 2 and 89 ± 3 mm/kyr for ^{10}Be and ^{26}Al-based determinations, respectively. These values are 1.4 times larger than the initial values reported by Shiroya et al. (2010), indicating that direct density measurements are indeed crucial for determining accurate denudation rates. These reanalyzed values are consistent with values obtained from sites ABK101 and ABK103 in the present study. Hence, the three sampling sites at topographic highs in the Abukuma Mountains show consistent values between 67 and 89 mm/kyr.

Comparison between depth profile and fluvial methods

Denudation rates deduced from analyses of fluvial sediments are 1.3 to 2.7 times larger than those determined from depth profiles ($p < 0.01$). Multiple depth profiles of cosmogenic nuclides provide the denudation rate of the summit plane, whereas fluvial sediments represent a basin-scale average. Thus, our measurements show that there are contrasts in denudation rates between topographic highs and the entire basin as a whole, primarily due to differences in surficial gradient. This supports previous studies that found that the slope angle of a basin significantly affects the overall denudation rate (Riebe et al. 2000; von Blanckenburg 2005). However, it should be noted that the mean basin slope is 13.9° and 12.0° for ABK-R1 and ABK-R3, respectively (Table 1), indicating that fluvial denudation rates are not sensitive to small differences in slope angles in the study area before the threshold for hill slope stability is reached. The slow upper stream denudation rate is also not attributed to the spatial distribution of the slope angle because the two basins exhibit a similar slope pattern (Figure 8). The

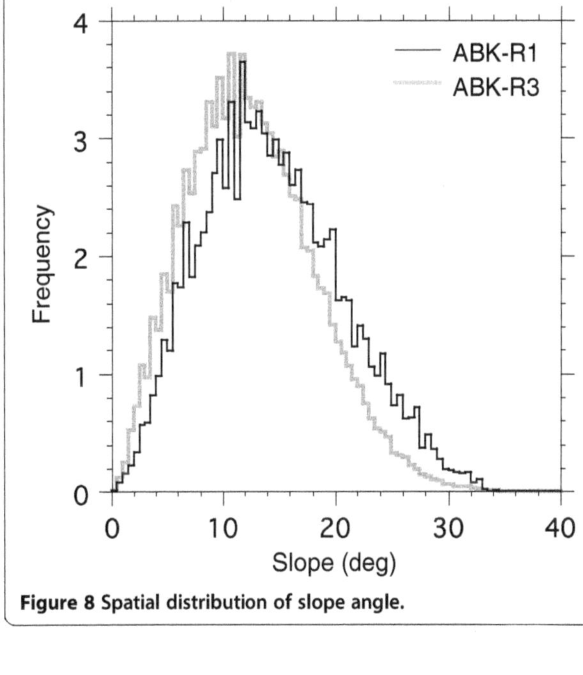

Figure 8 Spatial distribution of slope angle.

lower basin-scale denudation rate of ABK-R1 compared with ABK-R3 is likely derived from localized low denudation rates due in part to its location proximal to the ridgeline.

Subtraction of the site-specific denudation rate from the basin-scale rate (29 to 113 mm/kyr) represents the rate of change of local topographic relief. Overall, local topographic relief of the Abukuma Mountain is interpreted to be increasing. This is the first comparison of these two methods in the same lithological region in Japan, which allows us to quantitatively investigate regional landscape evolution. Yamamoto (2005) showed that fluvial incision rates, based on tephrostratigraphy, are 80 to 150 mm/kyr during the Late Quaternary according to the relative height of fluvial terraces. This is consistent with our results. Comparison between rates derived from depth profiles and those applicable to the entire basin is important for understanding landscape development. This requires careful site selection of topographic highs in order to obtain denudation rates that have minimal potential slope influence.

Conclusions

We compare denudation rates derived from cosmogenic nuclide data from the Abukuma Mountains between summits and catchments. This is the first comparison of these two methods in the same lithological region, where the entire study area is characterized by well-developed saprolite. Slope angle strongly affects the basin-scale averaged denudation rates by as much as 1.3 to 2.7 times, whereas rates derived from depth profiles at summit sites are unaffected. The importance of physical density measurements, as opposed to modeled values, to obtain accurate denudation rates is demonstrated. Denudation rates of topographic highs inferred from depth profiles of cosmogenic nuclides are 67 to 85 mm/kyr, whereas basin-scale denudation rates derived from concentration of cosmogenic nuclides in fluvial sediments indicate 114 to 180 mm/kyr.

Competing interests

The authors declare that they have no competing financial interests.

Authors' contributions

AN collected the samples, performed experiments, analyzed data, and wrote the paper. YY designed and directed the project, collected the samples, and contributed in writing the paper. KS and YM collected the samples with AN. HM conducted accelerator mass spectrometry experiments along with AN. All authors participated in the data analysis and interpretation of the results. All authors read and approved the final manuscript.

Acknowledgements

This work was supported by JSPS NEXT program GR031 and GCOE program. We thank Dr. L.L. Siame and an anonymous reviewer for their insightful reviews. We thank A. Suzuki, A. Horio, and S. Ochiai for their assistance and S.P. Obrochta for discussing the manuscript.

Author details

[1]Atmosphere and Ocean Research Institute, University of Tokyo, 5-1-5 Kashiwanoha, Kashiwa, Chiba 277-8564, Japan. [2]Department of Earth and Planetary Science, Graduate School of Science, University of Tokyo, 7-3-1 Hongo, Bunkyo-ku, Tokyo 113-0033, Japan. [3]Institute of Biogeoscience, Japan Agency of Marine Science and Technology (JAMSTEC), 2-15 Natsushima-cho, Yokosuka, Kanagawa 237-0061, Japan. [4]Geological Survey of Japan, National Institute of Advanced Industrial Science and Technology (AIST), AIST Tsukuba Central 7, 1-1-1 Higashi, Tsukuba 305-8567, Japan. [5]Department of Nuclear Engineering and Management, The University of Tokyo, Yayoi 2-11-16, Bunkyo-ku, Tokyo 113-0032, Japan.

References

Balco G, Stone JO, Lifton NA, Dunai TJ (2008) A complete and easily accessible means of calculating surface exposure ages or erosion rates from ^{10}Be and ^{26}Al measurements. Quat Geochronol 3:174–195

Berner RA, Lasaga AC, Garrels RM (1983) The carbonate-silicate geochemical cycle and its effect on atmospheric carbon-dioxide over the past 100 million years. Am J Sci 283:641–683

Bierman PR, Caffee M (2001) Slow rates of rock surface erosion and sediment production across the Namib Desert and escarpment, southern Africa. Am J Sci 301:326–358

Bierman PR, Steig EJ (1996) Estimating rates of denudation using cosmogenic isotope abundances in sediment. Earth Surf Processes Landforms 21:125–139

Braucher R, Del Castillo P, Siame L, Hidy AJ, Bourles DL (2009) Determination of both exposure time and denudation rate from an in situ-produced ^{10}Be depth profile: a mathematical proof of uniqueness. Model sensitivity and applications to natural cases. Quat Geochronol 4:56–67

Braucher R, Merchel S, Borgomano J, Bourles DL (2011) Production of cosmogenic radionuclides at great depth: A multi element approach. Earth Planet Sci Lett 309:1–9

Brown ET, Stallard RF, Larsen MC, Raisbeck GM, Yiou F (1995) Denudation rates determined from the accumulation of in situ-produced ^{10}Be in the Luquillo experimental forest, Puerto-Rico. Earth Planet Sci Lett 129:193–202

Clapp EM, Bierman PR, Schick AP, Lekach J, Enzel Y, Caffee M (2000) Sediment yield exceeds sediment production in arid region drainage basins. Geology 28:995–998

Cockburn HAP, Seidl MA, Summerfield MA (1999) Quantifying denudation rates on inselbergs in the central Namib Desert using in situ-produced cosmogenic ^{10}Be and ^{26}Al. Geology 27:399–402

Fukushimaken (1999) Futaba fault survey, earthquake-related basic research grants FY 1998 report. Doc 3826:109, Fukushima

Gosse JC, Phillips FM (2001) Terrestrial in situ cosmogenic nuclides: theory and application. Quat Sci Rev 20:1475–1560

Granger DE, Riebe CS (2007) Cosmogenic nuclides in weathering and erosion. In: Drever JI (ed) Surface and Ground Water, Weathering, Denudation and Soils, vol. 5. Treatise on Geochemistry (Holland HG, Turekian KK (exec eds). Elsevier, London, pp 1–43

Granger DE, Kirchner JW, Finkel R (1996) Spatially averaged long-term erosion rates measured from in situ-produced cosmogenic nuclides in alluvial sediment. J Geol 104:249–257

Hancock G, Kirwan M (2007) Summit erosion rates deduced from ^{10}Be: implications for relief production in the Central Appalachians. Geology 35:89–92

Heimsath AM, Dietrich WE, Nishiizumi K, Finkel RC (2001a) Stochastic processes of soil production and transport: erosion rates, topographic variation and cosmogenic nuclides in the Oregon Coast Range. Earth Surf Processes Landforms 26:531–552

Heimsath AM, Chappell J, Dietrich WE, Nishiizumi K, Finkel RC (2001b) Late quaternary erosion in southeastern Australia: a field example using cosmogenic nuclides. Quat Int 83–85:169–185

Heimsath AM, Chappell J, Finkel RC, Fifield K, Alimanovic A (2006) Escarpment erosion and landscape evolution in southeastern Australia. Geol Soc Am Spec Pap 398:173–190

Heimsath AM, DiBiase RA, Whipple KX (2012) Soil production limits and the transition to bedrock-dominated landscapes. Nat Geosci 5:210–214

Hein AS, Hulton NRJ, Dunai TJ, Schnabel C, Kaplan MR, Naylor M, Xu S (2009) Middle Pleistocene glaciation in Patagonia dated by cosmogenic-nuclide measurements on outwash gravels. Earth Planet Sci Lett 286:184–197

Heisinger B, Lal D, Jull AJT, Kubik P, Ivy-Ochs S, Neumaier S, Knie K, Lazarev V, Nolte E (2002a) Production of selected cosmogenic radionuclides by muons 1. Fast muons. Earth Planet Sci Lett 200:345 355

Heisinger B, Lal D, Jull AJT, Kubik P, Ivy-Ochs S, Knie K, Nolte E (2002b) Production of selected cosmogenic radionuclides by muons: 2. Capture of negative muons. Earth Planet Sci Lett 200:357–369

Henck AC, Huntington KW, Stone JO, Montgomery DR, Hallet B (2011) Spatial controls on erosion in the Three Rivers Region, southeastern Tibet and southwestern China. Earth Planet Sci Lett 303:71–83

Hidy AJ, Pederson JL, Cragun WS, Gosse JC (2005) Cosmogenic ^{10}Be exposure dating of Colorado river terraces at Lees Ferry, Arizona. Geol Soc Am Abstr 37–7:296

Hidy AJ, Gosse JC, Pederson JL, Mattern JP, Finkel RC (2010) A geologically constrained Monte Carlo approach to modeling exposure ages from profiles of cosmogenic nuclides: an example from Lees Ferry, Arizona. Geochem Geophys Geosyst 11:Q0AA10, doi:10.1029/2010GC003084

Kohl CP, Nishiizumi K (1992) Chemical isolation of quartz for measurement of in-situ-produced cosmogenic nuclides. Geochim Cosmochim Acta 56:3583–3587

Koike K (1968) Geomorphological development of the northern part of the Abukuma Mountains. Komazawa Geography 4–5:109–126

Lal D (1991) Cosmic ray labeling of erosion surfaces: in situ nuclide production rates and denudation models. Earth Planet Sci Lett 104:424–439

Larsen IJ, Almond PC, Eger A, Stone JO, Montgomery DR, Malcom B (2014) Rapid soil production and weathering in the Western Alps, New Zealand. Science 343:637–640

Matsumoto K, Yokoyama Y (2013) Atmospheric Δ^{14}C reduction in simulations of Atlantic overturning circulation shutdown. Global Biogeochem Cycles 27:1–9. doi:10.1002/gbc.20035

Matsushi Y, Wakasa S, Matsuzaki H, Matsukura Y (2006) Long-term denudation rates of actively uplifting hillcrests in the Boso Peninsula, Japan, estimated from depth profiling of in situ-produced cosmogenic ^{10}Be and ^{26}Al. Geomorphology 82:283–294

Meyer H, Hetzel R, Fügenschuh B, Strauss H (2010) Determining the growth rate of topographic relief using in situ-produced ^{10}Be: a case study in the Black Forest, Germany. Earth Planet Sci Lett 290:391–402

Moon S, Chamberlain CP, Blisniuk K, Levine N, Rood DH, Hilley GE (2011) Climatic control of denudation in the deglaciated landscape of the Washington Cascades. Nat Geosci 4:469–473

Niemi NA, Oskin ME, Burbank DW, Heimsath AM, Gabet EJ (2005) Effects of bedrock landslides on cosmogenically determined erosion rates. Earth Planet Sci Lett 237:480–498

Nishiizumi K (2004) Preparation of ^{26}Al AMS standards. Nucl Instrum Methods Phys Res B 223:388–392

Nishiizumi K, Lal D, Klein J, Middleton R, Arnold JR (1986) Production of ^{10}Be and ^{26}Al by cosmic rays in terrestrial quartz in situ and implications for denudation rates. Nature 319:134–136

Nishiizumi K, Imamura M, Caffee MW, Southon JR, Finkel RC, McAninch J (2007) Absolute calibration of ^{10}Be AMS standards. Nucl Instrum Methods Phys Res B 258:403–413

Portenga EW, Bierman PR (2011) Understanding Earth's eroding surface with ^{10}Be. GSA Today 21:4–10

Portenga EW, Bierman PR, Rizzo DM, Rood DH (2013) Low rates of bedrock outcrop erosion in the central Appalachian Mountains inferred from in situ ^{10}Be. Geol Soc Am Bull 125:201–215

Riebe CS, Kirchner JW, Granger DE, Finkel RC (2000) Erosional equilibrium and disequilibrium in the Sierra Nevada, inferred from cosmogenic ^{26}Al and ^{10}Be in alluvial sediment. Geology 28:803–806

Rixhon G, Braucher R, Bourles D, Siame L, Bovy B, Demoulin A (2011) Quaternary river incision in NE Ardennes (Belgium) - insights from ^{10}Be/^{26}Al dating of river terraces. Quat Geochronol 6:273–284

Rodés Á, Pallàs R, Braucher R, Moreno X, Masana E, Bourlés DL (2011) Effect of density uncertainties in cosmogenic ^{10}Be depth-profiles: dating a cemented Pleistocene alluvial fan (Carboneras Fault, SE Iberia). Quat Geochronol 6:186–194

Schaller M, von Blanckenburg F, Hovius N, Kubik PW (2001) Large-scale erosion rates from in situ-produced cosmogenic nuclides in European river sediments. Earth Planet Sci Lett 188:441–458

Shimizu F, Oyagi N (1988) Landslide map, vol 6. Technical Note No. 125, NIED, Tsukuba, Japan

Shiroya K, Yokoyama Y, Matsuzaki H (2010) Quantitative determination of long-term erosion rates of weathered granitic soil surfaces in western Abukuma, Japan using cosmogenic ^{10}Be and ^{26}Al depth profile. Geochem J 44:E23–E27

Shiroya K, Yokoyama Y, Obrochta SP, Harada N, Miyairi Y, Matsuzaki H (2013) Melting history of the patagonian ice sheet during termination I inferred from marine sediments. Geochem J 47:107–117

Siame LL, Bellier O, Braucher R, Sebrier M, Cushing M, Bourles D, Hamelin B, Baroux E, de Voogd B, Raisbeck G, Yiou F (2004) Local erosion rates versus active tectonics: cosmic ray exposure modelling in Provence (south-east France). Earth Planet Sci Lett 220:345–364

Siame LL, Angelier J, Chen RF, Godard V, Derrieux F, Bourles DL, Braucher R, Chang KJ, Chu HT, Lee JC (2011) Erosion rates in an active orogen (NE-Taiwan): a confrontation of cosmogenic measurements with river suspended loads. Quat Geochronol 6:246–260

Stone JO (2000) Air pressure and cosmogenic isotope production. J Geophys Res 105:23753–23759

von Blanckenburg F (2005) The control mechanisms of erosion and weathering at basin scale from cosmogenic nuclides in river sediment. Earth Planet Sci Lett 237:462–479

Wittmann H, von Blanckenburg F, Maurice L, Guyot JL, Kubik PW (2011) Recycling of Amazon floodplain sediment quantified by cosmogenic ^{26}Al and ^{10}Be. Geology 39:467–470

Yamamoto T (2005) The rate of fluvial incision during the Late Quaternary period in the Abukuma Mountains, northeast Japan, deduced from tephrochronology. Isl Arc 14:199–212

Yokoyama Y, Esat TM (2011) Global climate and sea level: enduring variability and rapid fluctuations over the past 150,000 years. Oceanography 24:54–69

Yokoyama Y, Aze T, Murasawa H, Matsuzaki H (2005) Terrestrial Cosmogenic Nuclides as a tool for studying earth surface processes. J Geol Soc Jpn 111:693–700

12

Geochemical approaches to the quantification of dispersed volcanic ash in marine sediment

Rachel P. Scudder[1,6*], Richard W. Murray[1], Julie C. Schindlbeck[2], Steffen Kutterolf[2], Folkmar Hauff[2], Michael B. Underwood[3], Samantha Gwizd[4], Rebecca Lauzon[5] and Claire C. McKinley[6]

Abstract

Volcanic ash has long been recognized in marine sediment, and given the prevalence of oceanic and continental arc volcanism around the globe in regard to widespread transport of ash, its presence is nearly ubiquitous. However, the presence/absence of very fine-grained ash material, and identification of its composition in particular, is challenging given its broad classification as an "aluminosilicate" component in sediment. Given this challenge, many studies of ash have focused on discrete layers (that is, layers of ash that are of millimeter-to-centimeter or greater thickness, and their respective glass shards) found in sequences at a variety of locations and timescales and how to link their presence with a number of Earth processes. The ash that has been mixed into the bulk sediment, known as dispersed ash, has been relatively unstudied, yet represents a large fraction of the total ash in a given sequence. The application of a combined geochemical and statistical technique has allowed identification of this dispersed ash as part of the original ash contribution to the sediment. In this paper, we summarize the development of these geochemical/statistical techniques and provide case studies from the quantification of dispersed ash in the Caribbean Sea, equatorial Pacific Ocean, and northwest Pacific Ocean. These geochemical studies (and their sedimentological precursors of smear slides) collectively demonstrate that local and regional arc-related ash can be an important component of sedimentary sequences throughout large regions of the ocean.

Keywords: Dispersed ash, Caribbean Sea, Equatorial Pacific Ocean, Northwest Pacific Ocean, Ash layers, Volcanic eruptions, DSDP, ODP, IODP

Introduction

In the modern era, marine scientists' interest in volcanic ash can be traced back at least to the *HMS Challenger* expedition (1872–1876). Indeed, "What is marine sediment made of?" is a question that has been asked for 100s of years, and while many aspects of the petrology, mineralogy, and geochemistry of marine sediment have been well characterized, it is remarkable how much remains unknown about some key aspects of ocean mud. In particular, the presence/absence and composition of volcanic ash is challenging given its fine grain size, broad classification as an "aluminosilicate", and common presence, especially since both oceanic and continental arc volcanism are often located physically adjacent to oceanic environments and are known to globally generate and transport widespread ash (Fig. 1). Certainly, civilization's natural curiosity about volcanism has driven interest in the marine records of ash for generations of scientists. Additionally, these studies considered how to link their presence with explosive volcanism, climate, arc evolution, biological productivity, and other geological processes, as described below. Although the term "ash" is normally used in volcanology as a grain size definition (32 μm to 2 mm, e.g., Fisher and Schmincke 1984) for explosive volcanic products (juvenile matter and lithic fragments), we here use the term "ash" to indicate a product (mostly volcanic glass) from explosive volcanism

* Correspondence: rscudder@tamu.edu
[1]Department of Earth & Environment, Boston University, Boston, MA 02215, USA
[6]Department of Oceanography, Texas A&M University, College Station, TX 77843, USA
Full list of author information is available at the end of the article

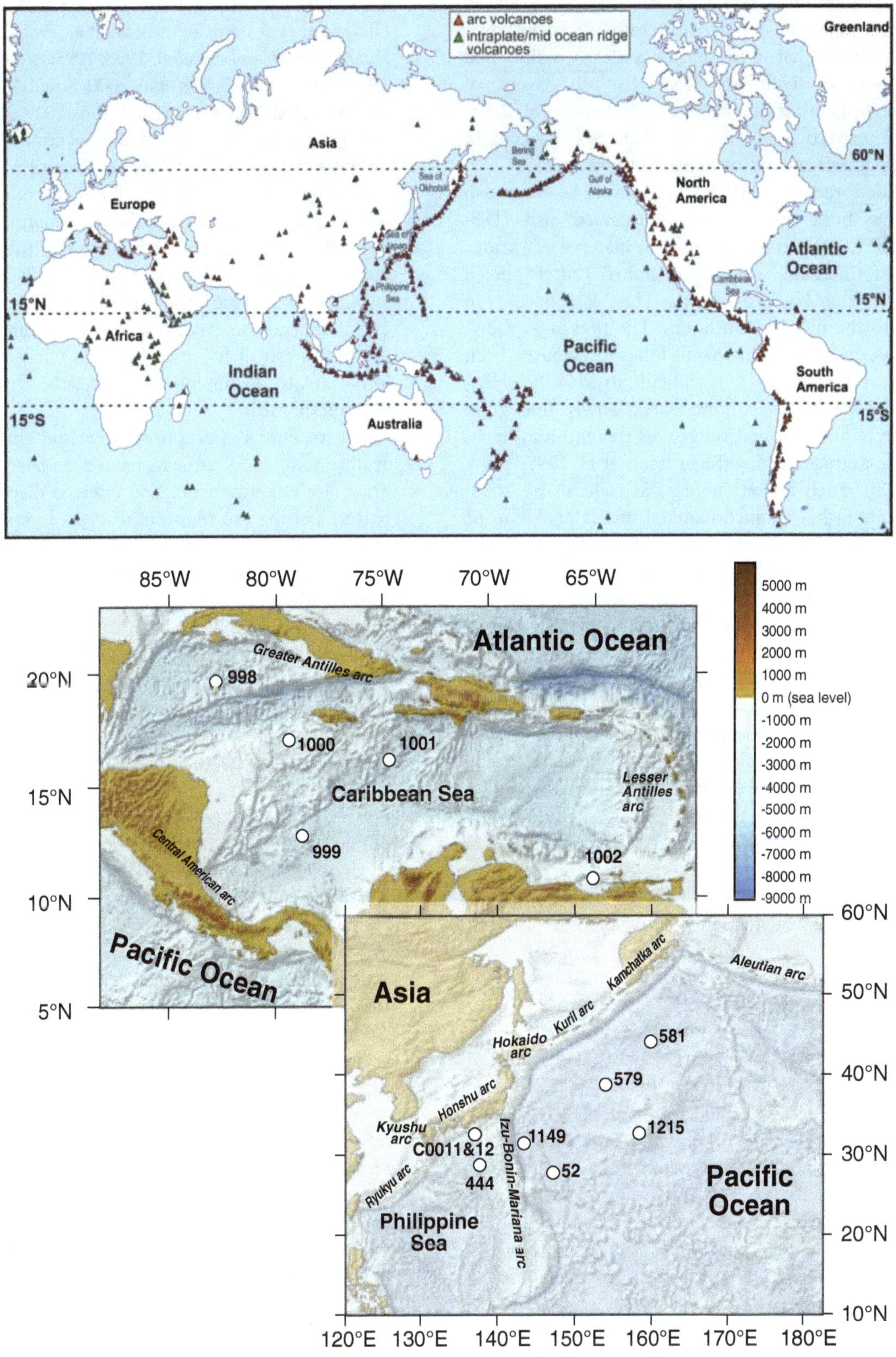

Fig. 1 (*Top*) Global distribution of subaerial volcanoes in different tectonic settings. Map has been modified after http://d-maps.com and volcano positions are taken from the Smithsonian Global Volcanism Program (http://www.volcano.si.edu). (*Bottom*) Locations of sites highlighted in this paper. Data from GeoMapApp (http://www.geomapapp.org); *GMRT* Global Multi-Resolution Topography (Ryan et al. 2009)

no matter which grain size a particle may have. Therefore, in this paper we refer to "discrete ash layers" as a volcanic product that is present as a clearly visible bed in the marine sediments (that is, ash layers of millimeter-to-centimeter or greater thickness, Fig. 2a–f) and their respective glass shards (Fig. 2g–l)) found in sequences at a variety of locations and timescales.

Less widely recognized than the discrete layers of ash is what has been referred to as "dispersed" ash. "Dispersed ash" is defined as volcanic ash material of various grain sizes (including single μm grains) that is mixed throughout the bulk sediment. This ash occurs in addition to the discrete ash layers. The presence of dispersed ash in the marine record has previously been relatively overlooked as it is difficult to identify petrographically due to its commonly extremely fine grain size (Rose et al. 2003, and references therein) and/or alteration to authigenic clay (Sigurdsson et al. 1997; Plank et al. 2000). Such alteration (e.g., Schacht et al. 2008) often results in fine-grained material that is unable to be differentiated from detrital terrigenous clay (non-ash, derived from land) on the basis of visual study alone.

Dispersed ash is the result from bioturbation of pre-existing discrete ash layers, the settling of airborne or subaqueous ash through the water column, transport by rivers and currents from terrestrially exposed ash deposits, and other processes (e.g., gravity flows). Because dispersed ash has often been only qualitatively addressed (e.g., via sedimentological smear slides), however it still plays as vital a role as discrete ash layers in many studies on multiple spatial and temporal scales, including (but not limited to) the following:

- Ash plays an important role for geochemical budgets (e.g., "Subduction Factory"), in that studies of geochemical recycling will benefit from an increased understanding of both input (subduction) and output (volcanism) fluxes (e.g., Straub 1997, 2008; Plank and Langmuir 1998; Straub and Schmincke 1998; Bryant et al. 1999, 2003; Straub and Layne 2002, 2003a,b; Hauff et al. 2003; Straub et al. 2004; Stern et al. 2006; Plank et al. 2007; Scudder et al. 2009, 2014; Tamura et al. 2009; Völker et al. 2011, 2014; Freundt et al. 2014). The amount, distribution, and composition of ash in sediment speaks to inputs as well as outputs in the budget, namely, knowing how much ash is entering the Subduction Factory is important for understanding recycling of geological material. Additionally, the ash record in the sediment itself provides an archive of local, regional, and global magmatic evolution.
- The marine repository of explosive volcanic eruptions is a vital record from which the geological history of arc evolution is recorded (e.g., Kennett et al. 1977; Ninkovich et al 1978; Sigurdsson et al. 1980, 2000; Huang 1980; Cambray et al. 1993, 1995; Lee et al. 1995; Bailey 1996; Peters et al. 2000; Straub 2003; Kutterolf et al. 2008a,b,c; Straub et al. 2009, 2010, 2015). An improved understanding of arc evolution has the potential to contribute significantly to knowledge regarding the birth, life, and death of arc magmatism. Furthermore, a full accounting of the ash record (that is, comparison of the dispersed ash record with that of the discrete layers) may provide information about arc evolution and erosional histories of landmasses. Studies of ash distribution are also vitally important to document risk associated with volcanic hazards.
- Reconstructions of eruption intensities and atmospheric wind patterns have been based on grain size characteristics and inferred dispersal patterns of the ash found in discrete layers, both at sea and on land (e.g., Huang et al. 1973, 1975; Shaw et al. 1974; Ledbetter and Sparks 1979; Carey and Sigurdsson 1980, 2000; Carey and Sparks 1986; Reid et al. 1996; Rose et al. 2003; Carey et al. 2010). These finer particles are probably the main constituent of dispersed ash, and therefore documenting the distribution of dispersed ash will constrain inferences of the transport pathways of volcanic material.
- The ability to quantify the amount of dispersed ash will increase knowledge of sedimentary physical properties (e.g., Peacock 1990; Kastner et al. 1991; Underwood and Pickering 1996; Chan and Kastner 2000; Saffer et al. 2008, 2012; Hüpers et al. 2015). Given that ash alters to clay in a hydration reaction, the amount of water bound in the sediment will increase, thus decreasing the shear strength of the mud. Additionally, the fluid budget of subducting sediment will be affected by these and, further, hydration/dehydration reactions. A quantification of the amount of dispersed ash will thus assist interpretation of physical properties of marine sediment and subduction zone modeling.
- Volcanic ash and eolian dust input to the surface ocean contributes nutrients to the surface ocean and are thus important for understanding trace metal cycling, biological productivity, climate change, volcano-climate interactions, and allied biogeochemical studies (e.g., Huang et al. 1974; Shaw et al. 1974; Carey 1997; Frogner et al. 2001; Jones and Gislason 2008; Robock et al. 2009; Duggen et al. 2010; Olgun et al. 2011; Kutterolf et al. 2013; Metzner et al. 2014).

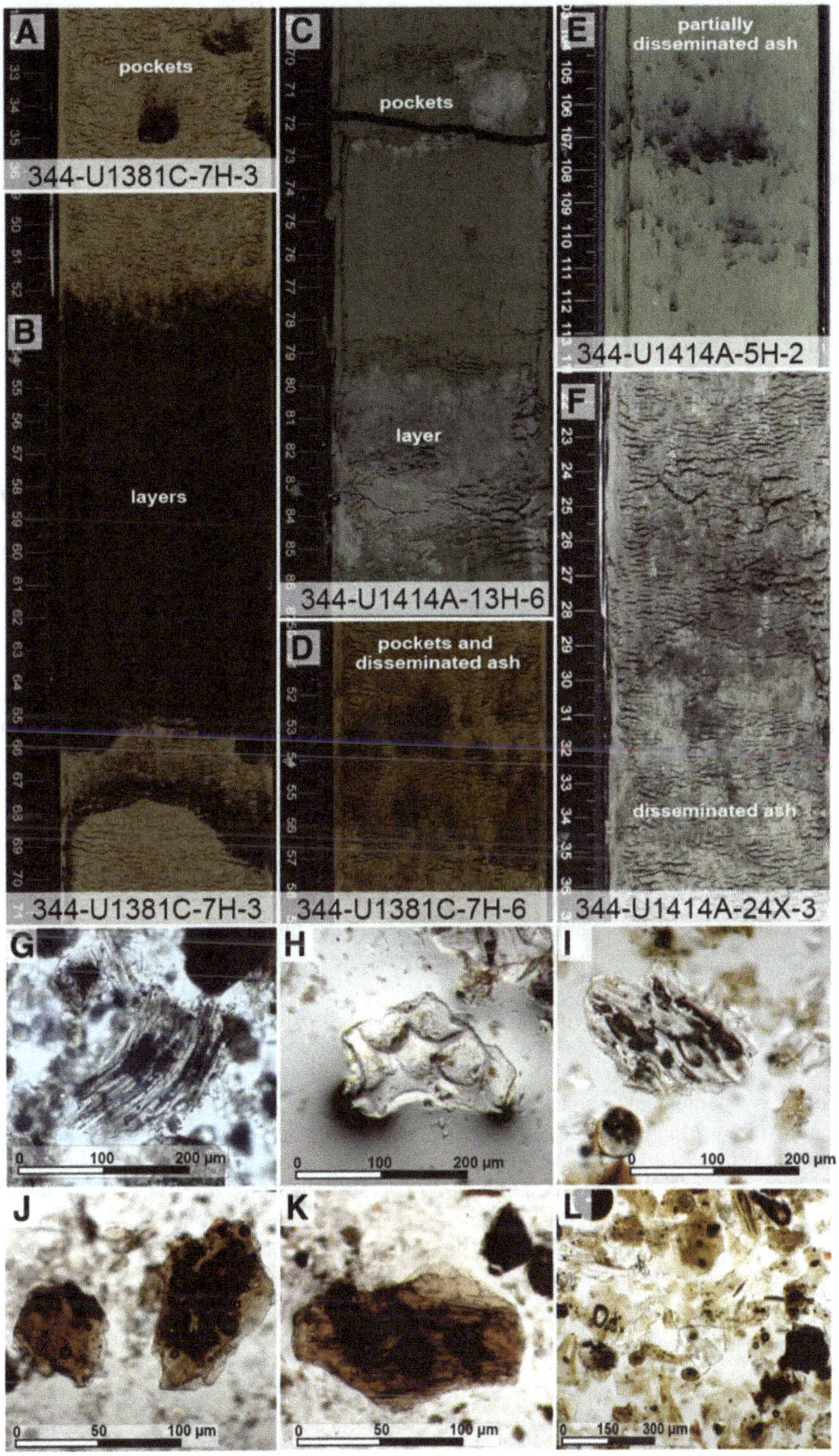

Fig. 2 (See legend on next page.)

(See figure on previous page.)

Fig. 2 Photographs of **a** mafic pockets, **b** discrete mafic ash layers, **c** felsic ash layer and pockets, **d** disseminated mafic ash pockets selected primary mafic ash layers, **e** partially disseminated ash pockets, and **f** disseminated ash. All examples from IODP expedition 344 (CRISP II) offshore Costa Rica. Example smear-slide microphotographs showing glass shard textures of felsic (G-I) and mafic (J-L) marine ash layers. **g** U1381-3H-7, 14–19 cm: felsic ash with tubular transparent to light brown shards. **h** M66-226/78 cm bsf: silicic shard from highly vesicular pumice. **i** M66-222/51 cm bsf: pumice fragment with moderately elongated vesicles. **j** M66-223/305 cm bsf: blocky sideromelane shard with and without vesicles. **k** M66-222/436-441 cm bsf: elongated vesicles in sideromelane shard. **l** M66-222/517–518 cm bsf: overview picture of different glass shards in one sample

Ash is commonly altered on and/or in the seafloor by diagenesis. In addition to the research cited above, for example, in a historically classic study, Gardner et al. (1986) observed many thin, pale green laminae (PGL) in the sediment cored during Deep Sea Drilling Project (DSDP) Leg 90 (Lord Howe Rise), which they identified as bentonites resulting from discrete ash fallout events. They found the temporal distribution of the PGLs to be similar to that of the discrete layers and logically concluded the PGLs were merely altered ash. For dispersed ash, however, the alteration increases the complexity for the application of techniques based on sedimentological and physical separation (e.g., grain size or sequential leachings to arrive at an aluminosilicate "residue"), because a fine-grained altered aluminosilicate ash is difficult to distinguish from, for example, fine-grained aluminosilicate eolian dust. The geochemical techniques summarized in this review, however, will still identify these alteration products as part of the original ash contribution to the sediment. For example, Hein et al. (1978) showed that ash beds in the Bering Sea become more altered with burial and form bentonite layers. They also documented that the chemistry of the bentonite beds reflects the chemistry of the parent ash, and differentiated authigenic smectite, which forms from the *in situ* breakdown of glass in ash layers or glass shards dispersed throughout the sediment, from detrital smectite derived from subaerial breakdown of recycled volcanic debris.

The use of tephra as a stratigraphic and chronologic tool has a long and wide-ranging history (e.g., Cambray et al. 1995; Straub and Schmincke 1998; Lowe 2011) and we do not intend here to repeat those efforts. Rather, in this paper, we (a) present a review of previous geochemical approaches to quantify dispersed ash in the Caribbean Sea and equatorial Pacific Ocean, (b) provide a summary of recent work applying a combined geochemical and multivariate statistical technique to identify dispersed ash in the northwest Pacific Ocean, and (c) describe preliminary results moving toward a regional assessment of inputs to the western Pacific Ocean (Fig. 1). We include some areas of research that are currently unresolved as they are likely to provide further insight into the nature and distribution of this important component of marine sediment. While radiogenic isotopes have proven helpful in studies of dispersed ash and the very fine detrital component (e.g., Ziegler et al. 2008; Scudder et al. 2014), in this paper we focus on major, trace, and rare earth elements since they are more commonly used in the various regions we discuss.

Review

Geochemical approaches: theory and practice

Composition is different from transport

A basic tenet of provenance research is that studies based on chemical, isotopic, or mineralogic composition, or many aspects of sedimentology (e.g., grain size), cannot alone infer the transport pathway (e.g., eolian vs. subaqueous transport) by which a grain of volcanic ash eventually becomes entrained in the sediment. Strictly speaking, transport pathway must be inferred based upon geological or geographic constraints. For example, in the context of eolian transport of eroded continental material, truly detrital aluminosilicate grains must be transported by wind to reach the deepest and most distal north Pacific (Rea and Leinen 1988; Kyte et al. 1993; Rea 1994).

For volcanic ash, however, the situation is more complicated. Subaqueous eruptions can lead to significant volcanic input (e.g., Fiske et al. 2001; Maicher and White 2001; White et al. 2003; Tani et al. 2008). Fiske et al. (2001, their p. 822) in particular note how the finest-grained fraction is missing from caldera deposits in the Izu-Bonin system. This "missing material" undoubtedly ends up as dispersed ash throughout the bulk sediment and due to its grain size has the potential to be advected great distances away from its point of origin. Determining this material's composition in the bulk sediment, a priori will not differentiate between subaerial (eolian) or subaqueous transport.

Subaqueous eruptions can also lead to pumice rafts, which may persist in the ocean for weeks, months, or years (Simkin and Fiske 1983; Risso et al. 2002; Bryan et al. 2004). Even in today's modern era of enhanced ocean and atmospheric observation, such occurrences are surprises (BBC News, 10 Aug. 2012). Yet, although these events may be uncommon on the human timescale, they are most likely significant pathways of ash to the deep sea over longer time frames.

Mixes of mixtures

In the broad context of geologic material, all non-biogenic rocks and sediments on the surface of Earth can be chemically described as falling along a mixing line or mixing curve between families of primordial materials (e.g., basalts) and families of fractionated "upper continental crust". Deviations from such a mixing line or curve are due to local provenance effects, authigenic/diagenetic resetting, weathering, and other processes, but the overall relationship holds true. While such a compositional range may seem relatively easy to differentiate analytically (e.g., felsic vs. mafic), the cases are rare when there are only two aluminosilicate sources and they are sufficiently compositionally distant from each other along the mixing line or curve. Far more common is the challenge of resolving multiple sources that are each at different fractional lengths along a theoretical mixing line or curve. For example, continental crust has been approximated as "granodiorite", "dacite", or other intermediate composition igneous/volcanic rocks (e.g., Taylor and McLennan 1985), and therefore distinguishing eroded continental crust from wind-blown dacitic ash is challenging. Or, still more challenging are scenarios whereby two disparate sources such as a marine basalt and a wind-blown dust of broad continental crust composition are both authigenically altered to, for example, assorted smectites, and thus have the potential to appear as a single source.

The above examples speak to the importance for geochemical studies of dispersed ash to address their attention to refractory elements that are appropriate to unravel mixing problems and that will also be relatively unaffected by diagenetic and authigenic processes. There are no chemical elements that unambiguously record any single geochemical provenance or domain. Within the spectrum of potential elements for use in general characterization as well as the multivariate statistics discussed below, inclusion of certain common major elements such as Al, Ti, and Fe, along with certain common trace elements such as La and Sc, proves to be beneficial. On a case-by-case basis, certain other key elements such as Th, Nb, and Cr can be advantageous as well (e.g., Plank 2005), although interpretations based on only a few elements are likely to be less representative than studies based on suites of many elements. However, common elements involved in reverse weathering reactions (e.g., Mg, Mackenzie and Garrels 1966; Presti and Michalopoulos 2008) should be avoided for such studies of provenance, given their labile behavior.

The contribution from geochemistry teamed with multivariate statistics

A variety of geochemical approaches can help resolve such mixing issues. In certain situations, such as in the Caribbean Sea example below, single element "normative calculations" (e.g., Leinen 1987) are sufficient, while in other cases multiple different analytical and/or computational approaches need to be simultaneously deployed to develop a more holistic understanding of the geochemical variability within datasets. We emphasize here a combined geochemical approach that generates a wide-ranging elemental suite that is examined with multivariate statistics.

The overall approach we summarize in this paper is based on Q-Mode Factor Analysis (QFA) and two forms of multiple linear regression, a Constrained Least Squares (CLS) technique and a Total Inversion (TI) technique (Pisias et al. 2013). The history of these techniques, as well as the specific MATLAB scripts to apply them to marine sediment chemistry (and recommendations for their use), are detailed in Pisias et al. (2013). The interested reader is also pointed toward previous papers by Dymond (1981), Leinen and Pisias (1984), Zhou and Kyte (1992), and Kyte et al. (1993). In our research group, we have successfully used these techniques in marine sediment from the Cariaco Basin, Arctic Ocean, equatorial Pacific, northwest Pacific Ocean, and South Pacific Gyre (Martinez et al. 2007, 2009, 2010; Ziegler and Murray 2007; Ziegler et al. 2007, 2008; Scudder et al. 2009, 2014; Dunlea et al. 2015a, b), demonstrating their utility throughout a variety of oceanic depositional regimes.

QFA is a terrific tool because—among other things—it is an "objective" technique (e.g., Leinen and Pisias 1984) that is minimally affected by inadvertent researcher bias. There are no inputs other than the data itself. It is limited, however, in many cases in that the resultant compositional factor scores may not be geologically reasonable or geologically precise enough across the spectrum of the wide element menus commonly generated in today's world of rapid geochemical analysis. Factor analysis is best used to help identify the number of independent components (also referred to as "end members") and their general composition, but not necessarily the extremely specific composition thereof. Therefore, one of the limitations of using QFA alone is that the user often cannot identify the specific geologic source(s) of the bulk sediment.

The number of end members necessary to best describe the variability of the dataset (that is, the number of factors indicated by the QFA) helps steer the CLS and TI multiple linear regressions. Specifically, in order to identify the exact sources identified by QFA, the CLS or TI linear regressions must be applied. In CLS and TI, the researcher inputs the composition of potential end members and mixes them until the best solution is reached (the differences between CLS and TI are detailed in Pisias et al. (2013)). Thus, we use the QFA to generate an objective "answer" as to the number of sources and their broad composition(s), and then the

CLS and/or TI, based on thoughtful selection of potential sources derived from knowledge of previous work and the local/regional geology, to generate the quantitative mix of specific sources. Libraries of potential end members are typically (a) constructed from the literature of all likely volcanic contributors (that is, individual papers about the geochemistry of particular volcano or region), (b) compiled from international global databases (e.g., EarthChem; http://earthchem.org), and/or (c) developed internally from the study's own data (e.g., using the composition of a discrete ash layer to assess whether it could potentially be a contributor to the dispersed ash component.).

Despite the quantitative attractiveness of applying statistical treatments to a well-measured geochemical dataset, there is no single diagnostic "magic bullet" to fulfill the task of identifying the dispersed ash component within bulk sediment. Instead, a series of interpretative techniques must be used prior to applying the statistical techniques. The use of standard geochemical approaches such as expressing data as concentrations, calculation of fluxes and/or elemental ratios, study of simple coefficient of determination (R^2) matrices, x-y diagrams, and other well-established approaches is extremely important. Such approaches define the boundaries for mass balance, allow comparison with lithologic descriptions and mineralogy, help ground the ensuing statistical treatments, and, in a way that is hard to quantify, ensure that the researcher is intimately familiar with the fundamental underpinnings of each dataset.

Finally, statistical treatments such as those discussed here often lead to non-unique solutions. Contextual oceanographic or geological knowledge plays a major role, however, and can eliminate some of the non-unique solutions. Our group takes a very labor-intensive approach to dealing with this broad subject for both factor analysis and multiple linear regression, strategically built around performing dozens to hundreds of statistical runs per dataset to assess sensitivity of, and variability in, the results (Scudder et al. 2009, 2014; Dunlea et al. 2015a,b). Each dataset is different, and we have found some in which any variation in one element will make a difference; but in another dataset varying a different element will exert a larger lever arm. We take great care to ensure that our results are not dependent upon, or unique to, one or two narrow statistical models nor to one or two specific elements. Doing so would be counter to the very nature of our task at hand—to develop a widely applicable, internally consistent understanding of "What is marine sediment made of?"

Previous work

In this section, we present a review of previously published studies focused on unraveling bulk marine sedimentary compositions. These case studies range from a simple three-component system (Caribbean Sea) in which element normative calculations alone are sufficient to determine the end members, to a very complex system in which the bulk sediment is composed entirely of aluminosilicates with very similar compositions that requires a combination of geochemical and statistical technique in order to identify the individual compositions (Izu-Bonin-Mariana Arc).

Caribbean sea: a simple case

The Caribbean Sea is ringed by volcanic arcs (Fig. 1; Sigurdsson 1999). To the west rises the Central American volcanic arc system, with a spectrum of volcanic edifices ranging in size from basaltic cinder cones to large stratovolcanos and calderas. To the east is the Antilles island arc, from which some of the most famous volcanic eruptions in the world originate (e.g., Lesser Antilles, Carey and Sigurdsson 1978). These combined volcanic systems contribute large amounts of ash to sediment in the Atlantic Ocean and Caribbean Sea from the Miocene on (Carey and Sigurdsson 1980, 2000; Peters et al. 2000; Jordan et al. 2006; Expedition 340 Scientists 2012; Le Friant et al. 2013) Ocean Drilling Program (ODP) Leg 165 cored throughout the Caribbean Sea (Fig. 1) with its most important objectives being related to the Cretaceous-Neogene bolide impact. One surprise that resulted from this research cruise was the startling number of discrete ash layers (Fig. 3, Carey and Sigurdsson 2000; Sigurdsson et al. 2000), their extent in time and space of which was unexpected. In addition to the discrete ash layers, shipboard study observed ash dispersed throughout the background sediment. Bulk chemical analyses of the bulk sediment allowed for the development of a single-element normative calculation scheme to quantify the abundance of this dispersed ash in these carbonate-rich sediments (Peters et al. 2000). Fortuitously, most of the sequence is a relatively simple three-component system (of $CaCO_3$, terrigenous, and ash), which facilitated the study of dispersed ash in parallel to the discrete layers. Petrographically, the bulk sediment appeared to consist of only carbonate and aluminosilicate clay—and the aluminosilicate clay was assumed to be entirely comprised of eroded terrigenous material (Peters et al. 2000).

Bulk chemical analyses, however, demonstrated that the sediment was vastly depleted in Cr beyond that, which could be explained by carbonate dilution. For example, in a sample with an equal (50 and 50 %) mix of $CaCO_3$ and terrigenous clay, one would predict there should be ~55 ppm Cr in the bulk composition (based on typical "average shale" of ~ 110 ppm Cr, using Post-Archean average Australian Shale [PAAS] from Taylor and McLennan (1985) to broadly represent an upper crustal terrigenous source). However, such a sample would instead present <10 ppm Cr, indicating that the brown aluminosilicate

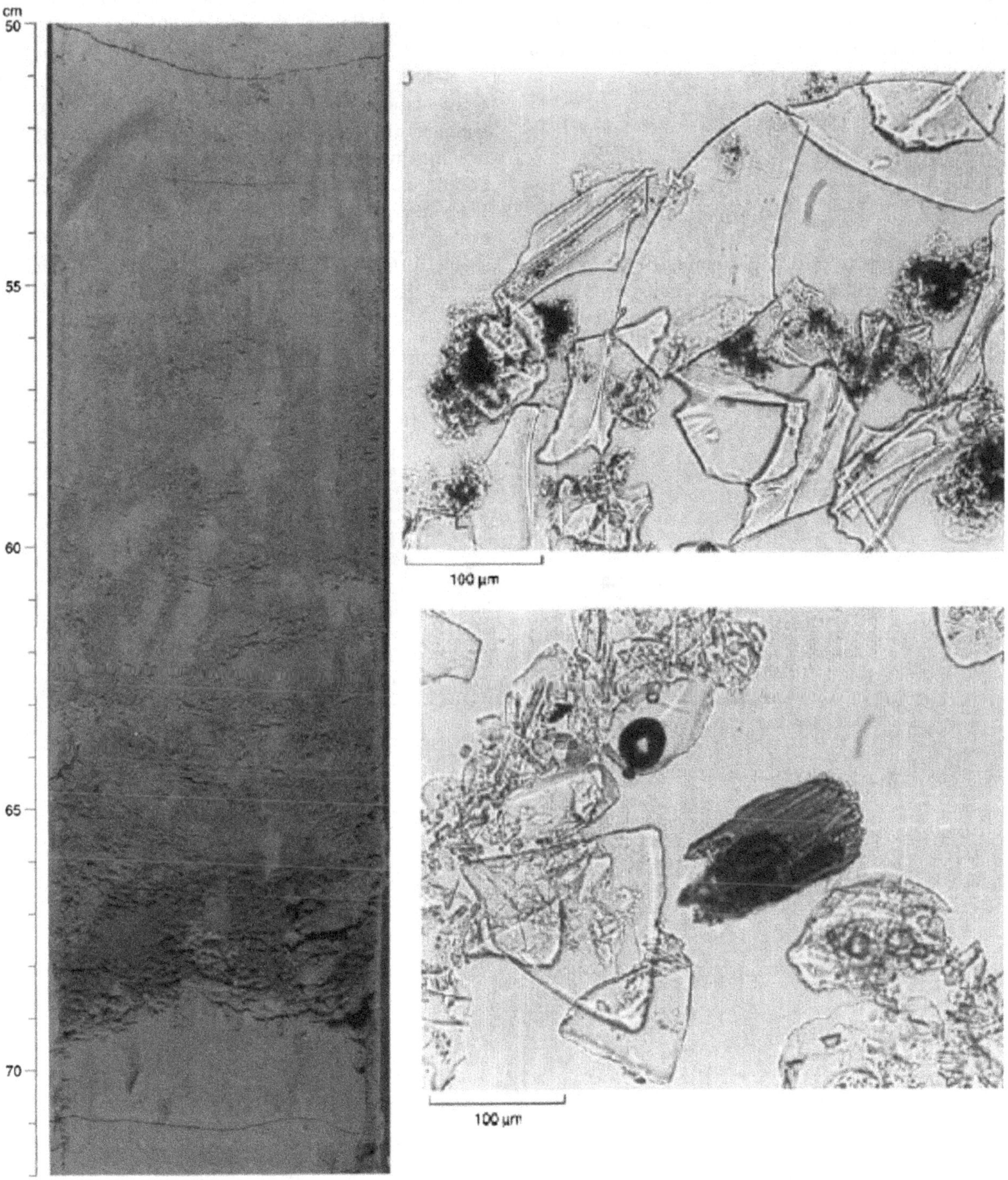

Fig. 3 Closeup photos and photomicrographs of ash layers from ODP Site 998, western Caribbean Sea. *(Left)* Volcanic ash layer in Section 165-998A-10H-4, 50–72 cm. Note the sharp base and transitional top that has been reworked by bioturbation. *(Right, top)* Section 165-998A-19X-3, 55–56 cm. Typical rhyolitic glass shards from a 13.8 Ma volcanic ash fall layer. The width of the field of view is 0.5 mm. *(Right, bottom)* Typical fresh *(clear)* and altered *(cloudy)* glass shards and an amphibole phenocryst in a 15 Ma volcanic ash fall layer in Section 165-998A-21X-3, 51–52 cm. From Sigurdsson, Leckie, Acton, et al. 1997

groundmass was not solely detrital terrigenous material (Peters et al. 2000).

Given the simple three-component system, normative calculations (Leinen 1987) were appropriate to identify the sources to the bulk sediment. Peters et al. (2000) based their calculations on Cr, given that the end member values of Cr in typical shale (110 ppm) and rhyolite (<3 ppm) are strongly divergent. Thus, in a three-component system of $CaCO_3$ (which is quantified precisely by coulometry), terrigenous material (calculated via Cr), and ash, the amount of ash can be calculated by difference, as follows:

$$(\%\,\text{Ash})_{\text{sample}} = 100 - (\%\,\text{CaCO}_3)_{\text{sample}} - (\%\,\text{Terrigenous})_{\text{sample}}$$

where

$$(\%\,\text{Terrigenous})_{\text{sample}} = 100 * \left(\text{Cr}_{\text{sample}} / \text{Cr}_{\text{PAAS}}\right)$$

Peters et al. (2000) thus documented on a sample-by-sample basis that dispersed ash comprises 15–20 wt.%, with a maximum of 45 wt.%, of the bulk sediment in the western Caribbean. These values were consistent with the sedimentological smear-slide analyses (Sigurdsson et al. 1997) but were more precise.

Peters et al. (2000) also observed that the timing of the accumulation rate of dispersed ash paralleled the sedimentation rate of the discrete layers, although the maxima in dispersed ash preceded the Miocene and Eocene maxima in discrete layers by ~2–4 Ma (Fig. 4). They interpreted the relative timing as recording arc evolution, with the dispersed ash being generated by smaller volcanoes characteristic of the more juvenile arc, and the larger discrete layers representing a mature arc characterized by larger stratovolcanoes and caldera

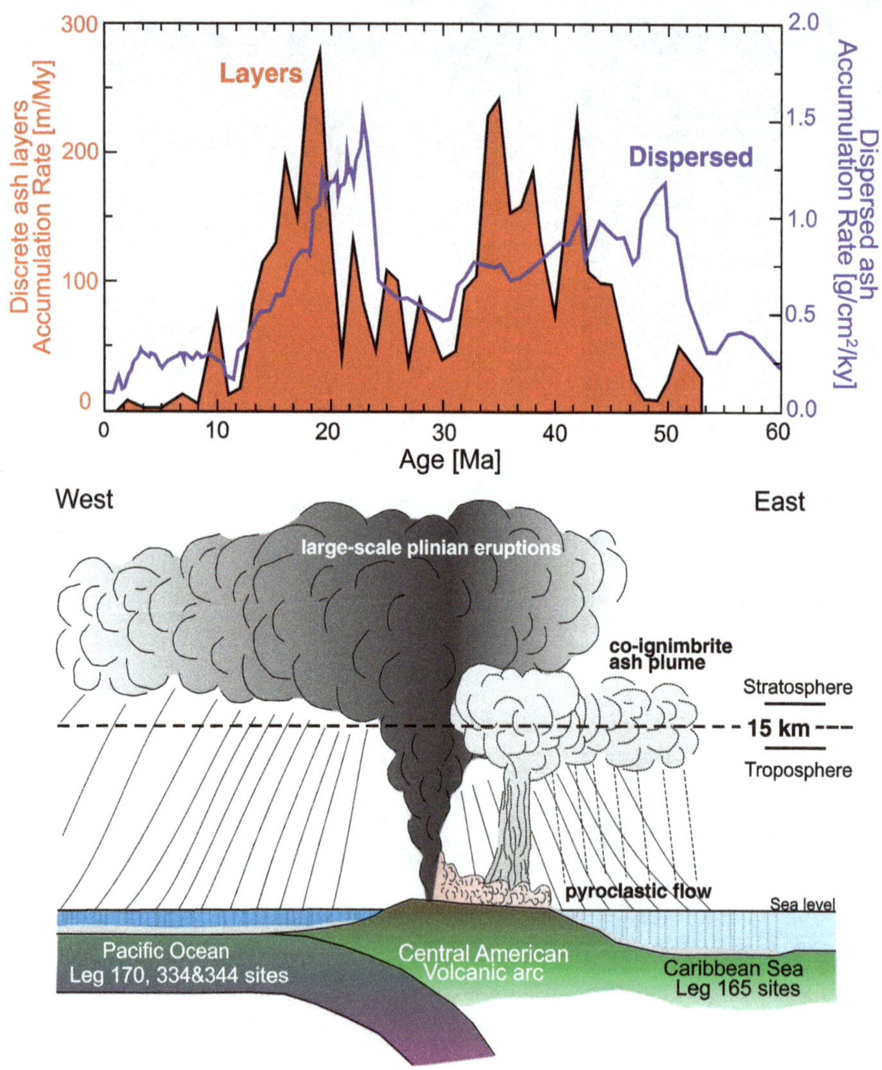

Fig. 4 (*Top*) Dispersed (*blue line*) vs. discrete (*red field*) ash accumulation in the Caribbean Sea. Accumulation of discrete ash layers (Carey and Sigurdsson 2000; Sigurdsson et al. 2000) and dispersed ash (Peters et al. 2000). ODP Site 998, western Caribbean Sea. (*Bottom*) Eruption transport and deposition patterns for recent to Miocene eruptions of the Central American Arc. Large plinian eruption columns (*light to dark gray*) reach deep into the stratosphere and are therefore transported to the west by the prevailing winds. Large ignimbrite-generating eruptions, predominantly in the Miocene, generate co-ignimbrite ash plumes (*pinkish to light gray*) reaching only into the upper troposphere/lower stratosphere and are therefore distributed to the east. Tropopause is shown with a *dashed line* at 15 km altitude. *Gray fields* below the sea level demonstrate submarine ash deposits

systems, as suggested by Carey and Sigurdsson (2000) and Sigurdsson et al. (2000). These are capable of injecting large plumes of Central American co-ignimbrite ash into the lower stratosphere and upper troposphere, that is, the ideal atmospheric height to facilitate west-to-east blowing wind transport (e.g., rather than transport from east-to-west in the higher stratosphere) (Fig. 4). This interpretation is consistent with the previous studies that have shown that Paleogene and Neogene Central American volcanism is predominantly characterized by larger ignimbrite-forming eruptions (Jordan et al. 2006) that may have frequently produced co-ignimbrite ash plumes reaching lower stratospheric heights (~15 to 20 km; e.g., Woods and Wohletz 1991; Bursik 2001), in contrast to the higher stratospheric eruption columns of Plinian eruptions (>>20 km; e.g., Kutterolf et al. 2008a).

Equatorial pacific ocean: a complex case

Our group also has a long-standing interest studying the equatorial Pacific (e.g., Murray and Leinen 1993, 1996; Murray et al. 1993, 1995, 2000; Kryc et al. 2003; Ziegler and Murray 2007; Ziegler et al. 2007, 2008, plus others). The non-biogenic component of these sediments includes terrigenous clay and ash, which are challenging to resolve in sediment with more than 98 % biogenic material (carbonate and biogenic silica). Some of our work specifically targeted how to resolve the eolian, ash, and authigenic components of these sediments and built upon the classic research of the long-standing Michigan group and their progeny (e.g., Hovan et al. 1991; Rea 1994; Rea et al. 1998).

For example, Ziegler et al. (2007) detailed chemical criteria that can be applied to differentiate authigenic, terrigenous, and volcanogenic aluminosilicates from each other. Sequential extractions were performed at ODP Site 1215 (Fig. 1) to remove non-aluminosilicate components. The material remaining upon completion of the extraction procedure was termed the "residual" component, consistent with established protocols. La-Sc-Th ternary diagrams and PAAS-normalized rare earth element (REE) patterns (Fig. 5a) and reference REE patterns (Fig. 5b) were used to aid in determination of the sedimentary components. In the residual component, Ziegler et al. (2007) quantified upper continental and lower crustal components using selected rare earth element (REE), Sc, and Th abundances. In old (>50 Ma) nannofossil ooze, the residual component exhibited a large La contribution. Additionally, the residual component of the nannofossil ooze fell outside the boundaries of a simple two component mixing between upper and lower crust components. REE patterns for these samples exhibit sharp, seawater-like negative Ce anomalies, providing a clear indication that an authigenic phase is present in the residual component (Fig. 5c). The old age of these sediments may contribute to the formation of the authigenic phase, given that such samples have likely experienced significant diagenesis. In this way, the geochemical approach clearly identified samples in which it is virtually impossible to differentiate eolian material from volcanic ash (Fig. 5c, d).

Izu-Bonin-Mariana Arc (ODP Site 1149 and DSDP Site 52)

One of the goals of ODP Leg 185 was to test whether subducted sediments control along-strike geochemical differences in Izu-Bonin-Mariana arc composition (Fig. 1, Plank et al. 2000). In addition to the potential influence of the terrigenous (non-ash) sedimentary component, the amount of ash in the sediment column is critical to constrain, not only for the determination of absolute geochemical fluxes into the "Subduction Factory" but also in order to determine how much of the input is "recycled" from the arc itself. Due to the long-standing interest in the region from a variety of geological perspectives, there is also a wealth of tectonic- and arc-related information to draw upon in the context of both ash layers and dispersed ash.

Preliminary work based on a Nb-based normative calculation showed that 33 ± 9 wt.% of the sediment at ODP Site 1149 is comprised of dispersed ash (Plank et al. 2000). As described in Scudder et al. (2009, 2014), in order to determine the sources to the bulk sediment, we followed the QFA and TI approaches and MATLAB scripts from Pisias et al. (2013). The elemental suite we targeted was composed of refractory elements predominantly associated with aluminosilicate components (Al, Ti, Sc, Cr, Ni, Nb, La, Th) as these are the most likely to identify differences between source components. The QFA results indicate that four factors (end members) explain 97 % of the variability of the bulk sediment (Scudder et al. 2014, Fig. 6). Given that QFA alone is not sufficient to identify the source of these factors, TI modeling was employed. The TI modeling identifies a combination of Chinese Loess (CL), rhyolitic ash from the Honshu arc (HR), mafic ash from the Izu-Bonin Front Arc (IBFA), and a second eolian dust (termed "Eolian 2") that best explain the total chemical composition of the bulk sediment.

Based on the TI results, the dispersed ash mass accumulation rate (MAR, $g/cm^2/ky$) at Site 1149 can be calculated and compared to the MAR of a series of ash layer parameters. At this site, the number of ash layers most closely tracks the total dispersed MAR (that is the sum of HR + IBFA). That the number of layers is most similar to the dispersed MAR suggests that eruption *frequency*, rather than eruption *size*, is the driving mechanism for the dispersed ash record (Scudder et al. 2014, Fig. 7).

The MAR patterns of the ash components are consistent with published eruption records of both the

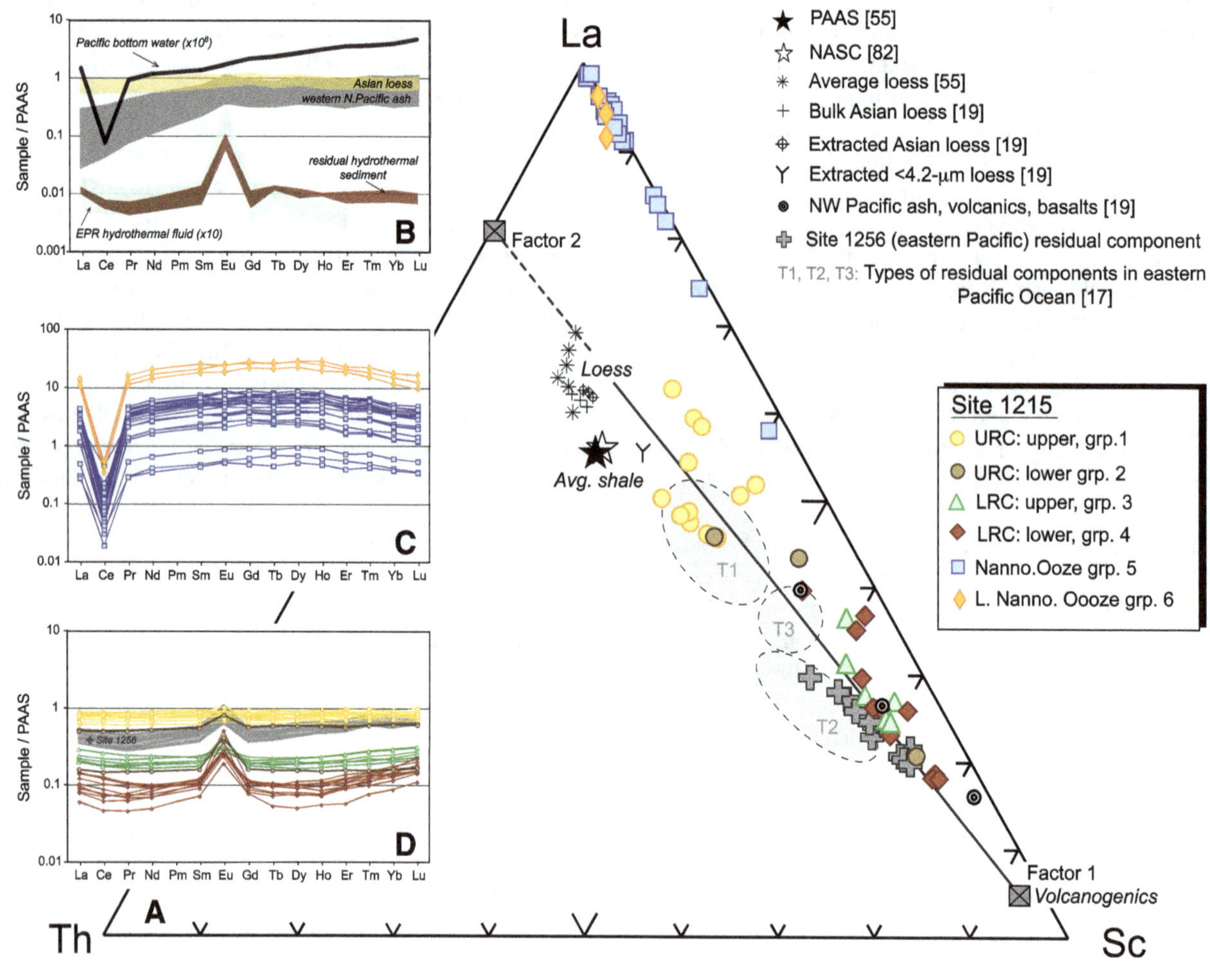

Fig. 5 a La-Sc-Th ternary diagram and PAAS-normalized rare earth element (REE) patterns from Site 1215 (central Pacific) and Site 1256 (eastern Pacific) from Ziegler et al (2007). **b** Reference REE patterns. Asian loess (Jahn et al. 2001), western Pacific ash (Bailey 1993), EPR and Pacific bottom water (Piepgras and Jacobsen 1992), hydrothermal residual sediment (Severmann et al. 2004), and hydrothermal fluid (Klinkhammer et al. 1994). **c** Residual component extracted from early-mid Eocene biogenic sediment at Site 1215 as defined in Ziegler et al. (2007). Note the significantly negative Ce anomaly indicating a seawater source. **d** Residual component extracted from the red clay unit from Site 1215 and the silty nannofossil ooze from Site 1256 as defined in Ziegler et al. (2007). See legend for symbol description. Groupings are based on REE patterns. Factor 1 is interpreted as lower crust island arc volcanic arc material and Factor 2 is similar to upper continental crust. T1, T2, and T3 are eolian dust samples classified based on REE patterns, Eu/Sm, and La/Sc ratios from Hyeong et al. (2005). *URC* upper red clay; *LRC* lower red clay

Izu-Bonin and Honshu arcs, and we can interpret these changes in accumulation rate in terms of arc history (Fig. 8). Focusing first on the Honshu Rhyolite, we observe that this component is the dominant dispersed ash, even though the Honshu arc is relatively far away. Given Site 1149's distal location, eruptions from the Honshu arc could have been large and yet resulted in only thin layers. These layers may subsequently be mixed into the bulk sediment to create the dispersed ash component. Alternately, the dispersed ash could simply represent an increase in deposition of ash from the atmosphere. The MAR of HR is consistent with the tectonic history of the Honshu arc, particularly through the younger portions of the record (e.g., Taira 2001). For the IBFA component, the models closely track the temporal changes in the tectonic record of Izu-Bonin (e.g., Sibuet et al. 1987). MAR patterns indicate that a gradual increase in IBFA volcanism occurred from ~18–10 Ma, with a burst of activity from ~4.5–3 Ma followed by a steady increase beginning at ~2 Ma (Scudder et al. 2014, Fig. 8).

We can compare and contrast Site 1149 to DSDP Site 52 in the northern Marianas arc (Fig. 1). Site 52 contains extremely high abundances of volcanic ash not directly linked to a specific eruption style (upwards of 50 % volcanic glass mixed in the brown clay; Fischer et al. 1971). Unfortunately, rotary drilling at Site 52 left only a few ash layers intact (all in the upper 20 mbsf; Fischer et al. 1971). Applying QFA with the same chemical suite as used at Site 1149 indicates that four factors (end

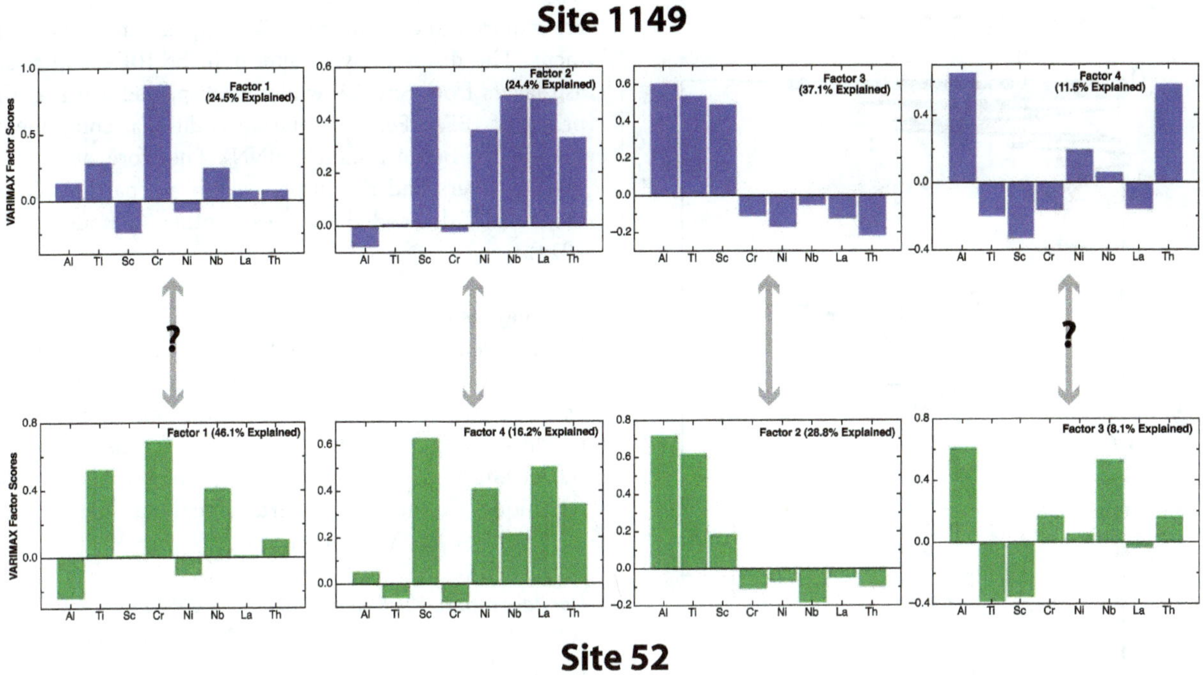

Fig. 6 VARIMAX factor scores from QFA, *(Top)* Site 1149, *(Bottom)* Site 52, from Scudder et al. (2014). Note the similarities between Site 1149 Factor 2 and Site 52 Factor 4 as well as Site 1149 Factor 3 and Site 52 Factor 2

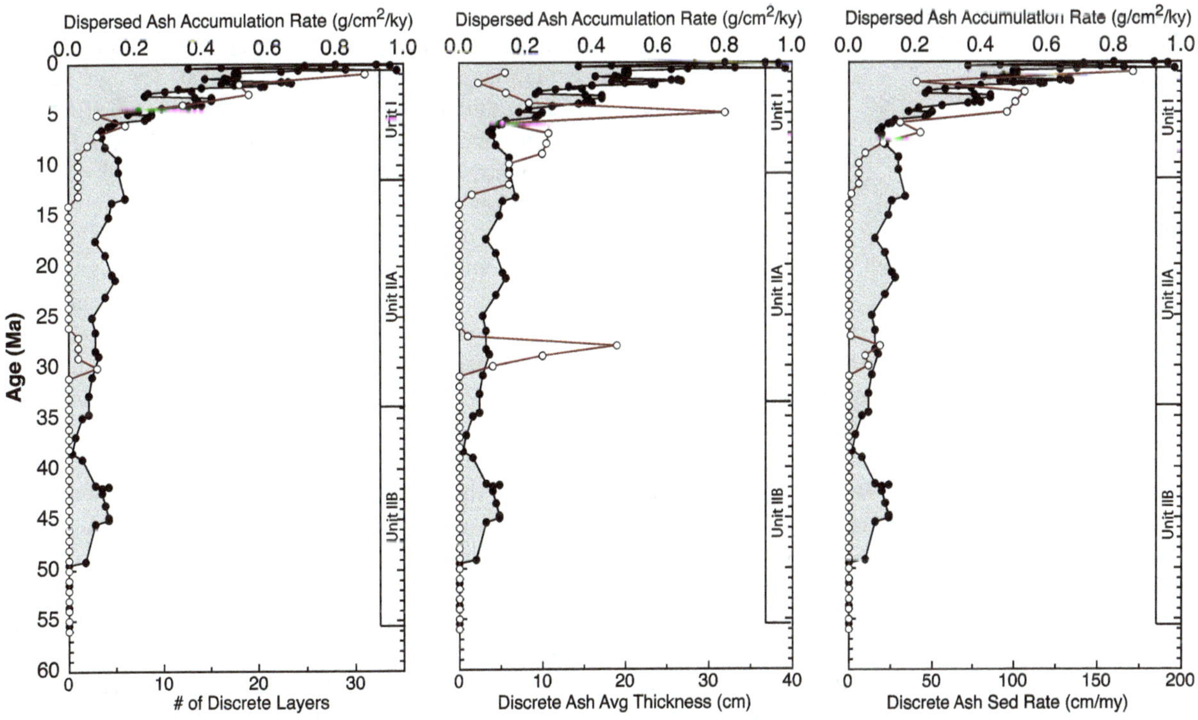

Fig. 7 Site 1149. The accumulation rate of total dispersed ash (*gray shaded* with *filled circles* in g/cm^2/ky) plotted against parameters associated with the sedimentation of discrete ash layers per 1 Ma (*open circles*) from Scudder et al. (2014)

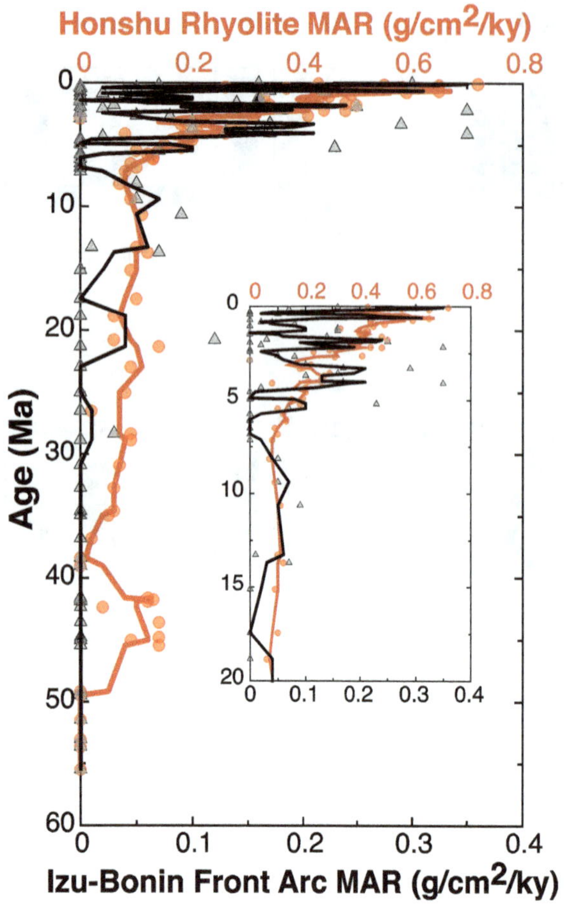

Fig. 8 Individual data and three-point averages of MARs of each dispersed ash end member, Site 1149. *Inset* focuses on the younger portion of each record. Note color-coded different *x*-axis scales and labels. From Scudder et al. (2014)

members) are present (Fig. 6). Two of these factors closely resemble those found at Site 1149, while the other two do not correspond well, if at all (Scudder et al. 2014). Given the limitations of QFA described previously, this interpretation is based on the factor scores (the weight of each element on the discrimination of a single factor) and the broad compositional scores produced by the modeling. TI linear modeling confirms this interpretation, identifying a mix of Chinese Loess, IBFA, dispersed boninite from the Izu-Bonin arc (BNN), and a dispersed felsic ash with the same composition as felsic layers from Site 52 (referred to as Felsic52) as best explaining the bulk sediment chemical composition.

Site 52 presents some key similarities and differences from Site 1149. First, the total amount of dispersed ash (regardless of composition) at both sites is very high, averaging 30 ± 17 wt.% at Site 1149 and 36 ± 18 wt.% at Site 52. Second, whereas at Site 1149 the two ash sources (HR, IBFA) are found both as layers and as dispersed ash, at Site 52 at least three volcanic sources are required to explain the bulk composition of the sediment. The discrete layers appear to be IBFA and felsic ash layers from Site 52, while the dispersed ash is comprised of IBFA, Felsic52, and an additional component, average Izu-Bonin Boninite (BNN). Therefore, at Site 52, the ash layers and the dispersed ash are partially compositionally decoupled from each other (Scudder et al. 2014).

Ongoing research

In this section, we present new data and ongoing observations from a variety of settings in the Northwest Pacific. While some of this research may still be developing, we include it here in order to show the strength of the combined geochemical/statistical technique in moving toward a regional perspective of inputs to the Western Pacific.

Nankai Trough, IODP Sites C0011 and C0012

IODP Sites C0011 and C0012 (Fig. 1), drilled as part of the Integrated Ocean Drilling Program (IODP) NanTroSEIZE transect, are located in the Shikoku Basin ~100 km southeast of the Kii Peninsula and ~160 km west of the Izu-Bonin arc on the Kashinosaki Knoll, a prominent bathymetric high. Located near the crest of the Kashinosaki Knoll, Site C0012 represents a condensed sediment section in comparison to Site C0011, which is located on the northwestern flank. The main goal of NanTroSEIZE is to drill across the up-dip limit of the seismogenic and tsunamigenic zone over the Nankai Trough subduction boundary, along which mega-thrust earthquakes are known to occur (Tobin and Kinoshita 2006). Shipboard sedimentological smear slide analyses at both Sites C0011 and C0012 estimate that on average dispersed ash constitutes ~25–30 wt.% through Units I and II, and decreases to ~7–15 wt.% in the hemipelagic facies of Unit III (Saito et al. 2010; Henry et al. 2012).

Statistical analyses of newly acquired data from Sites C0011 and C0012 were performed using the same element suite as that applied to Site 1149. At Site C0011, QFA indicates that three factors explain 99 % of the variability of the bulk sediment (Fig. 9). While we cannot determine the exact composition of the end members from QFA, one of the modeled sources appears to be intermediate (roughly equal contributions of most of the elements in the model, that is, of Sc, Cr, Ni, Nb, La, and Th) in composition while the other two are broadly felsic (contributions driven by Al, Ti, and Sc, with lesser covarying Nb, La, and Th). CLS multiple linear regression analysis suggest that the three sources at

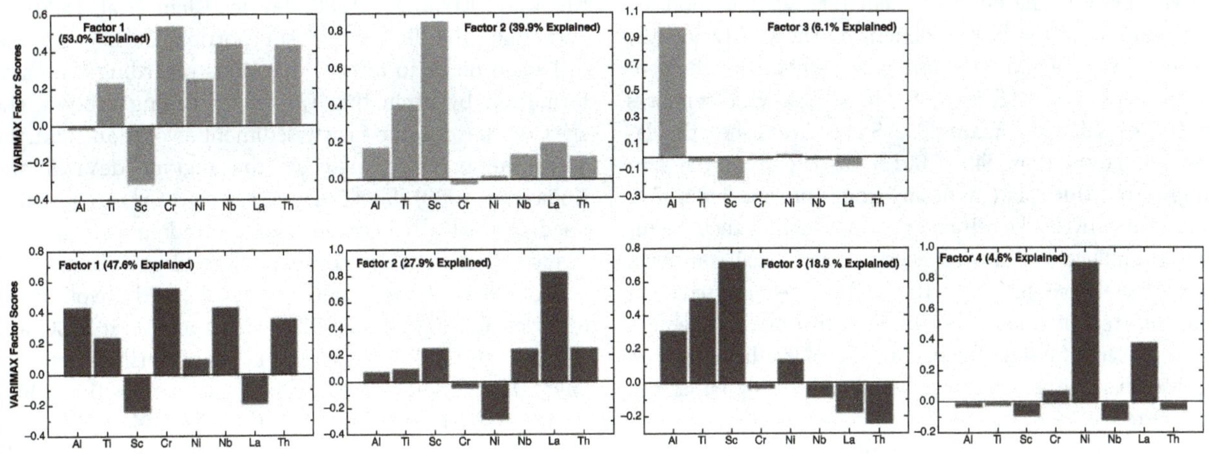

Fig. 9 VARIMAX factor scores from QFA. (*Top*) Site C0011. (*Bottom*) Site C0012

Site C0011 are consistent with one being PAAS, and the other two being ashes having compositions similar to a representative Rhyo-Dacite and Rhyolite from the Honshu Arc (HR) (Fig. 10). These latter two sources may in fact be dispersed ash but could also be ash that has since been altered and yet retained its refractory chemical signature. Regardless, these two types of dispersed ash and/or clay minerals with comparable geochemical signatures comprise 36 ± 8 wt.% of the bulk sediment. Evidence for multiple sources of volcanic material, and/or the chemically equivalent clay-alteration products, is consistent with studies of tuffaceous and volcaniclastic sandstones from this site, which calls for volcanic material from both the Izu-Bonin and Honshu Arcs (e.g., Pickering et al. 2013; Schindlbeck et al. 2013; Kutterolf et al. 2014).

Applying the same methods to Site C0012, QFA indicates that four factors (end members) explain 99 % of

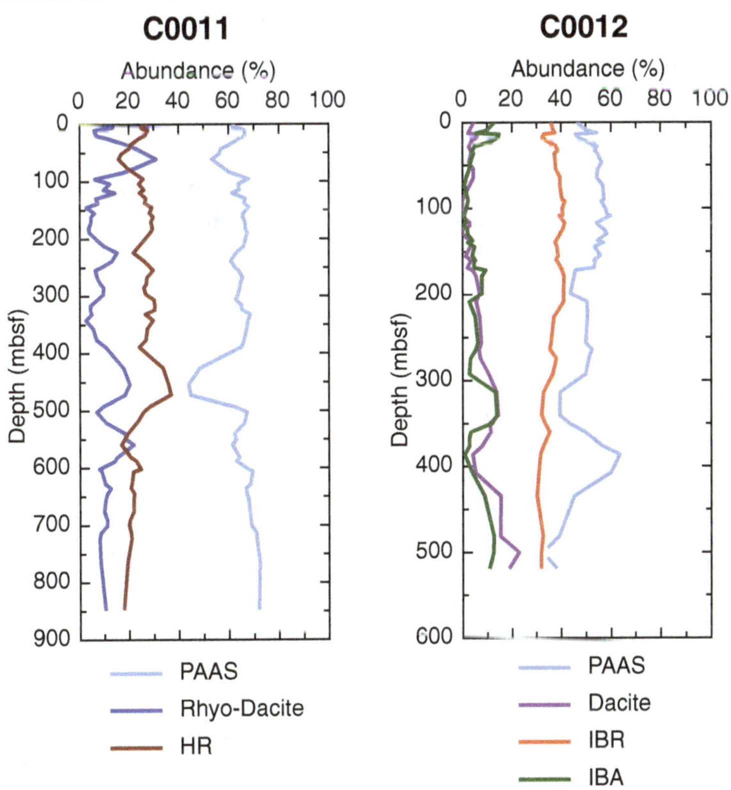

Fig. 10 Abundance (weight %) of sources identified by constrained least squares multiple linear regressions (CLS). (*Left*) Site C0011. (*Right*) Site C0012

the variability in the bulk sediment (Fig. 9). The modeled compositions of the bulk sediment at Site C0012 identify two intermediate and two felsic end members for the geochemical proxies. CLS indicates that these end members are best explained by mixing PAAS and three ashes: the dacitic ash layers from Site C0012, rhyolite from the Izu-Bonin Arc (IBR), and andesite from the Izu-Bonin Arc (IBA). Collectively, the ash compositions (and/or their chemically equivalent clay-mineral alteration products) contribute approximately one-half (49 ± 10 wt.%) of the bulk sediment (Fig. 10). Here, it is important to recall that our geochemical approach cannot resolve the pathway by which the ashes have reached these sites, as some of these ashes may be eroded from terrestrial deposits in the Japanese archipelago.

We compare the mass accumulation rate of the dispersed ash (or its clay alteration) component to a number of common ash layer parameters, including the number of ash layers per 0.2 Ma, the thickest ash layer per 0.2 Ma, and the total thickness of ash layers per 0.2 Ma. We binned the data by 0.2 Ma units, in order to generate a discrete ash layer dataset of approximately the same temporal resolution as that of our dispersed ash record (Fig. 11).

When compared to various ash layer parameters, we find that the MAR of the total dispersed ash component at Site C0011 correlates best with "Thickest Layer" in Unit I. This suggests that through this interval, eruption volume, rather than the frequency of explosive eruptions, drives the dispersed ash accumulation (Fig. 11). In contrast, in Unit I of Site C0012, the number of discrete ash layers rather than "thickness" is best correlated to the dispersed ash MAR. At both sites, the number and thickness of ash layers vastly decreases below the Unit I/II boundary while the dispersed ash component remains high as a geochemical proxy. This is reflective of time periods when the volcanism in SW Japan was reduced or eliminated (Mahony et al. 2011), when the sites were influenced by siliciclastic turbidites (Units II, IV, and V), or when dispersal paths from the main detrital sources were further away from eruptive fronts (Unit III). Changes in any of these factors may have resulted in either fewer and/or thinner layers that would be more susceptible to being homogenized (by bioturbation or otherwise). That the geochemical signature equivalent to dispersed ash is high at times when so few ash layers were deposited clearly demonstrates that clay minerals derived from volcanic ash and volcanic rock (e.g., smectite group) are a much greater contributor to the sedimentary sequence than is recorded only by the ash layers.

Shikoku Basin, DSDP Site 444

DSDP Site 444 was drilled during Leg 58 to test hypotheses about the origin of marginal basins (Fig. 1, deVries Klein and Kobayashi 1980; deVries Klein et al. 1980). In particular, the Shikoku Basin portion of Leg 58 was drilled in order to answer questions regarding the basin formation by evaluating various spreading models, the ages of the oceanic crust, sediment evolution, and the paleoceanographic origin of the region (deVries and Kobayashi 1980). Site 444 was selected for our research on dispersed ash because of its abundant volcanic ash layers (for example, in comparison to Sites 442 and 443).

Site 444 contains basaltic, andesitic, and rhyolitic ash (deVries and Kobayashi 1980; Furuta and Arai 1980; and our own studies). Most tephra in other northwestern Pacific DSDP holes are rhyolitic and andesitic, which makes the basaltic nature of the Site 444 tephra geologically and tectonically unusual (White et al. 1980). In the Pleistocene and Pliocene, the ash layers are non-alkali rhyolites. As felsic ash is believed to mainly be a product of explosive volcanic eruptions, these layers could have originated from relatively large-scale eruptions in an island arc setting, most likely located westward of Site 444 and transported in the prevailing westerlies. This direction of transport is evident as grain size increases toward the west and decreases toward the east (Furuta and Arai 1980).

In the Miocene, the tephra is of non-alkali tholeiitic basaltic as well as alkali basaltic composition. These basaltic layers imply that another volcanic source was involved in the Miocene. This source could have been closer to the drilling sites than those sources of rhyolitic and andesitic tephra (possibly the Kinan Seamounts and the Shichito-Iwo Jima volcanic arc), but it could also be the distal evidence for large mafic eruptions from a more oceanic arc source. The knowledge that the occurrence of such large mafic plinian eruptions with a wide dispersion is not limited to arc settings alone and can also be found at ocean island settings (e.g., the Galápagos archipelago, Schindlbeck et al. 2015). Such observations are especially important for future studies of dispersed mafic ash in marine sediments. We note, however, that at least some of these ashes may also be subaqueous in origin. For example, certain volcanic arcs near the Shikoku basin (e.g., Izu-Bonin) contain subaqueous volcanoes (Hochstaedter et al. 2001).

Ash layers were carefully identified throughout the 310 meters of sediment at Site 444 using core photos and the shipboard lithologic description (deVries and Kobayashi 1980). Site 444 was cored using the Rotary Core Barrel (RCB) system, as it was occupied during the early-mid stages of the DSDP prior to the advent of the advanced piston corer. Because of the RCB drilling, the recovered core is highly disturbed. Therefore, we classified the ash deposits as either an "ash layer" or an "ash pocket". Anything that is ash in distinct form but that is

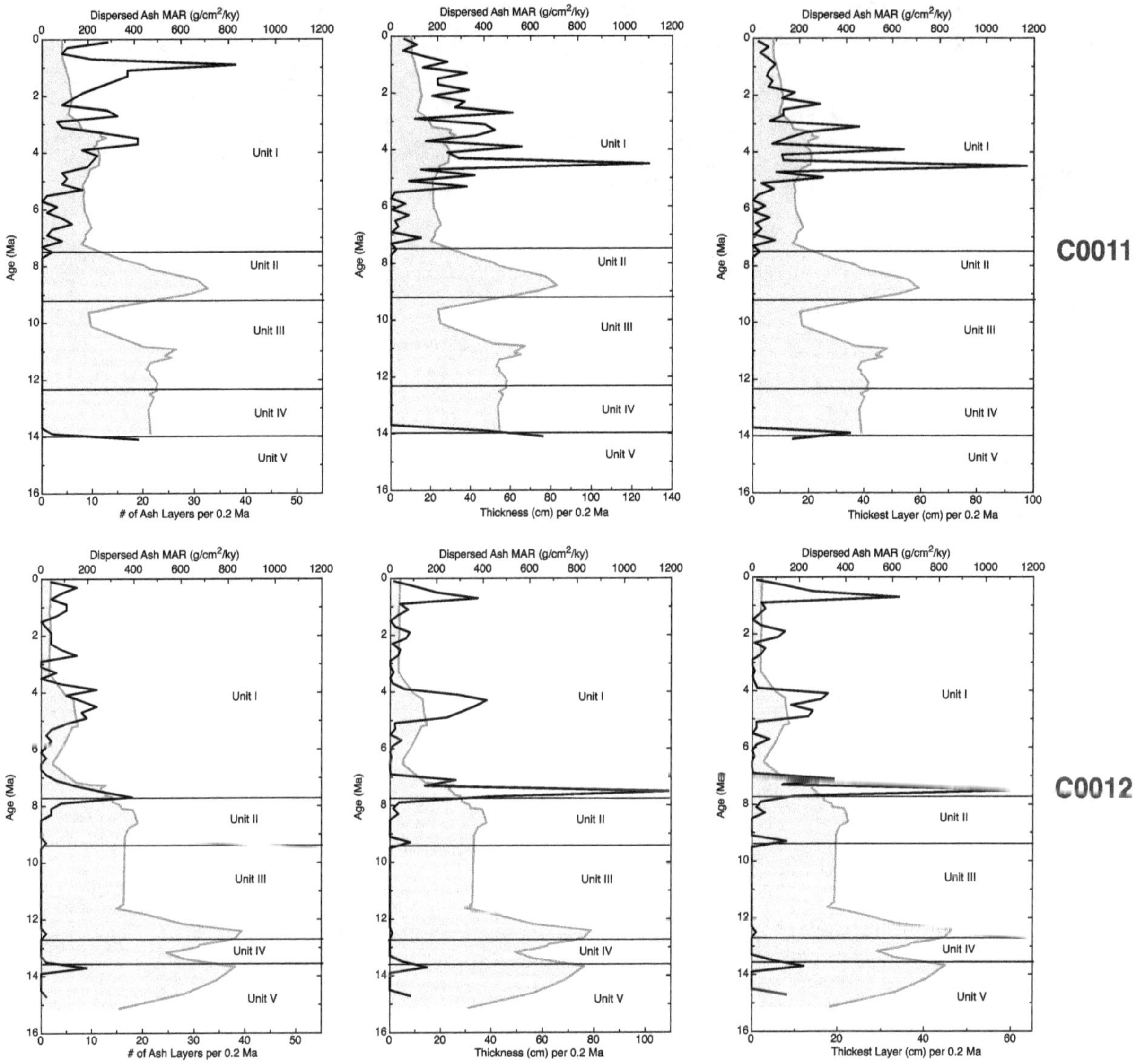

Fig. 11 The accumulation rate of total dispersed ash (*gray shaded line*, g/cm^2/ky, the same in each panel by site) plotted against parameters associated with the sedimentation of discrete ash layers per 0.2 Ma (*black line*, parameters are different in each panel as shown in the *bottom* *x*-axis labels). A moving 0.2 Ma window was chosen to bin each discrete ash layer parameter to approximate the resolution of the modeled dispersed ash record. (*Top*) Site C0011. (*Bottom*) Site C0012

not in the shape of a full horizontal layer (e.g., touching both sides of the core liner) is termed an "ash pocket". The pockets are often, though not always, circular in form. Ash layers and ash pockets correlate with each other with depth. Thus, we interpret that the pockets are likely to have originated from ash layers mixed by RCB drilling.

A total of 128 ash layers and 76 ash pockets were identified. Of the 28 ash layers chemically analyzed in this study, four (4) of them are categorized as ash pockets. This is important to note when assessing their compositions as any seemingly unusual compositions can be evaluated for mixing with non-ash sources (e.g., clays). The distribution of ash layers and pockets were parameterized by (1) thickness with depth, (2) number of layers per 10 meters, (3) thickest layer (cm) over successive 10 meter depth increments, and (4) total thickness of ash (cm) in that same 10 meter window (Fig. 12). These parameters collectively should yield a baseline understanding of the distribution of ash layers, and we acknowledge that any one of these parameters is not necessarily preferable to the others. Additionally, we note that there is a relatively thick 31-centimeter-thick ash layer at a depth of 30 mbsf (~0.66 Ma, Pleistocene), which may overly influence some of these parameters over that interval.

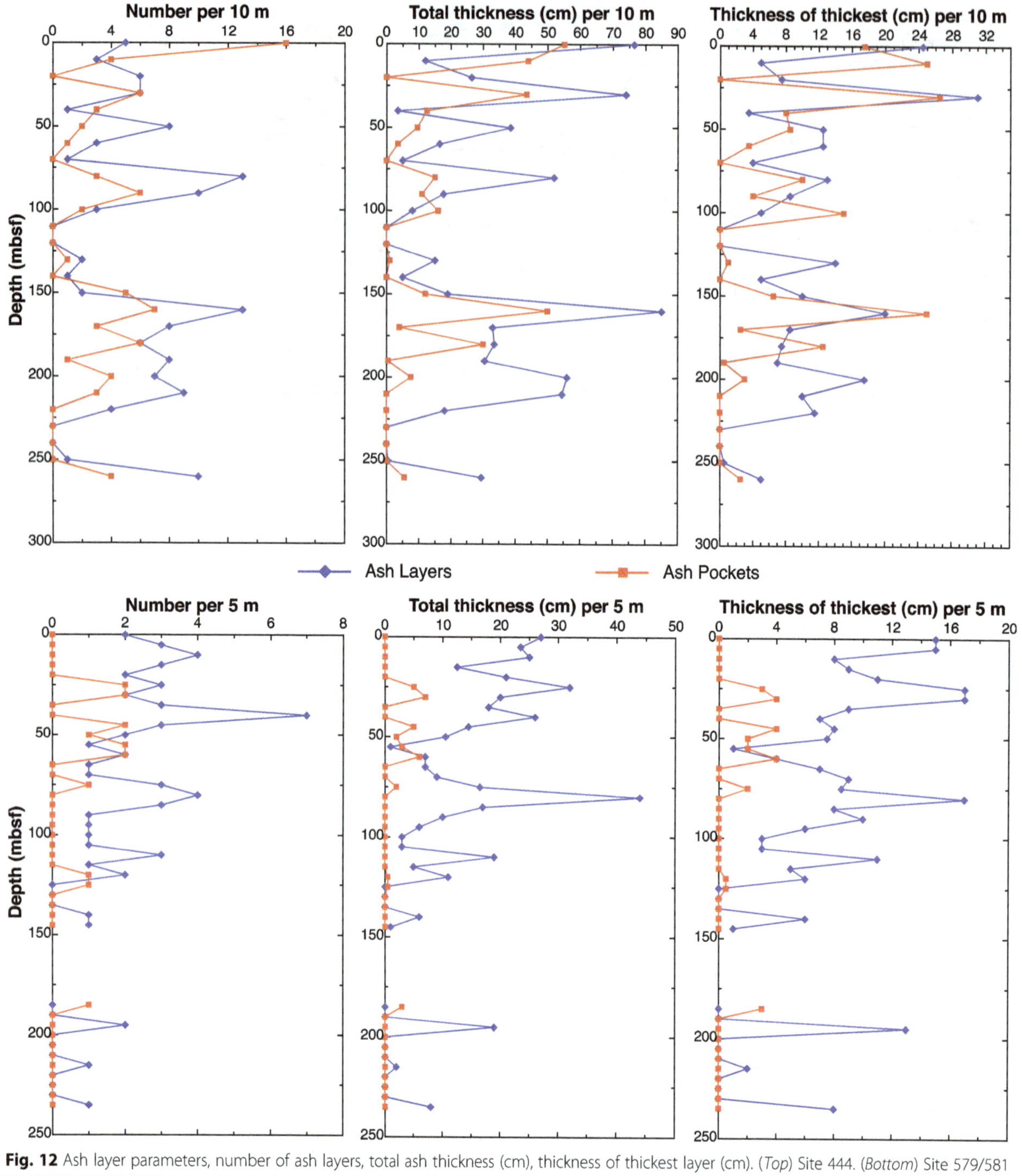

Fig. 12 Ash layer parameters, number of ash layers, total ash thickness (cm), thickness of thickest layer (cm). (*Top*) Site 444. (*Bottom*) Site 579/581

Ash layers were analyzed for major element glass shard compositions by electron microprobe (average of 15 single shard measurements), and additionally, the bulk ash layers were analyzed by ICP-ES and ICP-MS for major, trace, and REE compositions following the methods of Scudder et al. (2014). Even though the bulk ash data may be compromised by the inclusion of the terrigenous aluminosilicate material (e.g., adjacent pelagic clay mixed into the ash layer or ash pocket; Scudder et al. 2014), we can still infer other key aspects of the distribution of the ash layer/pocket chemistry.

The distribution of key major elements used to differentiate compositional variation, such as SiO_2 or TiO_2, (Fig. 13), suggests the presence of multiple ash populations ranging from mafic to intermediate to felsic in composition. A transition in ash layer composition is observed at ~160–170 mbsf in which the SiO_2 concentration of the bulk ash layers decreases with depth, while the TiO_2 concentration of the bulk ash layers increases with depth. At ~160–170 mbsf, the ash layers are enriched in elements such as Sc and V, and these elemental changes are indicative of a shift to ash of intermediate to mafic composition at this depth (Fig. 13). Ternary diagrams (Fig. 14) indicate at least two populations of ash spanning a mixing line between relatively more felsic (dacites and rhyolite sources) and more mafic (basalt sources) end members.

Other existing data supports the hypothesis of multiple ash populations. Some of the Site 444 cores possess visually black and chemically basaltic ash, which is a unique aspect for the Shikoku Basin. Refractive indices of volcanic glass and associated minerals have been documented as an effective means for correlating widespread tephra layers (Machida and Arai 1983). Based on the refractive indices of glass shards from ash layers at Site 444, Furuta and Arai (1980) (their Figure 12) show a linear relationship between refractive index and FeO

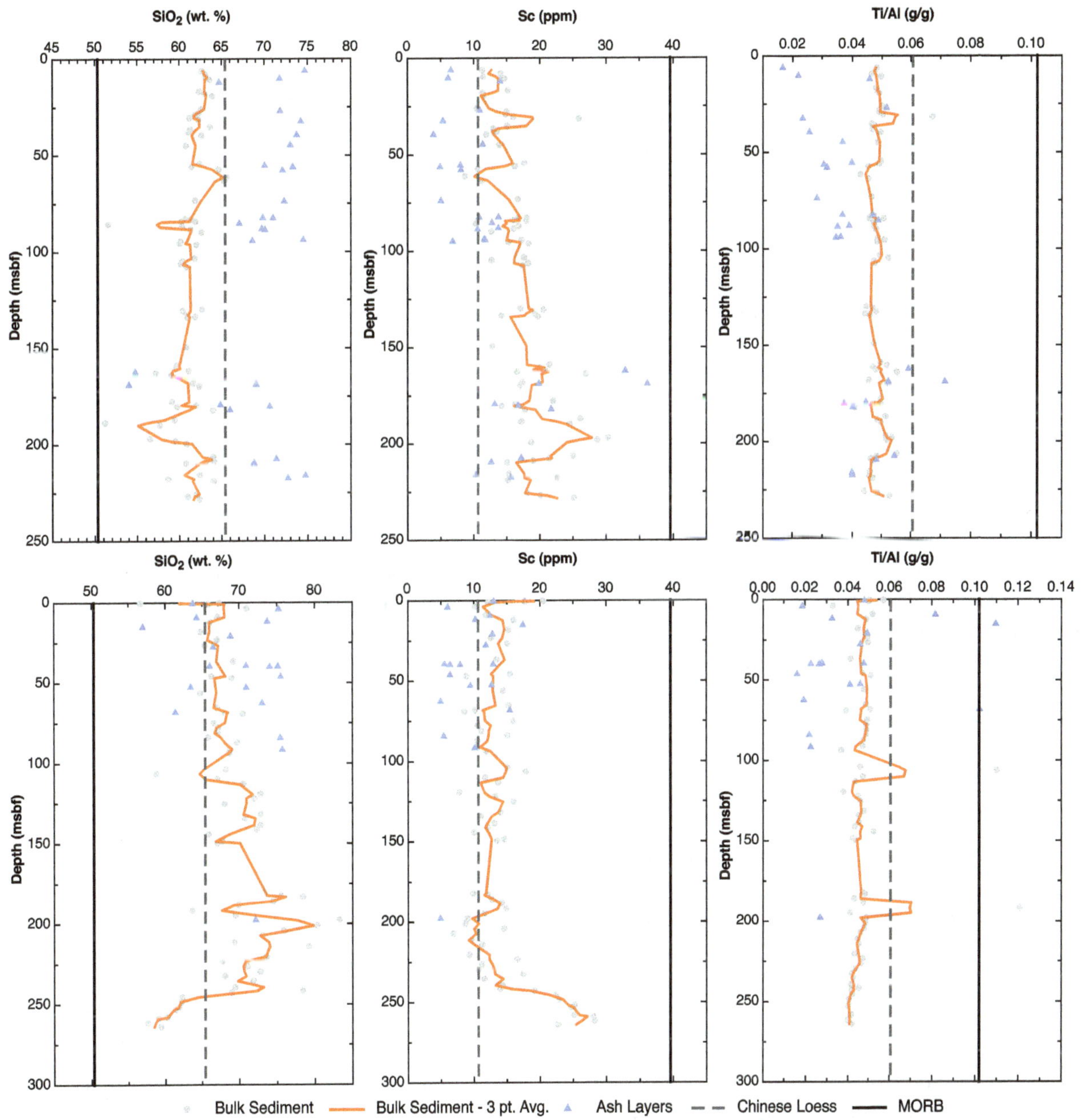

Fig. 13 Representative compositions and elemental ratios plotted with depth. Chinese Loess (Jahn et al. 2001), MORB (Arevalo and McDonough 2010). (*Top*) Site 444. (*Bottom*) Site 579/581

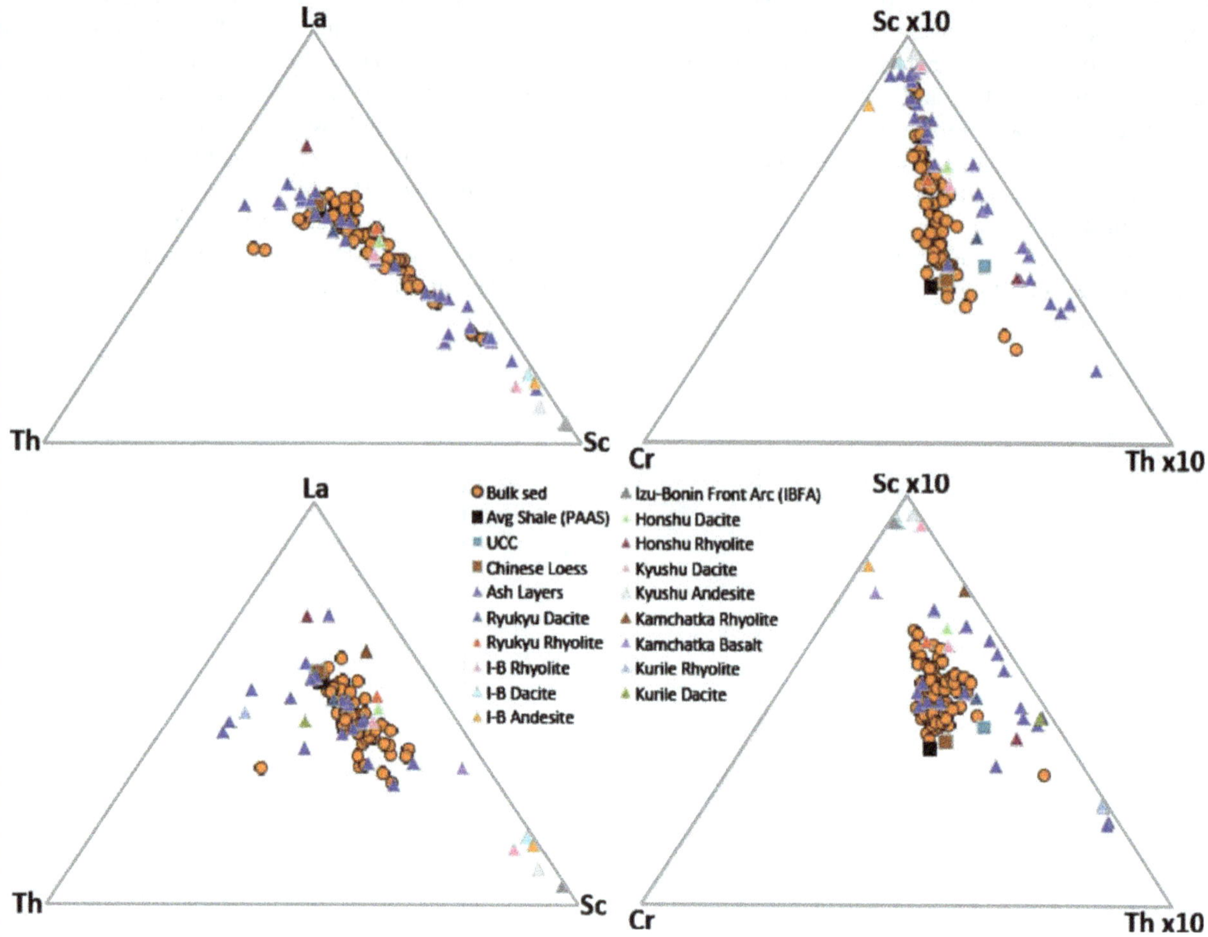

Fig. 14 Ternary diagrams showing potential end members and bulk sediment chemistry at Site 444 (*top*) and Site 579/581 (*bottom*)

content. They find that the refractivity of the transition-metal oxides (e.g., FeO, TiO_2, and MnO) at this site is higher than that of SiO_2 and other oxides, thus the refractive index seems to be driven primarily by the transition-metal oxides. The predominate transition-metal oxides in volcanic glass shards are iron-based (i.e., FeO or Fe_2O_3), thus at Site 444 the refractive index of volcanic glass shards can be determined by the content of the iron oxides and indicate that basaltic ashes are present. At Site 444, the ash layers are divided into several groups based on petrographic characteristics, chemical composition, and age. As discussed above, Furuta and Arai (1980) identify rhyolitic and basaltic ash layers at Site 444. Multiple ash populations are also apparent in our microprobe and bulk ash layer chemistry.

Downcore profiles of the bulk sediment do not necessarily provide clear distinction of a single source or two-component mixing (Fig. 13). Additionally, ternary diagrams show that the bulk sediment does not necessarily lie along a mixing line (Fig. 14) with an upper continental component (e.g., Chinese Loess or PAAS) influencing the sediment in addition to the ash components discussed above. Thus, there is likely at least one additional end member beyond the multiple ash populations that influences the bulk sediment composition. This will be explored further in our future studies.

Based on the above discussion of compositional trends, QFA was performed on the data at Site 444 using a modified element suite compared to that used for Site 1149 and Site 52 (Table 1). When performing factor analysis, there is a balance that must be maintained between the number of elements and the number of samples in the dataset. Reimann et al. (2002) outline a number of parameters by which to define the appropriate number of elements. We maintain a very conservative use of these rules, and as such during our study of Site 444, with 71 samples in the dataset, we restricted our element menu to seven (7) elements. Of the refractory elements chosen as representative of the aluminosilicate end members (Al, Ti, Sc, Cr, Ni, Nb, La, Th), Ni overall shows the least variability between potential sources and thus was removed from the element menu for our statistical analysis of Site 444.

Table 1 Bulk sediment elemental composition (ppm) used for QFA analysis, Site 444

Leg	Site	Hole	Core	Type	Section	Top depth (cm)	Bottom depth (cm)	Depth (msbf)	Al (ppm)	Ti (ppm)	Sc (ppm)	Cr (ppm)	Nb (ppm)	La (ppm)	Th (ppm)
58	444	*	2	R	1	24	26	5.24	79,996	3853	11.27	70.41	7.94	29.26	11.58
58	444	*	2	R	2	57	59	8.07	80,559	3832	14.09	68.73	10.92	36.56	10.75
58	444	*	2	R	3	53.5	55.5	9.54	83,039	3853	11.42	77.58	8.36	28.96	12.48
58	444	*	2	R	4	25	27	10.75	84,474	4190	14.67	88.30	12.79	31.16	11.36
58	444	*	3	R	1	124	126	6.74	81,860	3930	15.15	73.85	11.52	32.08	10.63
58	444	*	3	R	2	107.5	109.5	18.08	84,837	4085	11.31	78.40	7.94	30.28	23.13
58	444	*	3	R	3	86	88	19.36	79,428	3889	11.58	63.46	6.84	28.87	13.82
58	444	*	4	R	1	106	108	26.06	82,894	4157	10.43	80.53	6.65	32.58	26.08
58	444	*	4	R	2	108	110	27.58	80,804	3970	14.96	64.32	11.20	29.80	10.11
58	444	*	4	R	3	108	110	29.08	81,114	3965	14.79	66.16	10.00	33.88	9.93
58	444	*	4	R	4	109	111	30.59	78,989	3969	15.99	68.26	11.75	29.27	10.49
58	444	*	4	R	5	46	48	31.46	79,119	5323	25.88	15.44	3.20	10.26	2.67
58	444	*	5	R	1	62	64	35.12	82,536	3834	15.05	66.76	10.52	32.18	11.10
58	444	*	5	R	2	40.5	42.5	36.41	85,835	4087	13.03	81.69	8.47	32.84	11.82
58	444	*	5	R	3	53	55	38.03	76,601	3582	13.45	60.93	11.02	35.68	11.12
58	444	*	5	R	4	64	66	39.64	85,534	4233	11.88	77.77	7.62	32.26	15.33
58	444	*	6	R	1	76	78	44.76	76,374	3713	13.91	67.92	10.15	29.64	10.31
58	444	*	7	R	1	82.5	84.5	54.33	83,010	4126	17.62	68.63	11.54	32.11	10.78
58	444	*	7	R	2	82.5	84.5	55.83	85,068	4160	16.22	66.89	10.94	31.47	10.62
58	444	*	7	R	3	87	89	57.37	78,574	3564	10.59	50.80	5.41	28.56	9.71
58	444	*	7	R	6	86	88	61.31	77,641	3315	8.65	41.39	3.72	24.15	7.96
58	444	*	8	R	1	49	51	63.49	64,239	2914	11.20	21.17	3.33	11.96	4.33
58	444	*	9	R	1	73.5	75.5	73.24	83,958	3897	16.66	44.20	8.50	33.57	10.12
58	444	A	1	R	1	82.5	84.5	82.83	86,300	4026	17.98	65.34	9.35	29.67	10.09
58	444	*	10	R	1	100.5	102.5	83.01	81,435	3842	16.72	72.61	8.44	34.45	8.59
58	444	*	10	R	2	58	60	84.08	85,360	3903	16.11	69.87	11.43	32.10	11.45
58	444	A	1	R	2	86	88	84.36	82,646	4180	17.91	54.83	8.60	26.22	8.16
58	444	A	1	R	3	60	62	85.6	65,077	3067	10.31	47.18	4.61	27.01	7.58
58	444	A	1	R	4	53	55	87.03	84,916	4014	16.79	65.35	10.50	31.42	10.87
58	444	A	1	R	5	77	79	88.27	86,317	4087	16.05	63.91	10.71	30.28	10.35
58	444	A	1	R	6	82	84	89.78	81,209	3897	12.62	57.46	4.99	29.31	8.33
58	444	A	2	R	2	134	136	94.34	80,643	3913	17.20	61.23	8.65	28.27	9.64
58	444	A	2	R	3	96	98	95.46	80,589	4103	15.22	55.74	5.50	28.58	8.52

Table 1 Bulk sediment elemental composition (ppm) used for QFA analysis, Site 444 *(Continued)*

58	444	A	2	R	4	18	20	96.18	78,731	3810	18.30	49.13	8.16	27.02	8.03
58	444	A	3	R	2	74	76	103.24	82,638	4071	17.58	46.40	9.18	25.53	7.54
58	444	A	3	R	3	73	75	104.73	82,775	4320	12.68	55.50	3.79	26.30	7.31
58	444	A	3	R	4	79	81	106.29	85,566	4046	18.19	61.54	8.92	31.96	9.70
58	444	A	3	R	5	53	55	107.53	83,808	3974	17.42	55.99	8.49	30.56	8.95
58	444	A	6	R	1	34	36	129.84	72,532	3238	17.06	57.16	9.60	28.59	9.97
58	444	A	6	R	2	55	57	130.6	84,911	3983	20.53	44.81	7.98	43.70	8.98
58	444	A	6	R	3	94	96	132.49	84,151	4042	19.21	56.38	9.16	29.40	9.67
58	444	A	6	R	4	52	54	133.57	86,082	3811	14.22	65.93	7.13	30.96	11.47
58	444	A	6	R	5	48	50	134.24	87,133	3982	19.72	38.44	7.93	37.25	8.78
58	444	A	8	R	1	74	76	149.24	82,530	3904	12.79	62.99	4.02	27.01	9.03
58	444	A	9	R	1	95.5	97.5	158.96	82,663	4093	21.47	39.19	5.58	18.86	5.19
58	444	A	9	R	2	89	91	160.39	77,146	4000	19.96	47.48	4.25	25.95	6.92
58	444	A	9	R	3	48	50	161.48	76,262	3576	20.69	37.71	5.49	19.33	5.85
58	444	A	9	R	4	65.5	67.5	162.59	74,302	3564	16.57	47.67	4.26	21.70	6.72
58	444	A	9	R	5	37	39	163.8	76,727	4246	26.94	26.36	4.44	16.22	4.59
58	444	A	10	R	1	45	47	167.95	83,868	3826	17.41	49.19	8.77	39.44	9.45
58	444	A	10	R	2	54.5	56.5	169.55	77,551	4036	17.00	28.06	2.93	16.86	4.13
58	444	A	11	R	1	62.5	64.5	177.63	80,760	4019	22.00	43.95	6.23	22.21	5.90
58	444	A	11	R	2	102	104	179.52	77,568	3853	16.32	31.54	3.37	19.58	4.90
58	444	A	11	R	3	30.5	32.5	180.31	73,192	3451	14.22	30.30	2.33	20.26	5.12
58	444	A	11	R	4	68	70	182.18	73,600	3011	17.74	32.50	4.40	18.30	5.17
58	444	A	12	R	1	53.5	55.5	187.04	79,144	4070	26.12	41.96	6.10	19.34	5.91
58	444	A	12	R	2	60	62	188.6	65,137	3165	17.19	39.43	5.67	18.21	5.61
58	444	A	12	R	3	41	43	189.91	76,284	3793	24.56	47.45	6.42	20.07	6.47
58	444	A	13	R	1	95	97	196.95	81,168	4176	30.31	26.61	3.26	12.44	3.16
58	444	A	13	R	2	61	63	198.11	77,704	4165	28.66	39.02	5.47	18.67	5.39
58	444	A	13	R	3	36	38	199.36	76,949	3981	18.65	32.70	8.20	26.95	6.21
58	444	A	14	R	1	83	85	206.33	70,259	3887	25.01	14.61	2.19	9.13	2.06
58	444	A	14	R	2	62	64	207.62	77,583	3748	21.44	30.50	3.88	13.31	3.88
58	444	A	14	R	3	29	31	208.79	73,567	3401	19.02	34.00	4.49	15.12	4.56
58	444	A	14	R	4	22	24	209.64	79,436	3744	15.03	57.19	5.12	22.21	8.21
58	444	A	15	R	1	53	55	215.53	79,550	3659	15.30	31.16	2.72	11.85	4.16

Table 1 Bulk sediment elemental composition (ppm) used for Q-FA analysis, Site 444 *(Continued)*

58	444	A	15	R	2	125	27	217.75	82,603	3788	22.63	43.75	4.36	17.21	5.07
58	444	A	15	R	3	83	35	218.83	74,140	3484	15.16	37.63	2.89	16.34	4.93
58	444	A	16	R	1	106	108	225.56	79,839	3528	18.10	40.54	8.06	22.48	8.21
58	444	A	16	R	2	79	31	226.79	75,025	3649	20.13	45.88	7.72	20.26	7.33

The results of QFA at Site 444 indicate that three factors (end members) explain 98 % of the variability of the data. This is consistent with the geochemical trends we observe in the bulk data, and the three factors explain 53, 33, and 11 % of the variability of the data respectively (Fig. 15, left). Based on the preliminary modeled compositions of these factors, we interpret that there may be one intermediate crustal source, and at least one dacitic source exhibiting higher Ti/Al ratios, although we cannot differentiate further based on QFA alone. If we were to increase the number of factors, we find that four factors can explain 99 % of the variability of the data, with the four factors explaining 55, 34, 9, and 1 % of the variability of the data, respectively (Fig. 15, right). While this fourth factor is small, we find that it predominantly splits the third factor from the previous QFA and identifies an end member that is dominated by variation in Nb.

In terms of the relative downhole distributions of square factor loadings (Fig. 16), Factor 1 dominates over

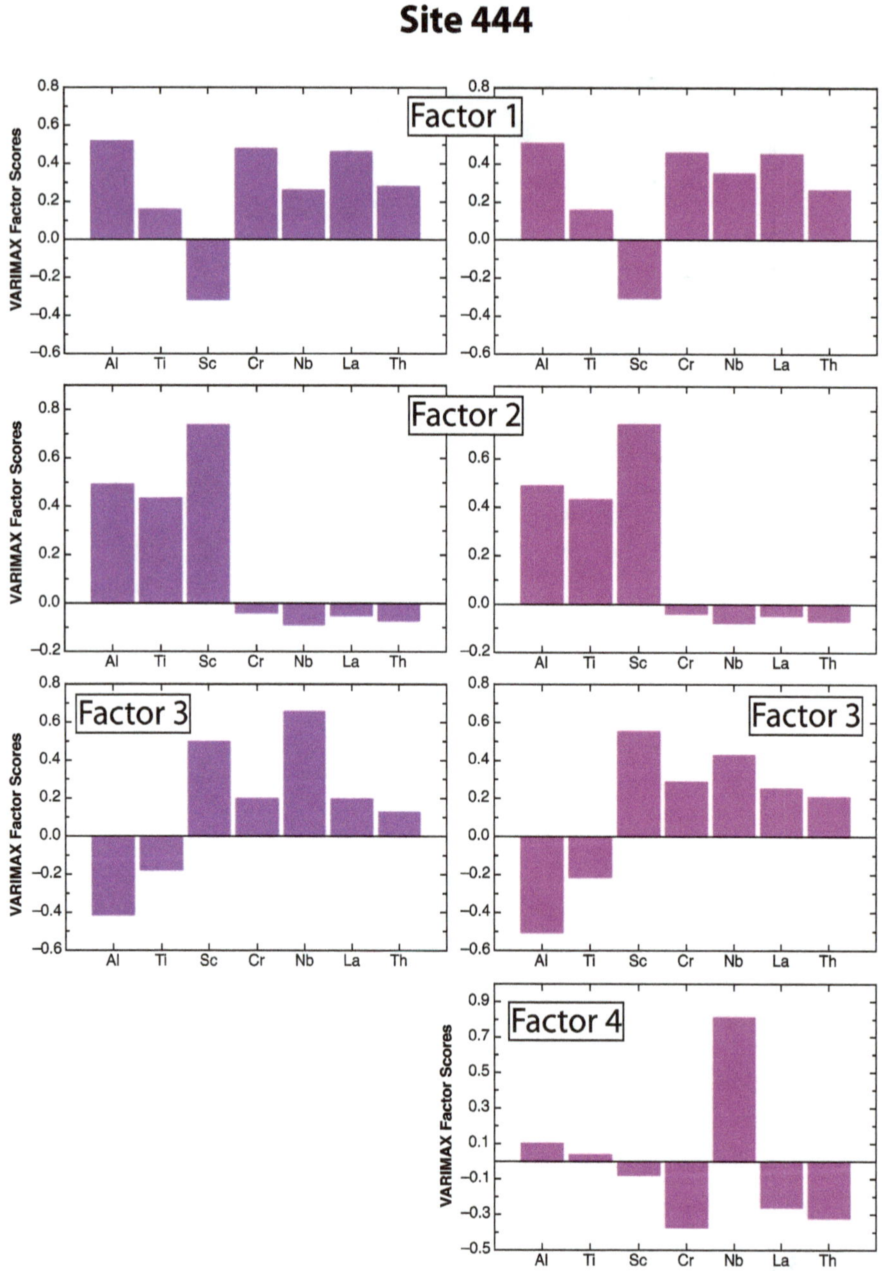

Fig. 15 Site 444, VARIMAX factor scores from QFA. (*Left*) Three-factor model. (*Right*) Four-factor model. As described in the text, Factor 1 is most likely an intermediate crustal source, Factors 2 and/or 3 are most likely dacitic source(s) and Factor 4 is a Nb-rich end member

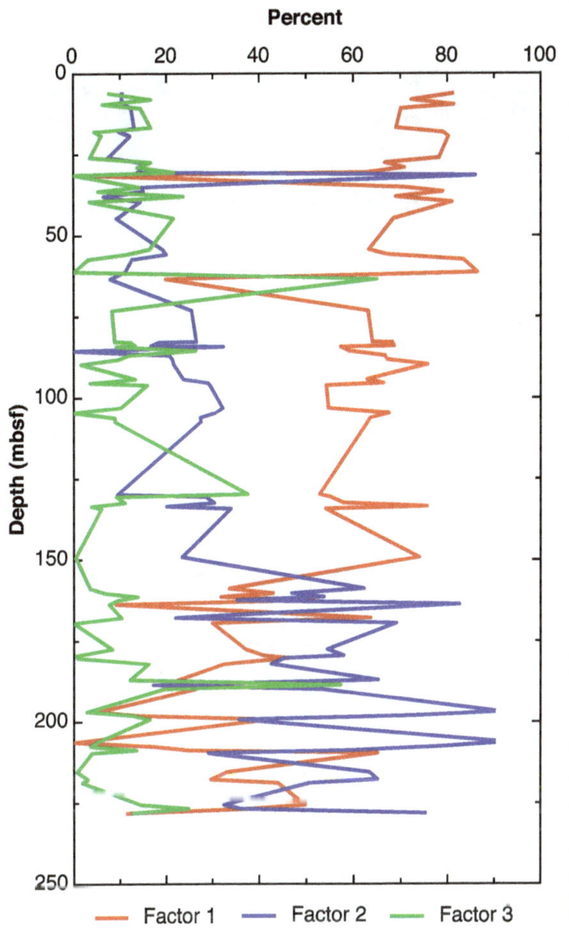

Fig. 16 Depth profiles of square factor loadings from QFA showing the relative contribution of each factor, Site 444

the upper ~150 mbsf and exhibits a broad decrease in importance with depth. Factor 2 is most important in the sequence below ~150 mbsf. Factors 2 and 3 have roughly equal relative contributions over the upper ~50 mbsf, below which Factor 2 exhibits a broad increase while Factor 3 generally decreases. Future statistical analysis by CLS will allow us to determine whether three or four factors are ultimately the appropriate number of end members and will allow us to further determine the sources to the sediment.

Kamchatka and the Kuriles, DSDP Site 579 and Site 581

DSDP Sites 579 and 581 (Fig. 1) were occupied in June 1982 as part of DSDP Leg 86 and were cored as part of a transect across the Kuroshio Current system designed to investigate the paleoceanography of the northwest Pacific. Although spaced relatively far apart, Sites 579 and 581 comprise an excellent composite section of the region (Plank and Langmuir 1998). Site 579 marks the southern margin of the transition between the sediments of the calcareous-siliceous subtropical gyre and the more siliceous subarctic gyre. It was selected for Leg 86 with the goal of compiling a paleoceanographic record of this transition zone for the late Neogene and Quaternary and comparing this record of migrations of the subarctic front with that of Asian eolian inputs. Site 581 is the northernmost of the transect sites drilled during Leg 86 and was designed to provide a high-resolution record of the late Neogene.

As with the other sites in this region, a number of ash layer parameters were examined (Fig. 12). There are no clear trends evident in the number of ash layers through the depth of the site. There are two peaks in the number of ash layers per 5 m, with seven ash layers from 40–45 mbsf and four layers from 80–85 mbsf. This peak at 80 mbsf also corresponds to a peak in total ash thickness (44 cm thick), meaning that not only are more layers present at this depth but also that those layers are thicker. This suggests a period of increased large-scale volcanic eruptions during the Pliocene. However, this is difficult to verify as late Pleistocene glaciations are likely to have obscured the terrestrial record of eruptive history in this area (Volynets et al. 1999). Ash pockets show good correlation with the ash layers in frequency, and we interpret that they originated from disturbance in the core due to the weather during core recovery. The cores had a high occurrence of "flow-in" zones due to heavy seas that may have affected coring and disturbed recovery of some ash layers (Heath et al. 1985).

Interpretation of the bulk ash layers and bulk sediment downhole is not straightforward (Fig. 13). In ternary diagrams, the bulk sediment commonly falls along a mixing line that can be explained by the bulk ash and a relatively more basaltic component, supporting the presence of dispersed ash in the sediment (Fig. 14). Much of the bulk sediment is similar in composition to Chinese Loess, which, while not necessary to explain the composition of the bulk sediment, can be inferred to be almost surely present due to the location of the sites.

Applying QFA to the combined data from these sites, we find that either five or six factors best explain the variability of the data (Table 2, Fig. 17). Comparison of the two sets of factor analysis indicates that Factors 1 through 4 correlate with each other whether there are five or, alternatively, six factors. If there are six factors, Factor 5 from the smaller factor model is split between Factors 5 and 6. Factor 6 explains 1.7 % of the variability of the data in this model, and thus it is interpreted as "real".

Focusing first on the five-factor model, the calculated compositions of the end members identifies Factor 4 (explaining 10.1 % of the variability) as a potential terrigenous source. This is unusual, as in all the other NW pacific sites discussed in this paper the crustal source is the first and/or second end member explaining the

Table 2 Bulk sediment elemental composition (ppm) used for QFA analysis, Sites 579 and 581

Leg	Site	Hole	Core	Type	Section	Top depth (cm)	Bottom depth (cm)	Depth (msbf)	Al (ppm)	Ti (ppm)	Sc (ppm)	Cr (ppm)	Nb (ppm)	La (ppm)	Th (ppm)
86	581	*	1	R	1	28	30	0.28	74,371	4244	20.49	46.23	4.02	16.27	5.36
86	581	*	1	R	1	36	38	0.36	64,845	3268	17.72	39.56	5.38	13.30	4.99
86	579	*	1	H	1	93	95	0.93	70,339	3518	12.17	72.45	6.68	25.01	9.40
86	579	*	1	H	3	75	77	3.75	66,319	2181	10.22	29.39	7.37	8.47	5.93
86	579	*	2	H	1	54	56	8.94	68,056	3532	11.81	58.75	8.02	14.70	6.78
86	579	*	2	H	3	73	75	12.13	73,188	3562	16.02	57.21	8.70	21.20	8.41
86	579	A	1	H	3	94	96	17.94	75,385	3367	15.23	44.72	7.02	15.57	6.65
86	579	A	1	H	7	6	8	23.06	68,144	3364	12.58	54.34	7.94	17.55	6.68
86	579	A	2	H	3	13	15	26.63	69,347	3438	15.07	63.72	9.20	21.34	8.62
86	579	A	3	H	3	73	75	36.73	66,624	2746	13.44	27.85	6.15	10.90	4.88
86	579	A	4	H	3	25	27	45.75	72,574	3398	15.46	59.67	9.37	15.94	7.88
86	579	A	4	H	4	4	6	47.04	43,351	2216	8.89	46.87	8.44	11.55	5.78
86	579	A	5	H	3	107	109	56.07	67,296	3207	14.39	55.31	8.70	21.52	8.26
86	579	A	6	H	3	76	78	65.26	70,026	3410	15.26	57.05	8.74	15.57	7.98
86	579	A	6	H	5	73	75	68.23	53,622	2724	10.20	49.76	6.55	19.78	6.26
86	579	A	6	H	6	10	12	69.1	62,451	2446	8.53	30.52	2.54	17.37	5.54
86	579	A	7	H	3	72	74	74.72	72,871	3585	15.95	59.25	10.44	23.68	9.41
86	579	A	7	H	5	72	74	77.72	68,002	3323	10.60	66.32	4.14	23.43	8.82
86	579	A	8	H	1	52	54	81.02	70,043	3426	11.11	59.48	6.03	23.35	8.69
86	579	A	8	H	3	30	32	83.8	73,144	3565	15.05	61.25	9.79	24.24	9.13
86	579	A	8	H	5	72	74	87.22	59,725	2829	10.54	52.81	6.70	27.02	8.18
86	579	A	9	H	1	110	112	91.1	65,280	3076	10.55	56.03	4.62	21.69	7.78
86	579	A	9	H	3	61	63	93.61	62,214	2294	11.87	38.56	7.77	21.26	8.69
86	579	A	10	H	3	131	133	103.81	64,853	2977	15.04	54.89	7.33	20.29	7.53
86	579	A	10	H	5	72	74	106.22	68,686	7575	17.89	51.60	6.58	15.94	6.28
86	579	A	11	H	1	94	96	109.94	67,054	3167	11.67	57.35	5.18	20.79	8.01
86	579	A	11	H	3	114	116	113.14	60,192	2612	13.67	40.83	6.95	14.15	6.71
86	579	A	12	H	1	83	85	119.33	54,488	2061	7.81	23.08	4.69	19.02	6.70
86	579	A	12	H	3	49	51	121.99	55,246	2474	13.26	41.86	6.21	14.32	6.26
86	579	A	12	H	5	76	78	125.26	63,097	2906	16.01	44.42	5.55	17.65	5.69
86	579	A	13	H	3	74	76	131.74	53,852	2582	13.99	43.75	5.69	12.81	5.30
86	579	A	13	H	5	27	29	134.27	57,088	2580	10.98	50.74	5.24	22.10	6.50
86	579	A	14	H	1	107	109	138.57	54,699	2471	13.11	47.12	6.56	18.01	7.14

Table 2 Bulk sediment elemental composition (ppm) used for QFA analysis, Sites 579 and 581 *(Continued)*

86	579	A	14	H	3	66	68	141.16	55,940	2423	12.06	44.14	6.49	12.61	5.58
86	579	A	14	H	5	70	72	144.2	73,974	3875	10.00	50.23	6.63	19.07	6.45
86	579	A	15	H	2	2	4	48.52	69,270	2925	14.56	55.19	7.95	22.95	8.46
86	579	A	15	H	CC	16	18	149.53	71,892	3146	13.49	62.63	9.86	18.70	9.01
86	581	*	2	R	1	62	64	182.12	42,056	1999	9.89	37.37	2.78	24.82	5.66
86	581	*	2	R	2	2	4	183.02	34,466	1662	12.08	27.39	4.28	18.16	5.07
86	581	*	2	R	3	137	139	185.87	50,664	2181	12.71	28.88	5.89	16.86	5.71
86	581	*	2	R	5	73	75	188.23	61,058	2893	14.22	38.94	5.38	25.91	6.30
86	581	*	3	R	1	58	60	191.58	56,561	6826	14.90	39.01	4.80	14.97	4.67
86	581	*	3	R	3	62	64	194.62	61,976	2550	11.56	33.12	5.22	12.46	4.74
86	581	*	3	R	5	74	76	197.74	27,691	1359	9.00	28.40	4.15	13.98	4.77
86	581	*	4	R	1	57	59	201.07	33,889	1641	8.65	24.95	2.31	15.01	4.71
86	581	*	4	R	3	76	78	204.26	46,571	2231	14.57	47.58	6.51	16.62	7.06
86	581	*	4	R	5	67	69	207.17	30,626	1412	6.83	26.45	1.97	8.75	4.53
86	581	*	5	R	1	126	128	211.26	60,537	2704	9.60	42.27	3.69	12.58	5.50
86	581	*	5	R	3	92	94	213.92	39,316	1781	11.23	32.04	5.22	7.86	4.84
86	581	*	6	R	1	86	88	220.36	51,086	2214	9.28	46.00	2.83	18.85	6.78
86	581	*	6	R	3	67	69	223.17	64,892	3061	16.55	54.76	8.36	18.97	8.96
86	581	*	6	R	5	74	76	226.24	58,477	2715	11.20	50.82	4.27	23.34	9.07
86	581	*	7	R	1	52	54	229.52	54,775	2401	10.65	50.44	3.28	21.27	8.15
86	581	*	7	R	3	36	38	232.36	70,340	2946	17.45	54.41	9.17	16.29	10.10
86	581	*	7	R	5	37	39	235.37	59,378	2529	11.56	50.07	3.26	23.01	8.75
86	581	*	8	R	1	77	79	239.27	63,987	2643	14.31	51.64	5.33	22.36	10.21
86	581	*	8	R	3	11	13	241.61	44,113	2001	14.00	38.89	6.23	17.51	7.56
86	581	*	8	R	3	115	117	242.65	64,594	2703	19.14	47.94	8.11	22.13	10.42
86	581	*	8	R	5	77	79	245.27	83,065	3497	22.35	61.47	10.04	33.19	14.07
86	581	*	9	R	1	16	18	248.16	83,298	3381	22.88	57.87	9.33	31.95	13.20
86	581	*	9	R	3	59	61	251.59	82,323	3316	25.21	57.11	10.80	20.48	13.59
86	581	*	9	R	4	46	48	252.96	85,210	3411	24.70	51.25	9.51	31.48	14.47
86	581	*	10	R	1	42	44	257.92	89,185	3707	25.16	60.99	11.65	34.16	16.84
86	581	*	10	R	1	124	126	258.74	86,271	3549	28.05	55.41	11.20	39.01	17.72
86	581	*	10	R	3	92	94	261.42	88,986	3546	28.26	55.69	11.63	17.12	11.30

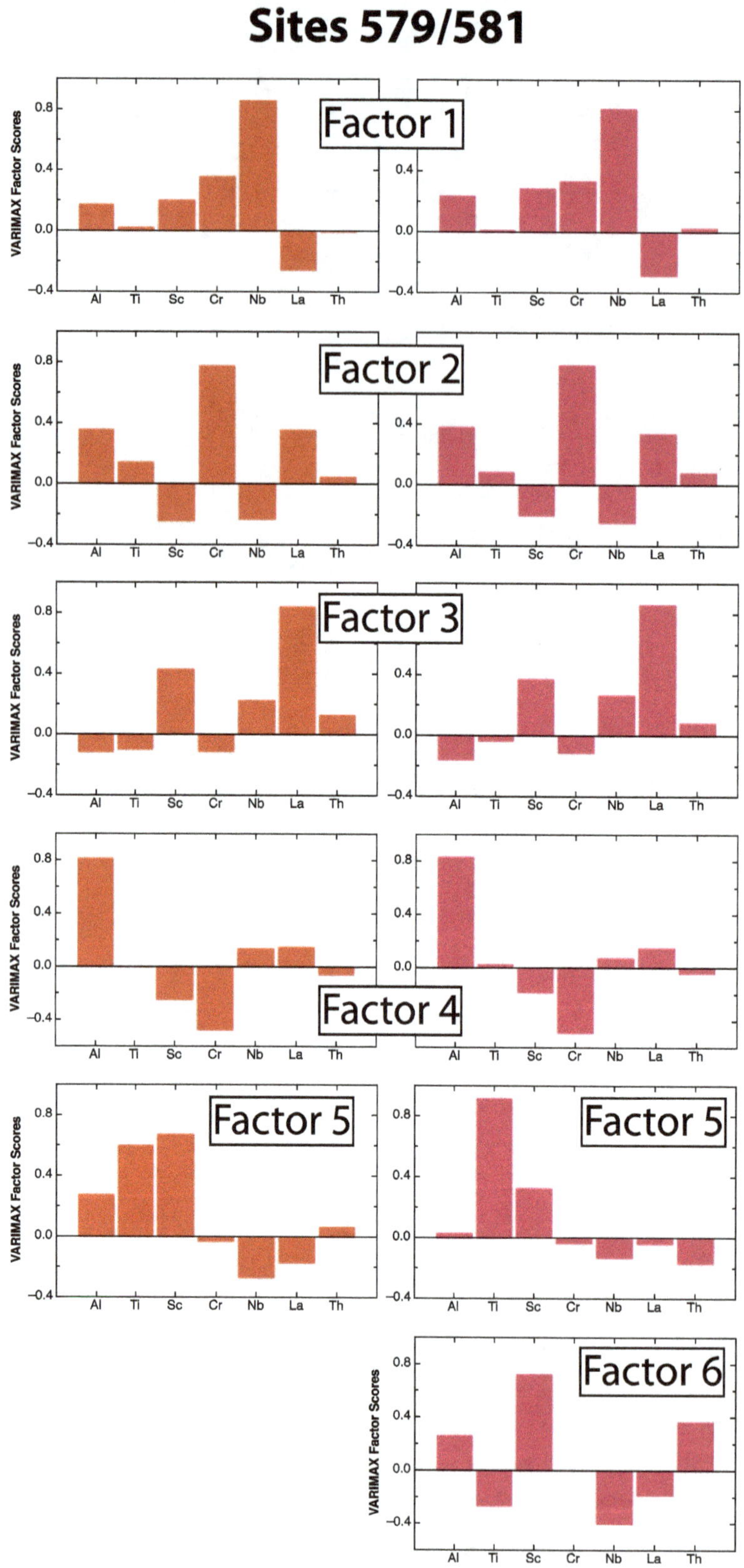

Fig. 17 Sites 579 and 581, VARIMAX factor scores from QFA

majority of the variability in the dataset. This speaks to the importance of dispersed ash at Site 579/581, since the dispersed ash component(s) appear to be controlling most of the variability. This will be further explored in future work. The six-factor model identifies a few factors that are dominated by a single element (Al in Factor 4, Ti in Factor 5) indicating that the division of factors is controlling variability as opposed to their specific concentrations.

As with Site 444, square factor loadings for these Kamchatka/Kurile sites can allow us to assess the relative contribution of the individual factors on a sample-by-sample basis (Fig. 18). In order to assess the relative contributions for this work, we will focus here on the five-factor model. We observe that for most of the sedimentary record, Factor 1 dominates the record, although from ~65 to ~93 mbsf, Factor 2 is the dominate component. Additionally, from ~150 to ~240 mbsf the variance in the record is dominated by Factor 3 (~150 to ~200 mbsf) or Factor 2 (~200 to ~240 mbsf). Factors 4 and 5 are remarkably similar in importance until ~200 mbsf below which Factor 4 increases in relative importance while Factor 5 remains fairly stable. Future statistical analysis by CLS will allow us to determine whether five or six factors are ultimately the appropriate number of end members and will allow us to further determine the sources to the sediment.

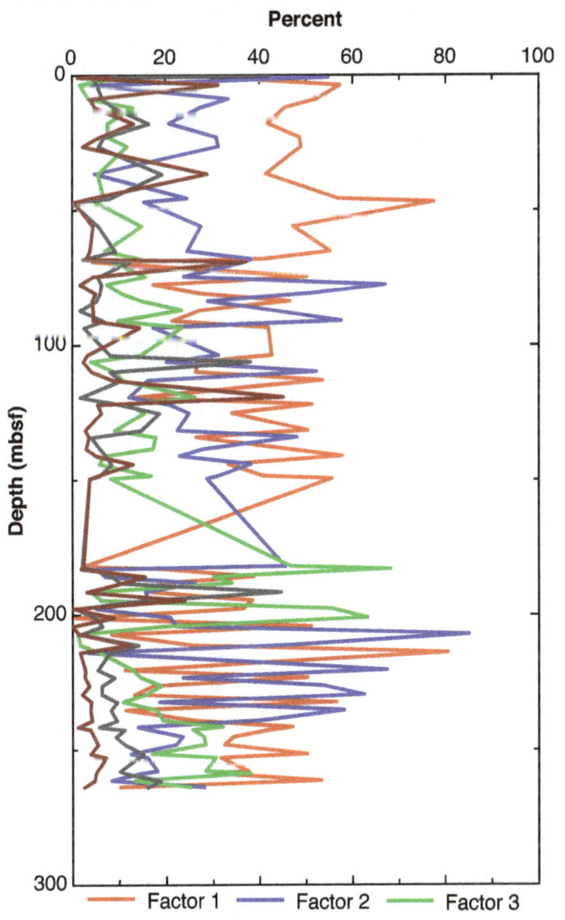

Fig. 18 Depth profiles of square factor loadings from QFA showing the relative contribution of each factor, Sites 579 and 581

Conclusions

Although the presence of dispersed ash in marine sediment has been long known by marine sedimentologists, the overall quantification of its abundance and its composition has recently received renewed attention. The dispersed ash, either altered or unaltered, is extremely difficult to differentiate from detrital/terrigenous/authigenic clay as they are all, fundamentally speaking, "aluminosilicates". The combined geochemical (major, trace, REEs) and statistical (Pisias et al. 2013) approach provides the chemical context for the determination of provenance and the distinctive resolution of the different aluminosilicate components based on their individual geochemical signature(s).

The case studies we present here show that this approach can be useful in multiple different arc systems. We have also shown that by adding the dispersed ash component to the study of discrete ash layers, many additional geological and oceanographic interpretations can be uncovered. Most obviously, we have shown how geochemical modeling of the bulk sediment composition can indicate the presence of an additional ash source that is not recorded in the discrete layer record. Furthermore, in cores where there are no remaining discrete layers (either due to natural bioturbation or man-made coring disturbance), we show that important ash contributions may still be recognized and studied for their composition and accumulation. We have also shown how the accumulation of dispersed ash will either best correlate with the *thickest* ash layer, or alternatively, the *frequency* of the discrete ash layers, which lead to dramatically different interpretations regarding the volcanic mechanisms of a given arc. Finally, when considering these studies throughout the northwest Pacific Ocean, or in marginal basins such as the Japan Sea, our approach may be used to develop a mass balance inventory of the total mass of volcanic ash that is more accurate than estimates based solely on the combined thicknesses of the discrete layers.

There are many other future research directions that are a natural extension of the work described above. There are a number of regions for which the characterization of the abundance and extent of dispersed ash would be of great scientific and societal importance (e.g., Tonga, Costa Rica, and Indonesia), and studying the dispersed ash component in these regions will most certainly lead to an improved understanding of explosive volcanism in those systems. Moreover, most marine volcanoes are beneath the sea surface, either in mid-ocean ridge settings or in convergent margins (e.g., Fiske et al. 2001; White et al. 2003). The vast majority of these eruptions go unobserved and their products are often unrecovered and thus are understudied. Often the only evidence of subaqueous eruptions are widespread pumice rafts. Little is known, however, about the fate of these rafts, yet they too are likely to contribute dispersed ash to the sediment.

Volcanic ash may have significant nutrient fertilization potential as well. Particulate matter from large volcanic eruptions may release enough iron and other nutrients to the surface ocean to stimulate primary production over short time scales (Watson 1997; Olgun et al. 2011). Thus, it would be important to develop an understanding of the effects of iron and other elements from volcanic ash on the biological productivity of the open ocean and, second, to consider the biological effects of volcanic ash-driven elemental fertilization in the open ocean through Earth's history. The use of the dispersed ash record can be applied to achieve these goals.

Additionally, ash, with its major component being volcanic glass, is a metastable phase with high potential biological availability to the subseafloor ecosystem, may provide both electron donors and electron acceptors important to microbial energetics. Characterizing the chemistry of the subseafloor sediment will provide vital information about the nutrients available to subseafloor microbial life as well as the pathways by which these nutrients are utilized.

Abbreviations

BNN: Izu-Bonin Boninite; CL: Chinese Loess; CLS: Constrained Least Squares; CRISP II: Costa Rica Seismogenesis Project Stage II; DSDP: Deep Sea Drilling Project; HR: Honshu Rhyolite; IBA: Izu-Bonin Andesite; IBFA: Izu-Bonin Front Arc; IBR: Izu-Bonin Rhyolite; ICP-ES: Inductively Coupled Plasma-Emission Spectrometer; ICP-MS: Inductively Coupled Plasma-Mass Spectrometer; IODP: Integrated Ocean Drilling Program; LRC: lower red clay MAR, mass accumulation rate; MORB: Mid-Ocean Ridge Basalt; ODP: Ocean Drilling Program; PAAS: Post-Archean average Australian Shale; PGL: pale green laminae; QFA: Q-Mode Factor Analysis; RCB: Rotary Core Barrel; REE: rare earth element; TI: Total Inversion; URC: upper red clay.

Competing interests

The authors' declare that they have no competing interests.

Authors' contributions

RPS is the lead author of the paper, performed the ICP-ES and ICP-MS analyses, and led the research and multiple conversations/interactions with the research team. RWM funded the research (see Acknowledgements), performed the oversight of the project, advised then-Ph.D. student RPS and then-undergraduate students SG, RL, and CCM, facilitated the discussion with co-authors and community, and contributed the text. JCS and SK performed microprobe analyses of volcanic glass from ash layers, discussed the research, contributed revisions of the text, compiled the reference list, created Figs. 1, 2, and 4, and provided the interpretation of the data. SK funded the research (see Acknowledgements) and advised Ph.D. student JCS regarding ash chemistry. FH reviewed and revised the manuscript. MBU discussed the research, contributed the revisions of the text, and provided the interpretation of the data. SG, RL, and CCM assisted in the ICP-ES and ICP-MS analyses, were involved in the preliminary statistics and interpretations of Sites 52, 444, and 579/581, and reviewed and commented on the manuscript. All authors read and approved the final manuscript.

Acknowledgements

This research used samples and/or data provided by the IODP. RPS and RWM thank the US Science Support Program (USSSP) of IODP, NSF OCE-0136855, and NSF OCE-0958002 for financial support, and T. Ireland, A. G. Dunlea, N. Murphy, and J. W. Sparks for laboratory assistance. JCS and SK thank the German Research Foundation for financial support (KU-2685/1-1, 2-1&2) and M. Thöner, K. Strehlow for laboratory assistance. Portions of this material are based upon work supported while RWM was serving at the National Science Foundation. Thanks to those who reviewed and commented on the manuscript.

Author details

[1]Department of Earth & Environment, Boston University, Boston, MA 02215, USA. [2]GEOMAR Helmholtz Centre for Ocean Research Kiel, 24148 Kiel, Germany. [3]Department of Earth and Environmental Science, New Mexico Institute of Mining and Technology, Socorro, NM 87801, USA. [4]Department of Geosciences, Princeton University, Princeton, NJ 08544, USA. [5]Division of Earth and Ocean Sciences, Nicholas School of the Environment, Duke University, Durham, NC 27708, USA. [6]Department of Oceanography, Texas A&M University, College Station, TX 77843, USA.

References

Arevalo R, McDonough WF (2010) Chemical variations and regional diversity observed in MORB. Chem Geol 271(1):70–85

Bailey JC (1993) Geochemical history of sediments in the northwestern Pacific Ocean. Geochem J 27:71–90

Bailey JC (1996) Role of subducted sediments in the genesis of Kurile-Kamchatka island arc basalts: Sr isotopic and elemental evidence. Geochem J 30(5):289–321

Bryan SE, Cook A, Evans JP, Colls PW, Wells MG, Lawrence MG et al (2004) Pumice rafting and faunal dispersion during 2001-2002 in the Southwest Pacific: record of a dacitic submarine explosive eruption from Tonga. Earth Planet Sci Lett 227(1-2):135–154

Bryant CJ, Arculus RJ, Eggins SM (1999) Laser ablation-inductively coupled plasmamass spectrometry and tephras: a new approach to understanding arc-magma genesis. Geology 27(12):1119–1122

Bryant CJ, Arculus RJ, Eggins SM (2003) The geochemical evolution of the Izu–Bonin arc system: a perspective from tephras recovered by deep-sea drilling. Geochem Geophys Geosyst 4(11):1094. doi:10.1029/2002GC000427

Bursik M (2001) Effect of wind on the rise height of volcanic plumes. Geophys Rese Lett 28(18):3621–3624

Cambray H, Cadet JP, Pouclet A (1993) Ash layers in deep-sea sediments as tracers of arc volcanic activity: Japan and Central America as case studies. Island Arc 2:72–86

Cambray H, Pubellier M, Jolivet L, Pouclet A. Volcanic activity recorded in deep-sea sediments and the geodynamic evolution of western Pacific island arcs in Active Margins and Marginal Basins of the Western Pacific. In B Taylor and J Natland, editors. Washington DC Geophys Monogr Ser 88:97–124 AGU Washington DC; 1995.

Carey SN (1997) Influence of convective sedimentation on the formation of widespread tephra fall layers in the deep sea. Geology 25(9):839–842

Carey SN, Sigurdsson H (1978) Deep-sea evidence for distribution of tephra from the mixed magma eruption of the Soufriére on St Vincent 1902: ash turbidites and airfall. Geology 6:271–274

Carey SN, Sigurdsson H (1980) The Roseau ash: Deep-sea tephra deposits from a major eruption on Dominica, Lesser Antilles arc. J Volcanol Geoth Res 7:67–86

Carey SN, Sigurdsson H (2000) Grain size of Miocene volcanic ash layers from Sites 998 999 and 1000: Implications for source areas and dispersal. Proc Ocean Drill Program Sci Results 165:101–110

Carey SN, Sparks RSJ (1986) Quantitative models of fallout and dispersal of tephra from volcanic eruption columns. Bull Volcanol 48:109–125

Carey RJ, Houghton BF, Thordarson T (2010) Tephra dispersal and eruption dynamics of wet and dry phases of the 1875 eruption of Askja Volcano Iceland. Bull 72(3):259_278

Chan LH, Kastner M (2000) Lithium isotopic compositions of pore fluids and sediments in the Costa Rica subduction zone: implications for fluid processes and sediment contribution to the arc volcanoes. Earth Planet Sci Lett 183(1):275–290

de Vries KG, Kobayashi K (1980) Initial Reports of the Deep Sea Drilling Project v. 58. US Goverment Printing Office, Washington DC, p 1022 deVries Klein G, McConville RL, Harris JM, Steffensen CK (1980) Petrology and Diagenesis of Sandstones Deep Sea Drilling Project Site 445 Daito Ridge. Init Rep DSDP 58:609–616. doi:10.2973/dsdpproc581121980

Duggen S, Olgun N, Croot P, Hoffmann LJ, Dietze H, Delmelle P et al (2010) The role of airborne volcanic ash for the surface ocean biogeo- chemical iron cycle: A review. Biogeosci 7:827–844. doi:10.5194/bg-7-827-2010

Dunlea AG, Murray RW, Sauvage J., Pockalny RA, Spivack AJ, Harris RN, et al. Cobalt-based age models of pelagic clay in the South Pacific Gyre. Geochem Geophys Geosys. 2015a;16. doi:10.1002/2015GC005892.

Dunlea AG, Murray RW, Sauvage J, Pockalny RA, Spivack AJ, Harris RN, et al. Dust, volcanic ash, and the evolution of the South Pacific Gyre through the Cenozoic. Paleocean. 2015b; 30. doi:10.1002/2015PA00282.

Dymond J (1981) Geochemistry of Nazca plate surface sediments: An evaluation of hydrothermal biogenic detrital and hydrogenous sources. Geol Soc Am Memoirs 154:133–174

Expedition 340 Scientists Lesser Antilles volcanism and landslides: Implications for hazard assessment and long-term magmatic evolution of the arc. Prelim Rep Integrated Ocean Drill Program 340. 2012. doi:10.2204/iodppr3402012.

Fischer AG et al. Site 52. Initial Rep Deep Sea Drill Proj 6:247–290. 1971. doi:10.2973/dsdpproc61101971.

Fisher RV, Schmincke H-U (1984) Pyroclastic Rocks. Springer-Verlag, New York, p 472

Fiske RS, Naka J, Iizasa K, Yuasa M, Klaus A (2001) Submarine silicic caldera at the front of the Izu-Bonin arc Japan: voluminous seafloor eruptions of rhyolite pumice. Geol Soc Amer Bull 113:813–824

Freundt A, Grevemeyer I, Rabbel W, Hansteen TH, Hensen C, Wehrmann H et al (2014) Volatile (H_2O, CO_2, Cl, S) budget of the Central American subduction zone. Int J Earth Sci 103(7):2101–2127. doi:10.1007/s00531-014-1001-1

Frogner P, Gíslason SR, Óskarsson N (2001) Fertilizing potential of volcanic ash in ocean surface water. Geology 29:487–490

Furuta T, Arai F (1980) Petrographic and geochemical properties of tephras in Deep Sea Drilling Project cores from the north Philippine Sea. Init Rep DSDP 58:617–627. doi:10.2973/dsdpproc581131980

Gardner JV, Nelson CS, Baker PA. Distribution and character of pale-green laminae in sediment from Lord Howe Rise: A probable late Neogene and Quaternary tephrostratigraphic record. In Blakeslee JH (ed) Initial Reports of the DSDP 90, Wahsington, 1986. p. 1145–1158.

Hauff F, Hoernle K, Schmidt A (2003) Sr-Nd-Pb composition of Mesozoic Pacific oceanic crust (Site 1149 and 801 ODP Leg 185): Implications for alteration of ocean crust and the input into the Izu-Bonin-Mariana subduction system. Geochem Geophys Geosyst 4(8):8913. doi:10.1029/2002GC000421

Heath GR, Kovar RB, Lopez C, Campi GL (1985) Elemental Composition of Cenozoic Pelagic Clays from Deep Sea Drilling Project Sites 576 and 578 Western North Pacific. Init Rep DSDP 86:605–648. doi:10.2973/dsdpproc861271985

Hein JR, Scholl DW, Barron JA, Jones MG, Miller J (1978) Diagenesis of late Cenozoic diatomaceous deposits and formation of the bottom simulating reflector in the southern Bering Sea. Sedimentol 25(2):155–181

Henry P, Kanamatsu T, Moe K, Expedition 333 Scientists. Proceedings IODP 333, Tokyo Integrated Ocean Drilling Program Management, International Inc Washington DC; 2012. doi:10.2204/iodpproc3332012.

Hochstaedter A, Gill J, Peters R, Broughton P, Holden P, Taylor B. Across-arc geochemical trends in the Izu-Bonin arc: Contributions from the subducting slab Geochem Geophys Geosys. 2001;2(7). doi:10.1029/2000GC000105.

Hovan SA, Rea DK, Pisias NG (1991) Late Pleistocene continental climate and oceanic variability recorded in northwest Pacific sediments. Paleoceanography 6(3):349–370

Huang TC (1980) A volcanic sedimentation model: Implications of processes and responses of deep-sea ashes. Mar Geol 38:103–122

Huang TC, Watkins ND, Shaw DM, Kennett JP (1973) Atmospherically transported volcanic dust in South Pacific deep sea sedimentary cores at distances over 3000 km from the eruptive source. Earth Planet Sci Lett 20(1):119 124

Huang TC, Fillon RH, Watkins ND, Shaw DM (1974) Volcanism and silicious microfaunal diversity in the southwest Pacific during the Pleistocene period. Deep-Sea Res 21:377–384

Huang TC, Watkins ND, Shaw DM (1975) Atmospherically transported volcanic glass in deep-sea sediments: Volcanism in sub-antarctic latitudes of the south Pacific during late Pliocene and pleistocene time. Bull Geol Soc Am 86(9):1305–1315

Hüpers A, Ikari MJ, Dugan B, Underwood MB, Kopf AJ (2015) Origin of a zone of anomalously high porosity in the subduction inputs to Nankai Trough. Mar Geol 361:47–162

Hyeong K, Park S, Yoo CM, Kim K. Mineralogical and geochemical compositions of the eolian dust from the northeast equatorial Pacific and their implications on paleolocation of the Intertropical Convergence Zone. Paleoceanography. 2005;20(0883-8305). doi: 10.1029/2004PA001053.

Jahn BM, Gallet S, Han JM (2001) Geochemistry of the Xining Xifeng and Jixian sections Loess Plateau of China: eolian dust provenance and paleosol evolution during the last 140 ka. Chem Geol 178(1–4):71–94

Jones MT, Gíslason SR (2008) Rapid releases of metal salts and nutrients following the deposition of volcanic ash into aqueous environments. Geochim Cosmochim Acta 72:3661–3680. doi:10.1016/jgca200805030

Jordan BR, Sigudsson H, Carey SN, Rogers R, Ehrenborg J. Geochemical correlation of Caribbean Sea tephra layers with ignimbrites in Central America. In: Siebe C, Macías JL, Aguirre-Díaz GJ (eds) Neogene-Quaternary Continental Margin Volcanism: A Perspective from México 2006. p. 175-208

Kastner M, Elderfield H, Martin JB (1991) Fluids in convergent margins: what do we know about their composition origin role in diagenesis and importance for oceanic chemical fluxes? Philos Trans R Soc Lond A 335(1638):243–259

Kennett JP, McBirney AR, Thunell RC (1977) Episodes of Cenozoic volcanism in the circum-Pacific region. J Volc Geo Res 2:145–163

Klinkhammer GP, Elderfield H, Edmond JM, Mitra A (1994) Geochemical implications of rare earth element patterns in hydrothermal fluids from mid-ocean ridges. Geochim Cosmochim Acta 58(23):5105–5113

Kryc KA, Murray RW, Murray DW (2003) Al-to-oxide and Ti-to-organic linkages in biogenic sediment: relationships to paleo-export production and bulk Al/Ti. Earth Planet Sci Lett 211(1):125–141

Kutterolf S, Freundt A, Peréz W, Mörz T, Schacht U, Wehrmann H, et al. The Pacific offshore record of Plinian arc volcanism in Central America part 1: Along-arc correlations. Geochem Geophys Geosyst. 2008a;9(2). doi:10.1029/2007GC001631.

Kutterolf S, Freundt A, Peréz W. The Pacific offshore record of Plinian arc volcanism in Central America part 2: Tephra volumes and erupted masses. Geochem Geophys Geosys. 2008b;9(2). doi:10.1029/2007GC001791.

Kutterolf S, Freundt A, Schacht U, Bürk D, Harders R, Mörz T, et al. The Pacific offshore record of Plinian arc volcanism in Central America part 3: Application to forearc geology. Geochem Geophys Geosys. 2008c;9(2). doi:10.1029/2007GC001826.

Kutterolf S, Jegen M, Mitrovica JX, Kwasnitschka T, Freundt A, Huybers P (2013) A detection of Milankovitch frequencies in global volcanic activity. Geology 41(2):227–230. doi:10.1130/G334191

Kutterolf S, Schindlbeck JC, Scudder RP, Murray RW, Pickering KT, Freundt A et al (2014) Large volume submarine ignimbrites in the Shikoku Basin: An example for explosive volcanism in the Western Pacific during the Late Miocene. Geochem Geophys Geosys 15(5):1837–1851

Kyte FT, Leinen M, Heath GR, Zhou L (1993) Cenozoic sedimentation history of the central North Pacific: Inferences from the elemental geochemistry of core LL44-GPC3. Geochim Cosmochim Acta 57:1719–1749

Le Friant A, Ishizuka O, Stroncik NA and the Expedition 340 Scientists (2013) Proceedings of Integrated Ocean Drilling Program vol 340, Tokyo. doi:10.2204/iodp.proc.340.104.2013

Ledbetter MT, Sparks RSJ (1979) Duration of large-magnitude explosive eruptions deduced from graded bedding in deep-sea ash layers. Geology 7:240–244

Lee J, Stern RJ, Bloomer SH (1995) Forty million years of magmatic evolution in the Mariana arc: the tephra glass record. J Geophys Res 100(B9):17671–17687

Leinen M (1987) The origin of paleochemical signatures in North Pacific pelagic clays: Partitioning experiments. Geochim Cosmochim Acta 51(2):305–319

Leinen M, Pisias N (1984) An objective technique for determining end-member compositions and for partitioning sediments according to their sources. Geochim Cosmochim Acta 48(1):47–62

Lowe DJ (2011) Tephrochronology and its application: A review. Quat Geochronol 6:107–153. doi:10.1016/j.quageo.2010.08.003

Machida H, Arai F (1983) Extensive ash falls in and around the Sea of Japan from large late Quaternary eruptions. J Volc Geotherm Res 18(1):151–164

Mackenzie FT, Garrels RM (1966) Chemical mass balance between rivers and oceans. Am J Sci 264(7):507–525

Mahony SH, Wallace LM, Miyoshi M, Villamor P, Sparks RSJ, Hasenaka T (2011) Volcano-tectonic interactions during rapid plate-boundary evolution in the Kyushu region SW Japan. Geol Soc Am Bull 123:2201–2223

Maicher D, White JDL (2001) The formation of deep-sea Limu o Pele. Bull Volcanol 63:482–496

Martinez NC, Murray RW, Thunell RC, Peterson LC, Muller-Karger F, Astor Y et al (2007) Modern climate forcing of terrigenous deposition in the tropics (Cariaco Basin Venezuela). Earth Planet Sci Lett 264(3):438–451

Martinez NC, Murray RW, Dickens GR, Kölling M. Discrimination of sources of terrigenous sediment deposited in the central Arctic Ocean through the Cenozoic. Paleoceanography. 2009;24(1). doi:10.1029/2007PA001567.

Martinez NC, Murray RW, Thunell RC, Peterson LC, Muller-Karger F, Lorenzoni L et al (2010) Local and regional geochemical signatures of surface sediments from the Cariaco Basin and Orinoco Delta Venezuela. Geology 38(2):159–162

Metzner D, Kutterolf S, Toohey M, Timmreck C, Niemeier U, Freundt A et al (2014) Radiative forcing and climate impact resulting from SO2 injections based on a 200,000 year record of Plinian eruptions along the Central American Volcanic Arc. Int J Earth Sci 103(7):2063–2079. doi:10.1007/s00531-012-0814-z

Murray RW, Leinen M (1993) Chemical transport to the seafloor of the equatorial Pacific Ocean across a latitudinal transect at 135 W: tracking sedimentary major trace and rare earth element fluxes at the Equator and the Intertropical Convergence Zone. Geochim Cosmochim Acta 57(17):4141–4163

Murray RW, Leinen M (1996) Scavenged excess aluminum and its relationship to bulk titanium in biogenic sediment from the central equatorial Pacific Ocean. Geochim Cosmochim Acta 60(20):3869–3878

Murray RW, Leinen M, Isern A (1993) Biogenic flux of Al to sediment in the central equatorial Pacific Ocean: Evidence for increased productivity during glacial periods. Paleoceanography 8(5):651–670

Murray RW, Leinen M, Murray DW, Mix AC, Knowlton CW (1995) Terrigenous Fe input and biogenic sedimentation in the glacial and interglacial equatorial Pacific Ocean. Global Biogeochem Cycles 9(4):667–684

Murray RW, Knowlton C, Leinen M, Mix AC, Polsky CH (2000) Export production and carbonate dissolution in the central equatorial Pacific Ocean over the past 1 Myr. Paleoceanography 15(6):570–592

Ninkovich D, Sparks RSJ, Ledbetter MT (1978) The exceptional magnitude and intensity of the Toba eruption Sumatra: An example of the use of deep-sea tephra layers as a geological tool. Bull Volcanol 41:286–298

Olgun N, Duggen S, Croot PL, Delmelle P, Dietze H, Schacht U, et al. Surface ocean iron fertilization: the role of subduction zone and hotspot volcanic ash and fluxes into the Pacific Ocean Global Biogeochemical Cycles. 2011;25. doi:10.1029/2009GB003761

Peacock SA (1990) Fluid processes in subduction zones. Science 248(4953):329–337

Peters JL, Murray RW, Sparks J, Coleman DS (2000) Terrigenous matter and dispersed ash in sediment from the CaribbeanSea: Results from Leg 165. Proc Ocean Drill Program Sci Results 165:115–124. doi:10.2973/odpprocsr1650032000

Pickering KT, Underwood MB, Saito S, Naruse H, Kutterolf S, Scudder RP et al (2013) Depositional architecture provenance and tectonic/eustatic modulation of Miocene submarine fans in the Shikoku Basin: Results from Nankai Seismogenic Zone experiment. Geochem Geophys Geosys 14:6. doi:10.1002/ggge.20107

Piepgras DJ, Jacobsen SB (1992) The behavior of rare earth elements in seawater: Precise determination of variations in the North Pacific water column. Geochim Cosmochim Acta 56(5):1851–1862

Pisias NG, Murray RW, Scudder RP (2013) Multivariate statistical analysis and partitioning of sedimentary geochemical data sets: General principles and specific MATLAB scripts. Geochem Geophys Geosys 14(10):4015–4020. doi:10.1002/ggge20247

Plank T (2005) Constraints from Thorium/Lanthanum on sediment recycling at subduction zones and the evolution of continents. J Petrol 46:921–944

Plank T, Langmuir CH (1998) The chemical composition of subducting sediment and its consequences for the crust and mantle. Chem Geol 145(3–4):325–394

Plank T, Ludden JN, Escutia C, et al. Proc Ocean Drill Program Init Results. 2000; 185:1–190. doi:10.2973/odpprocir1852000.

Plank T, Kelley KA, Murray RW, Stern LQ (2007) Chemical composition of sediments subducting at the Izu-Bonin trench. Geochem Geophys Geosyst 8: Q04I16. doi:10.1029/2006GC001444

Presti M, Michalopoulos P (2008) Estimating the contribution of the authigenic mineral component to the long-term reactive silica accumulation on the western shelf of the Mississippi River Delta. Cont Shelf Res 28(6):823–838

Rea DK (1994) The palcoclimatic record provided by eolian deposition in the deep sea: The geologic history of wind. Rev Geophys 32:159–195

Rea DK, Leinen M (1988) Asian aridity and the zonal westerlies: Late Pleistocene and Holocene record of eolian deposition in the northwest Pacific Ocean. Palaeogeog Palaeoclimatol Palaeoecol 66(1-2):1–8

Rea DK, Snoeckx H, Joseph LH (1998) Late Cenozoic eolian deposition in the North Pacific: Asian drying Tibetan uplift and cooling of the northern hemisphere. Paleoceanography 13(3):215–224

Reid P, Carey SN, Ross DR (1996) Late quaternary sedimentation in the Lesser Antilles island arc. Geol Soc Am Bull 108:78–100

Reimann C, Filzmoser P, Garrett RG (2002) Factor analysis applied to regional geochemical data: problems and possibilities. App Geochem 7(3):185–206

Risso C, Scasso RA, Aparicio A (2002) Presence of large pumice blocks on Tierra del Fuego and South Shetland Islands shorelines from 1962 South Sandwich Islands eruption. Mar Geol 186:413–422

Robock A, Ammann CM, Oman L, Shindell D, Levis S, Stenchikov G. Did the Toba volcanic eruption of ~ 74ka BP produce widespread glaciation? J Geophys Res: Atmospheres (1984–2012). 2009;114(D10). doi: 10.1029/2008JD011652.

Rose WI, Riley CM, Dartevelle S (2003) Sizes and shapes of 10-Ma distal fall pyroclasts in the Ogallala Group Nebraska. J Geol 111(1):115–124

Ryan WBF, Carbotte SM, Coplan JO, O'Hara S, Melkonian A, Arko R, et al. Global Multi-Resolution Topography synthesis. Geochem Geophys Geosys. 2009; 10(3). doi:10.1029/2008GC002332.

Saffer DM, Underwood MB, McKiernan AW (2008) Evaluation of factors controlling smectite transformation and fluid production in subduction zones: Application to the Nankai Trough. Island Arc 17(2):208–230

Saffer DM, Lockner DA, McKiernan A (2012) Effects of smectite to illite transformation on the frictional strength and sliding stability of intact marine mudstones. Geophys Res Lett 39:L11304. doi:10.1029/2012GL051761

Saito S, Underwood MB, Kubo Y. Expedition 322 Scientists. Proceedings IODP 322, Tokyo Integrated Ocean Drilling Program Management, International Inc Washington DC. 2010. doi:10.2204/iodpproc3221042010.

Schacht U, Wallmann K, Kutterolf S, Schmidt M (2008) Volcanogenic sediment–seawater interactions and the geochemistry of pore waters. Chem Geol 249(3–4):321

Schindlbeck JC, Kutterolf S, Freundt A, Scudder RP, Pickering KT, Murray RW (2013) Emplacement processes of submarine volcaniclastic deposits (IODP Site C0011 Nankai Trough). Mar Geol 343:115–124. doi:10.1016/j.margeo.2013.06.017

Schindlbeck JC, Kutterolf S, Freundt A, Straub SM, Wang K-L, Jegen M, et al. The Miocene Galápagos ash layer record of IODP Legs 334&344: Ocean-island explosive volcanism during plume-ridge interaction. Geology. 2015. doi:10.1130/G366451.

Scudder RP, Murray RW, Plank T (2009) Dispersed ash in deeply buried sediment from the northwest Pacific Ocean: An example from the Izu-Bonin arc (ODP Site 1149). Earth Planet Sci Lett 284(3–4):639–648

Scudder RP, Murray RW, Schindlbeck JC, Kutterolf S, Hauff F, McKinley CC (2014) Regional-scale input of dispersed and discrete volcanic ash to the Izu-Bonin and Mariana subduction zones. Geochem Geophys Geosys 15(11):4369–4379

Severmann S, Johnson CM, Beard BL, German CR, Edmonds HN, Chiba H et al (2004) The effect of plume processes on the Fe isotope composition of hydrothermally derived Fe in the deep ocean as inferred from the Rainbow vent site Mid-Atlantic Ridge 36 14 N. Earth Planet Sci Lett 225(1):63–76

Shaw DM, Watkins ND, Huang TC (1974) Atmospherically transported volcanic glass in deep-sea sediments: Theoretical considerations. J Geophysi Res 79(21):3087–3094

Sibuet JC, Letouzey J, Barbier F, Charvet J, Foucher JP, Hilde TW et al (1987) Back arc extension in the Okinawa Trough. J Geophys Res: Solid Earth (1978–2012) 92(B13):14041–14063

Sigurdsson H (1999) Volcanic episodes and rates of volcanism. In: Sigurdsson H, Houghten HB, McNutt SR, Ryme H, Stix J (eds) Encyclopedia of Volcanoes. Academic Press, New York, pp 271–279

Sigurdsson H, Sparks RSJ, Carey SN, Huang TC (1980) Volcanogenic sedimentation in the Lesser Antilles Arc. J Geol 88:523

Sigurdsson H, Leckie RM, Acton GD et al (1997) Proceedings of the Ocean Drilling Program: Initial reports Volume 165. Ocean Drilling Program, College Station Texas, p 865

Sigurdsson H, Kelley RM, Carey S, Bralower T, King J (2000) History of circum-Caribbean explosive volcanism: $^{40}Ar/^{39}Ar$ dating of tephra layers. Proc ODP Sci Results 165:299–314

Simkin T, Fiske RS (1983) Krakatau 1883: The Volcanic Eruption and its Effects. Smithsonian Institution Press, Washington DC, p 464

Stern RJ, Kohut EJ, Bloomer SH, Leybourne M, Fouch M, Vervoort J (2006) Subduction Factory processes beneath the Guguan Cross-chain Mariana Arc: no role for sediments are serpentinites important? Contrib Mineral Petrol 151:202–221

Straub SM (1997) Multiple sources of Quaternary tephra layers in the Mariana Trough. J Volcanol Geotherm Res 76(3–4):251–276

Straub SM. The evolution of the Izu Bonin–Mariana volcanic arcs (NW Pacific) in terms of major element chemistry. Geochem Geophys Geosyst. 2003;4. doi:10.1029/2002GC000357.

Straub SM. Timescales and causes of secular change in the Izu Bonin–Mariana volcanic arcs. Geochim Cosmochim Acta. 2008;72 (12). A905 Goldschmidt Conference Vancouver Canada.

Straub SM, Layne GD (2002) The systematics of boron isotopes in Izu arc front volcanic rocks. Earth Planet Sci Lett 198(1–2):25–39

Straub SM, Layne GD (2003b) The systematics of chlorine fluorine and water in Izu arc front volcanic rocks: Implications for volatile recycling in subduction zones. Geochim Cosmochim Acta 67(21):4179–4203

Straub SM, Layne GD. Decoupling of fluids and fluid-mobile elements during shallow subduction: evidence from halogen-rich andesite melt inclusions from the Izu arc volcanic front. Geochem Geophys Geosyst. 2003a;4. doi:10.1029/2002GC000349.

Straub SM, Schmincke HU (1998) Evaluating the tephra input into Pacific Ocean sediments: distribution in space and time. Geol Rundsch 87(3):461–476

Straub SM, Layne GD, Schmidt A, Langmuir CH (2004) Volcanic glasses at the Izu arc volcanic front: new perspectives on fluid and sediment melt recycling in subduction zones. Geochem Geophys Geosyst 5:Q01007. doi:10.1029/2002GC000408

Straub SM, Goldstein SL, Class C, Schmidt A (2009) Mid-ocean-ridge basalt of Indian type in the northwest Pacific Ocean basin. Nat Geosci 2(4):286–289

Straub SM, Goldstein SL, Class C, Schmidt A, Gomez-Tuena A (2010) Slab and Mantle Controls on the Sr-Nd-Pb-Hf Isotope Evolution of the Post 42Ma Izu-Bonin-Volcanic Arc. J Petrol 51(5):993–1026

Straub, SM, Woodhead, JD, Arculus, RJ. Temporal Evolution of the Mariana Arc: Mantle Wedge and Subducted Slab Controls Revealed with a Tephra Perspective. J Petrol. 2015. doi:10.1093/petrology/egv005.

Taira A (2001) Tectonic evolution of the Japanese island arc system. Annu Rev Earth Pl Sc 29(1):109–134

Tamura Y, Gill J, Tollstrup D, Kawabata H, Shukuno H, Chang Q et al (2009) Silicic Magmas in the Izu-Bonin Oceanic Arc and Implications for Crustal Evolution. J Petrol 50(4):685–723

Tani K, Fiske RS, Tamura Y, Kido Y, Naka J, Shukuno H et al (2008) Sumisu volcano Izu-Bonin arc Japan: site of a silicic caldera-forming eruption from a small open-ocean island. Bull Volcanol 70:547–562

Taylor SR, McLennan SM (1985) The Continental Crust: Its Composition And Evolution. Blackwell, Malden

Tobin HJ, Kinoshita M (2006) NanTroSEIZE: the IODP Nankai Trough seismogenic zone experiment. Sci Drill 2(2):23–27

Underwood MB, Pickering KT (1996) Clay-mineral provenance sediment dispersal patterns and mudrock diagenesis in the Nankai accretionary prism southwest Japan. Clays Clay Miner 44(3):339–356

Völker D, Kutterolf S, Wehrmann H (2011) Comparative mass balance of volcanic edifices at the Southern Volcanic Zone of the Andes between 33°S and 46°S. J Volcanol Geotherm Res 205:114–129. doi:10.1016/jjvolgeores201103011

Völker D, Wehrmann H, Kutterolf S, Iyer K, Geersen J, Rabbel W et al (2014) Constraining input and output fluxes of the southern Central Chile Subduction Zone: water chlorine sulfur. Int J Earth Sci 103(7):2129–2153. doi: 10.1007/s00531-014-1002-0

Volynets ON, Ponomareva VV, Braitseva OA, Melekestsev IV, Chen CH (1999) Holocene eruptive history of Ksudach volcanic massif South Kamchatka: evolution of a large magmatic chamber. J Volc Geotherm Res 91(1):23–42

Watson AJ (1997) Volcanic Fe CO2 ocean productivity and climate. Nature 385: 587–588. doi:10.1038/385587b0

White SM, Chamley H, Curtis DM, de Vries Klein G, Mizuno A (1980) Sediment synthesis: Deep Sea Drilling Project Leg 58 Philippine Sea. Init Rep DSDP 58: 963–1014. doi:10.2973/dsdpproc581451980

White JDL, Smellie JL, Clague DA (2003) Introduction: A deductive outline and topical overview of subaqueous explosive volcanism In: White JDL, Smellie JL, Clague DA (Ed) Explosive subaqueous volcanism, Geophys Monograph American Geophysical Union, Washington, p. 1-23.

Woods AW, Wohletz KH (1991) Dimensions and dynamics of co-ignimbrite eruption clouds. Nature 350:225–227

Zhou L, Kyte FT (1992) Sedimentation history of the South Pacific pelagic clay province over the last 85 million years inferred from the geochemistry of Deep Sea Drilling Project Hole 596. Paleoceanography 7(4):441–465

Ziegler CL, Murray RW. Geochemical evolution of the central Pacific Ocean over the past 56 Myr. Paleoceanography. 2007;22(2). doi:10.1029/2006PA001321.

Ziegler CL, Murray RW, Hovan SA, Rea DK (2007) Resolving eolian volcanogenic and authigenic components in pelagic sediment from the Pacific Ocean. Earth Planet Sci Lett 254(3):416–432

Ziegler CL, Murray RW, Plank T, Hemming SR (2008) Sources of Fe to the Equatorial Pacific Ocean from the Holocene to Miocene. Earth Planet Sci Lett 270:258–270

13

Unit cell determination of coexisting post-perovskite and H-phase in (Mg,Fe)SiO_3 using multigrain XRD: compositional variation across a laser heating spot at 119 GPa

Li Zhang[1*], Yue Meng[2] and Ho-kwang Mao[1,3]

Abstract

Multigrain X-ray diffraction (XRD) can be used to accurately calculate the unit cell parameters of individual mineral phases in a mineral assemblage contained in a diamond anvil cell (DAC). Coexisting post-perovskite (ppv) and H-phase were synthesized at 119 GPa and 2500 K from $(Mg_{0.85}Fe_{0.15})SiO_3$ in a laser-heated DAC. The unit cell parameters of the ppv and coexisting H-phase were determined using multigrain XRD with a 5 μm spatial resolution, close to the size of the X-ray beam, to understand compositional variations across the center area (20–30 μm) in a laser-heated sample. The ppv phase was Fe-depleted and the unit cell volume of ppv decreased by only 0.16 % (corresponding to ~3 % variation of $FeSiO_3$) from the heating center to 10 μm off the center, while the sample pressure remained at 119 GPa in a Ne quasi-hydrostatic environment. The unit cell volume of the H-phase decreased by 0.54 % (~10 % variation of FeSiO3 content) over the same 10 μm distance. Both phases were more Fe-enriched in the slightly hotter center. This observation suggests that thermal diffusion may not be the major driver for the compositional variations of ppv and H-phase in the center portion of a laser-heated sample. Instead, these variations could be caused by a temperature effect on the partitioning between the ppv and H-phase over the small gradient.

Keywords: Lower mantle, Multigrain XRD, Diamond anvil cell, Post-perovskite, H-phase, Unit cell, Synchrotron X-ray

Background

High-pressure (*P*) and high-temperature (*T*) experiments conducted in a laser-heated diamond anvil cell (DAC) have been widely used to simulate the extreme conditions in the Earth's interior. Integrating a laser-heated DAC with synchrotron X-ray diffraction (XRD) remains the dominant tool for studies of high-*P* mineral physics. In situ unit cell determination of minerals under high *P-T* conditions is a common task in the laser-heated DAC experiments. In an experiment involving element partitioning and/or chemical reaction, unit cell parameters often reflect compositional changes due to differences in the atomic radii of the elements involved. In the past decade, the combination of focused ion beam (FIB) with transmission electron microscopy (TEM) or other analytical techniques (Miyahara et al. 2008; Sinmyo et al. 2008) has been used to analyze the chemical compositions of samples recovered from a laser-heated DAC. This approach could allow the phase identification and detailed chemical analysis of materials under high *P-T* conditions, provided that all the phases and their chemical compositions remain unchanged after *T* quench from heating and during decompression to ambient conditions and that they remain unaffected by the FIB/TEM processes. On the other hand, in situ characterization is essential for obtaining information regarding phase changes and unit cell variations induced by pressure and composition.

In situ determination of unit cell parameters and crystal chemistry remains a challenge as pressure exceeds 100 GPa and temperature reaches several thousand kelvins (K).

* Correspondence: zhangli@hpstar.ac.cn
[1]Center for High Pressure Science and Technology Advanced Research (HPSTAR), Shanghai 201203, China
Full list of author information is available at the end of the article

Small sample sizes together with the re-crystallization of coexisting phases to coarse grains after laser heating significantly degrade the powder diffraction patterns required for accurate data analysis. In recent years, the multigrain method (Sørensen et al. 2012) has been successfully applied to study coarse-grained polycrystalline samples obtained under ultra-high *P-T* conditions in a DAC (Nisr et al. 2012; Zhang et al. 2013; Zhang et al. 2014; Rosa et al. 2015; Zhang et al. 2016). These studies have demonstrated that integration of the multigrain method with synchrotron XRD is suitable for treating each sub-micron-sized crystallite as a single crystal in a coarse-grained polycrystalline material in a DAC. In this study, we provide a detailed analysis of the unit cell parameters of Fe-depleted bridgmanite from multiple grains. Furthermore, the unit cell parameters of the coexisting post-perovskite (ppv) and H-phase, quenched from laser heating at 119 GPa, were calculated using multigrain XRD to determine the spatial variation of the compositions of individual phases across a laser heating spot.

Methods

An orthopyroxene (opx) sample with a composition of $(Mg_{0.85}Fe_{0.15})SiO_3$ (Fs15) was used as the starting material in the two experiments reported in this study. Diamond anvils with flat culet diameters of 150 μm beveled at 10° up to 300 μm were mounted in Boehler seats with up to a 60° opening. A piece of opx sample was pre-compressed into a disk of ~10 μm thickness, cut to 40–60 μm in diameter, and placed in a ~60 μm diameter Re gasket hole in a Mao-type symmetric DAC for each experiment. Compressed Ne was loaded as the pressure transmitting media for use as a pressure scale (Fei et al. 2007) and thermal insulation layer during laser heating. XRD measurements were conducted at the high-pressure synchrotron 16ID-B and 13ID-D beamlines of the advanced photon source (APS), Argonne National Laboratory.

To achieve the best possible uniform heating and data quality, and to minimize the temperature gradient in a relatively large sample area, samples were prepared with careful attention to details. Firstly, the heating spot size was adjusted to cover most of the sample in a DAC (Meng et al. 2015) and a heating spot with an almost flat area of ~30 μm was used for sample synthesis at the 16-ID-B beamline. Secondly, to achieve good coupling of the sample with the laser beam, efforts were made to control uniformity of the sample thickness, chemical homogeneity of the starting material, and the symmetric insulation layers of the quasi-hydrostatic Ne medium. The opx sample was prepared by compressing a single crystal, and thus, sample homogeneity and non-porosity was ensured. The sample was then successfully sandwiched between symmetric insulation layers of Ne from both sides. Note that the use of a quasi-hydrostatic medium such as Ne can speed up phase equilibrium. Spotty diffraction patterns were obtained, which indicated grain growth, after heating for only 2 min, in contrast to the 30–60 min required for experiments above 100 GPa reported in a previous study (Sinmyo et al. 2008). In situ XRD measurements indicated no obvious changes in the sample between heating for 2 and 10 min; therefore, a short duration of heating was used to avoid chemical segregation.

After successful sample synthesis, we aligned the sample to the ω-rotation center and collected a set of two-dimensional multigrain XRD patterns at small increments of the rotation angle over the maximum X-ray accessible range, similar to the rotation method used in conventional single-crystal crystallography. The FABLE package (Sørensen et al. 2012) was used to process the spotty XRD patterns, and the orientation matrices of individual grains were indexed using the GrainSpotter algorithm (Schmidt 2014).

Results and discussion

Unit cell determination of $(Mg,Fe)SiO_3$ bridgmanite coexisting with H-phase

Fe depletion in bridgmanite was observed as a result of the disproportionation of Fe-bearing bridgmanite and the appearance of an Fe-rich H-phase in a quasi-hydrostatic Ne medium (Zhang et al. 2014). XRD patterns were collected with an X-ray wavelength of 0.3738 Å by rotating the DAC from −21° to 21° in 0.1° steps at the 16ID-B beamline. The exposure time was 20 s/frame. A total of 51 grains of bridgmanite were indexed and the 16 of them with the most reflections available were selected for unit cell determination (Table 1). The 16 grains were then divided into three groups. A total of 150–180 reflections were merged from the five or six grains in each group and the unit cell parameters were refined from these reflections using UNITCELL (Holland and Redfern 1997), as shown in Table 1. The 5 grains in group #1 were orientated with reasonable constraints on all *a*, *b*, and *c* lattice parameters for individual grains, and the standard deviation was 0.23 $Å^3$ for the 5 unit cell volumes of each grain. In contrast, another 5 grains in groups #2 and 6 grains in group #3 showed weak constraints on certain directions of the lattice due to their orientations relative to the DAC opening. The unit cell volumes calculated from individual grains in group #2 and group #3 showed scattered values with standard deviations of 0.37 $Å^3$ for the 5 grains in group #2 and 0.89 $Å^3$ for the 6 grains in group #3. These results indicate that weak constraints on certain directions of the lattice of an individual grain could cause a large deviation in its unit cell volume, suggesting that unit cell determination from one crystal in conventional single-crystal XRD may suffer from a similar problem. Multigrain XRD overcomes this problem by merging enough reflections from several grains that have random orientations to determine the unit cell

Table 1 Unit cell parameters for Fe-depleted bridgmanite coexisting with H-phase at 102.5 GPa

Grain no.	*N*	*a* (Å)	*σ* (*a*)	*b* (Å)	*σ* (*b*)	*c* (Å)	*σ* (*c*)	*V* (Å^3)	*σ* (*V*)
Group #1									
12	39	4.6049	0.0006	6.3381	0.0018	4.3580	0.0008	127.19	0.03
17	30	4.6014	0.0019	6.3536	0.0010	4.3521	0.0010	127.24	0.04
32	38	4.6096	0.0009	6.3275	0.0012	4.3822	0.0011	127.82	0.03
35	38	4.6088	0.0014	6.3321	0.0013	4.3727	0.0006	127.61	0.02
43	36	4.6104	0.0008	6.3482	0.0011	4.3552	0.0011	127.47	0.03
Merged	181	4.6074	0.0004	6.3398	0.0005	4.3658	0.0004	127.52	0.01
Group #2									
6	38	4.6021	0.0018	6.3459	0.0018	4.3713	0.0004	127.66	0.03
16	51	4.6413	0.0020	6.3345	0.0011	4.3629	0.0003	128.27	0.04
23	33	4.5927	0.0015	6.3876	0.0039	4.3607	0.0004	127.93	0.05
24	33	4.5966	0.0024	6.3271	0.0024	4.3732	0.0004	127.19	0.03
29	31	4.6130	0.0029	6.3361	0.0019	4.3619	0.0006	127.49	0.05
Merged	185	4.6063	0.0007	6.3391	0.0007	4.3655	0.0002	127.47	0.01
Group #3									
33	28	4.6627	0.0066	6.3463	0.0008	4.3566	0.0005	128.92	0.17
36	30	4.5777	0.0027	6.3570	0.0017	4.3688	0.0005	127.13	0.05
41	33	4.5978	0.0011	6.4384	0.0061	4.3615	0.0004	129.11	0.10
46	27	4.5738	0.0027	6.3686	0.0024	4.3609	0.0005	127.03	0.05
51	18	4.6037	0.0010	6.3269	0.0017	4.3640	0.0018	127.11	0.04
1	23	4.5907	0.0024	6.3631	0.0013	4.3562	0.0011	127.25	0.04
Merged	158	4.6025	0.0005	6.3501	0.0005	4.3625	0.0002	127.50	0.01

N is the number of reflections. Sixteen grains were divided into three groups

parameters of each phase (Zhang et al. 2016). The refined unit cell parameters from merged reflections in each group did lead to consistent results, with a unit cell volume of 127.52(1) Å^3 in group #1, 127.47(1) Å^3 in group #2, and 127.50(1) Å^3 in group #3. The analysis demonstrated that the unit cell parameters of a high-*P* phase can be well constrained from enough merged reflections of several grains that have different orientations in a quasi-hydrostatic sample environment.

The unit cell parameter of Ne after *T* quench from 2300 K was calculated from three diffraction lines, 111, 200, and 220, and $a = 2.962$ (7) Å. The corresponding pressure was 102.5 GPa using the Vinet pressure scale of Ne (Fei et al. 2007). The Fe content in bridgmanite after disproportionation was characterized as $x = 0.02(2)$ in $(Mg_{1-x},Fe_x)SiO_3$ bridgmanite by ex situ FIB-TEM (Zhang et al. 2014). The unit cell volume of the Fe-depleted bridgmanite lies on the *P-V* curve of Fe-free $MgSiO_3$ bridgmanite (Lundin et al. 2008), as shown in Fig. 1, indicating consistency between the Vinet pressure scale of Ne (Fei et al. 2007) and the Au scale (Tsuchiya 2003). On the other hand, a previous study (Lundin et al. 2008) confirmed the consistency between the Au scale (Tsuchiya 2003) and another Au scale (Dewaele et al. 2004); the latter Au scale (Dewaele et al. 2004) agrees well with the internally consistent Au and Ne scales by Fei et al. (2007). However, the unit cell volume of $(Mg_{0.91}Fe_{0.09})SiO_3$ bridgmanite at 100 GPa (Dorfman et al. 2013) using the Au scale (Fei et al. 2007) is even slightly lower than that of Fe-free bridgmanite (Lundin et al. 2008) (Fig. 1), although the pressure scales used in

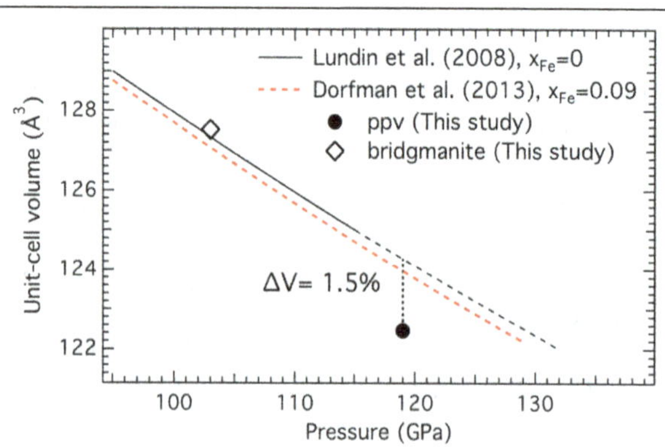

Fig. 1 Unit cell volumes of (Mg,Fe)SiO_3 bridgmanite and ppv versus pressure. The unit cell volumes of the Fe-depleted bridgmanite at 102.5 GPa and the ppv phase at 119 GPa coexisting with the H-phase are compared with the equations of state for Fe-free $MgSiO_3$ bridgmanite (Lundin et al. 2008) and $(Mg_{0.91}Fe_{0.09})SiO_3$ bridgmanite (Dorfman et al. 2013)

these two studies are consistent. The reason for this discrepancy is unclear. Multigrain XRD may have significant advantages over powder XRD for determining equations of state if one consistent pressure scale is used at all pressures and a quasi-hydrostatic sample environment is maintained.

Unit cell variations of coexisting ppv and H-phase across a laser heating spot

The coexisting ppv and H-phase were synthesized from $(Mg_{0.85}Fe_{0.15})SiO_3$ at 119 GPa (P after T quench) and 2500 K. After T quench and at 119 GPa, multigrain XRD data were collected for each sample position from the heating center to the margin, at steps of 5 μm across a laser heating spot. Diffraction patterns in step scans at each sample position were collected at ω increments of 0.25° over the DAC opening of 50°. The micro-focused XRD beam available at the 13ID-D beamline had a spot size of 3 by 4 μm with an X-ray wavelength of 0.3344 Å, and the exposure time was only 2 s/frame. The unit cell parameters of $(Mg,Fe)SiO_3$ ppv at each sample position were determined from ~150 reflections merged from several grains, as shown in Table 2 and Fig. 2a. The unit cell volume of ppv decreased by only 0.16 % from the heating center to 10 μm off the center, and the sample pressure remained at 119 GPa in the Ne quasi-hydrostatic environment. According to the previously determined relationship between the $FeSiO_3$ content in ppv and the equations of state (Zhang et al. 2012), the 0.16 % change in the unit cell volume of ppv corresponds to a ~3 % change of $FeSiO_3$ at 119 GPa, regardless of the absolute pressure determined by the different pressure scales. The unit cell volume of ppv in the heating center is also plotted in Fig. 1 and compared with the equation of state for Fe-free bridgmanite (Lundin et al. 2008). The unit cell volume of ppv is 1.5 % smaller than the value calculated from the equation of state for Fe-free bridgmanite and is consistent with the volume change from the bridgmanite to ppv transition in Fe-free $MgSiO_3$ (Tsuchiya et al. 2004). Fe depletion in ppv is confirmed when coexisting with H-phase.

H-phase was observed coexisting with ppv from the heating center to 10 μm off the center. The peak intensities of H-phase were comparable with those of the ppv phase. The unit cell volume of H-phase decreased by 0.54 % from the center to 10 μm off the center, as shown in Table 3 and Fig. 2b, suggesting a larger compositional variation in H-phase over the same distance. The relationship between the $FeSiO_3$ content and the equations of state for H-phase has not yet been established. Assuming that the effect of the $FeSiO_3$ content on the unit cell volume of H-phase is comparable to that of bridgmanite or ppv, the variation of $FeSiO_3$ content in H-phase is estimated as 9–14 %, based on the relationship between the $FeSiO_3$ content in ppv and their equations of state (Zhang et al. 2012) or 8–10 % based on the similar relationship to bridgmanite (Dorfman et al. 2013), corresponding to the 0.54 % change of the unit cell volume of H-phase.

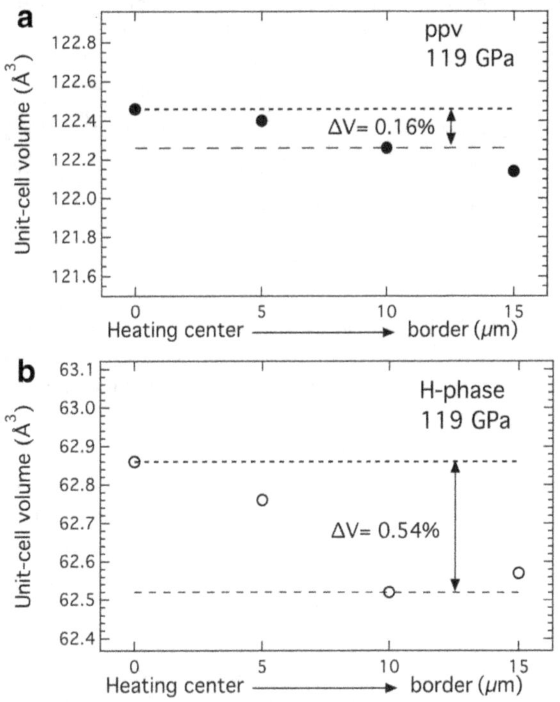

Fig. 2 Unit cell volumes of coexisting $(Mg,Fe)SiO_3$ ppv and H-phase across a laser heating spot. The unit cell volumes of **a** $(Mg,Fe)SiO_3$ ppv and **b** H-phase obtained using multigrain XRD are plotted as function of distance from the heating center to the margin at steps of 5 μm in a laser-heated sample at 119 GPa. The unit cell volumes of ppv and H-phase decrease by 0.16 and 0.54 %, respectively, from the center to 10 μm off the center, while the sample pressure remained the same in the Ne quasi-hydrostatic environment

Table 2 Variations of unit cell parameters for $(Mg,Fe)SiO_3$ ppv across a laser-heated spot at 119 GPa. The unit cell volume of the ppv phase slightly decreased by 0.16 % over the 10 μm distance from the center to the margin in the heated spot. *N* is the number of reflections from multiple grains at each sample position

Sample position	*N*	*a* (Å)	σ (*a*)	*b* (Å)	σ (*b*)	*c* (Å)	σ (*c*)	*V* (Å^3)	σ (*V*)	*a*Ne (Å)	σ (*a*)	*P* (GPa)
Heating center	136	2.4654	0.0002	8.1156	0.0005	6.1207	0.0004	122.46	0.01	2.923	0.005	119
5 μm off center	170	2.4647	0.0001	8.1180	0.0004	6.1173	0.0005	122.40	0.01	2.924	0.004	119
10 μm off center	155	2.4644	0.0001	8.1078	0.0008	6.1190	0.0004	122.26	0.01	2.924	0.006	119
15 μm off center	176	2.4644	0.0002	8.0986	0.0005	6.1200	0.0004	122.14	0.01	2.920	0.011	120

Table 3 Variations of unit cell parameters for H-phase across a laser-heated spot at 119 GPa. The unit cell volume of the H-phase coexisting with the ppv gradually decreased by 0.54 % over the same 10 μm distance from the center to the margin in the heated spot. *N* is the number of reflections from multiple grains at each sample position

Sample position	*N*	*a* (Å)	*σ* (*a*)	*c* (Å)	*σ* (*c*)	*V* (Å^3)	*σ* (*V*)	*a*Ne (Å)	*σ* (*a*)	*P* (GPa)
Heating center	71	5.0674	0.0003	2.8265	0.0003	62.86	0.01	2.923	0.005	119
5 μm off center	101	5.0665	0.0002	2.8232	0.0003	62.76	0.01	2.924	0.004	119
10 μm off center	71	5.0577	0.0004	2.8220	0.0003	62.52	0.01	2.924	0.006	119
15 μm off center	120	5.0624	0.0002	2.8192	0.0003	62.57	0.01	2.920	0.011	120

Previous studies using ex situ TEM analysis of recovered samples found that the compositions of ppv were approximately constant for the entire hot spot, while ferropericlase exhibited significantly larger spatial variations in iron content with FeO migration to the cold region (Sinmyo et al. 2008), indicating that the effect of thermal diffusion on an individual phase is related to its Mg-Fe interdiffusion rate. However, both ppv and H-phase in the present sample are more Fe-enriched in the slightly hotter center. The ppv phase is Fe-depleted with ~3 % variation of $FeSiO_3$ over the 10 μm distance, whereas the compositional variation in the coexisting H-phase is nearly three times of that in the ppv over the same distance. This observation suggests that thermal diffusion may not be the major driver for the compositional variations of ppv and H-phase in the center portion of the laser-heated sample. Instead, temperature effects on the partitioning behavior over a small gradient could explain the partitioning behavior between ppv and H-phase and their subsequent compositional variations. In addition, multigrain XRD has been demonstrated as effective for the detection of a minor phase by indexing a few individual grains within the sample (Zhang et al. 2016); however, no additional SiO_2 or other phases were detected in this sample.

Conclusions

Multigrain XRD provides accurate unit cell determination for individual phases in a multiphase system and may serve as an in situ probe for the spatial distribution of compositional changes and/or chemical variation within a sample region of interest, complementing ex situ chemical analysis by TEM or other analysis techniques. Through careful examination of the unit cell parameters of coexisting ppv and H-phase using multigrain XRD, we characterized the compositional variations of both phases across the center area of a laser-heated sample at 119 GPa. Chemical segregation due to a large temperature gradient can be problematic in some laser heating experiments. Previous ex situ TEM analyses of recovered samples revealed strong Fe depletion and Si enrichment in the hot area, and Fe migration to the cold margin through chemical diffusion due to a relatively large temperature gradient (Sinmyo et al. 2008; Andrault and Fiquet 2001). In this study, we observed relative Fe enrichment in both phases in the heating center, relative to the margin in a well-insulated and quasi-hydrostatic sample environment at 119 GPa. The temperature gradient in the center portion (20–30 μm) of the laser-heated sample was minimized by careful experimental preparation. Therefore, the relative Fe enrichment in the center may not be driven by thermal diffusion but could be caused by the effect of temperature on the partitioning. The temperature distribution in the laser heating spot should be measured in future experiments to quantitatively understand the effect of temperature on partitioning.

Abbreviations

APS: advanced photon source; DAC: diamond anvil cell; FIB: focused ion beam; K: kelvins; opx: orthopyroxene; *P*: pressure; ppv: post-perovskite; *T*: temperature; TEM: transmission electron microscopy; XRD: X-ray diffraction.

Competing interests

The authors declare that they have no competing interests.

Authors' contributions

LZ and HKM proposed the topic and conceived and designed the study. LZ and YM carried out the experimental study. LZ analyzed the data and wrote the manuscript. YM helped in the construction of the manuscript. All authors read and approved the final manuscript.

Acknowledgements

We thank V. Prakapenka for the assistance with the experiments. This work was supported by the Foundation of President of China Academy of Engineering Physics (Grant No: 201402032) and National Natural Science Foundation of China (NSFC) (Grant No: 41574080 and U1530402). This work was also partially supported by National Science Foundation (NSF) (Grant No: EAR-1345112 and EAR-1447438). The experiments were performed at HPCAT (Sector 16) and GeoSoilEnviroCARS (Sector 13), Advanced Photon Source (APS), Argonne National Laboratory. HPCAT operations are supported by the U.S. Department of Energy—National Nuclear Security Administration (DOE-NNSA) under award DE-NA0001974 and DOE—Basic Energy Sciences (BES) under award DE-FG02-99ER45775, with partial instrumentation funding by NSF. GeoSoilEnviroCARS is supported by the National Science Foundation—Earth Sciences (EAR-1128799) and Department of Energy—Geosciences (DE-FG02-94ER14466). Use of the APS facilities was supported by the U.S. Department of Energy, Office of Science, Office of Basic Energy Sciences, under Contract No. DE-AC02-06CH11357.

Author details

[1]Center for High Pressure Science and Technology Advanced Research (HPSTAR), Shanghai 201203, China. [2]High Pressure Collaborative Access Team (HPCAT), Geophysical Laboratory, Carnegie Institution of Washington, Argonne, IL 60439, USA. [3]Geophysical Laboratory, Carnegie Institution of Washington, Washington, DC 20015, USA.

References

Andrault D, Fiquet G (2001) Synchrotron radiation and laser heating in a diamond anvil cell. Rev Sci Instrum 72(2):1283. doi:10.1063/1.1343866

Dewaele A, Loubeyre P, Mezouar M (2004) Equations of state of six metals above 94 GPa. Phys Rev B 70(9):094112. doi:10.1103/PhysRevB.70.094112
Dorfman SM, Meng Y, Prakapenka VB, Duffy TS (2013) Effects of Fe-enrichment on the equation of state and stability of (Mg, Fe)SiO3 perovskite. Earth Planet Sci Lett 361:249–257. doi:10.1016/j.epsl.2012.10.033
Fei Y, Ricolleau A, Frank M, Mibe K, Shen G, Prakapenka V (2007) Toward an internally consistent pressure scale. Proc Natl Acad Sci USA 104(22):9182–9186. doi:10.1073/pnas.0609013104
Holland TJB, Redfern SAT (1997) UNITCELL refinement from powder diffraction data: the use of regression diagnostics. Mineral Mag 61:65–77
Lundin S, Catalli K, Santillán J, Shim SH, Prakapenka VB, Kunz M, Meng Y (2008) Effect of Fe on the equation of state of mantle silicate perovskite over 1Mbar. Phys Earth Planet Inter 168(1-2):97–102. doi:10.1016/j.pepi.2008.05.002
Meng Y, Hrubiak R, Rod E, Boehler R, Shen G (2015) New developments in laser-heated diamond anvil cell with in situ synchrotron x-ray diffraction at High Pressure Collaborative Access Team. Rev Sci Instrum 86(7):072201. doi:10.1063/1.4926895
Miyahara M, Sakai T, Ohtani E, Kobayashi Y, Kamada S, Kondo T, Nagase T, Yoo JH, Nishijima M, Vashaei Z (2008) Application of FIB system to ultra-high-pressure Earth science. J Mineral Petrol Sci 103(2):88–93. doi:10.2465/jmps.070612b
Nisr C, Ribárik G, Ungár T, Vaughan GBM, Cordier P, Merkel S (2012) High resolution three-dimensional X-ray diffraction study of dislocations in grains of $MgGeO_3$ post-perovskite at 90 GPa. J Geophys Res B: Solid Earth 117(B3): n/a-n/a. doi:10.1029/2011jb008401.
Rosa AD, Hilairet N, Ghosh S, Garbarino G, Jacobs J, Perrillat J-P, Vaughan G, Merkel S (2015) In situ monitoring of phase transformation microstructures at Earth's mantle pressure and temperature using multi-grain XRD. J Appl Crystallogr 48(5):1346–1354. doi:10.1107/s1600576715012765
Schmidt S (2014) GrainSpotter: a fast and robust polycrystalline indexing algorithm. J Appl Crystallogr 47(1):276–284. doi:10.1107/s1600576713030185
Sinmyo R, Hirose K, Nishio-Hamane D, Seto Y, Fujino K, Sata N, Ohishi Y (2008) Partitioning of iron between perovskite/postperovskite and ferropericlase in the lower mantle. J Geophys Res 113(B11):B11204. doi:10.1029/2008jb005730
Sørensen HO, Schmidt S, Wright JP, Vaughan GBM, Techert S, Garman EF, Oddershede J, Davaasambuu J, Paithankar KS, Gundlach C, Poulsen HF (2012) Multigrain crystallography. Z Kristallogr 227(1):63–78. doi:10.1524/zkri.2012.1438
Tsuchiya T (2003) First-principles prediction of the P-V-T equation of state of gold and the 660-km discontinuity in Earth's mantle. J Geophys Res 108(B10):2462. doi:10.1029/2003jb002446
Tsuchiya T, Tsuchiya J, Umemoto K, Wentzcovitch RM (2004) Phase transition in $MgSiO_3$ perovskite in the earth's lower mantle. Earth Planet Sci Lett 224(3-4): 241–248. doi:10.1016/j.epsl.2004.05.017
Zhang L, Meng Y, Mao WL (2012) Effect of pressure and composition on lattice parameters and unit-cell volume of (Fe, Mg)SiO3 post-perovskite. Earth Planet Sci Lett 317–318:120–125. doi:10.1016/j.epsl.2011.11.038
Zhang L, Meng Y, Dera P, Yang W, Mao WL, Mao HK (2013) Single-crystal structure determination of (Mg, Fe)SiO3 postperovskite. Proc Natl Acad Sci USA 110(16):6292–6295. doi:10.1073/pnas.1304402110
Zhang L, Meng Y, Yang W, Wang L, Mao WL, Zeng QS, Jeong JS, Wagner AJ, Mkhoyan KA, Liu W, Xu R, Mao HK (2014) Disproportionation of (Mg, Fe)SiO3 perovskite in Earth's deep lower mantle. Science 344(6186):877–882. doi:10.1126/science.1250274
Zhang L, Popov D, Meng Y, Wang J, Ji C, Li B, Mao HK (2016) In-situ crystal structure determination of seifertite SiO_2 at 129 GPa: studying a minor phase near Earth's core-mantle boundary. Am Mineral 101:231–234. doi:10.2138/am-2015-5525

14

Water and magmas: insights about the water solution mechanisms in alkali silicate melts from infrared, Raman, and ^{29}Si solid-state NMR spectroscopies

Charles Le Losq*, Bjorn O. Mysen and George D. Cody

Abstract

Degassing of water during the ascent of hydrous magma in a volcanic edifice produces dramatic changes in the magma density and viscosity. This can profoundly affect the dynamics of volcanic eruptions. The water exsolution history, in turn, is driven by the water solubility and solution mechanisms in the silicate melt. Previous studies pointed to dissolved water in silicate glasses and melts existing as molecules (H_2O_{mol} species) and hydroxyl groups, OH. These latter OH groups commonly are considered bonded to Si^{4+} but may form other bonds, such as with alkali or alkaline-earth cations, for instance. Those forms of bonding influence the structure of hydrous melts in different ways and, therefore, their properties. As a result, exsolution of water from magmas may have different eruptive consequences depending on the initial bonding mechanisms of the dissolved water. However, despite their importance, the solution mechanisms of water in silicate melts are not clear. In particular, how chemical composition of melts affects water solubility and solution mechanism is not well understood. In the present experimental study, components of such information are reported via determination of how water interacts with the cationic network of alkali (Li, Na, and K) silicate quenched melts. Results from ^{29}Si single-pulse magic-angle spinning nuclear magnetic resonance (^{29}Si SP MAS NMR), infrared, and Raman spectroscopies show that decreasing the ionic radius of alkali metal cation in silicate melts results in decreasing fraction of water dissolved as OH groups. The nature of OH bonding also changes as the alkali ionic radius changes. Therefore, as the speciation and bonding of water controls the degree of polymerization of melts, water will have different effects on the transport properties of silicate melts depending on their chemical composition. This conclusion, in turn, may affect volcanic phenomena related to the viscous relaxation of hydrous magmas, such as for instance the fragmentation process that occurs during explosive eruptions.

Keywords: Melt; Glass; Structure; Water bonding; Water speciation; NMR spectroscopy; Raman spectroscopy; Viscosity; Magma; Fragmentation

Background

Water is typically the major volatile component in magmas and can reach concentrations up to ~6–7 wt. % in silicic magma chambers (e.g., see Blake 1984; Clemens 1984). When magmas migrate from their source up towards the Earth surface, water exsolution can result in large (up to several orders of magnitude) increases in magma viscosity because of the strong effect of water on the viscosity of silicate melts (Leonteva 1940; Saucier 1952; Friedman et al. 1963; Shaw 1963; Dingwell et al. 1996; Richet et al. 1996). Water exsolution is also accompanied by the formation of vapor bubbles that participate in decreasing the bulk density of the magma, reinforcing, in this way, the buoyant force that pushes the magma out of the Earth crust. In silica-poor magmas with viscosities lower than 10^6 Pa s at eruptive temperature, such as with basalts, gas bubbles can coalesce and escape from the magma (Thomas et al. 1994; Gardner et al. 1996). This situation usually results in gentle (lava flows) to slightly explosive

* Correspondence: clelosq@carnegiescience.edu
Geophysical Laboratory, Carnegie Institution of Washington, 5251 Broad Branch Road NW, Washington, DC 20015, USA

(lava fountain, strombolian explosion) eruptions. In contrast, for silica-rich compositions, the magma viscosity can be considerably higher than 10^6 Pa s at eruptive temperature, so that bubbles are trapped in the melts (Thomas et al. 1994; Gardner et al. 1996). Exsolution of water from such magmatic liquids often results in foaming. In such case, brittle fragmentation of the magmatic foam can occur, for example, when the elongation strain rate of the foam ascending in the conduit exceeds its viscous relaxation time (Papale 1999). This situation leads to explosive eruptions.

The effects of water on magma properties, together with their possible consequences as described above, result from how water affects the disordered tetrahedral silicate network of melts. A simple and commonly proposed solution mechanism of water in such melts involves the presence of H_2O molecules (H_2O_{mol}), which can react with the oxygens O of the silicate network to form OH groups:

$$H_2O_{mol(melt)} + O_{(melt)} <=> 2\ OH_{(melt)} \quad (1)$$

The solution mechanism illustrated in Eq. (1) is documented by numerous nuclear magnetic resonance (NMR) and infrared (IR) spectroscopic studies (e.g., Scholtze 1960; Moulson and Roberts 1961; Bartholomew et al. 1980; Stolper 1982; Kohn et al. 1989a, b; Zotov and Keppler 1998), which also indicated that OH groups seem bonded predominantly to Si (see also Farnan et al. 1987; Cody et al. 2005; Mysen and Cody 2005). Therefore, in this simple model, H_2O_{mol} reacts predominantly with bridging oxygens (BO), breaking T-O-T bonds (with T = Si, Al), so that Eq. (1) can be rewritten as:

$$H_2O_{mol(melt)} + T{-}O{-}T_{(melt)} <=> 2\ T{-}OH_{(melt)} \quad (2)$$

Equation (2) implies formation of non-bridging oxygens (NBO) when the reaction shifts to the right. However, reaction (2) does not always proceed to completion. Indeed, both H_2O_{mol} and OH groups coexist in silicate melts, and hence glasses, in proportions defined by temperature (Nowak and Behrens 1995; Sowerby and Keppler 1999; Behrens and Yamashita 2008) and apparently water concentration (e.g., Stolper 1982).

However, Eq. (2) is just one reaction describing the reaction of H_2O with the silicate network of silicate melts. Other dissolution mechanisms have been proposed, such as the formation of MOH bonds (with $M^{+,2+}$ a metallic ion) (Xue and Kanzaki 2004, 2006, 2008; Moretti et al. 2014) or the protonation of T-O-T bonds (H^+ does not break T-O-T bonds but is attached to it; see for example Kohn et al. 1989b, 1992, 1998 and references therein). Moreover, Eq. (2) does not consider the Q^n speciation often employed to describe the silicate structure (where n denotes the number of bridging oxygen in a silicate complex, see Schramm et al. 1984). For example, water can react with different Q^n species in sodium silicate melts depending on their initial water content and Na/Si ratio (Zotov and Keppler 1998; Mysen and Cody 2005; Mysen 2007). In addition, it appears likely that the electronic properties of the alkali and alkaline earth metal cations will affect the reaction mechanisms of water with the silicate network (Xue and Kanzaki 2004).

To understand, quantitatively, how water interacts with the silicate structure of glasses and melts, we examined the structure of alkali (Li^+, Na^+, K^+) tetrasilicate glasses using Raman and ^{29}Si SP MAS NMR spectroscopy. Existing infrared spectroscopic information (Le Losq et al. 2015a, b) was also used. By studying such compositionally simple systems, it is possible to isolate effects of individual components on the water solution behavior. Furthermore, the chosen glasses have anhydrous glass transition temperature, *Tg*, within a ~30 K range: the *Tg* of the sodium tetrasilicate glass is 751 K (Neuville 2006), that of the potassium tetrasilicate glass to 762 K (Neuville 1992), and that of the LS4 glass to 740 K (Voigt et al. 2005). They also have very similar distributions of silicate Q^n species (e.g., Maekawa et al. 1991). Those two characteristics (comparable anhydrous T_gs and Q^n distributions) allow us to emphasize differences that characterize the hydrous quenched melts rather than differences related to the starting anhydrous glasses.

Methods

Sample preparation

The starting materials as well as the hydrous glasses used for this study are the same as those reported in Le Losq et al. (2015a, b). We refer readers to these references for details on sample synthesis. The chemical compositions of glass starting materials are reported in Table 1.

Table 1 Chemical composition of LS4, NS4, and KS4 anhydrous glasses

		SiO_2	Li_2O	Na_2O	K_2O
	nom. mol %	80.00	20.00	0.00	0.00
LS4	nom. wt. %	88.94	11.08	0.00	0.00
	meas. wt. %	88.77 (94)	11.23 (75)	*n.a.*	*n.a.*
	nom. mol %	80.0	0.00	20.00	0.00
NS4	nom. wt. %	79.50	0.00	20.50	0.00
	meas. wt. %	80.38 (36)	*n.a.*	19.62 (31)	*n.a.*
	nom. mol %	80.0	0.00	0.00	20.0
KS4	nom. wt. %	71.84	0.00	0.00	28.16
	meas. wt. %	71.16 (49)	*n.a.*	*n.a.*	28.84 (28)

nom. mol %, nom. wt. %, and meas. wt. % refer respectively to the nominal compositions in mol and wt. % and to the measured composition in wt. %. Chemical compositions are from Le Losq et al. (2015a). Errors are given at the 1σ confidence interval

The ^{29}Si NMR signal from the anhydrous glasses suffers from a very long T_1 relaxation time. To reduce this T_1 relaxation time, a portion of the starting glasses was synthesized with a few hundred ppm Fe_2O_3, as previously done by Maekawa et al. (1991) to study similar alkali silicate glasses. To ensure homogeneous distribution of this small amount of iron oxide in the materials, the glass synthesis was done in two steps. The first step consisted in enriching a proportion of the starting glasses to 15 wt. % Fe_2O_3. During the second step, the iron-rich glasses were crushed and then mixed with crushed iron-free starting materials, in order to produce iron-enrichment in the 100–1000 ppm concentration range.

Quantitative analysis of the glasses by energy-dispersive x-ray spectrometry (EDS) was performed with a JEOL field-emission scanning electron microscope (FE-SEM) equipped with an Oxford silicon drift detector and using a 15 kV accelerating voltage and a 1.00 nA current. Measurements were performed scanning 25 by 25 micron areas. The EDS results were calibrated using the following standards: pyrope for O and Si, scapolite for Na, K1597 GI for K, and cossyrite for Fe. Calibration was based on a set of glass standards. The EDS results are within error of standards analyzed as unknowns.

The compositions of the Fe-doped glasses are essentially the same as those of the Fe-free starting materials. The iron contents are below detection limit (~200 ppm) for Fe-doped LS4, 1500 ± 300 ppm for Fe-doped NS4, and 700 ± 400 ppm for Fe-doped KS4 (mean values and errors from >10 individual measurements made at different spots). These Fe-doped glasses were used only to perform the ^{29}Si NMR measurements of anhydrous end-members. Optically, they are clear and colorless with the same Raman signal as iron-free starting glasses, indicating a homogeneous distribution of Fe in their structure. This is supported by SEM observations that show no clustering of iron at the micrometer scale. The ^{29}Si NMR data of the hydrous glasses were obtained on iron-free glasses. Indeed, in those water-bearing glasses, the ^1H nuclei that have a strong magnetic moment serve to aid the relaxation of ^{29}Si nuclei, as Fe at the ‰ level does in the anhydrous glasses.

Solid-state NMR spectroscopy

Silicon-29 solid-state NMR experiments were carried out with a Chemagnetics CMX Infinity 300 solid-state NMR spectrometer. The static field strength of the magnet is ~7.05 T, and the Larmor frequency of ^{29}Si is ~59 MHz at this magnetic field. ^{29}Si spectra were referenced to tetramethylsilane (TMS), defined as 0 ppm.

Single-pulse ^{29}Si MAS spectra of each glass were obtained by spinning the samples at 11 kHz. ^{29}Si excitation pulse widths corresponded to 30° tip angles. One thousand acquisitions separated by a 100 s recycling time were co-added. Proton decoupling was performed with continuous RF irradiation at 62.5 kHz. As indicated in the "Sample preparation" section, the anhydrous starting materials had a long ^{29}Si spin lattice T_1 relaxation time. For the KS4 glass, this long relaxation time resulted in a signal from ^{29}Si in Q^4 polymerized units much too low compared to a prediction of a mixture of 50 % Q^3:50 % Q^4 units based on glass chemical composition. Therefore, a portion of the anhydrous samples was doped with iron (see "Sample preparation" section) in order to decrease of the ^{29}Si T_1. The resulting signal from KS4 glass was much more intense than without the iron dopant and the spectra yielding Q^3–Q^4 proportion near the nominal 50:50. In hydrous glasses, which were iron-free, the protons played a similar role as iron, thus reducing the ^{29}Si spin–lattice relaxation time, so that spectra are directly quantitative. More details on the ^{29}Si single-pulse NMR experiments can be found in Cody et al. (2005).

Raman spectroscopy

Raman spectra were recorded with a Jasco NRS 3100 spectrometer, equipped with holographic gratings, a single monochromator, and a 1024 × 128 Andor DV401-F1 CCD Peltier-cooled detector operating at −70 °C. The 490 nm line of a coherent solid-state laser was used for sample excitation, with a power of ~44 mW on sample. The laser was focused on sample through a 50× Olympus lens. The groove density of the gratings was 2400 lines/mm, which results in an accuracy of ±3 cm^{-1}. Precision was ±1 cm^{-1}. Alignment of the Raman spectrometer was tested against the 520.7 cm^{-1} Raman peak of pure Si. All spectra reported in this study are unpolarized. The spectra were not corrected from temperature and excitation line effects (e.g., see Neuville et al. 2014, for further discussion about the relevance of such corrections) because thermal effects mostly affect the Raman intensity at frequency below 500 cm^{-1}, which does not contain much information relevant to the focus of the present study, and because excitation line effects are very small and in all cases will not affect our results since we used the same laser line for recording all spectra.

The Raman spectra were fitted using Gaussian peaks. A two-step fitting procedure was employed. As both the number of peaks and the frequency range of the fitted region are large, a first fitting step was carried out by using an equal peak width for all peaks. This fit was accomplished by using the *nlinfit* non-linear least-square optimization function of Matlab®. Then, in order to refine the models produced by *nlinfit*, the parameters and associated errors from those fits were injected in the pseudo-Newtonian algorithm described in Tarantola (2005) and used in Le Losq et al. (2014). This algorithm has the advantage of accounting of the a priori errors on the initial models (e.g., the errors on the estimations provided by the user or by *nlinfit* in our case). Errors in peak areas were calculated by propagating the peak parameter errors.

Water concentration

Water concentrations $[H_2O^{TOT}]$ of hydrous glasses were determined from IR absorption spectroscopy data of the NS4 compositions (for which molar absorption coefficients are available; Yamashita et al. 2008) and an FTIR-based calibration of the relative area of the OH fundamental stretch Raman band for the LS4 and KS4 glasses as described by Le Losq et al. (2015a). The resulting water contents together with glass synthesis conditions are reported in the Table 2.

The molar absorption coefficients of OH groups and H_2O_{mol} species for LS4 glasses have been estimated by using the IR data reported in Le Losq et al. (2015b) and the data treatment reported in Le Losq et al. (2015a), which follows the protocol described in Yamashita et al. (2008). A six-order polynomial baseline was fitted on spectra to derive a baseline under the 4500 and 5200 cm^{-1} IR absorption peaks (see Fig. 1, top left) and subtracted from spectra. Then, areas of the 4500 (A4500) and 5200 (A5200) cm^{-1} IR peaks, assigned to OH groups and H_2O_{mol} species respectively (Scholtze 1960; Moulson and Roberts 1961; Stolper 1982), have been measured (Table 2). The absorption coefficients of these two peaks (ε_{5200} and ε_{4500}, respectively) were calculated from the following relationship (Yamashita et al. 2008):

$$\frac{1820 \times A5200}{d \times \rho \times [H_2O^{TOT}]} = \varepsilon_{5200} - \frac{\varepsilon_{5200}}{\varepsilon_{4500}} \times \left[\frac{1820 \times A4500}{d \times \rho \times [H_2O^{TOT}]}\right] \tag{3}$$

with d, the thickness of the double-polished glass piece, and ρ, its density. The density of anhydrous LS4 glass was taken from Tischendorf et al. (1998). The density of hydrous glasses was then estimated by assuming a linear mixing between the molar volume of water and that of the glasses. In the water content range of the present study, such an assumption seems valid in light of the results of Yamashita et al. (2008) for sodium silicate glasses, and, on a larger dataset, with the results of Richet et al. (2000). According to results from those two studies, a partial molar volume of 12 $cm^3\ mol^{-1}$ for dissolved water was chosen. Such a value may overestimate slightly the actual molar volume of water in lithium silicate glasses, as inferred from results from ^1H NMR spectroscopy on these same glasses (Le Losq et al. 2015a). However, a change of the partial molar volume of water of 4 $cm^3\ mol^{-1}$ for instance can result in a maximum change of 4 $L\ mol^{-1}\ cm^{-2}$ of the absorption coefficients. Such variations are incorporated in the reported error bars. By using Eq. (3), $\varepsilon_{5200} = 56 \pm 8\ (1\sigma)\ L\ mol^{-1}\ cm^{-2}$ and $\varepsilon_{4500} = 70 \pm 7\ (1\sigma)\ L\ mol^{-1}\ cm^{-2}$ for the LS4 glass.

Use of the IR combination peaks at 4500 and 5200 cm^{-1} to determine water content of glasses is only possible when those peaks are well defined. In water-bearing alkali silicate glasses, an increase of the ionic radius of alkali (decreasing their ionic field strength) produces a decrease of the O-O distances around protons dissolved as H_2O_{mol} species and OH groups (Le Losq et al. 2015a). This perturbs significantly the frequency and intensity of the signal assigned to O-H stretching vibrations in IR and Raman spectra of hydrous alkali silicate glasses (Le Losq et al. 2015b). For example, in IR spectra of KS4 glasses, this signal extends to frequencies much lower than 3000 cm^{-1}, presenting several intensity maxima near ~2350, ~2800, and ~3600 cm^{-1}. This extended frequency range results in a poorly defined 4500 cm^{-1} combination band (Le Losq et al. 2015b) so that this

Table 2 Water concentration of glasses

Glass	nom. mol % H_2O	Conditions of synthesis	FTIR values					RAMAN
			A_{4200}	A_{5200}	[OH]	$[H_2O_{mol}]$	$[H_2O^{TOT}]$	$[H_2O^{TOT}]$
LS4	3.28	1650 °C	151 (12)	93 (9)	2.11 (35)	1.04 (22)	3.15 (41)	3.08 (60)
	9.40	1.5 GPa	325 (18)	452 (21)	4.56 (70)	5.06 (42)	9.61 (81)	10.09 (60)
	17.64		447 (21)	1109 (33)	6.26 (94)	12.42 (55)	18.68 (1.09)	18.74 (60)
NS4	3.28	1500 °C	155 (12)	3 (2)	3.85 (40)	0.07 (3)	3.92 (40)	3.44 (67)
	9.40	1.5 GPa	301 (17)	86 (9)	7.35 (63)	1.89 (20)	9.24 (66)	9.46 (67)
	17.64		403 (20)	440 (21)	9.52 (80)	9.35 (60)	18.87 (1.00)	17.54 (67)
KS4	3.28	1550 °C	-	-	3.98 (70)[a]	0.00 (0)[a]	-	3.98 (74)
	9.40	1.5 GPa	-	-	8.89 (70)[a]	0.73 (7)[a]	-	9.62 (74)
	17.64		-	-	13.65 (70)[a]	2.30 (24)[a]	-	15.95 (74)

Nom. refers to the nominal water concentrations. $[H_2O_{mol}]$, [OH], and $[H_2O^{TOT}]$ concentrations are in mol %. A_{4500} and A_{5200} are the integrated intensities of the IR peaks at ~4500 and ~5200 cm^{-1} that have been used to quantify the content of [OH] and $[H_2O_{mol}]$ for the LS4 and NS4 glasses. FTIR and Raman values are from Le Losq et al. (2015a). Errors are given at the 1σ confidence interval

[a]Calculated with the calibration linking ^1H MAS NMR sideband area to H_2O_{mol} content shown in Fig. 1; the concentration of OH is estimated by difference to $[H_2O_{mol}]$ measured with using Raman spectroscopy

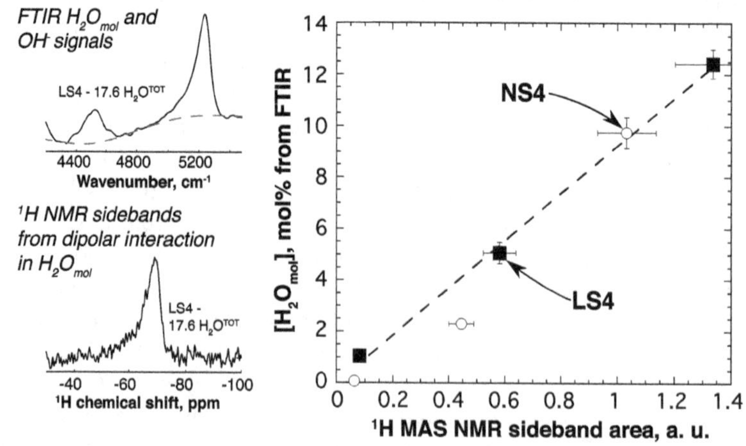

Fig. 1 Relationship between the area of ^{1}H MAS NMR sidebands, which arise from ^{1}H-^{1}H dipolar interaction in H_2O_{mol} species, and the concentration of the latters as measured by IR spectroscopy in the LS4 (plain squares) and NS4 (open circles) glasses. On the *left*, examples of the 4500 and 5200 cm^{-1} IR combination bands, assigned to OH and H_2O_{mol} species, respectively, and of one ^{1}H NMR sideband are shown for the LS4 glass containing 17.6 nom. mol% water. On the FTIR example, the *plain line* represents the experimental spectrum and the *dashed line* is a polynomial baseline defining the background (see text)

combination band cannot be used for measuring the concentration of OH groups. In such cases, an alternative to derive H_2O_{mol}/OH is to use the sidebands in ^{1}H MAS NMR spectra. These sidebands are residual signals from ^{1}H-^{1}H dipolar coupling, this interaction being strongly averaged by the fast magic angle spinning in the present experiments. The ^{1}H-^{1}H dipolar coupling is very strong in the case of H_2O_{mol} species and negligible in the case of isolated OH groups (Eckert et al. 1988). Therefore, in normalized ^{1}H NMR spectra acquired under similar conditions, sideband area would express the fraction of H_2O_{mol} species. Following this idea, Fig. 1 shows that the area of the sidebands ($A_{sidebands}$) of the ^{1}H MAS NMR spectra of LS4 and NS4 glasses, from Le7#146;Losq et al. (2015a), seem to be a linear function of the glasses H_2O_{mol} concentrations measured by IR. Particularly, the relationship seems nearly perfect for LS4 glasses. Slight deviations are observed for NS4 glasses. This can be rationalized with considering that most of the signal from O-H stretching vibrations is located above 3000 cm^{-1} in the IR spectra of LS4 glasses whereas it slightly extends toward lower frequency in the IR spectra of NS4 glasses (Le Losq et al. 2015b). This results in better defined ~ 4500 and ~ 5200 cm^{-1} combination bands in IR spectra of LS4 glasses than in those of NS4 glasses (Le Losq et al. 2015b), and hence, in higher accuracy on the calculated [H_2O_{mol}] concentrations. Because of that, we can take the IR-based H_2O_{mol} concentrations and $A_{sidebands}$ from LS4 glasses as standards. Doing so, we obtain the relationship: mol % $H_2O_{mol} = 9.21 \pm 0.21 \times A_{sidebands}$. This equation allows us to determine the concentration of H_2O_{mol} species and of OH groups (by difference to the total water content) in the KS4 glasses (Table 2).

Results

Raman spectroscopy

The Raman-active vibrational modes of the silicate tetrahedral network give rise to numerous bands between 400 and 1400 cm^{-1} in the Raman spectra of silicate glasses (Fig. 2). A strong band near ~500 cm^{-1} is always observed. Smaller peaks and shoulders near 550 and 590–600 cm^{-1} are also present. Their intensity and frequency depend on glass composition. The respective intensity and frequencies of the contributions located between 450 and 650 cm^{-1} depend on the water content of the glasses and on the type of alkali metal (Fig. 2). Those signals arise from the bending/breathing modes of Si-O-Si bonds that link Q^n units (noted "Q^n-Q^n modes" in Fig. 2; see for instance Bell et al. 1968; Sen and Thorpe 1977; Furukawa et al. 1981 or McMillan 1984).

A band near 800 cm^{-1} is also observed in all of the Raman spectra (Fig. 2). For a given alkali, the intensity of this peak changes with water content, but not its shape. At given water content, increasing ionic radius of alkali results in decreasing the intensity of this peak. This band may arise from vibrations involving the motion of the Si in its oxygen cage (McMillan 1984), this being possibly related to the threefold—degenerate "rigid cage" vibrational mode of SiO_2 tetrahedra (Sen and Thorpe 1977; Galeener 1979). However, its origin is not firmly established.

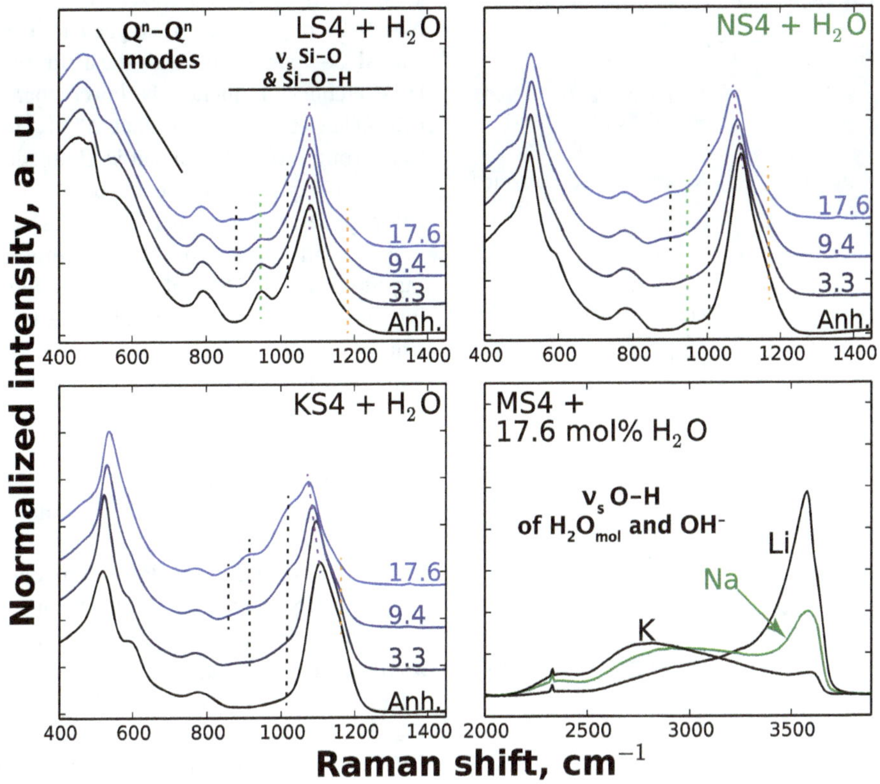

Fig. 2 Raman spectra of the lithium (LS4), sodium (NS4), and potassium (KS4) tetrasilicate glasses with 0.0 (Anh.), 3.3, 9.4, and 17.6 nom. mol % [H_2O^{TOT}]. Raman intensities are offset vertically for a better visibility, with the exception of the figure in *bottom right*. *Dashed lines* are used as visual markers of the signals contributing in the 820–1350 cm^{-1} portion of the spectra (see text). The small sharp peak located near 2330 cm^{-1} in the *lower right* figure is a signal from atmospheric N_2

Signals assigned to Si-O stretch vibrations in various Q^n units (ν_s Si-O in Fig. 2) form the broad envelope of bands between 850 and 1300 cm^{-1} (Brawer and White 1975, 1977; Virgo et al. 1980; Mysen et al. 1982; McMillan 1984; Neuville et al. 2014 and references therein). In hydrous alkali silicate glasses, a band often assigned to Si-O-H stretch vibrations contributes to the intensity in the 900–970 cm^{-1} frequency range (Stolen and Walrafen 1976; McMillan et al. 1993; Zotov and Keppler 1998; Malfait 2009). At frequencies higher than 1300 cm^{-1}, bands are observed between 2000 and 4000 cm^{-1} (Fig. 2, bottom right). These are assigned to O-H stretching (ν_s O-H) in H_2O_{mol} and OH groups (Scholtze 1960, Bartholomew et al. 1980; Wu 1980; Zotov and Keppler 1998). Their intensity and frequency depend on the chemical composition of the alkali silicate glasses. The larger the alkali ionic radius *r*, the lower the O-H stretch frequency (Fig. 2; see also Le Losq et al. 2015a; Le Losq et al. 2015b).

The topology of the 850–1300 cm^{-1} frequency region of the Raman spectra systematically varies as a function of both water content and ionic field strength of alkali in the glass (Fig. 2). This variation likely reflects changes of the Q^n species fractions. To quantify those changes, this portion of Raman spectra was fitted with Gaussian peaks. Such peak shape was used because it returns the best fitting results compared to other peak shapes (e.g., Lorentzian or Pseudo-Voigt shapes). This conclusion also is coherent with an expected pseudo Boltzmann distribution of Si-O vibrators in silicate melts in regards of their high thermal agitation and structural disorder at the glass transition temperature (Mysen et al. 1982). Table 3 summarizes the peaks used to perform the fits as well as their assignments and the references that substantiate those assignments.

The peak near 1100 cm^{-1} is assigned to Si-O stretching in Q^3 species. It is the most intense signal in the Raman spectra of all anhydrous glasses, whatever the alkali metal (Figs. 3 and 4). Areas of the peaks assigned to Si-O stretching in Q^2 and Q^4 species (near 950 and 1150–1200 cm^{-1}) respectively decrease and increase with increasing ionic radius of the alkali cation. These intensity variations indicate variations in the Q^n distribution as a function of the alkali ionic radius/field strength, similar

Table 3 Assignment of the peaks from curve-fitted Raman spectra

Raman shift, cm^{-1}	Attribution	Source
~850 cm^{-1}	v_s Si-O in Q^1	Furukawa et al. (1981), Virgo et al. (1980), Mysen et al. (1982), McMillan (1984)
~880–910 cm^{-1}	v_s Si-O-H	McMillan et al. (1993), Zotov and Keppler (1998), Malfait (2009), Spiekermann et al. (2012)
~960 cm^{-1}	v_s Si-O in Q^2	Brawer and White (1975), Brawer and White (1977), Virgo et al. (1980), Furukawa et al. (1981), Mysen et al. (1982), McMillan (1984)
~1020 cm^{-1}	v_s Si-BO in Q^n	Mysen et al. (1982), Spiekermann et al. (2012), Le Losq and Neuville (2013)
~1090 cm^{-1}	v_s Si-O in Q^3	Brawer and White (1975), Virgo et al. (1980), Furukawa et al. (1981), Mysen et al. (1982)
~1140 cm^{-1}	v_s Si-O in $Q^{4,II}$ (a)	Mysen et al. (1982), Seifert et al. (1982), Mysen (1990), Neuville and Mysen (1996), Le Losq and Neuville (2013), Le Losq et al. (2014)
~1190 cm^{-1}	v_s Si-O in $Q^{4,I}$ (a)	Mysen et al. (1982), Seifert et al. (1982), Mysen (1990), Neuville and Mysen (1996), Le Losq and Neuville (2013), Le Losq et al. (2014)

[a]Two Gaussians named $Q^{4,II}$ and $Q^{4,I}$ are used for fitting the Si-O stretch of Q^4 because of its non-Gaussian shape. See Mysen et al. (1982), Seifert et al. (1982), Neuville and Mysen (1996), and Le Losq and Neuville (2013) for further discussion on this topic. In more depolymerized glasses, the signal of Q^4 becomes weak and only one band is needed (see in Fig. 2 the spectrum of KS4 + 17.6 mol % H_2O for instance)

to the conclusions drawn from previous ^{29}Si NMR (e.g., Maekawa et al. 1991) and Raman (e.g., Matson et al. 1983; Mysen 1990) spectra of binary alkali silicate glasses.

Solution of water in the melts quenched to glasses results in the appearance in the Raman spectra of a new peak near 900 cm^{-1} (Fig. 3). Its intensity increases with increasing water content, but the extent of this change depends on ionic radius of alkali (Fig. 3). The greater the alkali ionic radius, the larger the intensity of the ~900 cm^{-1} peak. Moreover, it appears that the type of alkali in the glass affects the frequency of this peak. For example, in the Raman spectra of glasses with 17.6 mol % H_2O, it shifts from 882 cm^{-1} for LS4 to 902 cm^{-1} for NS4 and 918 cm^{-1} for KS4. According to the study of Zotov and Keppler (1998) on sodium silicate glasses, this ~900 cm^{-1} peak may be assigned to Si-O-H stretching vibrations in Q^n species with Si-O-H bonds, regardless their number of bridging oxygens n. In pure hydrous silica, a peak near 976 cm^{-1} has been assigned to the signal of Si-O-H stretching vibrations (Stolen and Walrafen 1976). In hydrous aluminosilicate glasses, a peak located near 900 cm^{-1} has been assigned to a mixture of Si-O-H and Al-O-H vibrations, with the assumption that the Al-O-H stretching frequency is lower than that of Si-O-H (McMillan et al. 1993; Malfait 2009). However, recent results from molecular dynamic simulations (Spiekermann et al. 2012) questioned the attribution of the 976 cm^{-1} signal in pure hydrous silica and suggested that this peak may actually arise from the asymmetric Si-O stretching vibration of Q^2 units and that the Si-O-H stretching signal may form a shoulder near 920 cm^{-1} on the side of the 976 cm^{-1} peak. The fact that a peak near 900 cm^{-1} is detected and grows with solution of water in silicate glasses (Figs. 2 and 3, see also Zotov and Keppler 1998) may support such an interpretation, at least for silicate glasses.

The area of the ~950 cm^{-1} peak, assigned to Si-O stretching vibrations in Q^2, increases with solution of water in KS4 and NS4 glasses but decreases slightly in spectra of LS4 glasses (Fig. 4). The area of the ~1100 cm^{-1} peak, assigned to Si-O stretching vibrations in Q^3, seems to increase slightly with solution of water, but this effect is small (≤6 % with addition of 17.6 mol% H_2O, Fig. 4). Finally, the area of the two peaks at ~1150 and 1200 cm^{-1}, assigned to Si-O stretching vibrations in Q^4, varies differently as a function of water concentration and alkali ionic radius (Fig. 4). In Raman spectra of the NS4 and KS4 glasses, the area of these two peaks decreases with increasing water content (~10–20 % with addition of 17.6 mol % H_2O). In contrast, their area increases slightly (~3 %) with increasing water content from 0.0 up to 17.6 mol % in LS4 glasses.

There is another peak, near 850 cm^{-1}, in spectra of the KS4 glasses at high water concentration (Figs. 2 and 3). Its origin could be related to the presence of Q^1 species (e.g., see Mysen et al. 1982 or McMillan 1984), but Q^1 species are not detected by ^{29}Si NMR spectroscopy (see below and Fig. 5). This could be explained by an enhanced Raman signal of Q^1 species compared to other more polymerized Q^2, Q^3, or Q^4 species because of the increase of the Raman cross section of Q^n units that results from their depolymerization (decreasing n; see for instance Furukawa et al. 1981). Such effect has been observed for Q^2 species in sodium silicate glasses (Le Losq et al. 2014). Therefore, a very small proportion of Q^1 species (<1 %) may be present in the KS4 + 17.6 mol % H_2O and undetected by ^{29}Si NMR spectroscopy. This interpretation is speculative but does not affect the global observations because the 850 cm^{-1} signal intensity is negligible compared with those of the other signals.

^{29}Si single-pulse (SP) MAS NMR spectra

The ^{29}Si SP MAS NMR spectra of anhydrous glasses comprise two main peaks near −95 and −105 ppm

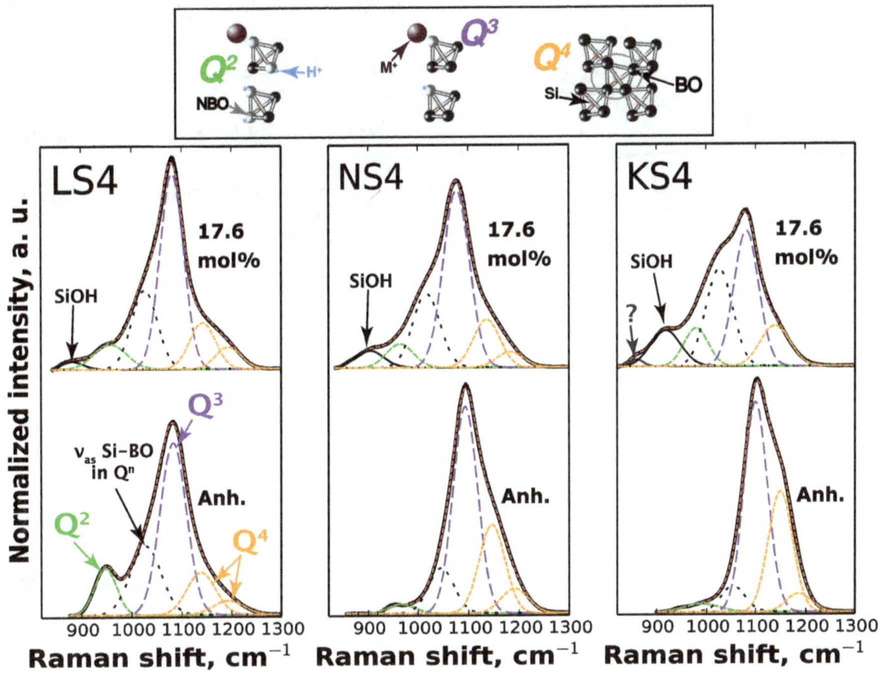

Fig. 3 Examples of curve-fitted Raman spectra of anhydrous and 17.6 mol% H_2O LS4 (*left*), NS4 (*middle*), and KS4 (*right*) glasses. Addition of water in the glasses clearly produces an increasing intensity near 900 cm^{-1}. It could be related to the stretching signal of Si-O-H bonds regardless of the Q^n specie carrying those bonds. Signal near 850 cm^{-1} in the Raman spectrum of KS4 + 17.6 mol % H_2O glass may be related to Q^1 species in the glass in negligible quantities. The various Q^n species are schematically represented at the top of the figure (ionic size is not at scale). *BO* bridging oxygen (Si-O-Si bonds), *NBO* non-bridging oxygen (Si-O-M or Si-O-H bonds, with M^+ an alkali)

(Fig. 5). A third contribution, located near −85 ppm, forms a shoulder in the spectrum of the anhydrous LS4 glass. The ~ −105, −95, and −85 ppm contributions are assigned to the signals of ^{29}Si in Q^4, Q^3, and Q^2 species, respectively (Lippmaa et al. 1980; Murdoch et al. 1985; Maekawa et al. 1991). The Q^3 and Q^4 contributions dominate the ^{29}Si SP MAS NMR spectra of anhydrous glasses (Fig. 5), in agreement with interpretation of Raman data (Figs. 3 and 4). The signal of Q^2 species is visible only in the spectrum of the LS4 anhydrous glass (Fig. 5). The relative intensities of Q^3 and Q^4 ^{29}Si NMR signals vary slightly with ionic radius of alkali in anhydrous glasses. The global width of the ^{29}Si NMR signal also decreases with increasing ionic radius of alkali metal, and the frequency difference between Q^3 and Q^4 signals narrows (Fig. 5). The Q^4 frequency is the lowest in the spectrum of LS4 glass compared with those in the spectra of the NS4 and KS4 glasses. Such ^{29}Si NMR frequency changes can result from a modification of the shielding of the ^{29}Si nuclei with changing ionic radius of alkali. Maekawa et al. (1991) suggested that such shielding evolution may reflect changes in the third neighbor environments (e.g., Olivier et al. 2001), bond length, or bridging Si-O-Si angle as a function of ionic radius of alkali in silicate glasses.

In order to highlight the perturbations induced by the water addition in the glasses networks, the ^{29}Si SP MAS NMR spectra of the anhydrous glasses have been subtracted from those of hydrous glasses, giving residual ^{29}Si NMR spectra where positive and negative peaks can be observed (Fig. 5, bottom). Those peaks represent the changes of the proportions in the various Q^n species upon water addition. In all glasses, solution of water results in a reduction in intensity of the Q^4 peak, whereas that of the Q^3 peak increases (Fig. 5). However, the extent of this intensity change depends on ionic radius of alkali metal. The Q^3/Q^4 ratio changes only slightly for LS4 glass, whereas it increases strongly for KS4 glass. In the NMR spectra of NS4 and KS4 glasses, the intensity near −80/−85 ppm also increases upon water solution but not in spectra of LS4 glasses. It follows, therefore, that upon solution of water in the silicate glasses, Q^4 units are converted to Q^3 and Q^2 units, and that this effect gets more pronounced with increasing ionic radius of alkali.

Quantification of the effect of water on glass polymerization

The relative areas, A_{Q^n} $\left(\sum_{n=0}^{4} A_{Q^n} = 1\right)$, of the Raman peaks assigned to Si-O stretching in Q^n units (Fig. 4) can be related to the fraction of Q^n units through (Mysen 1990; Mysen and Frantz 1993):

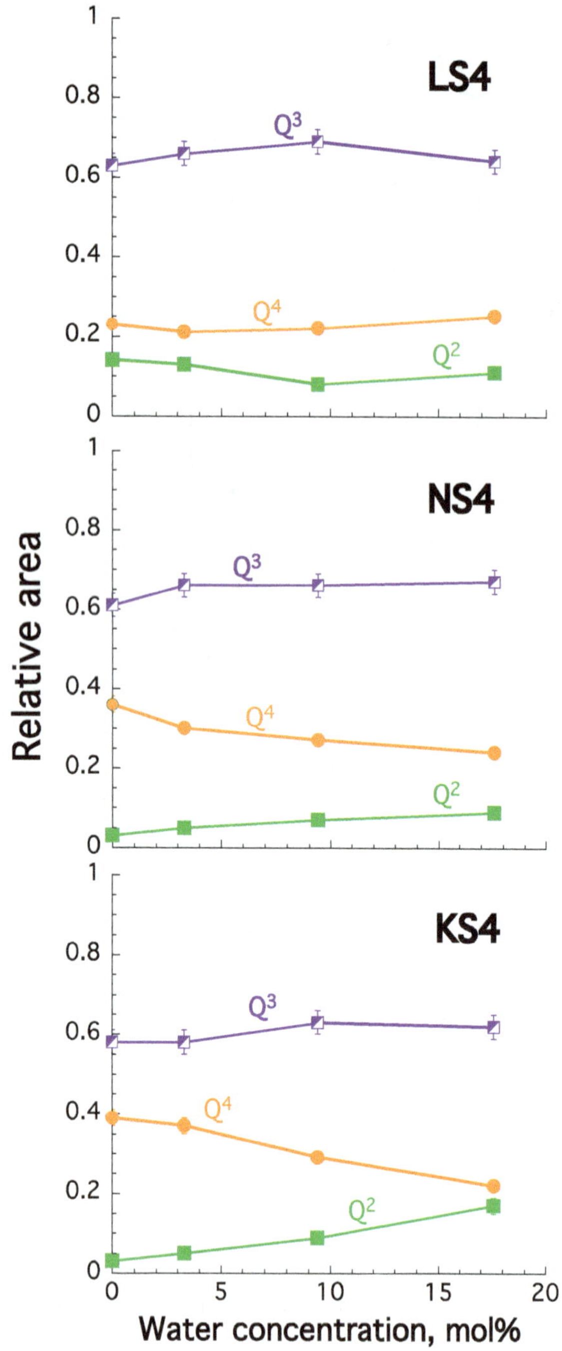

Fig. 4 Relative area of the Raman signals assigned to Q^2, Q^3, and Q^4 units composing the LS4 (*top*), NS4 (*middle*), and KS4 (*bottom*) glasses as a function of their nominal water concentration. Errors are given at the 1σ confidence interval. If not visible, they are smaller than the symbol sizes

$$[Q^n] = \theta^n A_{Q^n} \tag{4}$$

with θ^n a conversion factor that directly depends on the Raman cross section of the Si-O stretching vibrations in the Q^n units. The polymerization of a glass that contains a mixture of Q^2, Q^3 and Q^4units can be quantified by calculating the ratio of the mean number of NBOs other tetrahedral BOs as:

$$\mathrm{NBO/T} = 2[Q^2] + [Q^3] = 2 \times \theta^2 A_{Q^2} + \theta^3 A_{Q^3} \tag{5}$$

Between the anhydrous and hydrous glasses, we have the following equation:

$$\Delta\mathrm{NBO/T} = \mathrm{NBO/T^{hyd}}\text{–}\mathrm{NBO/T}^{anh} = 2[\Delta\mathrm{Q}^2] + [\Delta\mathrm{Q}^3], \tag{6}$$

where ΔQ^2 and ΔQ^3 are the variations of the Q^2 and Q^3 fractions between the anhydrous and hydrous glasses. The θ^2 and θ^3 values are near unity for sodium silicates and the effect of water solution on those values is negligible (Mysen 2007). Therefore, as a first approximation, we can neglect the θ^2 and θ^3 factors in Eqs. (5) and (6). It follows from the Raman data that addition of 17.6 mol % water in the LS4, NS4, and KS4 glasses results in *ΔNBO/T* of –0.04 ± 0.05, 0.18 ± 0.05, and 0.30 ± 0.05, respectively (Table 4).

Fitting ^{29}Si SP MAS NMR spectra with a set of Gaussian lines usually allows quantification of the Q^n fractions in silicate glasses (e.g., Maekawa et al. 1991). However, fitting the ^{29}Si SP MAS NMR spectra of the hydrous glasses is difficult because of the poor resolution of the shoulder associated to the signal of ^{29}Si in Q^2 units. Several solutions can be returned by the fitting procedure, and it is not clear how to discriminate between them. Fortunately, spectral fitting is not a requirement for quantifying the polymerization change induced by solution of water in glasses. Indeed, the areas of the peaks assigned to Q^2, Q^3, and Q^4 units in the residual ^{29}Si SP MAS NMR spectra (Fig. 5, bottom) are equivalent to the fractions of transformed Q^n units (ΔQ^n) upon solution of water. For the LS4 glasses, the ΔQ^3 signal area was measured between −87 and −102 ppm and the ΔQ^4 signal area between −102 and −122 ppm. In the spectra of the NS4 glasses, areas of the ΔQ^2, ΔQ^3, and ΔQ^4 signals were measured from −70 to −88, −88 to −101, and −101 to −120, respectively. For the spectra of KS4 glasses, the areas were measured from −70 to −91, −91 to −100, and −100 to −120, respectively. In order to evaluate the errors affecting the area determinations, the boundaries reported above were varied in a ±2–3 ppm uncertainty range. We have chosen such an uncertainty range because it is similar to the variations that affect the frequency of the peaks observed on the ^{29}Si MAS NMR spectra of the various glasses, and because it produces residual intensity variations higher than the noise level. As a result, such uncertainty range appears to be the best estimate for estimating the errors affecting the areas of residual signals.

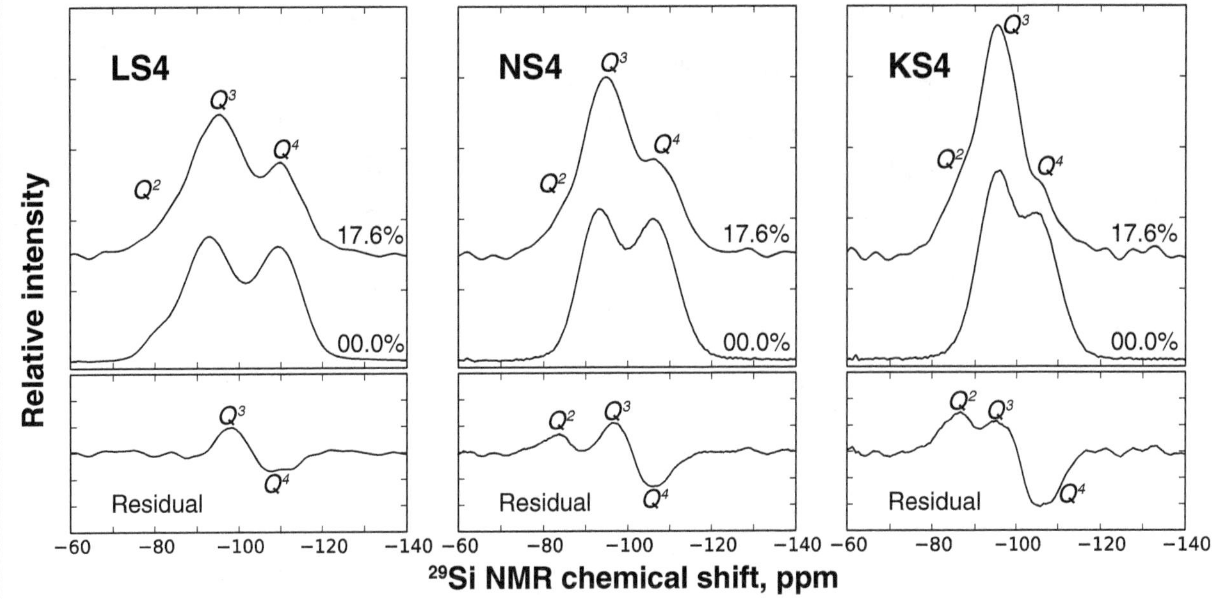

Fig. 5 *Top*: ^{29}Si single-pulse MAS (5 kHz) NMR spectra of the anhydrous and 17.6 nom. mol % H_2O LS4 (*left*), NS4 (*middle*), and KS4 (*right*) glasses. Intensities on spectra are normalized such that the total areas of the spectra are equal to unity. Changes in relative intensity and areas are thus directly proportional to the changes in the chemical concentrations of ^{29}Si in particular Q^n species. *Bottom*: residuals from the subtraction of the ^{29}Si NMR signals from hydrous glasses minus those from anhydrous glasses. Areas under the positive and negative peaks are quantitative and represent the fraction of transformed Q^n species upon hydration

The ΔQ^2 and ΔQ^3 values thus obtained can be used to compute $\Delta NBO/T$ with Eq. (6). The $\Delta NBO/T$ calculated from Raman data and from ^{29}Si SP MAS NMR difference spectra are in good agreement (Fig. 6 and Table 4) except for LS4 glass. Indeed, for the latter, the $\Delta NBO/T$ values obtained with the two different spectroscopic methods are different: −0.04 ± 0.05 (Raman) and 0.07 ± 0.01 (NMR). The Q^n speciation derived from the Raman spectra of LS4 glasses suggests almost no depolymerization of those glasses upon water solution. This conclusion differs somewhat from that drawn from the ^{29}Si NMR data, which indicate that there is a weak, but real, effect of water on the polymerization of LS4 glasses (Fig. 5). The errors affecting the fitting of Raman spectra might be too large to allow detection of the weak effect of water solution on the structure of LS4 glasses. Indeed, the Raman-derived $\Delta NBO/T$ values are affected by errors inherent to the fitting procedure (optimization error) as well as by errors inherent to the initial assumptions made for fitting (e.g., perfect Gaussian shape for the peaks). In this regards, NMR-derived $\Delta NBO/T$ values are more robust than the Raman-derived $\Delta NBO/T$ values because they do not rely on any fitting procedure but directly on calculations from raw data. This notwithstanding, both sets of values indicate a small effect of water addition on the degree of polymerization of the LS4 composition.

Table 4 ΔNBO/T, concentrations of total OH groups, SiOH, and *unknown* OH groups

	H_2O^{TOT} (Raman)	[OH]	ΔNBO/T (Raman)	[SiOH] (Raman)	[Unknown OH] (Raman)	ΔNBO/T (NMR)	[SiOH] (NMR)	[Unknown OH] (NMR)
LS4	3.08 (60)	2.11 (35)	0.02 (5)	0.63 (1.25)	1.47 (3.41)	-	-	-
	10.09 (60)	4.56 (70)	−0.05 (5)	0.00 (–)	4.56 (1.27)	-	-	-
	18.74 (60)	6.26 (94)	−0.04 (5)	0.00 (–)	6.26 (1.57)	0.07 (1)	2.77 (16)	3.49 (95)
NS4	3.44 (67)	3.85 (40)	0.08 (5)	3.24 (66)	0.61 (77)	-	-	-
	9.46 (67)	7.35 (63)	0.13 (5)	5.02 (43)	2.33 (76)	-	-	-
	17.54 (67)	9.52 (80)	0.18 (5)	7.28 (30)	2.24 (85)	0.18 (1)	7.17 (5)	2.35 (80)
KS4	3.98 (74)	3.98 (70)	0.04 (4)	1.42 (1.22)	2.56 (40)	-	-	-
	9.46 (74)	8.89 (70)	0.15 (5)	5.88 (32)	3.01 (77)	-	-	-
	15.95 (74)	13.65 (70)	0.30 (5)	12.08 (19)	1.57 (72)	0.34 (2)	13.63 (7)	0.02 (70)

Concentrations of OH, H_2O^{TOT}, SiOH, and *unknown* OH are in mol %

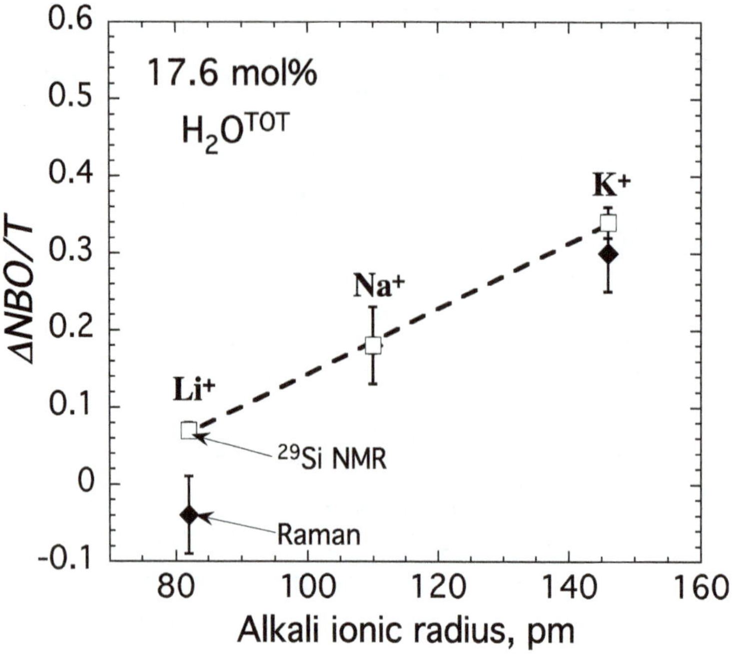

Fig. 6 Variation of the polymerization of the glasses (ΔNBO/T) upon addition of 17.6 nom. mol % water represented as a function of the ionic radius of alkali. ΔNBO/T (= $NBO/T_{anhydrous} - NBO/T_{17.6\ mol\%\ water}$) is calculated from the variations of the Q^n species as observed in Raman spectra (*plain symbols*) or in ^{29}Si NMR spectra (*open symbols*). Ionic radii of alkalis are from Whittaker and Muntus (1970), assuming sixfold oxygen-coordinated alkalis. The *black dotted line* is a linear fit of the ΔNBO/T vs ionic radius relationship for NMR data. For the Na-bearing silicate glass, Raman and ^{29}Si NMR give the same value, so that symbols are difficult to distinguish. *Error bars* are given at the 1σ confidence interval

Discussion

The dissolution of water into alkali silicate glasses results in their depolymerization, as shown by ^{29}Si NMR and Raman data (Figs. 4, 5, and 6). The Q^4 units transform to Q^3 and Q^2 units according to the reaction.

$$Q^4 \iff \phi Q^2 + (1-\phi) Q^3, \quad (7)$$

where ϕ is a constant reflecting the partitioning of the NBOs (formed upon H_2O addition) between Q^3 and Q^2. The equilibrium constant K_7 of reaction 7 depends on ionic radius of alkali. The larger the radius, the greater the *ΔNBO/T* and, thus, the higher the K_7 (Fig. 6). This dependence likely is linked to the influence of ionic radius of alkali metal on the OH/H_2O_{mol} ratio. At given water content, the larger the ionic size of alkali cation, the higher this ratio (Fig. 7). The variations of ϕ seem primarily related to the extent of depolymerization of the silicate network because the ratio $\Delta Q^2/\Delta Q^3$ is correlated with *ΔNBO/T*. At low *ΔNBO/T*, and hence for *NBO/*T close to 0.5, ϕ is low and Q^3 units are produced by reaction (7). At high *ΔNBO/T*, and hence for *NBO/T* > 0.5, ϕ is high and Q^2 units formation may dominate.

The concentration of OH groups that form SiOH bonds upon hydration of the glasses can be determined by using the *ΔNBO/T* values derived from the Q^n-species abundances. This is possible because oxygen in the

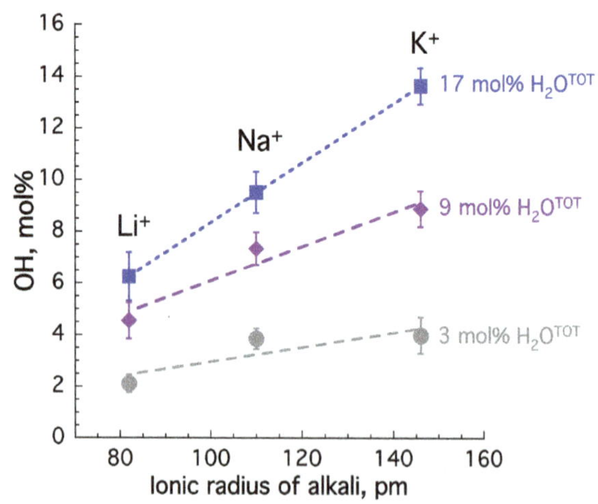

Fig. 7 Total OH group contents (mol %) of the alkali silicate glasses represented as a function of the alkali ionic radius (in picometers) for different water concentrations. Total OH group contents in KS4 glasses were determined by subtracting the H_2O_{mol} species contents determined with using ^{1}H MAS NMR spectra (see text and Fig. 1) to the total water content measured by Raman spectroscopy (see values in Table 2). The *dotted lines* are linear fits of the data. *Error bars* are given at the 1σ confidence interval

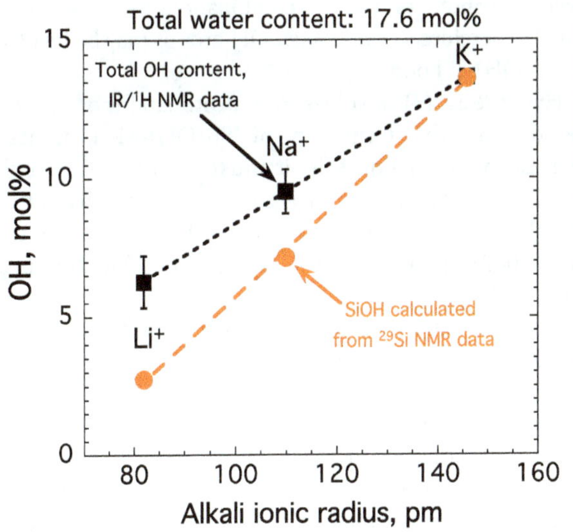

Fig. 8 Comparison between the concentration of OH groups measured with IR (LS4 and NS4 glasses) and ^{1}H NMR (KS4 glass) data and that of OH groups bonded to Si determined with using ^{29}Si NMR data for glasses with 17.6 mol % total H_2O. The *dotted lines* are linear fits of the data. *Error bars* are given at the 1σ confidence interval

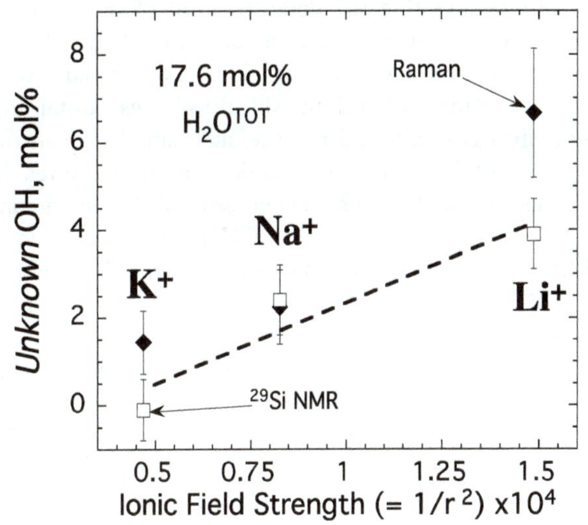

Fig. 9 Concentrations of *unknown* OH groups that do not participate in the depolymerization of the silicate network in the 17.6 mol % water glasses, represented as a function of the ionic field strength (IFS) of alkali. IFS = Z/r^2 with Z the electric charge and r the ionic radii of the alkalis (from Whittaker and Muntus 1970 assuming sixfold oxygen-coordinated alkalis). IFS was multiplied by 10^4 for clarity purpose. The *black dashed line* is a linear fit of the mol% unknown OH vs alkali IFS relationship calculated using NMR data. *Error bars* are given at the 1σ confidence interval

SiOH groups is non-bridging, so that SiOH group formation testifies to the depolymerization of the silicate network upon water solution (Eq. (2)). Figure 8 presents a comparison between the SiOH content determined with using *ΔNBO/T* derived from ^{29}Si SP MAS NMR data and the concentration of OH groups determined by using IR and ^{1}H MAS NMR data for glasses with 17.6 mol % water. At this water content, it appears that there are significant differences between these two sets of values for hydrous NS4 and LS4 glasses. For those glasses, the concentrations of SiOH groups are lower than those of OH groups detected by IR and ^{1}H MAS NMR spectroscopy. For the KS4 glass with 17.6 mol % water, the SiOH and the OH values obtained by the different methods agree. The SiOH fractions calculated from Raman-derived *ΔNBO/T* values confirm the above observations by showing a very similar trend following changes in ionic radius of alkali metal for glasses with 17.6 nom. mol % [H_2O^{TOT}] (Table 4).

The comparison in Fig. 8 and Table 4 between the SiOH concentrations retrieved through the ^{29}Si NMR data, validated by Raman results, and the total OH concentration from IR spectroscopy indicates that there is a significant fraction of OH groups in the NS4 and LS4 glasses that do not reflect depolymerization of the silicate network and for which we do not know the nature of their bonding. At 17.6 nom. mol % water, the concentration of those "unknown" OH groups increases with decreasing ionic radius, and hence, increasing ionic field strength of alkali metal (Fig. 9). It is negligible for the KS4 glass whereas it is maximum for the LS4 glass, which also has a negligible *ΔNBO/T*. It follows, therefore, that the chemical composition of the melt, in terms of ionic radius of the alkali metal, not only affects the speciation of water between OH groups and H_2O_{mol} species (Fig. 7), but also affects the bonding of the OH groups with the quenched melt structure (Figs. 8 and 9). The effect of the water concentration is less clear. From the Raman-derived SiOH contents, the fraction of *unknown* OH groups may increase in LS4 and NS4 with addition of water, whereas it may slightly decrease in the case of the KS4 glasses (Table 4). However, those values are difficult to interpret with confidence because they have not been validated by ^{29}Si NMR spectroscopy. This validation seems necessary in regards of the errors that can affect the *ΔNBO/T* determined from Raman spectroscopy, as shown by the data for LS4 (see Fig. 6).

Previous studies of hydrous sodium silicate glasses indicated that, at 80 mol% SiO_2, Eq. (2) appears to describe the mechanism of water solution in glasses (Zotov and Keppler 1998; Cody et al. 2005; Mysen and Cody 2005). The content of *unknown* OH is low in the NS4 glasses, near 2 mol % for a nominal total water content of 17.4 mol % (Table 4 and Fig. 9). The errors affecting the Q^n fractions determined by Raman or ^{29}Si NMR spectroscopy combined with those affecting the determination of the concentrations of total OH groups render difficult the detection of such a low amount. This may explain why previous studies of sodium silicate glasses containing high silica contents did not see such small concentrations of OH groups not bonded to Si. However, in alkaline-earth and alkali silicate glasses with low silica contents (NBO/T > ~ 1), presence of OH groups not bonded to Si has been inferred (e.g., Xue and Kanzaki 2004; Cody et al. 2005; Mysen and Cody 2005).

Present data indicate that Eq. (2) alone cannot explain the solution mechanism of water in silicate glasses containing small alkali metals, such as Li and Na. Another mechanism implying a different bonding of OH group must be considered. The concentration of total OH groups measured by IR in LS4 glasses is much higher than that of OH bonded to Si (Figs. 8 and 9). This indicates that the *unknown* OH groups contribute to the 4100–4500 cm^{-1} portion of IR spectra, where the peak assigned to the combination of the OH groups vibrational modes is usually observed (e.g., see Fig. 1). To our knowledge, three hypotheses can explain observations: (1) the formation of MOH groups, where M is an alkali; (2) the protonation of bridging Si-O-Si bonds; or (3) the protonation of Si-O-M bonds.

Infrared spectra of crystalline alkali hydroxide show absorption bands in the 4000–4400 cm^{-1} portion of IR spectra. It has been proposed that they arise from the combination of the O-H fundamental stretching vibrations with the lattice modes (Buchanan 1959; Wickersheim 1959; Snyder et al. 1960). Therefore, contribution from MOH species in the 4100–4500 cm^{-1} range of the IR spectra of the present glasses seems possible, so that the observed *unknown* OH could be MOH species. In this line of thinking, we note that the IR peak near 4500–4600 cm^{-1} is also observed in hydrous soda-lime-borate glasses and assigned to the combination of the OH stretching with lattice vibrational modes (Bauer et al. 2015). Therefore, the 4500 cm^{-1} IR peak in hydrous glasses not only contains contributions of OH groups bonded to Si, as it has been previously inferred (e.g., Malfait 2009 and references therein), but also possible contributions from other OH species bonded to metallic elements or even boron, as shown by the data of Bauer et al. (2015).

Hypothesis (2) involves the protonation of Si-O-Si bonds and thus the formation of Si-(OH)-Si protonated bonds. A similar hypothesis has been proposed to account for water solution in glasses along the joined SiO_2-$NaAlO_2$ because of the apparently negligible changes of ^{29}Si and ^{27}Al NMR signals upon water addition to these glasses (Kohn et al. 1989b, 1992, 1998). Molecular orbital calculations indicate that both Si-OH and Si-(OH)-Si bonds give signals near 900 cm^{-1} (Kubicki and Sykes 2004), so that the OH groups involved in Si-(OH)-Si bonds will contribute to the IR combination peak near 4500 cm^{-1}. Therefore, the *unknown* OH groups may be involved in Si-(OH)-Si bonds.

Hypothesis (3) involves the protonation of Si-O-M bonds and the formation of Si-(OH)-M protonated bonds. In this hypothesis, it must be inferred that the Si-(OH)-M bonds give a signal similar to that of Si-(OH)-Si protonated bonds near 900 cm^{-1}, such that those bonds will contribute to the IR combination peak near 4500 cm^{-1}.

To discriminate between those hypotheses, several points have to be considered. First, the link between the size/ionic field strength (abbreviated IFS in the following) of the alkali and the concentration of *unknown* OH (Fig. 9) indicates that alkalis that form short (and thus strong) M-O bonds promote *unknown* OH groups. Considering hypothesis (3), it is unlikely that protonation of Si-O-M bonds occurs when the M-O bond strength is high, such as for Li-O bonds, because protonation of Si-O-M bonds may be more likely to occur for stabilizing Si-O-M bonds with low M-O bond strengths. Considering now hypothesis (2), protonation of Si-O-Si bonds would result in the appearance of positively charged $Si\text{-}(OH)^{+}\text{-}Si$ bonds that would require charge balance by a negatively charged ion, i.e., by NBO oxygens or "free" oxygens O^{2-}. Such situation would require specific structural positions for the $Si\text{-}(OH)^{+}\text{-}Si$ bonds and the compensating oxygens and thus seems unlikely to occur at high temperature in silicate melts that lack long- and medium-range orders. It should be noted that protonation of Si-O-Al bonds has been inferred in aluminosilicate compositions because it may help the electrical charge compensation of AlO_2 tetrahedral units, but it also has been strongly questioned (see introduction of Malfait and Xue 2010 and references cited therein). Indeed, the most recent data point to formation of SiOH and AlOH groups in aluminosilicate glasses and not to protonation of bridging oxygen (e.g., see Xue and Kanzaki 2009, Malfait and Xue 2010 and references cited therein).

Therefore, the most plausible hypothesis to explain the observations of a fraction of OH groups not bonded to Si in the alkali silicate glasses is the formation of MOH groups. In particular, this hypothesis explains the link between the alkali IFS and the concentration of *unknown* OH groups (Fig. 9): metal cations with strong M-O bonds may help stabilizing MOH complexes, in agreement with interpretations of Xue and Kanzaki (2004). In addition, this hypothesis is supported by ^{1}H MAS and $^{1}H\text{->}^{29}Si\text{->}^{1}H$ CP MAS NMR data on Ca and Mg silicate glasses that indicate the present of MOH groups in such compositions (Xue 2009). Therefore, from the above discussion, the *unknown* OH groups likely are bonded to alkali elements, forming MOH molecular complexes.

We propose the following framework to explain the spectroscopic observations. A condition for the reaction of H_2O_{mol} species with the network of silicate melts is the self-ionization of H_2O_{mol} species (Moretti et al. 2014):

$$H_2O_{mol\,(melt)} <=> H^+_{(melt)} + OH^-_{(melt)} \quad (8)$$

When H_2O_{mol} reacts with Si-O-Si bonds, H^+ and OH^- are consumed to form SiOH bonds, leading to the process described by Eq. (2). In the present silica-rich glasses, H^+ and OH preferentially react with Q^4 bonds to form Q^2 and Q^3 species (Fig. 5), so that the depolymerization mechanism can be described by Eq. (7), confirming previous observations in similar composition (Zotov and Keppler 1998; Cody et al. 2005; Xue and Kanzaki 2004; Mysen and Cody 2005). Decreasing the ionic radius, and hence, increasing the IFS of alkali metal in silicate melts inhibits the self-ionization reaction of water (Fig. 7) but also promotes reaction of H^+ and OH^- with the non-bridging oxygen NBO and the alkalis of the Si-O-M bonds (Fig. 9) following:

$$H_2O_{mol} + SiOM <=> SiOH + MOH \quad (9)$$

Such a mechanism was proposed by Xue and Kanzaki (2004), who suggested the formation of MOH complexes in highly depolymerized Ca, Mg, and Na silicate melts (NBO/T > 1) on the basis of molecular dynamic calculations and ^{1}H MAS NMR observations. They also proposed the following reaction (see also Mysen and Cody 2005):

$$2\ SiOM + H_2O_{mol} <=> SiOSi + 2\ MOH, \quad (10)$$

Formation of MOH species can result in no change (Eq. (9)) or even in an increase (Eq. (10)) of the melt polymerization. Xue and Kanzaki (2004) indicated that such mechanisms mostly concerned depolymerized melts (NBO/T > ~1) (see also the discussion in Xue and Kanzaki 2009), in agreement with the findings of Cody et al. (2005) and Mysen and Cody (2005). However, the water solution mechanism for the LS4 glass is best described by the combination of Eqs. (8) and (9). Therefore, the present data support the solution mechanism described with Eq. (9) in melts with relatively high silica contents (NBO/T ~ 0.5) that contain network modifier elements with high IFS (Fig. 9).

According to the present data (Figs. 7, 8, and 9), IFS of network modifying alkali metals plays an important role in defining the equilibrium constants of Eqs. (8) (self-ionization of water), (2) (SiOH formation), (9) and (10) (MOH formation). The origin of such role could arise from electronic/steric considerations, hydration shell formation, or volume effects around alkalis governed by their size/IFS (Le Losq et al. 2015a). For instance, electronic effects may influence the hydrolysis energy of Si-O-Si and Si-O-M bonds, and hence, the equilibrium constants of reactions (2), (9), and (10) (i.e., see discussion in Xue and Kanzaki 2004). Volume effects cannot be excluded. For instance, large alkalis may inhibit the presence of cavities in which H_2O_{mol} species might be contained, and hence, will result in a large equilibrium constant for reaction (8). Formation of hydration shell around small alkalis may also help explaining the speciation of water (Fig. 7). Finally, IFS of metal cations may also affect the stability of MOH molecular complexes (Xue and Kanzaki 2004), explaining the link between the IFS of metal cations and the concentration of observed MOH groups (Fig. 9). However, the present data do not allow us to discriminate between those hypotheses. In addition, it cannot be excluded that several mechanisms play roles at the same time. Despite that, it is clear that the influence of the alkali IFS on the speciation, environment, and bonding of water in silicate quenched melts results in differences in water solubility (Mysen and Acton 1999; Mysen and Cody 2004) and partial molar volume (Mysen and Cody 2004). Beyond such concerns, the differences of bonding and speciation of water in hydrous glasses as a function of their chemical composition probably also result in variations in their physical properties such as their glass transition temperature and density.

Conclusions

The water speciation in glasses is that of the melts at their glass transition temperature, *Tg* (see for instance Nowak and Behrens 1995). Present results show that IFS of alkali affects the speciation of dissolved water between H_2O_{mol} species and OH groups, as well as the OH bonding mechanisms in glasses, and hence, in melts at their *Tg*. Therefore, the chemical composition of silicate melts affects the solution mechanisms of water (Figs. 7, 8, and 9), resulting in different effects on melt structures at *Tg* (Figs. 3, 4, 5, and 6). In addition, variations in the solubility of water with ionic radius of the alkali in tetrasilicate melts (see data for NS4 and KS4 in Mysen and Cody 2004) at given temperature and pressure indicate that such interplay between the melt composition, water solution mechanism, and the melt structure affects the solubility of water. As a result, such interplay probably drives the water exsolution history of magmas.

In addition to such effect, this complex interplay implies that water will affect differently the rheological properties of magmas depending on their composition. Indeed, the strong changes in melt structure (Figs. 4, 5, and 6) that accompanies variations in water solution mechanism (Figs. 7, 8, and 9) probably result in different glass transition temperatures for the hydrous alkali tetrasilicate glasses at given $[H_2O^{TOT}]$, and, as a corollary, in different viscosity vs temperature paths. The present data do not allow us to make inferences on the extent of those differences, but they are expected to occur in view of the intimate link between the structure and the viscosity of alkali silicate melts (e.g., see Mysen and Richet 2005 for further details). Therefore, magmas with different chemical compositions will not only present different degassing paths, but also they will present different changes in their viscosity as a function of water content and temperature. Such consideration might be important for assessing about volcanic processes intimately linked to the viscous relaxation of magmas, such as the fragmentation that occurs in explosive eruptions (Papale 1999). Indeed, following the above ideas, the complex interplay between the magma composition, molecular structure, viscous properties, and the water solution mechanisms and solubility may influence the fragmentation threshold of magmas.

Abbreviations

EDS: energy-dispersive x-ray spectroscopy; SP: single pulse; MAS: magic angle spinning; NMR: nuclear magnetic resonance; IR: infrared; IFS: ionic field strength.

Competing interests

The authors declare that they have no competing interests.

Authors' contributions

BOM and GDC proposed the study. CLL made the experimental study, treated and interpreted data, and drafted the manuscript. BOM helped CLL for the experiments and lead the study. GDC performed the NMR experiments with the help of CLL. BOM and GDC collaborated with CLL in the construction of the manuscript. All authors read and approve the final manuscript.

Authors' information

CLL performed this study as a Postdoctoral Fellow at the Geophysical Laboratory, Carnegie Institution of Science, Washington DC, USA, under the mentorship of BOM and GDC, respectively Senior Scientist at, and Acting Director of the Geophysical Laboratory.

Acknowledgements

This research was partially supported by grants EAR1212754 and EAR1250449 to BOM and the NASA Astrobiology Institute. All NMR experiments were performed at the W. M. Keck Solid State NMR Facility at the Geophysical Laboratory, supported in part by the W. M. Keck Foundation, the NSF, and the Carnegie Institution of Washington.
This paper belongs to the Solid Earth Sciences section.

References

Bartholomew RF, Butler BL, Hoover HL, Wu CK (1980) Infrared spectra of a water-containing glass. J Am Ceram Soc 63:481–485

Bauer U, Behrens H, Fechtelkord M, Reinsch S, Deubener J (2015) Water- and boron speciation in hydrous soda-lime-borate glasses. J Non Cryst Solids 423–424:58–67

Behrens H, Yamashita S (2008) Water speciation in hydrous sodium tetrasilicate and hexasilicate melts: constraint from high temperature NIR spectroscopy. Chem Geol 256:306–315

Bell RJ, Bird NF, Dean P (1968) The vibrational spectra of vitreous silica, germania and beryllium fluoride. J Physics C (Proc phys soc) 1:299–303

Blake S (1984) Volatile oversaturation during the evolution of silicic magma chambers as an eruption trigger. J Geophys Res 89:8237–8244

Brawer SA, White WB (1975) Raman spectroscopic investigation of the structure of silicate glasses. I. The binary alkali silicates*. J Chem Phys 63:2421–2432

Brawer SA, White WB (1977) Raman spectroscopic investigation of the structure of silicate glasses (II). Soda-alkaline earth-alumina ternary and quaternary glasses. J Non Cryst Solids 23:261–278

Buchanan RA (1959) Near infrared spectra of crystalline alkali hydroxides. J Chem Phys 31:870–874

Clemens JD (1984) Water contents of silicic to intermediate magmas. Lithos 17:273–28

Cody GD, Mysen BO, Lee SK (2005) Structure vs. composition: a solid-state ^1H and ^{29}Si NMR study of quenched glasses along the Na_2O-SiO_2-H_2O join. Geochim Cosmochim Acta 69:2373–2384

Dingwell DB, Romano C, Hess KU (1996) The effect of water on the viscosity of a haplogranitic melt under P-T-X conditions relevant to silicic volcanism. Contribut Mineral Petrol 124:19–28

Eckert H, Yesinowski JP, Silver LA, Stolper EM (1988) Water in silicate glasses: quantification and structural studies by ^1H solid echo and MAS-NMR methods. J Phys Chem 92:2055–2064

Farnan I, Kohn SC, Dupree R (1987) A study of the structural role of water in hydrous silica glass using cross-polarisation magic angle spinning NMR. Geochim Cosmochim Acta 51:2869–2873

Friedman I, Long W, Smith RL (1963) Viscosity and water content of rhyolite glass. J Geophys Res 68:6524–6535

Furukawa T, Fox KE, White WB (1981) Raman spectroscopic investigation of the structure of silicate glasses. III. Raman intensities and structural units in sodium silicate glasses. J Chem Phys 75:3226–3237

Galeener FL (1979) Band limits and the vibrational spectra of tetrahedral glasses. Phys Rev B 19:4292–4397

Gardner JE, Thomas RME, Jaupart C, Tait S (1996) Fragmentation of magma during Plinian volcanic eruptions. Bull Volcanol 58:144–162

Kohn SC, Dupree R, Mortuza MG (1992) The interaction between water and aluminosilicate magmas. Chem Geol 96:399–409

Kohn SC, Dupree R, Smith ME (1989a) Proton environments and hydrogen-bonding in hydrous silicate glasses from proton NMR. Nature 337:539–541

Kohn SC, Dupree R, Smith ME (1989b) A multinuclear magnetic resonance study of the structure of hydrous albite glass. Geochim Cosmochim Acta 53:2925–2935

Kohn SC, Smith ME, Dirken PJ, Van Eck ERH, Kentgens APM, Dupree R (1998) Sodium environment in dry and hydrous albite glasses: Improved ^{23}Na solid state NMR data and their implications for water dissolution mechanisms. Geochim Cosmochim Acta 62:79–87

Kubicki JD, Sykes DG (2004) Ab initio calculation of ^1H, ^{17}O, ^{27}Al and ^{29}Si NMR parameters, vibrational frequencies and bonding energetics in hydrous silicate and Na-aluminosilicate glasses. Geochim Cosmochim Acta 68:3909–3918

Le Losq C, Neuville DR (2013) Effect of the Na/K mixing on the structure and the rheology of tectosilicate silica-rich melts. Chem Geol 346:57–71

Le Losq C, Cody GD, Mysen BO (2015a) Alkali influence on the water speciation and the environment of protons in silicate glasses revealed by ^1H MAS NMR spectroscopy. Am Mineral 100:466–473

Le Losq C, Cody GD, Mysen BO (2015b) Complex stretching signal of OH groups in silicate glasses: implication for the use of the 4500 cm^{-1} IR peak as a marker of OH^- groups concentration. Am Mineral 100:945–950

Le Losq C, Neuville DR, Florian P, Henderson GS, Massiot D (2014) The role of Al^{3+} on rheology and structural changes of sodium silicate and aluminosilicate glasses and melts. Geochim Cosmochim Acta 126:495–517

Leonteva A (1940) Measurements of the viscosity of obsidian and of hydrated glasses. *Izvestiya Akademii Nauk SSSR*. Ser Geol 2:44–54

Lippmaa E, Mägi M, Samoson A, Engelhardt G, Grimmer A-R (1980) Structural studies of silicates by solid-state high-resolution ^{29}Si NMR. J Am Chem Soc 102:4889–4893

Maekawa H, Maekawa T, Kawamura K, Yokohama T (1991) The structural groups of alkali silicate glasses determined from ^{29}Si MAS-NMR. J Non Cryst Solids 127:53–64

Malfait WJ (2009) The 4500 cm–1 infrared absorption band in hydrous aluminosilicate glasses is a combination band of the fundamental {Si, Al}-OH and O-H vibrations. Am Mineral 94:849–852

Malfait WJ, Xue X (2010) The nature of hydroxyl groups in aluminosilicate glasses: Quantifying Si-OH and Al-OH abundances along the SiO_2-$NaAlSiO_4$ join by ^1H, ^{27}Al-^1H and ^{29}Si-^1H NMR spectroscopy. Geochim Cosmochim Acta 74:719–737

Matson DW, Sharma SK, Philpotts JA (1983) The structure of high-silica alkali-silicate glasses. A Raman spectroscopic investigation. J Non Cryst Solids 58:323–352

McMillan PF (1984) Structural studies of silicate glasses and melts—applications and limitations of Raman spectroscopy. Am Mineral 69:622–644

McMillan PF, Poe BT, Stanton TR, Remmele RL (1993) A Raman spectroscopic study of H/D isotopically substituted hydrous aluminosilicate glasses. Phys Chem Min 19:454–459

Moretti R, Le Losq C, Neuville DR (2014) The amphoteric behavior of water in silicate melts from the point of view of their ionic-polymeric constitution. Chem Geol 367:23–33

Moulson AJ, Roberts JP (1961) Water in silica glass. Trans Faraday Soc 57:1208–1216

Murdoch JB, Stebbins JF, Carmichael ISE (1985) High-resolution ^{29}Si NMR study of silicate and aluminosilicate glasses: the effect of network-modifying cations. Am Mineral 70:332–343

Mysen BO (1990) Role of Al in depolymerized, peralkaline aluminosilicate melts in the systems Li_2O-Al_2O_3-SiO_2, Na_2O-Al_2O_3-SiO_2, and K_2O-Al_2O_3-SiO_2. Am Mineral 75:120–134

Mysen BO (2007) The solution behavior of H_2O in peralkaline aluminosilicate melts at high pressure with implications for properties of hydrous melts. Geochim Cosmochim Acta 71:1820–1834

Mysen BO, Acton M (1999) Water in H_2O-saturated magma-fluid systems: solubility behavior in K_2O-Al_2O_3-SiO_2-H_2O to 2.0 GPa and 1300 °C. Geochim Cosmochim Acta 63:3799–3815

Mysen BO, Cody G (2005) Solution mechanisms of H_2O in depolymerized peralkaline melts. Geochim Cosmochim Acta 69:5557–5566

Mysen BO, Cody G (2004) Solubility and solution mechanism of H_2O in alkali silicate melts and glasses at high pressure and temperature. Geochim Cosmochim Acta 68:5113–5126

Mysen BO, Frantz JD (1993) Structure and properties of alkali silicate melts at magmatic temperatures. Eur J Mineral 5:393–407

Mysen BO, Richet P (2005) Silicate glasses and melts—properties and structure. Developments in Geochemistry 10. Elsevier B.V, Amsterdam

Mysen BO, Finger LW, Virgo D, Seifert FA (1982) Curve-fitting of Raman spectra of silicate glasses. Am Mineral 67:686–695

Neuville, D. R. (1992) Etude des propriétés thermodynamiques et rheologiques des silicates fondus. Ph.D. Thesis of the Université Paris 7, 274.

Neuville DR (2006) Viscosity, structure and mixing in (Ca, Na) silicate melts. Chem Geol 229:28–41

Neuville DR, Mysen BO (1996) Role of aluminium in the silicate network: in situ, high-temperature study of glasses and melts on the join SiO_2-$NaAlO_2$. Geochim Cosmochim Acta 60:1727–1737

Neuville DR, de Ligny D, Henderson GS (2014) Advances in Raman spectroscopy applied to Earth and Material Sciences. Rev Mineral Geochem 78:509–541

Nowak M, Behrens H (1995) The speciation of water in haplogranitic glasses and melts determined by in situ near-infrared spectroscopy. Geochim Cosmochim Acta 59:3445–3450

Ollvier L, Yuan X, Cormack AN, Jäger C (2001) Combined ^{29}Si double quantum NMR and MD simulation studies of network connectivities of binary $Na_2O.SiO_2$ glasses: new prospects and problems. J Non Cryst Solids 293–295:53–66

Papale P (1999) Strain-induced magma fragmentation in explosive eruptions. Nature 397:425–428

Richet P, Lejeune A-M, Holtz F, Roux J (1996) Water and the viscosity of andesite melts. Chem Geol 128:185–197

Richet P, Whittington AG, Holtz F, Behrens H, Ohlhorst S, Wilke M (2000) Water and the density of silicate glasses. Contribut Mineral Petrol 138:337–347

Saucier H (1952) Quelques expériences sur la viscosité à haute température de verres ayant la composition d'un granite. Influence de la vapeur d'eau sous pression. Bull de la Société Française de Minéralogie et de Cristallographie 75:1–45

Scholtze H (1960) Zur Frage der unterscheidung zwischen H_2O-Molekeln und OH-gruppen in gläsern und mineralen. Naturwissenschaften 47:226–227

Schramm CM, de Jong BHWS, Parziale VE (1984) ^{29}Si magic angle spinning NMR study on local silicon environments in amorphous and crystalline lithium silicates. J Am Chem Soc 106:4396–4402

Seifert F, Mysen BO, Virgo D (1982) Three-dimensional network structure of quenched melts (glass) in the systems SiO_2-$NaAlO_2$, SiO_2-$CaAl_2O_4$ and SiO_2-$MgAl_2O_4$. Am Mineral 67:696–717

Sen PN, Thorpe MF (1977) Phonons in AX_2 glasses: from molecular to band-like modes. Phys Rev B 15:4030–4038

Shaw HR (1963) Obsidian-H_2O viscosities at 1000 and 2000 bars in the temperature range 700° to 900 °C. J Geophys Res 68:6337–6343

Snyder RG, Kumamoto J, Ibers JA (1960) Vibrational spectrum of crystalline potassium hydroxide. J Chem Phys 33:1171–1177

Sowerby JR, Keppler H (1999) Water speciation in rhyolitic melt determined by in-situ infrared spectroscopy. Am Mineral 84:1843–1849

Spiekermann G, Steel-McInnis M, Schmidt C, Jahn S (2012) Vibrational mode frequencies of silica species in SiO_2-H_2O liquids and glasses from ab initio molecular dynamics. J Chem Phys 136:154501

Stolen RH, Walrafen GE (1976) Water and its relation to broken bond defects in fused silica. J Chem Phys 64:2623–2631

Stolper A (1982) Water in silicate glasses: an infrared spectroscopic study. Contribut Mineral Petrol 81:1–17

Tarantola, A. (2005) Inverse problem theory and methods for model parameter estimation. Society of Industrial and Applied Mathematics

Thomas N, Jaupart C, Vergniolle S (1994) On the vesicularity of pumice. J Geophys Res 99:15633–15644

Tischendorf B, Ma C, Hammersten E, Venhuizen P, Peters M, Affatigato M, Feller S (1998) The density of alkali silicate glasses over wide compositional ranges. J Non Cryst Solids 239:197–202

Virgo D, Mysen BO, Kushiro I (1980) Anionic constitution of silicate melts quenched at 1 atm from Raman spectroscopy: implications for the structure of igneous melts. Science 208:1371–1373

Voigt U, Lammert H, Eckert H, Heuer A (2005) Cation clustering in lithium silicate glasses: quantitative description by solid-state NMR and molecular dynamics simulations. Phy Rev B 72:064207–1–11

Whittaker EJW, Muntus R (1970) Ionic radii for use in geochemistry. Geochim Cosmochim Acta 34:945–956

Wickersheim KA (1959) Infrared absorption spectra of lithium hydroxide. J Chem Phys 31:863–869

Wu C-K (1980) Nature of incorporated water in hydrated silicate glasses. J Am Ceram Soc 63:453–457

Xue X (2009) Water speciation in hydrous silicate and aluminosilicate glasses: direct evidence from ^{29}Si-^{1}H and ^{27}Al-^{1}H double-resonance NMR. Am Mineral 94:395–398

Xue X, Kanzaki M (2004) Dissolution mechanisms of water in depolymerized silicate melts: constraints from ^{1}H and ^{29}Si NMR spectroscopy and *ab initio* calculations. Geochim Cosmochim Acta 68:5027–5057

Xue X, Kanzaki M (2006) Depolymerization effect of water in aluminosilicate glasses: direct evidence from ^{1}H-^{27}Al heteronuclear correlation NMR. Am Mineral 91:1922–1926

Xue X, Kanzaki M (2008) Structure of hydrous aluminosilicate glasses along the diopside—anorthite join: a comprehensive one- and two-dimensional ^{1}H and ^{27}Al NMR study. Geochim Cosmochim Acta 72:2331–2348

Xue X, Kanzaki M (2009) Proton distributions and hydrogen bonding in crystalline and glassy hydrous silicates and related inorganic materials: insights from high-resolution solid-state nuclear magnetic resonance spectroscopy. J Am Ceram Soc 92:2803–2830

Yamashita S, Behrens H, Schmidt BC, Dupree R (2008) Water speciation in sodium silicate glasses based on NIR and NMR spectroscopy. Chem Geol 256:231–241

Zotov N, Keppler H (1998) The influence of water on the structure of hydrous sodium tetrasilicate glasses. Am Mineral 83:823–834

15

High-resolution Quaternary record of marine organic carbon content in the hemipelagic sediments of the Japan Sea from bromine counts measured by XRF core scanner

Arisa Seki[1*], Ryuji Tada[2], Shunsuke Kurokawa[2] and Masafumi Murayama[3]

Abstract

The marine organic carbon content in sediments is a useful tool for reconstructing past productivity in the ocean. Bromine has been proposed as a useful proxy for marine organic carbon, since bromine is more concentrated in marine organic matter compared to terrestrial organic matter.

Here, we present a high-resolution Quaternary record of marine organic carbon in the hemipelagic sediments of the Japan Sea obtained during Integrated Ocean Drilling Program (IODP) Expedition 346. We measured the bromine content in the sediments using an XRF core scanner (ITRAX). The total organic carbon content, total nitrogen, and stable carbon isotope values were also measured for discrete samples from the same sediments. The total organic carbon/total nitrogen ratio and the stable carbon isotope ratio of the total organic carbon are used to estimate the marine organic carbon content within the total organic carbon.

The Br counts and marine organic carbon content show a high correlation, which we used to construct a calibration equation of the marine organic carbon content from the Br counts. Using the calibration equation, we reconstructed the changes in the marine organic carbon content during the Quaternary with a time resolution of ~ 50 years using sediment cores from IODP Site U1424 in the east central Japan Sea. The method to estimate the marine organic carbon content from the Br counts measured by the XRF core scanner proposed in this paper will be a useful tool to reconstruct the organic carbon content with high resolution and high speed.

Keywords: Marine organic carbon, Japan Sea, XRF core scanner, ITRAX, Bromine, IODP, Expedition 346, Site U1424

Introduction

Br as a potential proxy of marine organic carbon

Br (Bromine) is one of the major elements in seawater. The concentration of Br in seawater is 6.7×10^7 ng/L (0.84 mM), which is the eighth highest in seawater (National Astronomical Observatory of Japan 2016). The seawater/crust concentration ratio of Br is second highest next to Cl (The Geochemical Society of Japan 2012). The average residence time of Br in seawater is 100×10^6 years (Gamo 2014). The vertical profiles of the Br concentration in the water column are flat, and Br is thus classified as a conservative element in seawater (Nozaki 1992; National Astronomical Observatory of Japan 2016). However, a recent study on the biogeochemical cycling of Br casts doubt on the classification of Br as a conservative element (Leri et al. 2010).

Br is known to be present in organic materials of both marine and terrestrial origin in the form of organobromine compounds (Leri et al. 2010). More than 1600 naturally occurring organobromine compounds, such as bromoalkanes, bromopyrroles, and bromophenols have

* Correspondence: a_seki@shinshu-u.ac.jp
[1]Faculty of Science, Shinshu University, 3-1-1 Asahi, Matsumoto, Nagano 390-8621, Japan
Full list of author information is available at the end of the article

been identified, and many of them are of marine origin (Gribble 2000). It is known that some microalgae and heterotrophic organisms produce brominated organic compounds (Gribble 1998). Bromoperoxidase enzymes, which catalyze the formation of organobromine compounds, have been found in organisms such as marine algae (red, green, and brown), bacteria, and a marine annelid (Gribble 2000). However, the structures and identities of the organobromine compounds in the deep marine sediments have not been well characterized (Berg and Solomon 2016).

In the Japan Sea, Masuzawa et al. (1988) measured the Br concentrations of zooplankton and found that Br is 3 to 20 times more concentrated in zooplankton compared to seawater. Therefore, Br is expected to be concentrated in the Japan Sea sediments, especially in those with high concentrations of marine organic matter.

As a result of the formation of organobromine compounds by biobromination, good correlations between the bromine content and total organic carbon (TOC) content are reported for estuarine and marine sediments (Ziegler et al. 2008; Leri et al. 2010). Since the Br concentration in seawater (0.84 mM) is higher than that in terrestrial water (National Astronomical Observatory of Japan 2016), the Br content in marine organic matter is higher than that in terrestrial organic matter (Berg and Solomon 2016). The reported average Br/TOC molar ratio for marine organic matter is $0.9–1.8 \times 10^{-3}$, while that for terrestrial organic matter is $0.06–0.3 \times 10^{-3}$ (Berg and Solomon 2016; Mayer et al. 2007; Leri et al. 2010).

Owing to the large difference in Br/TOC ratios between marine and terrestrial organic matter, Br/TOC ratios have been used as a proxy to distinguish organic matter of marine and terrestrial origin in estuary sediments (Mayer et al. 1981, 2007; Malcolm and Price 1984). Additionally, based on the high correlation between Br and TOC, Br is used as a proxy for TOC in deep sea sediments (Ziegler et al. 2008, 2010).

Ziegler et al. (2008) established a method to estimate the TOC content from Br profiles measured by an X-ray fluorescence (XRF) core scanner, using sediments from the Arabian Sea and a Mediterranean brine basin. They showed the usefulness of the XRF core scanner to measure the variability of Br and estimate the TOC with high speed and high resolution. They also pointed out that a calibration for each XRF core scanner is necessary when applying their method because the peak area counts of elements are sensitive to the XRF core scanner settings. They also suggest the possibility of estimating the marine organic carbon (MOC) content from the Br counts by removing samples where the $\delta^{13}C$ values are less than − 20‰ (the typical value for marine organic carbon) to improve the correlation between the Br counts and the TOC contents of Arabian Sea sediments. However, the relationship between the Br and MOC content has not been evaluated.

XRF core scanner for high-resolution measurement

In recent decades, XRF core scanners have become intensively used for the analysis of sediment cores in many academic fields including paleoceanographic studies (Rothwell and Croudace 2015a). The advantages of the XRF core scanner are its nondestructive measurements, high speed, and high resolution.

In previous studies, the measurement results of XRF core scanners have generally been expressed as downcore profiles of the element peak area count calculated from XRF spectra. Rarely have these peak counts been converted to the concentration of an element in sediments (Rothwell and Croudace 2015b, and references therein). This fact is because measurement by an XRF core scanner is a nondestructive method that can analyze half-round cores or u-channels directly; thus, the result is easily affected by the sediment surface condition or water content of the sediments (Croudace et al. 2006; Kido et al. 2006; Weltje and Tjallingii 2008; Weltje et al. 2015). Therefore, the data processing is more complicated even though the measurement itself is simple and easy. Some studies have made efforts to construct calibration equations to calculate the concentrations of elements from their peak area counts, but these efforts have had limited success, and XRF core scanner measurements are generally used in a qualitative way.

The ITRAX XRF core scanner is one of the most popular XRF core scanners and is manufactured and sold by COX Analytical Systems (Croudace et al. 2006; Rothwell and Croudace 2015a). Owing to the flat-beam techniques developed for the ITRAX core scanner, it has advantages for high-resolution measurement, with a highest resolution of 0.2 mm. Since XRF measurements using the ITRAX are conducted under the ordinary atmospheric environment, elements lighter than Al cannot be measured because the absorption effect of X-rays by air is more significant for the lighter elements. Although the organic carbon content in sediments is important for many paleoceanographic studies, the organic carbon cannot be measured by XRF core scanners.

Aim of this study

In this study, the TOC, total nitrogen (TN), and organic matter $\delta^{13}C$ of selected discrete samples taken from IODP Expedition 346 Site U1424 are measured and compared with the Br counts measured by the ITRAX XRF core scanner to evaluate the possibility of estimating MOC contents from Br counts in the Japan Sea sediment. Based on the results obtained, the downcore organic carbon variability in the Japan Sea sediments will be reconstructed at Site U1424 to demonstrate the usefulness of the method.

Oceanographic settings

Oceanographic and geological settings

The Japan Sea is a semiclosed marginal sea located along the northwest margin of the Pacific Ocean. The sea is connected with the surrounding seas through four shallow channels, the Tsushima Strait (sill depth ~ 130 m), the Tsugaru Strait (sill depth ~ 130 m), the Soya Strait (sill depth ~ 55 m), and the Mamiya (Tatar) Strait (sill depth ~ 15 m) (Fig. 1). The average depth of the Japan Sea is more than 1600 m, with a maximum depth of more than 3700 m (National Astronomical Observatory of Japan 2016). The Japan Sea is composed of three basins, the Japan Basin (~ 3800 m), Yamato Basin (~ 3000 m), and Tsushima (Ulleung) Basin (~ 2600 m), divided by the Yamato Rise (Gamo 2016; Fig. 1).

At present, the Tsushima Warm Current (TWC) is the only current flowing into the Japan Sea through the

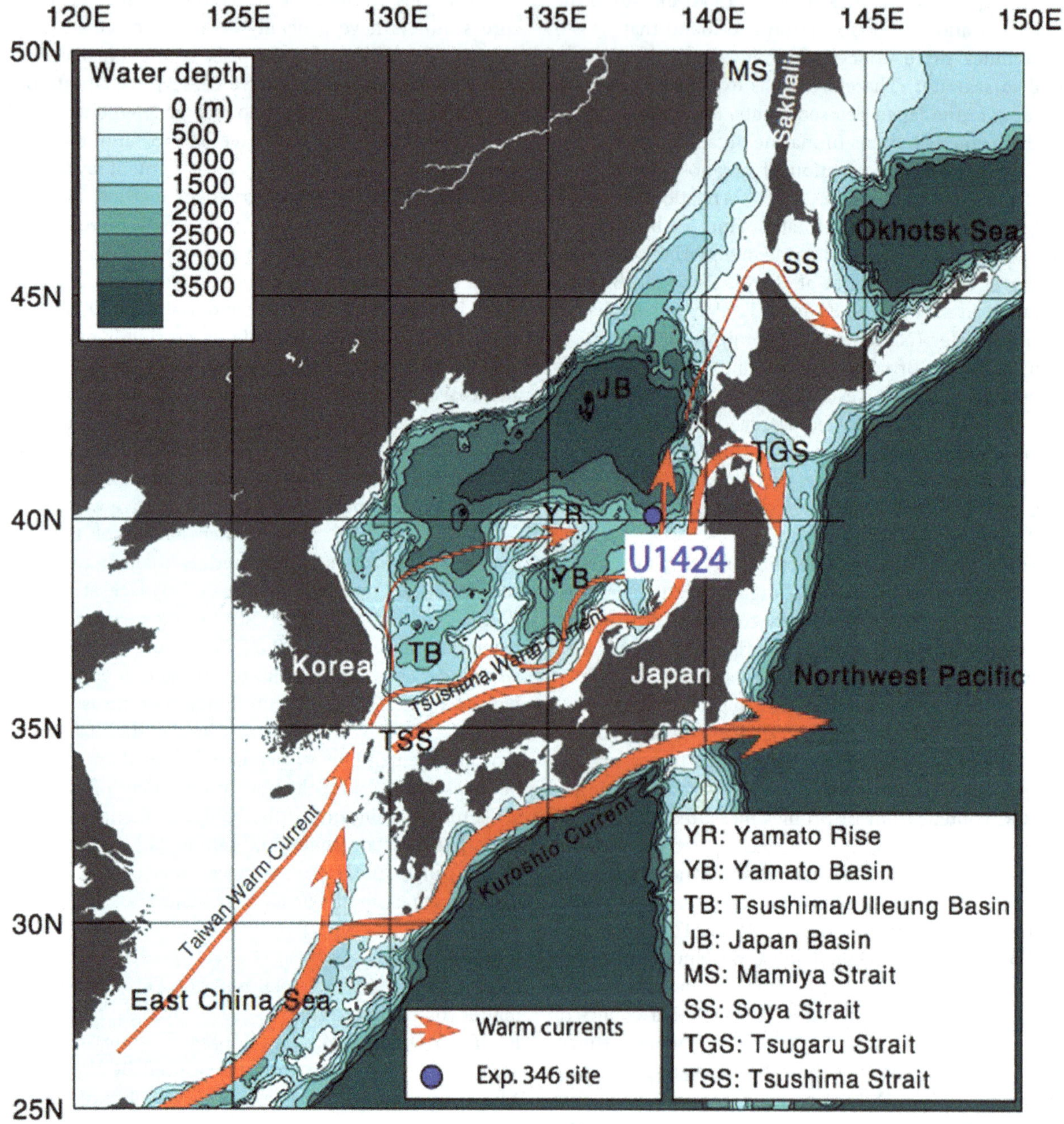

Fig. 1 A map of the Japan Sea with the study site, IODP Site U1424. Main surface currents and name of basins and straits are also indicated. (Modified after Tada et al. 2015a)

Tsushima Strait. The TWC is composed of two components, the Kuroshio Warm Current (KWC) and the Taiwan Warm Current (TWWC). The TWC originates from the KWC that flows into the Japan Sea through the southern channel, whereas the TWC originates from the TWWC flows into the Japan Sea through the northern channel of the Tsushima Strait. The TWWC flows from the South China Sea into the East China Sea (ECS), and then flows northeastward on the ECS shelf where it is mixed with river water supplied from the continent to become lower in salinity and higher in nutrients compared to the KWC flowing northeastward in the Okinawa Trough. Therefore, the surface water salinity in the Japan Sea is easily influenced by the salinity of the TWWC, which, in turn, is affected by the fresh water flux from the continent (Yanagi 2002), and the relative contributions of the TWWC and the KWC to the influx into the Japan Sea.

The modern Japan Sea produces its own deep water called the Japan Sea Proper Water (JSPW), which is characterized by a low potential temperature of 0–1 °C and high oxygen content (200–300 μmol/kg) throughout the deeper (> 200 m) part of the Japan Sea (Gamo et al. 2014; Gamo 2016). The dissolved oxygen levels in the Japan Sea deep water are more than 200 μmol/kg in all three basins, and this value is higher than that in the Northwestern Pacific at the same latitude (~ 150 μmol kg^{-1}) (Gamo et al. 2014). The potential temperature deeper than 2000–2500 m is uniform in all three basins with < 0.001 °C fluctuations, suggesting active vertical convection in the deep water (Gamo and Horibe 1983; Gamo et al. 2014). The current that flows along the western margin of the Japan Sea is cooled by the winter monsoon wind and the JSPW is formed especially in the area off Vladivostok where strong northwesterly winter monsoon winds are focused because of a topographic effect due to a mountain gap.

Since the sill depths of the modern straits are shallow (< 130 m) and the tectonic influences are minimal, especially in the Tsushima Strait (Itaki 2016), sea level changes should be the major factor that controls the paleoceanographic conditions of the Japan Sea during the Quaternary (Tada et al. 1999). During the Last Glacial Maximum (LGM), when the sea level dropped to around − 120 m, the surface water inflow from the Tsushima Strait dramatically decreased, and the sea became a closed system (Tada et al. 1999). The surface water salinity in the sea also decreased due to the excess precipitation plus fresh water inflow from the surrounding rivers over evaporation (Matsui et al. 1998). Consequently, the sea became stratified and the deep water became euxinic because the production of the dense deep water was reduced due to the low-salinity surface water (Tada et al. 1999).

The Japan Sea is also characterized by its low nutrient content (Yanagi 2002; Tada 2012). In the modern Japan Sea, the phosphorus content in the JSPW is low (2.3 μM), and the residence times of phosphorus and nitrogen are short (2.2 years and 1.6 years in the uppermost 200 m, respectively) (Yanagi 2002, Tada 2012). The phosphorous budget in the Japan Sea reveals that the phosphorus input from the ECS through the Tsushima Strait is by far the dominant source (Yanagi 2002). However, 90% of the phosphorus flowing into the sea via the Tsushima Strait flows out from the Tsugaru Strait in the modern Japan Sea because the subsurface water, which has a higher nutrient content than the surface water, flows out from the Tsugaru Strait, whose sill depth (~ 130 m) is deeper than the thermocline (Tada 2012). Thus, the nutrient budget in the Japan Sea is also thought to have been affected significantly by changes in the sill depth controlled by sea level changes (Tada 2012).

Study site

In the summer of 2013, the Integrated Ocean Drilling Program (IODP) Expedition 346 "Asian Monsoon" drilled hemipelagic sediments at seven sites in the Japan Sea (Tada et al. 2015a). Among the seven sites, sediments recovered from Site U1424 (water depth 2808 m) are selected for this study.

Site U1424 is located ~ 150 km to the west of northern Honshu Island and ~ 200 km to the southwest of the Tsugaru Strait (40° 11.40′ N, 138° 13.90′ E), close to the boundary of the Japan Basin and Yamato Basin, with a water depth of 2808 m (Fig. 1; Tada et al. 2015c).

The hemipelagic sediments of the Japan Sea are known to have distinct alternations of dark and light layers (e.g., Tada et al. 1999; Tada 2004). During IODP Expedition 346, these dark and light layers were recovered at six sites in the Japan Sea at water depths deeper than 900 m (Tada et al. 2015a). The deposition of dark and light layers started at approximately 2.6 Ma (million years ago) and became increasingly distinct at approximately 1.45 Ma (Tada et al. 2015a; Tada et al. 2018). Distinct dark layers have been correlated among the six sites (Irino et al. 2018), and their deposition was synchronous basinwide, suggesting that the Japan Sea has been acting as a single system responding to climatic-oceanographic change in climatic-tectonic coupling in Asia throughout the Quaternary (e.g., Tada et al. 2016: Clift 2017).

Three holes were drilled at Site U1424 to recover a continuous sequence of sediments. The sediments back to 5 Ma were continuously recovered. These mainly consist of clay and diatom ooze, with a minor component of volcaniclastic materials (Tada et al. 2015c).

Splice intervals were defined onboard, and a continuous composite column made of each splice interval was constructed. The depth scale of this composite column is expressed as "CCSF-D (m)" (Tada et al. 2015b). The splice intervals were further revised after the cruise and the new depth scale for U1424 is called "U1424_CCSF-D_Patched_rev20170308 (m)" (Irino et al. 2018). The difference between the shipboard depth scale (CCSF-D) and the new depth scale is less than 1.5 m throughout the splice (Irino et al. 2018). The new depth scale is employed in this study unless otherwise noted. The thickness of the Quaternary sediment is approximately 90 m based on the shipboard age model (Tada et al. 2015c) and ~83 m based on the revised age model described below (Tada et al. 2018). Since this site is at the same location as Ocean Drilling Program (ODP) Site 794, the sediment sequence was correlated with ODP Site 794 cores (Irino et al. 2018).

One reason to choose Site U1424 is because no sediment disturbance (such as slumps) or hiatus is observed at this site (Tada et al. 2015c). In addition, the color alternations of dark and light layers are distinct at this site compared to sites drilled from shallower water depths (such as Site U1426 drilled at 903 m; Tada et al. 2015a; Irino et al. 2018). Another reason to choose Site U1424 is that a high-precision and high-resolution age model has been constructed at this site (Fig. 2; Tada et al. 2018).

In the Japan Sea hemipelagic sediments, especially the sediments deposited at a deep-water depth with alternations of dark and light layers, foraminifer shells are not continuously found from all horizons (Tada et al. 2015c). Thus, it is difficult to construct a $\delta^{18}O$ stratigraphy of foraminifer shells for the Japan Sea hemipelagic sediments. Tada et al. (2018) constructed a "LR04-tuned age model" for Site U1424 by tuning the gamma ray attenuation (GRA) bulk density of the Site U1424 composite section to the LR04 $\delta^{18}O$ stack (Lisiecki and Raymo 2005) because the diatom content, which is the major reason for the density changes in the Japan Sea sediments, has been changing in association with glacial-interglacial cycles (Tada et al. 1999). Since their age model has a higher resolution and reliability covering the entire Quaternary compared to the shipboard age model, their age model (Fig. 2) is employed in this study.

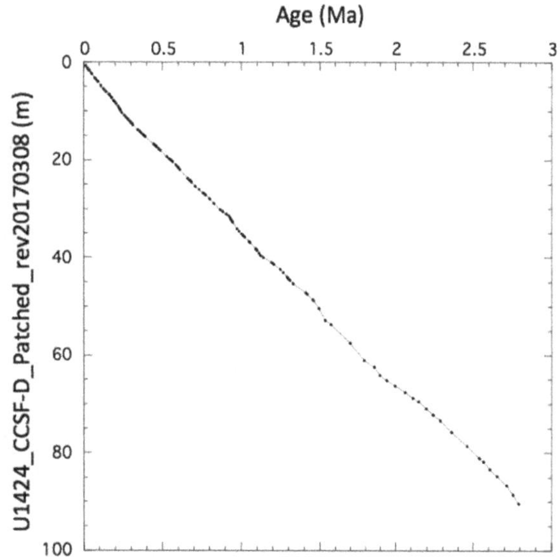

Fig. 2 Age model of IODP Site U1424. Data are from Tada et al. (2018)

Methods/Experimental

TOC, TN, and $\delta^{13}C$ measurements

Discrete samples of U1424 cores were subsampled from working halves in August 2016 (Table 1). Since the TOC content is different in the dark and light layers (Tada et al. 1999), every sample was obtained from the middle of either a dark or light layer and represents a 1-cm stratigraphic interval. Although tephra layers sometimes occur at Site U1424, no discrete sample contained visual tephra identified from the shipboard visual core description (VCD) and core photographs. In total, 23 samples were used in this study (Table 1).

The discrete samples were dried at 50 °C for more than 3 days and were gently ground to powder using an agate mortar. Then, 5- to 50-mg powdered samples were weighed in an Ag cup (dimensions: ϕ5, 9 mm) and treated with 100 μl of 3 M HCl for 3 days at room temperature to remove carbonate carbon. During the HCl treatment, the samples were stored in the Ag cups inside a desiccator with NaOH and P_2O_5 to remove HCl and H_2O. After the HCl treatment, each Ag cup was enfolded by a Sn cup (dimensions: ϕ9, 10 mm) and stored in a desiccator until the measurement of the TOC, TN, and $\delta^{13}C$. To estimate any carbon contamination from the HCl, blank Ag cups with HCl treatment that were enfolded by Sn cups were also prepared.

The TOC, TN, and $\delta^{13}C$ of the samples were measured by an elemental analyzer (Flash EA 1112) coupled to a ConFlo III and a Thermo Finnigan DeltaPlus Advantage isotope ratio mass spectrometer at the Center for Advanced Marine Core Research, Kochi University. Sulfanilamide was used as a standard for the TOC and TN measurements, and its analytical error (σ) was within 0.03 wt% for both TOC and TN. L-Alanine ($\delta^{13}C$ = 19.6‰ ± 0.2) was used as a standard for the $\delta^{13}C$ measurement, and its analytical error (σ) is 0.89‰.

Table 1 List of discrete samples taken from IODP Site U1424 and their analytical results

Sample No.	Sample interval								Sample top depth U1424_ CCSF-D_ Patched_ rev20170308 (m)	Age (Ma)	Marine isotope stage	Layer type	TOC (%)	TOC error (%)	TN (%)	TN error (%)	TOC/(TN-0.035)	$\delta^{13}C$ (‰ VPDB)	$\delta^{13}C$ error ‰)	Br (counts)	Br error (counts)	MOC (%)	MOC error (%)
	Site	Hole	Core	Type	Sect	A/W	Offset Top (cm)	Offset Bottom (cm)															
1	U1424	A	1	H	1	W	58	59	0.58	0.011	1	Dark	1.32	0.01	0.16	0.02	10.3	−22.95	0.89	2223	97	0.76	0.17
2	U1424	A	1	H	1	W	75	76	0.75	0.014	1	Light	0.58	0.00	0.08	0.00	12.0	−23.47	0.89	1593	82	0.29	0.07
3	U1424	A	1	H	1	W	87	88	0.87	0.016	2	Light	0.38	0.00	0.07	0.00	10.9	−23.80	0.89	1472	79	0.17	0.05
4	U1424	B	1	H	1	W	105	106	1.02	0.019	2	Dark	1.59	0.01	0.16	0.01	12.5	−26.63	0.89	1443	78	0.08	0.20
6	U1424	B	1	H	1	W	116	117	1.13	0.022	2	Dark	1.40	0.01	0.14	0.01	12.8	−26.94	0.89	1130	69	0.01	0.18
7	U1424	A	1	H	1	W	117	118	1.17	0.023	2	Dark	1.54	0.01	0.15	0.01	13.0	−26.73	0.89	1206	72	0.06	0.20
8	U1424	A	1	H	1	W	120	121	1.20	0.024	2	Dark	1.64	0.01	0.17	0.01	12.5	−26.40	0.89	1422	78	0.14	0.21
9	U1424	A	1	H	1	W	125	126	1.25	0.025	2	Light	0.48	0.00	0.07	0.00	13.4	−23.43	0.89	1045	67	0.25	0.06
10	U1424	A	1	H	2	W	10	11	1.57	0.032	3	Dark	2.42	0.02	0.26	0.02	10.7	−24.16	0.89	2286	98	0.98	0.31
11	U1424	A	1	H	2	W	98	99	2.38	0.052	3	Dark	1.10	0.01	0.14	0.01	10.7	−23.17	0.89	1615	83	0.60	0.14
12	U1424	A	1	H	3	W	25	26	3.15	0.070	4	Dark	2.49	0.01	0.26	0.01	10.9	−24.96	0.89	1819	88	0.72	0.32
14	U1424	A	1	H	3	W	79	80	3.69	0.083	5	Dark	1.54	0.01	0.18	0.01	10.7	−22.75	0.89	1955	91	0.94	0.20
16	U1424	A	1	H	3	W	106	107	3.96	0.089	5	Dark	2.15	0.01	0.23	0.01	11.1	−24.41	0.89	2462	102	0.80	0.27
17	U1424	A	1	H	4	W	14	15	4.54	0.102	5	Dark	1.25	0.01	0.16	0.01	9.8	−21.85	0.89	1777	87	0.92	0.16
18	U1424	A	1	H	4	W	40	41	4.80	0.108	5	Dark	3.50	0.02	0.35	0.02	11.0	−23.46	0.89	2500	103	1.77	0.44
19	U1424	B	2	H	1	W	129	130	6.04	0.139	6	Dark	1.16	0.01	0.11	0.01	15.5	−27.57	0.89	831	59	−0.09	0.15
20	U1424	B	2	H	1	W	148	149	6.23	0.149	6	Light	0.30	0.00	0.06	0.00	11.9	−23.99	0.89	872	61	0.13	0.04
21	U1424	B	2	H	2	W	15	16	6.40	0.157	6	Dark	0.98	0.01	0.10	0.01	14.4	−26.64	0.89	992	65	0.05	0.12
22	U1424	B	2	H	2	W	23	24	6.48	0.159	6	Dark	1.23	0.01	0.12	0.01	14.3	−27.76	0.89	1036	66	−0.13	0.16
23	U1424	B	2	H	3	W	5	6	7.80	0.190	6	Dark	3.24	0.01	0.33	0.01	10.9	−25.10	0.89	2366	100	0.88	0.41
24	U1424	B	2	H	3	W	30	31	8.05	0.196	7	Dark	3.96	0.04	0.40	0.04	11.0	−23.28	0.89	3857	128	2.11	0.50
25	U1424	B	2	H	3	W	126	127	9.01	0.219	7	Dark	1.57	0.01	0.18	0.01	11.1	−23.23	0.89	1819	88	0.84	0.20
26	U1424	B	2	H	4	W	32	33	9.57	0.230	7	Light	0.26	0.00	0.06	0.00	12.3	−24.70	0.89	797	58	0.09	0.03

Ages are calculated based on Tada et al. (2018)

Br measurement using XRF core scanner (ITRAX)

Archive halves of the U1424 cores from the Quaternary intervals were measured by ITRAX, using a Mo X-ray tube with settings of 30 kV, 55 mA, a 2-mm step size, and a 10-s measurement time. Outliers in the ITRAX analytical results due to core surface disturbances (such as cracks) were eliminated after the measurement. Before the sample measurement, five repeated measurements were conducted using standard samples to evaluate the analytical precision of the measurement (Seki 2017); the analytical precision of five times repeated measurement for Br was determined to be 2.1*√N (N: peak area counts; Seki 2017). To check the relationship between the Br counts and Br concentration, standard reference materials (Additional file 1: Table S1) were measured by ITRAX (Seki 2017). Since the concentration of Br in the standard reference material is not sufficiently high, KBr was added to standard reference materials and these KBr-supplemented standard reference materials were also measured by ITRAX (Seki 2017). Based on the measurement, the Br counts show a high correlation with the Br concentrations within a range between 0 and 500 ppm (Additional file 2: Figure S1; Seki 2017). Thus, the Br counts can be used as a semiquantitative measure of the Br concentration in the sediments.

Br counts measured on exactly the same stratigraphic intervals as the discrete samples used for the TOC, TN, and $\delta^{13}C$ measurements (Table 1) were used for comparison with the TOC, TN, and $\delta^{13}C$ results. Since discrete samples represent 1-cm intervals and the step size of the ITRAX measurement was 2 mm, the ITRAX results from five adjacent measurements for the same stratigraphic intervals as the discrete samples were averaged (Table 1).

Results and discussion

Results of TOC, TN, and $\delta^{13}C$ measurements

Table 1 shows the results of the TOC, TN, and $\delta^{13}C$ measurements for discrete samples from Site U1424. In general, the TOC is high (typically 1–6%) in dark layers and low (< 1%) in light layers (Table 1). The TN values ranged between 0.06 and 0.4% (Table 1), and the stable carbon isotope values ranged between − 20 and − 28‰ (Table 1).

It is known that the terrestrial organic matter and marine organic matter have different TOC/total organic nitrogen (TON) ratios (Lamb et al. 2006). Therefore, the TOC/TON ratio is commonly used to estimate the relative contributions of terrestrial and marine organic matter in sediments (Lamb et al. 2006). However, many paleoceanographic studies use the TOC/TN ratio instead of the TOC/TON ratio to evaluate the relative contributions because the analysis of the TN is much easier than the TON measurement (Schubert and Calvert 2001). In the Arctic Ocean, Schubert and Calvert (2001) demonstrated that a significant amount of inorganic nitrogen binds to clay minerals (especially illite) and is contained in the total nitrogen values. According to their results, inorganic nitrogen significantly affects the TOC/TN ratios (Schubert and Calvert 2001). Since the Japan Sea sediments contain significant amounts of clay, including illite (Irino and Tada 2000; Tada et al. 2015c), we investigated the possible influence of inorganic nitrogen on the TOC/TN ratio. Figure 3 shows the relationship between the TOC/TN ratio and the TOC. The TOC/TN ratio increases with the TOC, but the rate of increase gradually decreases. This relation can be explained by the presence of inorganic nitrogen binding to clay minerals. Schubert and Calvert (2001) calculated that the inorganic nitrogen content in the sediments of the central Arctic Ocean is 0.035%, based on TON and inorganic nitrogen content measurements in sediments taken from the central Arctic Ocean. Since the reported illite contents in the Japan Sea sediment (~ 14–20%; Irino and Tada 2000) are comparable to those of the central Arctic Ocean sediments (~ 20%, Schubert and Calvert 2001), we assumed an inorganic nitrogen content in the sediment of 0.035% (the same amount as in the central Arctic Ocean) for the Japan Sea sediment used in this study. Based on this assumption, theoretical TOC/TN ratios were calculated and are shown in Fig. 3. Figure 3 reveals that the TOC/TON ratio of the U1424 samples analyzed in this study is between 9 and 16 when the correction for the inorganic nitrogen content (estimated as 0.035%) is made. The calculated TOC/(TN-0.035) ratios have no dependency on the TOC, suggesting that the assumed content of inorganic nitrogen of 0.035% is reasonable for the U1424 samples.

The C/N ratio and $\delta^{13}C$ are both used as parameters to evaluate the relative ratio of marine versus terrestrial

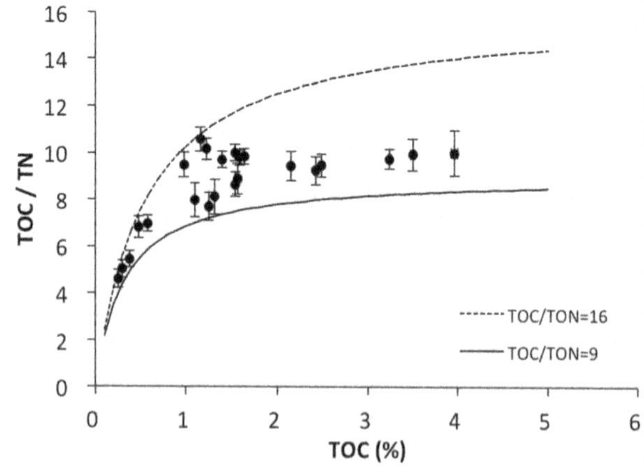

Fig. 3 TOC versus TOC/TN ratio of samples from IODP Site U1424 measured in this study. The TOC/TN ratio decreases as the TOC decreases, suggesting a probable effect of inorganic nitrogen, especially when the TOC is low. Solid and dashed lines are theoretical lines for the TOC/TN ratio versus the TOC when the TOC/TON ratio is assumed as 9 and 16 and inorganic nitrogen content is assumed as 0.035% (see text for details)

organic matter because the two parameters show distinctly different values for marine and terrestrial organic matter. Marine organic matter has a lower C/N ratio and higher $\delta^{13}C$ compared to terrestrial organic carbon (Lamb et al. 2006). The TOC/(TN-0.035) ratio and $\delta^{13}C$ obtained in this study show a high correlation of $r^2 = 0.64$ (Fig. 4), with samples with a lower C/N ratio having higher $\delta^{13}C$ values. This result suggests that the organic matter in the Japan Sea sediments is a mixture of marine and terrestrial organic matter. These results suggest that both the TOC/(TN-0.035) ratio and $\delta^{13}C$ in the Japan Sea sediment are good proxies to distinguish marine from terrestrial organic matter. In this study, we use $\delta^{13}C$ to calculate the percentages of marine and terrestrial organic carbon, because $\delta^{13}C$ is more reliable than the TOC/(TN-0.035) ratio, which has uncertainty in the inorganic nitrogen estimate. We assume that the $\delta^{13}C$ of marine organic carbon is −20‰, and that of terrestrial organic carbon is −27‰ (Minoura et al. 1997; Ikehara et al. 2009). Then, the MOC can be calculated as follows (Eq. 1), where "marine organic carbon fraction" means the percentage of marine organic carbon within the total organic carbon.

$$\mathrm{MOC}\ (\%) = \mathrm{TOC}\ (\%) \times < \text{marine organic carbon fraction} > - \mathrm{TOC}(\%) \times (\delta^{13}C_{terrestrial} - \delta^{13}C_{sample}) / (\delta^{13}C_{terrestrial} - \delta^{13}C_{marine}) \quad (1)$$

Correlation between Br and marine organic carbon

The relationship between the Br and TOC is shown in the cross-plot of Br and TOC in Fig. 5. It shows a high correlation ($R^2 = 0.69$) and a y-intercept of 739 (Br counts) (Fig. 5).

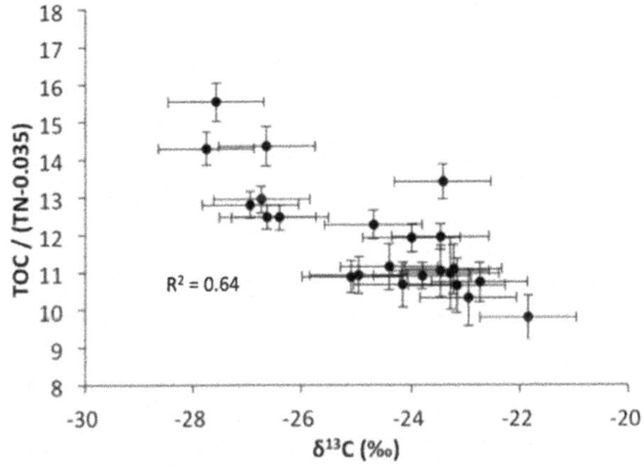

Fig. 4 Correlation between TOC/(TN-0.035) ratio and $\delta^{13}C$ of the TOC for samples analyzed in this study. (TN-0.035) represents the estimated organic nitrogen content of the sediments (see text for details)

Since wet archive halves of cores were used for the ITRAX measurement, the Br counts include the Br in the interstitial water. As bromine does not accumulate in solution as dissolved organobromine compounds (Berg and Solomon 2016), the Br in organic matter and Br in the interstitial water should behave independently. The Br concentration in the interstitial water is reported from a shipboard analysis as being 0.91–1.05 (± 0.03) mM for the uppermost 10 m of U1424 (Tada et al. 2015b, 2015c). Using water contents measured onboard (Tada et al. 2015c) and the relationship between the Br counts and concentration, the contribution of the Br in the interstitial water to the total Br counts of the wet archive halves is estimated as being between 470 and 880 counts (minimum and maximum estimates). Thus, the intercept of the Br counts in Fig. 5 (739 counts) is interpreted as representing the background Br in the interstitial water.

Since the Br is mainly contained in the MOC, the relation between the Br and MOC (calculated from TOC and $\delta^{13}C$) is further examined. Figure 6 shows a cross-plot of Br and MOC calculated from Eq. 1. For U1424, the Br counts show a better correlation with the MOC ($R^2 = 0.85$) compared to the TOC ($R^2 = 0.69$) (Figs. 5 and 6).

Since the Br is mainly contained in the MOC, the calibration equation obtained from Fig. 6 is used to estimate the MOC content from the Br counts in this study.

$$\mathrm{Br\ (counts)} = 1143 \times \mathrm{MOC}\ (\%) + 1060 \quad (2)$$

$$\mathrm{MOC}\ (\%) = (\mathrm{Br\ (counts)} - 1060)/1143 \quad (3)$$

Since the analytical error of Br is small (less than 150 counts when the peak area count is less than 5000 counts), the error of this equation mainly comes from deviations from the regression curve in Fig. 6. The estimated error of the MOC calculation is approximately

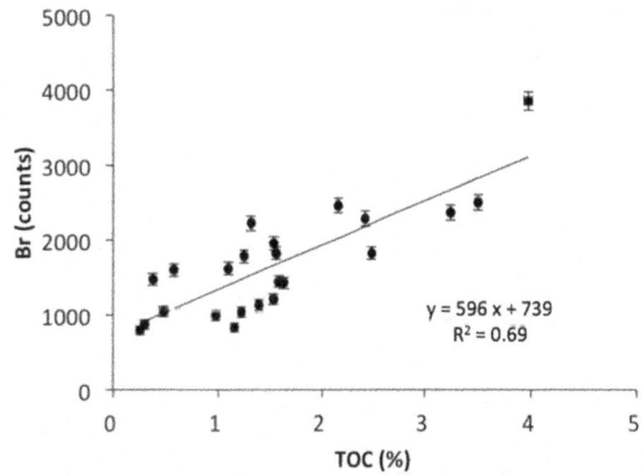

Fig. 5 Cross-plot of Br counts by XRF core scanner and TOC measured in this study

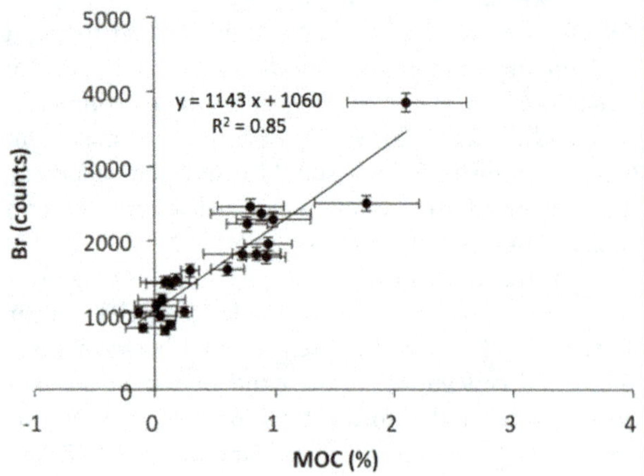

Fig. 6 Cross-plot of Br counts by XRF core scanner and MOC calculated from TOC and $\delta^{13}C$ in this study

0.5% (Fig. 6). It should be noted that Eqs. 2 and 3 are based on the assumption that the Br in terrestrial organic matter is negligible because the Br/TOC ratio is much lower in terrestrial organic matter ($0.06–0.3 \times 10^{-3}$) compared with marine organic matter ($0.9–1.8 \times 10^{-3}$) (Berg and Solomon 2016; Mayer et al. 2007; Leri et al. 2010). Since analytical data on Br/MOC ratios in marine organic matter for individual seas are not available, the applicability of the equations for the MOC calculation (Eqs. 2 and 3) for sediments from other seas (such as the open ocean) should be checked in the future. Even given the limitations described above, this method to estimate the MOC content by the high-resolution, high-speed, and nondestructive measurement of Br using ITRAX is a useful tool for paleoceanographic studies.

MOC variability in the Japan Sea during the Quaternary

Using Eq. 3, high-resolution temporal variations of the MOC during the Quaternary were reconstructed for Site U1424 from the Br counts measured by ITRAX (Fig. 7; Additional file 3 and Additional file 4: Table

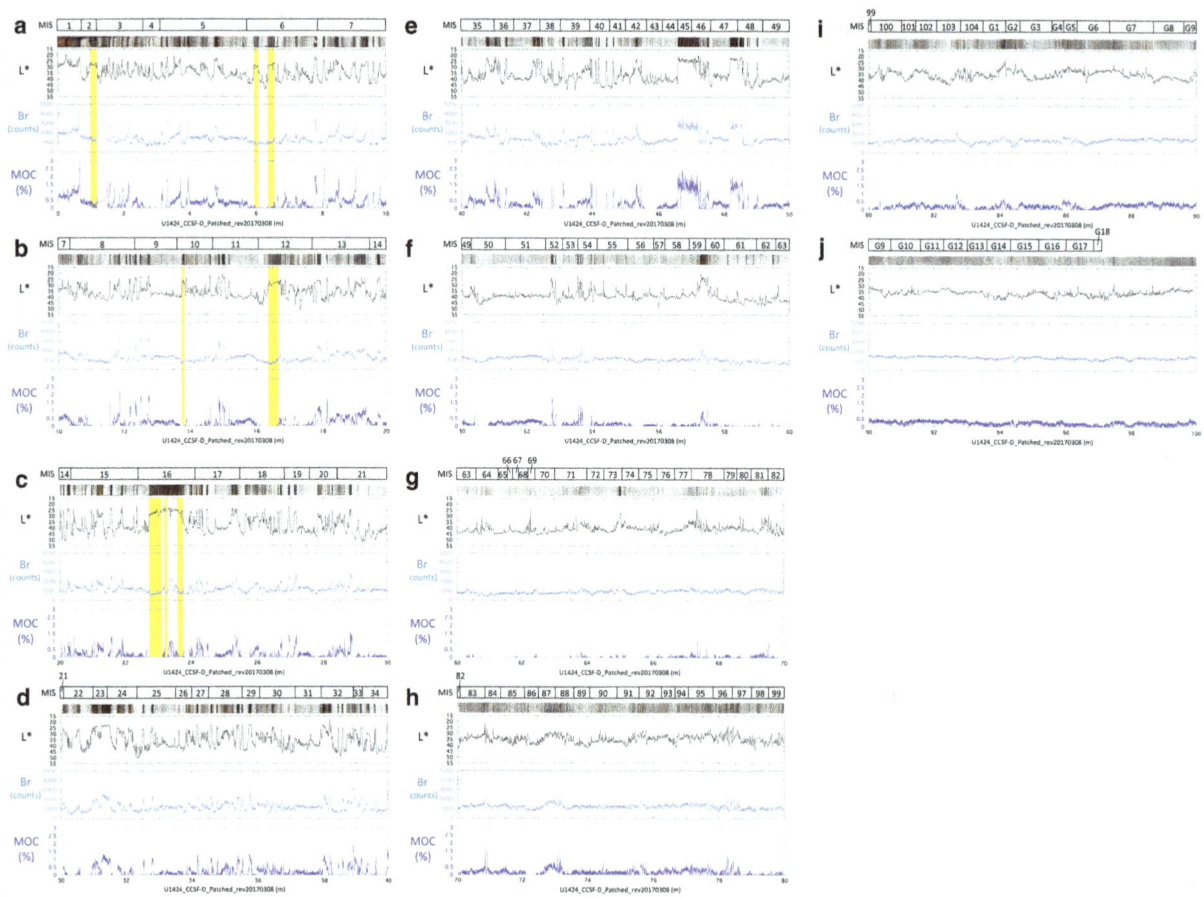

Fig. 7 Reconstructed MOC for Site U1424. Enhanced photograph, lightness (L^*, black; data from Irino et al. 2018) and Br counts (light blue) are also shown for comparison. The hatched intervals are of a dark color (low L^* value) but are low in MOC content (see text for details). **a** 0–10 m, **b** 10–20 m, **c** 20–30 m, **d** 30–40 m, **e** 40–50 m, **f** 50–60 m, **g** 60–70 m, **h** 70–80 m, **i** 80–90 m, **j** 90–100 m

S2). Enhanced core photographs, lightness (L^*; data from Irino et al. 2018), and Br counts measured by ITRAX are shown with the MOC in (Fig. 7; Additional file 3). The MOC values tend to be high (> 1%) in the dark layers (low L^* values) and low (< 0.5%) in the light layers (high L^* values). The concentration of the MOC in the dark layers occurs throughout the sequence, but more distinct in the uppermost 50 m (Fig. 7; Additional file 3), which corresponds to the last 1.5 Myr (Fig. 2; Tada et al. 2018). Below 50 m (before 1.5 Ma), dark layers are rare, and thus the low-MOC layers prevail (Fig. 7; Additional file 3).

The reconstructed MOC shows cm-scale variability associated with the dark layers (Fig. 7; Additional file 3), which correspond to millennial-scale variability (Fig. 2; Tada et al. 2018). These millennial-scale variations are more distinct during the last 1.5 Myr (uppermost 50 m (Fig. 7; Additional file 3). The MOC in the sediments reflects the surface productivity, degradation of marine organic matter within the water column and during early burial, and dilution by terrigenous and biogenic materials. Thus, the reconstructed MOC enables us to discuss possible oceanic condition changes. In the glacial period during the last 200 kyr, a millennial-scale variation in the TOC was reported (Tada et al. 1999). This millennial-scale alternation of dark and light layers in the Japan Sea sediments was proposed to reflect the East Asian summer monsoon (EASM) variability associated with Dansgaard-Oeschger cycles (Tada et al. 1999; Tada 2004). Tada et al. (1999) proposed a hypothesis on dark and light layer deposition. Namely, when the EASM became stronger during interstadials, more nutrients was provided to the Japan Sea via the Tsushima Strait, and the surface productivity in the Japan Sea was enhanced. The reconstructed MOC variability revealed that millennial-scale variability occurred not only during the last 200 kyr but also during most of the glacial periods during the last 1.5 Ma. This suggests the occurrence of climate perturbations such as Dansgaard-Oeschger cycles in the earlier period. Since exploring the relationship between the Asian monsoon and the oceanography of the Japan Sea is important (Tada and Murray 2016), the detailed mechanism and onset timing of the millennial-scale variability of the MOC should be addressed in the future.

As described above, MOC values tend to be high (> 1%) in the dark layers throughout the sequence (Fig. 7; Additional file 3). However, there are some exceptions to the relationship between the lightness (L^*) of the sediment and the MOC content. For example, the dark layers in the intervals of 5.9–6.1 m and 6.3–6.6 m are of a dark color (low L^* value) but are low in MOC content (Fig. 7; Additional file 3). These layers were deposited during marine isotope stage (MIS) 6, when the eustatic sea level was more than ~ 100 m lower than it is today (Tada et al. 2018). Similar trends are obvious in the layers deposited during MIS 2, 10, 12, and 16 (Fig. 7; Additional file 3). These layers have high concentrations of pyrite and are considered to have been deposited under euxinic deep-water conditions (Tada et al. 1999; Kido et al. 2007). From the results of the $\delta^{13}C$ of organic carbon (sample No. 4, 6, 7, 8, 19, 21, and 22 in Table 1), these layers are also characterized by high terrestrial/marine organic matter ratios. Since organic matter is the main cause of the dark color in the Japan Sea sediments (Tada et al. 1992), L^* has been used to estimate the TOC content in the Japan Sea because it is easy to measure compared to the conventional TOC measurement methods (Kido et al. 2007). Thus, the Br counts have an advantage over L^* as a parameter to reconstruct the paleoproductivity. On the other hand, it may be possible to estimate the fraction of terrestrial organic carbon within the TOC using a combination of L^* and Br counts, which will be tested in the future.

From the results of this study, it is clear that MOC estimation from Br counts measured by ITRAX has a high potential to reconstruct the paleoproductivity with high resolution and high speed and in a nondestructive way, which allows us to examine past marine productivity, the carbon cycle in the ocean, and oceanographic changes.

Conclusions

In this paper, a method to estimate the MOC content from Br as measured by an XRF core scanner is described. For establishing the method, the TOC, TN, and $\delta^{13}C$ were measured on discrete sediment samples. The TOC/TN ratio, which is usually used to distinguish marine organic matter from terrestrial organic matter, is significantly influenced by inorganic nitrogen bound to clay minerals. The inorganic nitrogen content of the sediment is assumed to be a constant 0.035% based on a previously reported value for sediments from the Arctic Ocean (Schubert and Calvert 2001). Both the TOC/(TN-0.035) ratio and $\delta^{13}C$ of the Quaternary Japan Sea sediments suggest a wide range of mixing of terrestrial and marine organic matter. The $\delta^{13}C$ of the organic matter is used to calculate the marine organic carbon fraction, because $\delta^{13}C$ is more reliable than the TOC/(TN-0.035) ratio because the latter contains uncertainty in the inorganic nitrogen content assumption. Then, the MOC is calculated from the TOC and the marine organic carbon fraction, which is calculated from the $\delta^{13}C$ value, and the calculated MOC is compared with the Br counts. A high positive correlation is found between the MOC and Br counts, and a calibration equation to estimate the MOC from the Br counts is obtained. To demonstrate the usefulness of the equation, it is applied to the Quaternary sediments of the Japan Sea. The MOC variability was reconstructed for the entire Quaternary sediment sequence at IODP Site U1424. The reconstructed MOC record reveals that millennial-scale variations exist since 1.5 Ma. Since the millennial-scale variations reflect climate perturbations similar to Dansgaard-Oeschger cycles, the reconstructed MOC may provide useful information to investigate past climate and oceanographic changes. L^* is commonly used as a proxy for the TOC and used to semiquantitatively estimate the surface productivity. However, the TOC sometimes contains nonnegligible amounts of terrestrial organic carbon, the effect of which should be corrected to estimate the surface productivity. Consequently, Br counts are superior to L^* if one wants to obtain more quantitative information on the surface productivity. Therefore, the MOC estimation from Br counts measured by ITRAX has a high potential to reconstruct the paleoproductivity with high resolution and high speed and in a nondestructive way.

Additional files

Additional file 1: Table S1. Standard reference material used in this study.

Additional file 2: Figure S1. Relationship between Br counts and Br concentration. Results of standard reference material (Additional file 1: Table S1) and KBr-supplemented standard reference materials are shown.

Additional file 3: Reconstructed MOC for Site U1424. Enhanced photograph, lightness (L*, black; data from Irino et al. 2018) and Br counts (light blue) are also shown for comparison. The hatched intervals are of a dark color (low L* value) but are low in MOC content (see text for details). **a** 0–10 m, **b** 10–20 m, **c** 20–30 m, **d** 30–40 m, **e** 40–50 m, **f** 50–60 m, **g** 60–70 m, **h** 70–80 m, **i** 80–90 m, **j** 90–100 m.

Additional file 4: Table S2. Raw data of Br counts and MOC content shown in Fig. 7.

Abbreviations

EASM: East Asian summer monsoon; ECS: East China Sea; GRA: Gamma ray attenuation; IODP: Integrated Ocean Drilling Program; JSPW: Japan Sea Proper Water; KWC: Kuroshio Warm Current; LGM: Last Glacial Maximum; MIS: Marine isotope stage; MOC: Marine organic carbon; ODP: Ocean Drilling Program; TN: Total nitrogen; TOC: Total organic carbon; TON: Total organic nitrogen; TWC: Tsushima Warm Current; TWWC: Taiwan Warm Current; VCD: Visual core description; XRF: X-ray fluorescence

Acknowledgements

We appreciate Kaz Mitake of the University of Tokyo and Minoru Ikehara, Takuya Matsuzaki, and the staff of the Kochi Core Center for laboratory assistance. We appreciate Larry Peterson and an anonymous reviewer for their constructive comments and helpful suggestions.
This research used samples provided by the International Ocean Discovery Program (IODP). This study was performed under the cooperative research program of Center for Advanced Marine Core Research (CMCR), Kochi University (16A035, 16B031).

Funding

Funding for this research was provided by JSPS KAKENHI Grant Numbers 16H01765 for RT and 25•9053 for AS. This work is also supported by the Sasakawa Scientific Research Grant from the Japan Science Society for AS.

Authors' contributions

RT and AS proposed the topic and conceived and designed the study. AS, SK, and MM carried out the experimental study. AS and SK analyzed the data. All authors took part in discussions about the data analysis and interpretation. AS drew the figures and wrote a draft of the manuscript. All authors read and approved the final manuscript.

Authors' information

RT was cochief scientist of IODP Expedition 346. AS and SK were students of RT.

Competing interests

The authors declare that they have no competing interests.

Author details

[1]Faculty of Science, Shinshu University, 3-1-1 Asahi, Matsumoto, Nagano 390-8621, Japan. [2]Graduate School of Science, The University of Tokyo, 7-3-1 Hongo, Bunkyo-ku, Tokyo 113-0033, Japan. [3]Faculty of Agriculture and Marine Science, Kochi University, 200 Monobe, Nankoku-shi, Kochi 783-8502, Japan.

References

Berg RD, Solomon EA (2016) Geochemical constraints on the distribution and rates of debromination in the deep subseafloor biosphere. Geochim Cosmochim Acta 174:30–41.

Clift PD (2017) Cenozoic sedimentary records of climate-tectonic coupling in the Western Himalaya. Prog Earth Planet Sci 4:39. https://doi.org/10.1186/s40645-017-0151-8.

Croudace IW, Rindby A, Rothwell RG (2006) ITRAX: description and evaluation of a new multi-function X-ray core scanner. In: Rothwell RG (ed) New techniques in sediment Core analysis, Special publication, vol 267. Geological Society, London, pp 51–63.

Gamo T (2014) Marine geochemistry. Kodansha Ltd., Japan (In Japanese).

Gamo T (2016) Tha Japan Sea—things happen in the deep sea. Kodansha Ltd., Japan (In Japanese).

Gamo T, Horibe Y (1983) Abyssal circulation in the Japan Sea. J Oceanogr Soc Jpn 39:220–230.

Gamo T, Nakayama N, Takahata N, Sano Y, Zhang J, Yamazaki E, Taniyasu S, Yamashita N (2014) The Sea of Japan and its unique chemistry revealed by time-series observations over the last 30 years. Monogr Environ Earth Planets 2(1):1–22. https://doi.org/10.5047/meep.2014.00201.0001.

Gribble GW (1998) Naturally occurring organohalogen compounds. Acc Chem Res 31:141–152.

Gribble GW (2000) The natural production of organobromine compounds. Environ Sci Pollut Res 7(1):37–49. https://doi.org/10.1065/espr199910.002.

Ikehara M, Akita D, Matsuda A (2009) Enhanced marine productivity in the Kuroshio region off Shikoku during the last glacial period inferred from the accumulation and carbon isotopes of sedimentary organic matter. J Quat Sci 24(8):848–855.

Irino T, Tada R (2000) Quantification of aeolian dust (Kosa) contribution to the Japan Sea sediments and its variation during the last 200 ky. Geochem J 34:59–93.

Irino T, Tada R, Ikehara K, Sagawa T, Karasuda A, Kurokawa S, Seki A, Lu S (2018) Construction of perfectly continuous records of physical properties for dark-light sediment sequences collected during IODP expedition 346. Prog Earth Planet Sci 5:19. https://doi.org/10.1186/s40645-018-0167-8l.

Itaki T (2016) Transitional changes in microfossil assemblages in the Japan Sea from the Late Pliocene to Early Pleistocene related to global climatic and local tectonic events. Prog Earth Planet Sci 3:11. https://doi.org/10.1186/s40645-016-0087-4.

Kido Y, Koshikawa T, Tada R (2006) Rapid and quantitative major element analysis method for wet fine-grained sediments using an XRF mixroscanner. Mar Geol 229:209–225.

Kido Y, Minami I, Tada R, Fujine K, Irino T, Ikehara K, Chun JH (2007) Orbital-scale stratigraphy and high-resolution analysis of biogenic components and deep water oxygenation conditions in the Japan Sea during the last 640 kyrs using XRF microscanner. Palaeogeogr Palaeoclimatol Palaeoecol 247:32–49. https://doi.org/10.1016/j.palaeo.2006.11.020.

Lamb AL, Wilson GP, Leng MJ (2006) A review of coastal palaeoclimate and relative sea-level reconstructions using $\delta^{13}C$ and C/N ratios in organic material. Earth-Sci Rev 75:29–57.

Leri AC, Hakala JA, Marcus MA, Lanzirotti A, Reddy CM, Myneni SCB (2010) Natural organobromine in marine sediments: new evidence of biogeochemical Br cycling. Glob Biogeochem Cycles 24:GB4017.

Lisiecki LE, Raymo ME (2005) A Pliocene-Pleistocene stack of 57 globally distributed benthic $\delta^{18}O$ records. Paleoceanography 20(1):PA1003. https://doi.org/10.1029/2004PA001071.

Malcolm SJ, Price NB (1984) The behaviour of iodine and bromine in estuarine surface sediments. Mar Chem 15:263–271.

Masuzawa T, Koyama M, Terazaki M (1988) A regularity in trace element contents of marine zooplankton species. Mar Biol 97:587–591. https://doi.org/10.1007/BF00391056.

Matsui H, Tada R, Oba T (1998) Low-salinity isolation event of the Japan Sea in response to eustatic sea-level drop during the LGM: reconstruction based on salinity-balance model. Quatern Res (Daiyonki-kenkyu) 37:221–233 (in Japanese with English abstract).

Mayer LM, Macko SA, Mook WH, Murray SM (1981) The distribution of bromine in coastal sediments and its use as a source indicator for organic matter. Org Geochem 3:37–42.

Mayer LM, Schick LL, Allison MA, Ruttenberg KC, Bentley SJ (2007) Marine vs. terrigenous organic matter in Louisiana coastal sediments: the uses of bromine:organic carbon ratios. Mar Chem 107:244–254.

Minoura K, Hoshino K, Nakamura T, Wada E (1997) Late Pleistocene-Holocene palcoproductivity circulation in the Japan Sea: sea-level control on $\delta^{13}C$ and $\delta^{15}N$ records of sediment organic material. Palaeogeogr Palaeoclimatol Palaeoecol 135:41–50.

National Astronomical Observatory of Japan (ed) (2016) Rika Nenpyo (chronological scientific tables). Maruzen Publishing Co., Ltd., Japan (In Japanese).

Nozaki Y (1992) Trace elements in sea water: their mean concentrations and North Pacific profiles. Chikyukagaku (Geochemistry) 26:25–39 (In Japanese with English abstract).

Rothwell RG, Croudace IW (2015a) Micro-XRF studies of sediment cores: a perspective on capability and application in the environmental sciences. In: Croudace IW, Rothwell RG (eds) Micro-XRF Studies of sediment cores, developments in Paleoenvironmental research 17. https://doi.org/10.1007/978-94-017-9849-51.

Rothwell RG, Croudace IW (2015b) Twenty years of XRF Core scanning marine sediments: what do geochemical proxies tell us? In: Croudace IW, Rothwell RG (eds) Micro-XRF studies of sediment cores, developments in Paleoenvironmental research 17. https://doi.org/10.1007/978-94-017-9849-52.

Schubert CJ, Calvert SE (2001) Nitrogen and carbon isotopic composition of marine and terrestrial organic matter in Arctic Ocean sediments: implications for nutrient utilization and organic matter composition. Deep Sea Res Part I 48:789–810.

Seki A (2017) Reconstruction of marine organic matter content in the sediments and deep-water redox condition variability of the Japan Sea during the quaternary using high-resolution XRF core scanner. Dissertation, the University of Tokyo, Tokyo.

Tada R (2004) Onset and evolution of millennial-scale variability in the Asian Monsoon and its impact on paleoceanography of the Japan Sea. Millennial-Scale Variability in the Asian Monsoon, American Geological Union, U.S.A, pp 1–16.

Tada R (2012) The Japan Sea sediments and variability of East Asian Monsoon: toward the IODP drilling of the Japan Sea and East China Sea. Quatern Res 51:151–164 (In Japanese).

Tada R, Irino T, Ikehara K, Karasuda A, Sugisaki S, Xuan C, Sagawa T, Itaki T, Kubota Y, Lu S, Seki A, Murray RW, Alvarez-Zarikian C, Exp. 346 Scientists (2018) High-resolution and -precision correlation of dark and light layers in the Quaternary hemipelagic sediments of the Japan Sea recovered during IODP expedition 346. Prog Earth Planet Sci 5:19. https://doi.org/10.1186/s40645-018-0167-8.

Tada R, Irino T, Koizumi I (1999) Land-ocean linkages over orbital and millennial timescales recorded in late Quaternary sediments of the Japan Sea. Paleoceanography 14(2):236–247.

Tada R, Koizumi I, Cramp A, Rahman A (1992) Correlation of dark and light layers, and the origin of their cyclicity in the Quaternary sediments from the Japan Sea. In: Pisciotto KA, Ingle JC, von Breymann J et al (eds) Proceedings of ocean drilling program scientific results, vol 127/128. Ocean Drilling Program, TX, pp 577–601.

Tada R, Murray RW (2016) Preface for the article collection "Land–Ocean linkages under the influence of the Asian monsoon". Prog Earth Planet Sci 3:24. https://doi.org/10.1186/s40645-016-0100-y.

Tada R, Murray RW, Alvarez Zarikian CA, Anderson WT Jr, Bassetti MA, Brace BJ, Clemens SC, da Costa Gurgel MH, Dickens GR, Dunlea AG, Gallagher SJ, Giosan L, Henderson ACG, Holbourn AE, Ikehara K, Irino T, Itaki T, Karasuda A, Kinsley CW, Kubota Y, Lee GS, Lee KE, Lofi J, Lopes CICD, Peterson LC, Saavedra-Pellitero M, Sagawa T, Singh RK, Sugisaki S, Toucanne S, Wan S, Xuan C, Zheng H, Ziegler M (2015a) Expedition 346 summary. In: Tada R, Murray RW, Alvarez Zarikian CA, the Expedition 346 Scientists (eds) Proc. IODP, 346. Integrated Ocean Drilling Program, College Station. https://doi.org/10.2204/iodp.proc.346.101.2015.

Tada R, Murray RW, Alvarez Zarikian CA, Anderson WT Jr, Bassetti MA, Brace BJ, Clemens SC, da Costa Gurgel MH, Dickens GR, Dunlea AG, Gallagher SJ, Giosan L, Henderson ACG, Holbourn AE, Ikehara K, Irino T, Itaki T, Karasuda A, Kinsley CW, Kubota Y, Lee GS, Lee KE, Lofi J, Lopes CICD, Peterson LC, Saavedra-Pellitero M, Sagawa T, Singh RK, Sugisaki S, Toucanne S, Wan S, Xuan C, Zheng H, Ziegler M (2015b) Methods. In: Tada R, Murray RW, Alvarez Zarikian CA, the Expedition 346 Scientists (eds) Proc. IODP, 346. Integrated Ocean Drilling Program, College Station. https://doi.org/10.2204/iodp.proc.346.102.2015.

Tada R, Murray RW, Alvarez Zarikian CA, Anderson WT Jr, Bassetti MA, Brace BJ, Clemens SC, da Costa Gurgel MH, Dickens GR, Dunlea AG, Gallagher SJ, Giosan L, Henderson ACG, Holbourn AE, Ikehara K, Irino T, Itaki T, Karasuda A, Kinsley CW, Kubota Y, Lee GS, Lee KE, Lofi J, Lopes CICD, Peterson LC, Saavedra-Pellitero M, Sagawa T, Singh RK, Sugisaki S, Toucanne S, Wan S, Xuan C, Zheng H, Ziegler M (2015c) Site U1424. In: Tada R, Murray RW, Alvarez Zarikian CA, the Expedition 346 Scientists (eds) Proc. IODP, 346. Integrated Ocean Drilling Program, College Station. https://doi.org/10.2204/iodp.proc.346.105.2015.

Tada R, Zheng H, Clift PD (2016) Evolution and variability of the Asian monsoon and its potential linkage with uplift of the Himalaya and Tibetan plateau. Prog Earth Planet Sci 3:4. https://doi.org/10.1186/s40645-016-0080-y.

The geochemical society of Japan (2012) Encyclopedia of chemistry in earth and space. Asakura Publishing Co., Ltd., Japan ISBN978-4-254-16057-4 C3544 (in Japanese).

Weltje GJ, Bloemsma MR, Tjallingii R, Heslop D, Röhl U, Croudace IW (2015) Prediction of geochemical composition from XRF-core-scanner data: a new multivariate approach including automatic selection of calibration samples and quantification of uncertainties. In: Croudace IW, Rothwell RG (eds) Micro-XRF Studies of Sediment Cores, Developments in Paleoenvironmental Research 17. https://doi.org/10.1007/978-94-017-9849-5_21.

Weltje GJ, Tjallingii R (2008) Calibration of XRF core scanners for quantitative geochemical logging of sediment cores: theory and application. Earth Planet Sci Lett 274:423–438.

Yanagi T (2002) Water, salt, phosphorus and nitrogen budgets of the Japan Sea. J Oceanogr 58:797–804.

Ziegler M, Jilbert T, de Lange GJ, Lourens LJ, Reichart GJ (2008) Bromine counts from XRF scanning as an estimate of the marine organic carbon content of sediment cores. G-cubed 9:Q05009.

Ziegler M, Lourens LJ, Tuenter E, Reichart GJ (2010) High Arabian Sea productivity conditions during MIS 13—odd monsoon event or intensified overturning circulation at the end of the mid-Pleistocene transition? Clim Past 6:63–76.

PERMISSIONS

All chapters in this book were first published in PEPS, by Springer; hereby published with permission under the Creative Commons Attribution License or equivalent. Every chapter published in this book has been scrutinized by our experts. Their significance has been extensively debated. The topics covered herein carry significant findings which will fuel the growth of the discipline. They may even be implemented as practical applications or may be referred to as a beginning point for another development.

The contributors of this book come from diverse backgrounds, making this book a truly international effort. This book will bring forth new frontiers with its revolutionizing research information and detailed analysis of the nascent developments around the world.

We would like to thank all the contributing authors for lending their expertise to make the book truly unique. They have played a crucial role in the development of this book. Without their invaluable contributions this book wouldn't have been possible. They have made vital efforts to compile up to date information on the varied aspects of this subject to make this book a valuable addition to the collection of many professionals and students.

This book was conceptualized with the vision of imparting up-to-date information and advanced data in this field. To ensure the same, a matchless editorial board was set up. Every individual on the board went through rigorous rounds of assessment to prove their worth. After which they invested a large part of their time researching and compiling the most relevant data for our readers.

The editorial board has been involved in producing this book since its inception. They have spent rigorous hours researching and exploring the diverse topics which have resulted in the successful publishing of this book. They have passed on their knowledge of decades through this book. To expedite this challenging task, the publisher supported the team at every step. A small team of assistant editors was also appointed to further simplify the editing procedure and attain best results for the readers.

Apart from the editorial board, the designing team has also invested a significant amount of their time in understanding the subject and creating the most relevant covers. They scrutinized every image to scout for the most suitable representation of the subject and create an appropriate cover for the book.

The publishing team has been an ardent support to the editorial, designing and production team. Their endless efforts to recruit the best for this project, has resulted in the accomplishment of this book. They are a veteran in the field of academics and their pool of knowledge is as vast as their experience in printing. Their expertise and guidance has proved useful at every step. Their uncompromising quality standards have made this book an exceptional effort. Their encouragement from time to time has been an inspiration for everyone.

The publisher and the editorial board hope that this book will prove to be a valuable piece of knowledge for researchers, students, practitioners and scholars across the globe.

LIST OF CONTRIBUTORS

Bjorn Mysen
Geophysical Laboratory, Carnegie Institution of Washington, 5151 Broad Branch Rd., NW, Washington DC 20015, USA

Mitsuyo Saito
Graduate School of Environmental and Life Science, Okayama University, 3-1-1 Tsushima-naka, Kita-ku, Okayama 700-8530, Japan

Shin-ichi Onodera
Graduate School of Integrated Arts and Sciences, Hiroshima University, 1-7-1 Kagamiyama, Higashi-Hiroshima 739-8521, Japan

Guangzhe Jin
Graduate School of Integrated Arts and Sciences, Hiroshima University, 1-7-1 Kagamiyama, Higashi-Hiroshima 739-8521, Japan
Guangdong Ocean University, Huguangyan, Zhanjiang, Guandong Province 524-088, China

Yuta Shimizu
Western Region Agricultural Research Center, National Agriculture and Food Research Organization, 6-12-1 Nishi-Fukatsu-cho, Fukuyama 721-8514, Japan

Masanobu Taniguchi
Faculty of Systems Engineering, Wakayama University, Sakaedani 930, Wakayama 640-8510, Japan

Yuya Tsunazawa
Department of Earth and Planetary Science, Graduate School of Science, The University of Tokyo, 7-3-1 Hongo, Bunkyo-ku, Tokyo 113-0033, Japan

Tadashi Yokoyama
Graduate School of Integrated Arts and Sciences, Hiroshima University, 1-7-1 Kagamiyama, Higashi-Hiroshima, Hiroshima 739-8521, Japan

Naoki Nishiyama
Research Center for Functional Materials, National Institute for Materials Science, 1-1Namiki, Tsukuba, Ibaraki 305-0044, Japan

Razvan Caracas
CNRS, Ecole Normale Supérieure de Lyon, Université Claude Bernard Lyon 1, Laboratoire de Géologie de Lyon, CNRS UMR 5276, Lyon, France

Wendy R. Panero
School of Earth Sciences, Ohio State University, Columbus, OH 43210, USA

Takuya Sagawa
Institute of Science and Engineering, Kanazawa University, Kakuma-machi, Kanazawa, Ishikawa 920-1192, Japan

Yoshitaka Nagahashi
Faculty of Symbiotic Systems Science, Fukushima University, 1 Kanayagawa, Fukushima, Fukushima 960-1296, Japan

Yasufumi Satoguchi
Lake Biwa Museum, 1091 Oroshimo, Kusatsu, Shiga 525-0001, Japan

Ann Holbourn
Institute of Geosciences, Christian-Albrechts-University, Ludewig-Meyn-Str. 10, 24118 Kiel, Germany

Takuya Itaki and Ken Ikehara
Geological Survey of Japan, National Institute of Advanced Industrial Science and Technology, 1-1-1 Higashi, Tsukuba, Ibaraki 305-8567, Japan

Stephen J. Gallagher
School of Earth Sciences, The University of Melbourne, Melbourne, Victoria 3010, Australia

Mariem Saavedra-Pellitero
Department of Geosciences, FB5, University of Bremen, Klagenfurter Strasse, 28359 Bremen, Germany

Tomohisa Irino
Faculty of Environmental Earth Science, Hokkaido University, N10W5, Sapporo, Hokkaido 060-0810, Japan

Ryuji Tada
Department of Earth and Planetary Science, The University of Tokyo, 7-3-1 Hongo, Bunkyo-ku, Tokyo 113-0033, Japan

Keiji Horikawa and Jing Zhang
Graduate School of Science and Engineering for Research, University of Toyama, Gofuku 3190, Toyama 930-8555, Japan

Tomohiro Kodaira
Graduate School of Science and Engineering for Education, University of Toyama, Gofuku 3190, Toyama 930-8555, Japan

Masafumi Murayama
Center for Advanced Marine Core Research, Kochi University, B200 Monobe, Nankoku 783-8502, Japan

Harue Masuda
Graduate School of Science, Osaka City University, 3-3-138, Sugimoto, Sumiyoshi-ku, Osaka 558-8585, Japan

Rachel Smith and Hugh St.C. O'Neill
Research School of Earth Sciences, The Australian National University, Building 142, Mills Road, Canberra, ACT 2601, Australia

Peter M. E. Tollan and Jörg Hermann
Research School of Earth Sciences, The Australian National University, Building 142, Mills Road, Canberra, ACT 2601, Australia
Institute of Geological Sciences, Universität Bern, Bern 3012, Switzerland

Ryuji Asami
Institute of Geology and Paleontology, Graduate School of Science, Tohoku University, Aobayama, Sendai 980-8578, Japan

Akira Kinjo, Daiki Ohshiro and Ryuichi Shinjo
Department of Earth Science, Faculty of Science, University of the Ryukyus, 1 Senbaru, Nishihara, Okinawa 903-0213, Japan

Tohru Naruse
Tropical Biosphere Research Center, Iriomote Station, University of the Ryukyus, 870 Uehara, Taketomi, Okinawa 907-1541, Japan

Masaru Mizuyama
Department of Chemistry, Biology, and Marine Science, Faculty of Science, University of the Ryukyus, 1 Senbaru, Nishihara, Okinawa 903-0213, Japan

Ryu Uemura
Division of Earth and Environmental Sciences, Graduate School of Environmental Studies, Nagoya University, Furo-cho, Chikusa-ku, Nagoya 464-8601, Japan

Yuji Ise
Centre for Marine and Coastal Studies (CEMACS), Universiti Sains Malaysia, 11800 USM, Penang, Malaysia

Yoshihisa Fujita
Okinawa Prefectural University of Arts, 1–4 Shuri-Tounokura, Naha, Okinawa 903-8602, Japan

Takashi Sakamaki
Department of Civil and Environmental Engineering, Graduate School of Engineering, Tohoku University, 6-6-06 Aramaki-Aza-Aoba, Sendai 980-8579, Japan

Tomoyo Okumura
Laboratory of Ocean-Earth Life Evolution Research (OELE), Japan Agency for Marine-Earth Science and Technology (JAMSTEC), 2-15 Natsushima-cho, Yokosuka 237-0061, Japan
Department of Subsurface Geobiological Analysis and Research (D-SUGAR), Japan Agency for Marine-Earth Science and Technology (JAMSTEC), 2-15 Natsushima-cho, Yokosuka 237-0061, Japan

Shinsuke Kawagucci and Ken Takai
Laboratory of Ocean-Earth Life Evolution Research (OELE), Japan Agency for Marine-Earth Science and Technology (JAMSTEC), 2-15 Natsushima-cho, Yokosuka 237-0061, Japan
Department of Subsurface Geobiological Analysis and Research (D-SUGAR), Japan Agency for Marine-Earth Science and Technology (JAMSTEC), 2-15 Natsushima-cho, Yokosuka 237-0061, Japan
Research and Development Center for Submarine Resources, Japan Agency for Marine-Earth Science and Technology (JAMSTEC), 2-15 Natsushima-cho, Yokosuka 237-0061, Japan

Yohei Matsui and Hiroyuki Imachi
Department of Subsurface Geobiological Analysis and Research (D-SUGAR), Japan Agency for Marine-Earth Science and Technology (JAMSTEC), 2-15 Natsushima-cho, Yokosuka 237-0061, Japan
Research and Development Center for Submarine Resources, Japan Agency for Marine-Earth Science and Technology (JAMSTEC), 2-15 Natsushima-cho, Yokosuka 237-0061, Japan

Yayoi Saito
Department of Subsurface Geobiological Analysis and Research (D-SUGAR), Japan Agency for Marine-Earth Science and Technology (JAMSTEC), 2-15 Natsushima-cho, Yokosuka 237-0061, Japan
Department of Environmental Systems Engineering, Nagaoka University of Technology, 1603-1 Kamitomiokamachi, Nagaoka 940-2188, Japan

Yosuke Miyairi
Atmosphere and Ocean Research Institute, University of Tokyo, 5-1-5 Kashiwanoha, Kashiwa, Chiba 277-8564, Japan

Atsunori Nakamura
Atmosphere and Ocean Research Institute, University of Tokyo, 5-1-5 Kashiwanoha, Kashiwa, Chiba 277-8564, Japan
Department of Earth and Planetary Science, Graduate School of Science, University of Tokyo, 7-3-1 Hongo, Bunkyo-ku, Tokyo 113-0033, Japan

Yusuke Yokoyama
Atmosphere and Ocean Research Institute, University of Tokyo, 5-1-5 Kashiwanoha, Kashiwa, Chiba 277-8564, Japan
Department of Earth and Planetary Science, Graduate School of Science, University of Tokyo, 7-3-1 Hongo, Bunkyo-ku, Tokyo 113-0033, Japan
Institute of Biogeoscience, Japan Agency of Marine Science and Technology (JAMSTEC), 2-15 Natsushima-cho, Yokosuka, Kanagawa 237-0061, Japan

Kazuyo Shiroya
Geological Survey of Japan, National Institute of Advanced Industrial Science and Technology (AIST), AIST Tsukuba Central 7, 1-1-1 Higashi, Tsukuba 305-8567, Japan

Hiroyuki Matsuzaki
Department of Nuclear Engineering and Management, The University of Tokyo, Yayoi 2-11-16, Bunkyo- ku, Tokyo 113-0032, Japan

Richard W. Murray
Department of Earth & Environment, Boston University, Boston, MA 02215,USA

Rachel P. Scudder
Department of Earth & Environment, Boston University, Boston, MA 02215, USA
Department of Oceanography, Texas A&M University, College Station, TX 77843, USA

Julie C. Schindlbeck, Steffen Kutterolf and Folkmar Hauff
GEOMAR Helmholtz Centre for Ocean Research Kiel, 24148 Kiel, Germany

Michael B. Underwood
Department of Earth and Environmental Science, New Mexico Institute of Mining and Technology, Socorro, NM 87801, USA

Samantha Gwizd
Department of Geosciences, Princeton University, Princeton, NJ 08544, USA

Rebecca Lauzon
Division of Earth and Ocean Sciences, Nicholas School of the Environment, Duke University, Durham, NC 27708, USA

Claire C. McKinley
Department of Oceanography, Texas A&M University, College Station, TX 77843, USA

Li Zhang
Center for High Pressure Science and Technology Advanced Research (HPSTAR), Shanghai 201203, China

Ho-kwang Mao
Center for High Pressure Science and Technology Advanced Research (HPSTAR), Shanghai 201203, China
Geophysical Laboratory, Carnegie Institution of Washington, Washington, DC 20015, USA

Yue Meng
High Pressure Collaborative Access Team (HPCAT), Geophysical Laboratory, Carnegie Institution of Washington, Argonne, IL 60439, USA

Charles Le Losq, Bjorn O. Mysen and George D. Cody
Geophysical Laboratory, Carnegie Institution of Washington, 5251 Broad Branch Road NW, Washington, DC 20015, USA

Arisa Seki
Faculty of Science, Shinshu University, 3-1-1 Asahi, Matsumoto, Nagano 390-8621, Japan

Ryuji Tada and Shunsuke Kurokawa
Graduate School of Science, The University of Tokyo, 7-3-1 Hongo, Bunkyo-ku, Tokyo 113-0033, Japan

Masafumi Murayama
Faculty of Agriculture and Marine Science, Kochi University, 200 Monobe, Nankoku-shi, Kochi 783-8502, Japan

Index

www.ingramcontent.com/pod-product-compliance
Lightning Source LLC
LaVergne TN
LVHW080921210926
848485LV00004B/700